Edited by
David J. Kwiatkowski,
Vicky Holets Whittemore, and
Elizabeth A. Thiele

Tuberous Sclerosis Complex

Related Titles

Gires, O., Seliger, B. (eds.)

Tumor-Associated Antigens

Identification, Characterization, and Clinical Applications

2009

ISBN: 978-3-527-32084-4

Allgayer, H., Rehder, H., Fulda, S. (eds.)

Hereditary Tumors

From Genes to Clinical Consequences

2009

ISBN: 978-3-527-32028-8

Nothwang, H., Pfeiffer, E (eds.)

Proteomics of the Nervous System

2008

ISBN: 978-3-527-31716-5

Müller, H. W. (ed.)

Neural Degeneration and Repair

Gene Expression Profiling, Proteomics and Systems Biology

2008

Hardcover

ISBN: 978-3-527-31707-3

Edited by
David J. Kwiatkowski, Vicky Holets Whittemore, and Elizabeth A. Thiele

Tuberous Sclerosis Complex

Genes, Clinical Features, and Therapeutics

WILEY-BLACKWELL

The Editors

Dr. David J. Kwiatkowski
Brigham & Women's Hospital
Dana Farber Cancer Institute
Harvard Medical School
1 Blackfan Circle
Boston, MA 02115
USA

Dr. Vicky Holets Whittemore
Tuberous Sclerosis Alliance
801 Roeder Road
Silver Spring, MD 20910
USA

Dr. Elizabeth A. Thiele
Carol & James Herscot Center For TCS
Massachusetts General Hospital
Department of Neurology
175 Cambridge Street
Boston, MA 02114
USA

Cover: Tuberous sclerosis complex (TSC) affects people of all races, ages, and sexes. The cover shows photographs of individuals with TSC, provided by Rick Guidotti, New York, NY (www.positiveexposure.org) and MGH Photography (www.massgeneral.org/photography), Boston, Massachusetts.

Library of Congress Card No.: applied for

British Library Cataloguing-in-Publication Data
A catalogue record for this book is available from the British Library.

Bibliographic information published by the Deutsche Nationalbibliothek
The Deutsche Nationalbibliothek lists this publication in the Deutsche Nationalbibliografie; detailed bibliographic data are available on the Internet at http://dnb.d-nb.de.

Wiley-Blackwell is an imprint of John Wiley & Sons, formed by the merger of Wiley's global Scientific, Technical, and Medical business with Blackwell Publishing.

Cover Design Adam-Design, Weinheim
Typesetting Thomson Digital, Noida, India
Printing and Binding Strauss GmbH, Mörlenbach

Printed in the Federal Republic of Germany
Printed on acid-free paper

ISBN: 978-3-527-32201-5

Contents

Tuberous Sclerosis Complex: Genes, Clinical Features, and Therapeutics
Edited by David J. Kwiatkowski, Vicky Holets Whittemore, and Elizabeth A. Thiele

ISBN: 978-3-527-32201-5

Preface

It is a great pleasure and honor to present this book, *Tuberous Sclerosis Complex: From Genes to Therapeutics*, for your thoughtful reading. This book was conceived in the spring of 2007, by David and Vicky, as we realized that the traditional Tuberous Sclerosis Complex (TSC) book edited by Manuel Gomez was eight years old, and was already outdated then in several respects. We recruited Elizabeth as a third Editor, and began serious work at that time in developing the chapter outlines and recruiting the best authors for the chapters from TSC clinicians and investigators from around the world.

We have sought to make the presentation in this book both scholarly and scientifically accurate, and understandable to the average TSC family member. We hope that it will find use to research scientists interested in the clinical details of this syndrome, clinicians caring for individuals with TSC, and individuals with TSC patients and their family members. We apologize in advance if the presentation is too technical in some areas.

TSC clinical and basic investigation has made great strides in the past 10 years. The identification of the two genes, TSC1 and TSC2, and the discovery of the main signaling pathway in which they play a important role, the mTOR pathway, has opened up an increasing flood of investigation into their role in cellular growth control and the mechanism by which inactivation of either gene leads to hamartoma development in individuals with TSC. Although there remain many unanswered questions of great importance, these findings have led to the introduction of rational therapy for TSC lesions, directed at the abnormal activation of the mTORC1 complex, in the form of rapamycin and analogues. Although there is much hope for these compounds, they are the subject of current clinical trials and ongoing investigation, so it is not yet clear what their long term benefits versus side-effects and toxicities will be. Fortunately, even if these compounds fail to work as well as desired, many related compounds have been or will be generated in the coming years, based upon our expanding knowledge of this pathway, providing additional therapeutic molecules to be tested in the clinic. These developments, combined with the general current concept of personalized medicine, provide much optimism about the long-term reduction in both morbidity and mortality due to TSC.

Tuberous Sclerosis Complex: Genes, Clinical Features, and Therapeutics
Edited by David J. Kwiatkowski, Vicky Holets Whittemore, and Elizabeth A. Thiele

ISBN: 978-3-527-32201-5

We have divided the book into 6 sections: Basics, Genetics, Basic Science, Brain Involvement, Other Organ Systems, and Family Impact. The Basics section provides information on the history of TSC clinical description and research, an overview of the clinical manifestations of TSC, and diagnostic criteria. The Genetics section covers the two TSC genes in great detail, as well as correlations between different mutations and clinical features. The Basic science section describes the biochemical function of the TSC1 and TSC2 proteins and their role in mTOR regulation, as well as insights from the fly mouse and rat models of TSC. The Brain Involvement section covers the many different aspects of brain involvement in TSC, including pathological and clinical. The Other Organs Section covers all the other organs commonly involved by TSC. Finally, the Family Impact chapter describes effects of TSC on the family and the importance of genetic counseling in TSC.

Our literature review for this book, as well as our own experience, has made it clear that there are many issues in regard to TSC management in the family for which there has been both relatively little investigation and little well-founded guidance. These issues fall largely in the neurocognitive sphere, and include: attention deficit hyperactive disorder (ADHD), autism spectrum disorder, tantrums and behavioral outbursts, intellectual disability, and sleep disturbance. In some instances, these issues are understood to be due in part to chronic seizures. However, this is not the case for all individuals with TSC. This is an area of great importance to TSC individuals and their families, and we hope to be able to report in a revised edition of this book in the future that there has been significant progress in both understanding and management of these issues.

Boston and Silver Spring
February 2010

David J. Kwiatkowski
Elizabeth A. Thiele
Vicky H. Whittemore

Acknowledgements

The Editors give many thanks to: all of the chapter authors for their contributions to this book; our families for their perseverance and understanding; our grant support enabling this work (DJK- NIH/NCI 1P01CA120964, NIH NINDS 2R37NS031535, NIH NINDS 1P01NS24279; ET- NIH NINDS 1P01NS24279; the Carol and James Herscot Center for TSC); the continuing support of the Tuberous Sclerosis Alliance, and other TSC support groups worldwide; and individuals with TSC and families who have not only permitted but facilitated, encouraged, and even funded in part many studies on this condition for several decades.

List of Contributors

Shivi Agrawal
Boston Children's Hospital
and Harvard Medical School
Boston, MA 02115
USA

Kit S. Au
The University of Texas Medical School
at Houston
Division of Medical Genetics
Department of Pediatrics
Houston, TX 77030
USA

M. Gregory Balko
Wright State University Boonshoft
School of Medicine
Dayton, OH
USA

John J. Bissler
University of Cincinnati
College of Medicine
Cincinnati Children's Hospital Medical
Center
Division of Nephrology and
Hypertension
Cincinnati, OH 45435
USA

Peter B. Crino
University of Pennsylvania
PENN Epilepsy Center
Philadelphia, PA 19104
USA

Petrus J. de Vries
University of Cambridge
Cambridgeshire & Peterborough NHS
Foundation Trust
Developmental Psychiatry Section
Douglas House
Cambridge CB2 8AH
UK

Thomas N. Darling
Uniformed Services University of the
Health Sciences
Department of Dermatology
Bethesda, MD 20814
USA

David Neal Franz
University of Cincinnati College of
Medicine
Cincinnati Children's Hospital Medical
Center
Cincinnati, OH 45229
USA

Tuberous Sclerosis Complex: Genes, Clinical Features, and Therapeutics
Edited by David J. Kwiatkowski, Vicky Holets Whittemore, and Elizabeth A. Thiele

ISBN: 978-3-527-32201-5

Anne B. Fulton
Boston Children's Hospital
and Harvard Medical School
Boston, MA 02115
USA

Elizabeth P. Henske
Harvard Medical School
Brigham and Women's Hospital
Center for LAM Research and Patient Care
Boston, MA 02115
USA

Sergiusz Jóźwiak
The Children's Memorial Health Institute
Department of Pediatric Neurology and Epileptology
Warsaw
Poland

Darcy A. Krueger
University of Cincinnati College of Medicine
Cincinnati Children's Hospital Medical Center
Cincinnati, OH 45229
USA

David J. Kwiatkowski
Brigham & Women's Hospital
Dana Farber Cancer Institute
Harvard Medical School
Boston, MA 02115
USA

Janine Lewis
The Genetic and Rare
Disease Information Center
National Institute of Health
Gaithersburg, MD 20898
USA

Brendan D. Manning
Harvard University, School of Public Health
Department of Genetics and Complex Diseases
Boston, MA 02115
USA

Mark Mausner
Mausner Plastic Surgery Center
Bethesda, MD 20817
USA

Francis X. McCormack
The University of Cincinnati
Division of Pulmonary, Critical Care and Sleep Medicine
Cincinnati, OH 45219
USA

Rupal Mehta
David Geffen School of Medicine at UCLA
Department of Pathology & Laboratory Medicine
Los Angeles, CA 90095
USA

Joel Moss
National Institutes of Health
National Heart, Lung, and Blood Institute
Translational Medicine Branch
Bethesda, MD 20892
USA

Hope Northrup
The University of Texas Medical School of Houston
Division of Medical Genetics
Department of Pediatrics
Houston, TX 77030
USA

Finbar J. O'Callaghan
University of Bristol
Institute of Child Life and Health,
Education Centre
Bristol
UK

John P. Osborne
University of Bath
UK

Duojia Pan
Johns Hopkins University School of
Medicine
Howard Hughes Medical Institute
Department of Molecular Biology and
Genetics
Baltimore, MD 21205
USA

Maria Respondek-Liberska
Medical University of Łódź and
Research Institute Polish Mother's
Memorial Hospital
Department for Diagnosis and
Prevention of Fetal Malformations
Łódź
Poland

E. Steve Roach
Ohio State University College
of Medicine
Division of Child Neurology
Columbus, OH 43205
USA

Steven P. Sparagana
Texas Scottish Rite Hospital for
Children
Dallas, TX 75219
USA

Elizabeth A. Thiele
Massachusetts General Hospital
Carol & James Herscot Center for TSC
Department of Neurology
Boston, MA 02114
USA

Harry V. Vinters
David Geffen School of Medicine at
UCLA
Department of Pathology & Laboratory
Medicine
Los Angeles, CA 90095
USA

Howard L. Weiner
Massachusetts General Hospital
Carol & James Herscot Center
Boston, MA 02114
USA

Vicky H. Whittemore
Tuberous Sclerosis Alliance
Silver Spring, MD 20910
USA

Part I
Basics

Tuberous Sclerosis Complex: Genes, Clinical Features, and Therapeutics
Edited by David J. Kwiatkowski, Vicky Holets Whittemore, and Elizabeth A. Thiele

ISBN: 978-3-527-32201-5

1
The History of Tuberous Sclerosis Complex

Vicky H. Whittemore

There are very few rare genetic disorders where the research has moved from clinical descriptions and case reports to identification of the disease-causing genes, to an understanding of the underlying mechanisms of disease, and finally to clinical trials in just 12 years. Research on tuberous sclerosis complex (TSC) has done just that with the identification of the *TSC1* and *TSC2* genes in 1993 and 1997, respectively, identification of the role of the genes in an important cell signaling pathway, and launching of clinical trials with drugs that specifically target the molecular defect in individuals with TSC.

1.1
Definition

Tuberous sclerosis complex is a genetically determined multisystem disorder that may affect any human organ system. Skin, brain, retina, heart, kidneys, and lungs are most frequently involved with the growth of noncancerous tumors, although tumors can also be found in other organs such as the gastrointestinal tract, liver, and reproductive organs. There may also be manifestations of TSC in the central nervous system (CNS), including tubers (disorganized areas of the cerebral cortex that contain abnormal cells), scattered abnormal cells throughout the CNS, and other lesions. The majority of individuals with TSC have learning disabilities that range from mild to severe, and may include severe intellectual disability and autism spectrum disorder. In addition, the majority of individuals with TSC will have epilepsy beginning in early childhood or at any point in the individual's life. Psychiatric issues including attention deficit, depression, and anxiety disorder may significantly impair the life of an individual with TSC and their family, and may impair their ability to live an independent life. However, there are many very able individuals with TSC who can carry on healthy and productive lives.

TSC can be inherited in an autosomal dominant manner, but the majority of cases are thought to be sporadic mutations with no family history of the disease. As our clinical understanding of the disease has improved over the last century, it is clear

Tuberous Sclerosis Complex: Genes, Clinical Features, and Therapeutics
Edited by David J. Kwiatkowski, Vicky Holets Whittemore, and Elizabeth A. Thiele

ISBN 978-3-527-32201-5

that the disease is variably expressed, even in the same family and even in two individuals from different families who have the same genetic mutation in one of the two TSC genes.

1.2
The History of Tuberous Sclerosis Complex

The first documented descriptions of TSC date back to the early 1800s. Rayer [1] illustrated the skin lesions on a young man's face in his atlas in 1835. These skin lesions had the characteristic distribution and appearance of the facial angiofibromas frequently seen in individuals with TSC. The pathological findings of a newborn who died shortly after birth was provided by von Recklinghausen in 1862, and is the first documented report of a child with cardiac tumors (called "myomata") and a "great number of scleroses" in the brain [2] (Table 1.1).

The first detailed description of the neurological symptoms and the gross pathology in the central nervous system of three individuals with TSC was provided by Bourneville in 1880 [3]. He used the term "tuberous sclerosis of the cerebral convolutions" to describe the CNS pathology in a child with seizures and learning disability [3]. Moolten first used the term "tuberous sclerosis complex" to describe the multisystem genetic disorder that may predominantly include involvement of the skin, heart, brain, kidneys, lungs, eyes, and liver, but can also involve other organ systems (e.g., the gastrointestinal tract and reproductive organs) [4].

In 1881, Bourneville and Brissaud [5] described a 4-year-old boy with seizures, limited verbal skills, and a cardiac murmur who subsequently stopped eating and drinking and died. At autopsy, the brain showed sclerotic, hypertrophic convolutions, and they described many small sclerotic tumors covering the lateral walls of the ventricles – the first description of what later became known as subependymal nodules. They also described small yellowish-white tumors in the kidneys and proposed the association between the CNS and renal manifestations of TSC. Balzer and Menetrier [6] and then Pringle [7] described the facial lesions illustrated much earlier by Rayer and called them "congenital adenoma sebaceum." It was not until 1962 that Nickel and Reed [8] showed that the sebaceum glands were not enlarged in the facial lesions in TSC, but that they were often absent or atrophic. However, these lesions were only renamed facial angiofibromas after additional pathological descriptions of the lesions showed that the term adenoma sebaceum was a misnomer [9].

For many years, Vogt's triad of seizures, learning disability, and "adenoma sebaceum" (facial angiofibromas) was used to diagnose TSC [10]. Vogt also noted that cardiac and renal tumors were part of the disease.

In 1920, van der Hoeve coined the term phakomatoses to describe disorders that were characterized by the presence of circumscribed lesions or phakomas that had the potential to enlarge and form a tumor [11]. The three phakomatoses included TSC, neurofibromatosis, and von Hippel–Lindau disease. All three diseases have a spotty distribution of the lesions and the lesions can grow as benign tumors.

Table 1.1 Historical milestones of the tuberous sclerosis complex.

Clinicopathological developments	
1835	First illustration of facial angiofibromas in atlas [1]
1862	Cardiac "myomata" described in newborn [2]
1879	Cortical "tuberosities" identified [3]
1885	Report of "adenoma sebaceum" [6]
1908	Diagnostic triad proposed [10]
1910	Hereditary nature of TSC described [20]
1912	Hereditary nature of TSC [21]
1913	Forme fruste with normal intelligence [22]
1920	Retinal phakoma identified [11]
1932	Review of clinical aspects and discovery of hypomelanotic macules [12]
1942	First use of the term "tuberous sclerosis complex" [4]
1967	Significant number of individuals with TSC found to have average (normal) intelligence [17]
1979	New criteria for diagnosis of TSC, decline of Vogt's triad [18]
1987	Full spectrum of psychiatric issues described [14–16]
1988	Revised diagnostic criteria for TSC [18]
1998	Diagnostic criteria revised [19]
1999	Phenotype/genotype correlations [30]
2001	Phenotype/genotype correlations [31]
2007	Phenotype/genotype correlations [32]
Genetic and scientific developments	
1987	Positional cloning: mapping of the *TSC1* gene to chromosome 9q34.3 [25]
1992	Finding of nonlinkage to chromosome 9 [26]; mapping of the *TSC2* gene to chromosome 16p13.3 [27]
1993	Cloning of the *TSC2* gene; its protein product is called tuberin [28]
1997	Cloning of the *TSC1* gene; its protein product is called hamartin [29]
2001	*Drosophila* homologues *Tsc1* and *Tsc2* involved in regulation of cell and organ size [33–35]
2002	Tuberin found as a target of the PI3k/akt pathway [36]; TSC1/2 protein complex described [37]
2002	Activation of mTOR pathway in TSC described [38]
2003	mTOR activation confirmed in renal angiomyolipomas from individuals with TSC [39]
2005	Rapamycin (mTOR inhibitor) reduces renal tumors in Eker rats [40] and mouse models [41]
2006	Rapamycin shown to reduce the size of subependymal giant cell astrocytomas [42]
2008	Rapamycin reduces size of renal angiomyolipomas [43]

It was not until 1932 that the significance of the white spots (hypomelanotic macules) on the skin of individuals was noted as helpful in the diagnosis of TSC [12]. They also described autistic behavior in some of the 29 individuals with TSC they observed. Kanner [13] described "early infantile autism" 11 years later, but it was not until far more recently that the link between TSC and autism spectrum disorder was truly recognized [14–16].

A very important shift in our understanding and diagnosis of TSC occurred in 1967 when Lagos and Gomez [17] reported their findings from a family with 71 affected individuals in which five generations were affected by TSC. In this family, 38% of the 69 individuals, where information on their intellectual abilities was known, had average intelligence, while 62% had learning disabilities. These data led to the new diagnostic criteria that were first published in 1988 [18], although many clinicians still used Vogt's triad to diagnose TSC for many years, incorrectly and inappropriately referring to individuals with TSC as persons with "fits, zits and who are nitwits." The diagnostic criteria were revised again in 1998 [19] and will continue to be revised as more knowledge is gained about the clinical and genetic aspects of the disease.

The hereditary nature of TSC was recognized in the early 1900s through the observation of families that had multiple affected individuals in two or more generations [20, 21]. Schuster [22] confirmed that TSC was a hereditary disease, but also described individuals with only the "adenoma sebaceum" component of Vogt's triad, with no seizures or intellectual disability. Initially, these individuals were described as having *forme fruste* TSC (from the French fluster, or defaced), a term that was not clearly defined but was used for individuals with "incomplete" phenotypes who did not meet diagnostic criteria.

With the improvement of technology to image the human body starting in the mid-1970s, it became possible to diagnose individuals with TSC who had manifestations of the disease but who were clinically asymptomatic. The development of computed tomography (CT) of the head allowed the imaging of subependymal nodules, subependymal giant cell tumors (SGCTs), and calcified tubers starting in 1974. This was followed by echocardiography to image cardiac rhabdomyomas and renal ultrasound to image renal tumors in individuals with TSC. However, the development of magnetic resonance imaging (MRI) in 1982 provided the means to much more accurately and explicitly image cortical tubers and other manifestations of TSC. As new technologies are developed and applied to the study of the clinical manifestations of TSC, our knowledge of the disease and our ability to diagnose TSC will significantly improve.

1.3
Hereditary Nature of TSC

Kirpicznik [20] first recognized TSC as a genetic condition after reporting on a family with affected individuals in three generations, including identical and fraternal twins. Adenoma sebaceum (correctly termed facial angiofibromas) were reported to be inherited in families [6, 7]. Berg [21] also described the hereditary nature of TSC in 1913, and Schuster [22] confirmed this and noted the exceptional individual with only the facial lesions without intellectual disability.

The dominant inheritance of TSC and its high mutation rate were demonstrated [23, 24], but very little progress was made until genetic linkage analysis identified a probably TSC gene on chromosome 9q34 in 1987 [25], identified as the *TSC1* locus. Numerous linkage analysis publications narrowed the search for the TSC gene(s), with a group in the United States showing that there some families with TSC had a linkage to chromosome 9, but that there were certainly one or more additional loci [26]. This led to the identification of a second linkage to chromosome 16p13 [27], designated as the *TSC2* locus. The *TSC2* gene was cloned first by the European Chromosome 16 Consortium [28] in 1993, with the *TSC1* gene cloned in 1997 [29].

A molecular diagnostic test for TSC was launched in the early 2000s, and is used today for confirmation of a clinical diagnosis of TSC, to assist in the diagnosis of TSC, and for reproductive decision making, including prenatal diagnosis and preimplantation genetic diagnosis combined with *in vitro* fertilization. Several studies have attempted to correlate the phenotype (the clinical manifestations of the disease expressed) with the genotype (the specific genetic mutation) for individuals with TSC, with reinforcement of the notion that TSC is variably expressed even in individuals with the exact genetic mutation [30–32].

1.4 Molecular Mechanisms in TSC

Little was known about the cause of TSC prior to identification of the *TSC1* and *TSC2* genes in the 1990s. A naturally occurring rat mutation in *Tsc2*, the Eker rat model, had been used extensively to study TSC, but it was not until the *Drosophila* homologues, *Tsc1* and *Tsc2*, were found to be involved in regulation of cell and organ size [33–35] that significant progress could be made. Finding that the *TSC2* gene product, tuberin, was a target in an important cell signaling pathway [36] and the identification that the *TSC1* and *TSC2* gene products worked together in a complex [37] led to finding the critical role of the TSC genes in regulation of the mTOR pathway [38]. mTOR activation has been confirmed in renal angiomyolipomas from individuals with TSC [39], and an mTOR inhibitor, rapamycin, has been shown to reduce renal tumors in Eker rats [40] and TSC mouse models [41] and, more recently, to reduce the size of subependymal giant cell astrocytomas [42] and renal angiomyoloipomas [43] in individuals with TSC.

1.5 The Future of TSC

Significant progress has been made in TSC research, but there are still many questions left unanswered. The clinical trials look promising, but may or may not be effective for treatment of both the CNS manifestations and tumor growth in various organ systems without very early treatment and/or chronic drug therapy. Yet another revision of the diagnostic criteria is needed to include those individuals who do not meet criteria for a diagnosis based on the previous criteria, but are found to have a disease-causing variation in either the *TSC1* or *TSC2* gene. The future holds much promise for improving the quality of life for individuals with TSC, and for reaching an even more complete understanding of the underlying mechanisms that result in the many and variable manifestations of the disease.

References

1 Rayer, P.F.O. (1835) *Traité Theorique et Pratique des Maladies de la Peau*, 2nd edn, JB Baillière, Paris.

2 von Recklinghausen, F. (1862) Ein Herz von einem Neugeborene welches mehrere theils nach aussen, theils nach den Höhlen prominirende Tumoren (Myomen) trug. *Monatschr. Geburtsheklkd.*, **20**, 1–2.

3 Bourneville, D.M. (1880) Sclérose tubéreuse des circonvultions cérébrales: idiotie et épilepsie hémiplégique. *Arch. Neurol. (Paris)*, **1**, 81–91.

4 Moolten, S.E. (1942) Hamartial nature of the tuberous sclerosis complex and its bearing on the tumor problem: report of one case with tumor anomaly of the kidney and adenoma sebaceum. *Arch. Intern. Med.*, **69**, 589–623.

5 Bourneville, D.M. and Brissaud, E. (1881) Encéphalite ou sclérose tubéreuse des circonvultions cérébrales. *Arch. Neurol. (Paris)*, **1**, 390–410.

6 Balzer, F. and Menetrier, P. (1885) Étude sur un cas d'adénomes sébacés de la face et du cuir chevelu. *Arch. Physiol. Norm. Pathol. (série III)*, **6**, 564–576.

7 Pringle, J.J. (1890) A case of congenital adenoma sebaceum. *Br. J. Dermatol.*, **2**, 1–14.

8 Nickel, W.R. and Reed, W.B. (1962) Tuberous sclerosis. *Arch. Dermatol.*, **85**, 209–226.

9 Sanchez, N.P., Wick, M.R., and Perry, H.O. (1981) Adenoma sebaceum of Pringle: a clinicopathologic review, with a discussion of related pathologic entities. *J. Cutan. Pathol*, **8** (6), 395–403.

10 Vogt, H. (1908) Zur Pathologie und pathologishcen Anatomie der verschiedenen Idiotieform. *Monatsschr. Psychiatr. Neurol.*, **24**, 106–150.

11 van der Hoeve, J. (1920) Eye symptoms in tuberous sclerosis of the brain. *Trans. Ophthalmol. Soc. UK*, **20**, 329–334.

12 Critchley, M. and Earl, C.J.C. (1932) Tuberose sclerosis and allied conditions. *Brain*, **55**, 311–346.

13 Kanner, L. (1943) Autistic disturbances of affective contact. *J. Pediatr.*, **2**, 217–250.

14 Hunt, A. and Dennis, J. (1987) Psychiatric disorders among children with tuberous sclerosis. *Dev. Med. Child Neurol.*, **29**, 190–198.

15 Smalley, S., Smith, M., and Tanguay, P. (1991) Autism and psychiatric disorders in tuberous sclerosis. *Ann. N.Y. Acad. Sci.*, **615**, 382–383.

16 Curatolo, P., Cusmai, R., Cortesi, F., Chiron, C., Jambaque, I., and Dulac, O. (1991) Neuropsychiatric aspects of tuberous sclerosis. *Ann. N.Y. Acad. Sci.*, **615**, 8–16.

17 Lagos, J.C. and Gomez, M.R. (1967) Tuberous sclerosis: reappraisal of a clinical entity. *Mayo Clin. Proc.*, **42**, 26–29.

18 Gomez, M.R. (1988) Criteria for diagnosis, in *Tuberous Sclerosis*, 2nd edn (ed. M.R. Gomez), Raven Press, New York.

19 Roach, E.S., Gomez, M.R., and Northrup, H. (1998) Tuberous sclerosis complex consensus conference: revised clinical diagnostic criteria. *J. Child Neurol.*, **13**, 624–628.

20 Kirpicznik, J. (1910) Ein Fall von Tuberoser Sklerose und gleichzeitigen multiplem Nierengeschwülsten. *Virchow Arch. Pathol. Anat.*, **202** (3), 358.

21 Berg, H. (1913) Vererbung der Tuberösen Sklerose durch zweigzu drie Generationen. *Z. Ges. Neurol. Psychiatr.*, **19**, 528–539.

22 Schuster, P. (1914) Beiträge zur Klinik der tuberösen Sklerose des Gehirns. *Dtsch. Z. Nervenheilkd.*, **50**, 96–133.

23 Gunther, M. and Penrose, L.S. (1935) The genetics of epiloia. *J. Genet.*, **31**, 413–430.

24 Nevin, N.C. and Pearce, W.G. (1968) Diagnostic and genetical aspects of tuberous sclerosis. *J. Med. Genet.*, **5**, 273–280.

25 Fryer, A.E., Chalmers, A.H., Connor, J.M., Fraser, I., Povey, S., Yates, A.D., Yates, J.R., and Osborne, J.P. (1987) Evidence that the gene for tuberous sclerosis is on chromosome 9. *Lancet*, **1**, 659–661.

26 Northrup, H., Kwiatkowski, D.J., Roach, E.S., Dobyns, W.B., Lewis, R.A., Herman, G.E., Rodriguez, E., Daiger, S., and Blanton, S.H. (1992) Evidence for genetic heterogeneity in tuberous sclerosis: one locus on chromosome 9 and at least one

locus elsewhere. *Am. J. Hum. Genet.*, **51**, 709–720.

27 Kandt, R.S., Haines, J.K., Smith, M., Northrup, H., Gardner, R.J., Short, M.P., Dumars, K., Roach, E.S. *et al.* (1992) Linkage of an important gene locus for tuberous sclerosis to a chromosome 16 marker for polycystic kidney disease. *Nat. Genet.*, **2**, 37–41.

28 The European Chromosome 16 Consortium (1993) Identification and characterization of the tuberous sclerosis gene on chromosome 16. *Cell*, **75**, 1305–1315.

29 Van Slegenhorst, M., de Hoogt, R., Hermans, C., Nellist, M., Janssen, B., Verhoef, S., Lindhout, D., van den Ouweland, A. *et al.* (1997) Identification of the tuberous sclerosis gene *TSC1* on chromosome 9q34. *Science*, **277**, 805–808.

30 Jones, A.C., Shyamsundar, M.M., Thomas, H.W., Maynard, J., Idziaszczyk, S. Tomkins, S., and Sampson, J.R. (1999) Comprehensive mutation analysis of TSC1 and TSC2 and phenotypic correlations in 150 families with tuberous sclerosis. *Am. J. Hum. Genet.*, **2**, 217–250.

31 Dabora, S.L., Jozwiak, S., Franz, D.N., Roberts, P.S., Nieto, A., Chung, J., Choy, Y.S., Reeve, M.P. *et al.* (2001) Mutational analysis in a cohort of 224 tuberous sclerosis patients indicates increased severity of TSC2 compared with TSC1 disease in multiple organs. *Am. J. Hum. Genet.*, **68**, 64–80.

32 Au, K.S., Williams, A.T., Roach, E.S., Batchelor, L., Sparagana, S.P., Delgado, M.R., Wheless, J.W., Baumgartner, J.E. *et al.* (2007) Genotype/phenotype correlation in 325 individuals referred for a diagnosis of tuberous sclerosis complex in the United States. *Genet. Med.*, **9**, 88–100.

33 Gao, X. and Pan, D. (2001) TSC1 and TSC2 tumor suppressors antagonize insulin signaling in cell growth. *Genes Dev.*, **15**, 1383–1392.

34 Potter, C.J., Huang, H., and Xu, T. (2001) Drosophila Tsc1 functions with Tsc2 to antagonize insulin signaling in regulating cell growth, cell proliferation, and organ size. *Cell*, **105**, 357–368.

35 Tapon, N., Ito, N., Dickson, B.J., Treisman, J.E., and Hariharan, I.K. (2001) The *Drosophila* tuberous sclerosis complex gene homologs restrict cell growth and cell proliferation. *Cell*, **105**, 345–355.

36 Manning, B.D., Tee, A.R., Logsdon, M.N., Blenis, J., and Cantley, L.C. (2002) Identification of the tuberous sclerosis complex-2 tumor suppressor gene product tuberin as a target of the phosphoinositide 3-kinase/akt pathway. *Mol. Cell*, **10**, 151–162.

37 Tee, A.R., Fingar, D.C., Manning, B.D., Kwiatkowski, D.J., Cantley, L.C., and Blenis, J. (2002) Tuberous sclerosis complex-1 and -2 gene products function together to inhibit mammalian target of rapamycin (mTOR)-mediated downstream signaling. *Proc. Natl. Acad. Sci. USA*, **99**, 13571–13576.

38 Kenerson, H.L., Aicher, L.D., True, L.D., and Yeung, R.S. (2002) Activated mammalian target of rapamycin pathway in the pathogenesis of tuberous sclerosis complex renal tumors. *Cancer Res.*, **62**, 5645–5650.

39 El-Hashemite, N., Zhang, H., Henske, E.P., and Kwiatkowski, D.J. (2003) Mutation in TSC2 and activation of mammalian target of rapamycin signalling pathway in renal angiomyolipoma. *Lancet*, **361**, 1348–1349.

40 Kenerson, H., Dundon, T.A., and Yeung, R.S. (2005) Effects of rapamycin in the Eker rat model of tuberous sclerosis complex. *Pediatr. Res.*, **57**, 67–75.

41 Lee, L., Sudentas, P., Donohue, B., Asrican, K., Worku, A., Walker, V., Sun, Y., Schmidt, K. *et al.* (2005) Efficacy of a rapamycin analog (CCI-779) and IFN-gamma in tuberous sclerosis mouse models. *Genes Chromosomes Cancer*, **42**, 213–227.

42 Franz, D.N., Leonard, J., Tudor, C., Chuck, G., Care, M., Sethuraman, G., Dinopoulos, A., Thomas, G., and Crone, K.R. (2006) Rapamycin causes regression of astrocytomas in tuberous sclerosis complex. *Ann. Neurol.*, **59**, 490–498.

43 Bissler, J.J., McCormack, F.X., Young, L.R., Elwing, J.M., Chuck, G., Leonard, J.M., Schmithorst, V.J., Laor, T. *et al.* (2008) Sirolimus for angiomyolipoma in tuberous sclerosis complex or lymphangioleiomyomatosis. *N. Engl. J. Med.*, **358**, 140–151.

2
Natural History of Tuberous Sclerosis Complex and Overview of Manifestations

Elizabeth A. Thiele and Sergiusz Jóźwiak

Tuberous sclerosis complex (TSC) is a genetic disorder that is characterized by multisystem involvement and wide phenotypic variability. Originally thought to be a rare disorder, it is now known that TSC affects at least 1 in 6000 individuals worldwide, with no recognized ethnic predilection. TSC has an autosomal dominant inheritance, and there is no known effect of paternal or maternal age or of birth order on disease severity. However, approximately two thirds of individuals diagnosed with TSC develop it as the result of an apparent spontaneous DNA mutation not found in either parent (although mosaicism in a parent is possible).

Tuberous sclerosis complex can affect nearly every organ system, with various manifestations occurring at various times throughout the individual's lifetime. Unfortunately, there is very limited information available regarding the natural history of many aspects of TSC. Longitudinal clinical information on large populations of individuals with TSC has not been available, especially characterizing the behaviors of various manifestations over time. As this information is now being collected by many groups, we will hopefully have significant advances in our understanding of the natural history of TSC. This will likely have a profound impact not only on clinical care but also on our understanding of the pathogenesis of TSC.

In addition, an increasing number of "mildly affected" individuals with TSC are now being diagnosed, including many older adults who have never experienced a seizure and are cognitively normal. They are typically diagnosed with TSC after the diagnosis of a child or grandchild, or after they experience renal or other symptoms. This will undoubtedly impact our understanding of and appreciation for the wide phenotypic variability of the disorder and will likely expand the recognized clinical spectrum of TSC.

At present, we also have limited understanding of the impact of an individual's age on the various clinical manifestations. It is known that some features are more frequently seen or almost exclusively seen during early childhood, such as cardiac rhabdomyomas or the onset of epilepsy, while other features have been observed to occur only following puberty, such as pulmonary lymphangioleiomyomatosis (LAM). We know that some manifestations of TSC such as renal angiomyolipoma (AML) and facial angiofibroma can continue to progress throughout an individual's life. We also

Tuberous Sclerosis Complex: Genes, Clinical Features, and Therapeutics
Edited by David J. Kwiatkowski, Vicky Holets Whittemore, and Elizabeth A. Thiele

ISBN: 978-3-527-32201-5

know that others appear to either lose their growth potential, such as subependymal giant cell tumors, or even regress, such as cardiac rhabdomyoma.

Our understanding of the impact of gender on TSC is also limited, and despite knowing that LAM in TSC occurs almost exclusively in women, we do not know why. It is known that gender also matters for other features of TSC. Renal angiomyolipoma have a tendency to grow much larger in women with TSC than men. Autism spectrum disorders occur equally in boys and girls with TSC, differing from the strong male preponderance in autism due to other causes.

At present, we are also limited in our understanding of factors affecting the phenotypic variability of TSC. We know that a mutation in the TSC2 gene is more likely to result in a more severe phenotype than a TSC1 mutation, and although there is some speculation, we do not know why (Chapter 5). Although we are able to identify a disease-causing mutation in either TSC1 or TSC2 in 85% of individuals meeting clinical criteria for TSC, we cannot find a mutation in the other 15%–either because our techniques are not sophisticated enough, or because there are other genetic mechanisms involved (Chapters 4 and 5). And it is not clear what makes individuals with TSC sharing the same mutation "different" with regard to severity of various organ involvement, even within the same family.

Despite much uncertainty however, advances in our understanding of TSC over the past two decades have had a profound impact on individuals living with the disorder. Identification of the two genes associated with the disorder, TSC1 and TSC2, has allowed genetic testing, including prenatal testing, to become available. Recent advances in the understanding of the molecular biology of the TSC protein products tuberin and hamartin, and in particular their involvement in the phosphatidyloinositol-3-kinase/Akt/mTOR signal transduction pathway (Chapter 6), have made specific drug therapy a possibility and clinical trials a reality.

In addition, advances in medical technologies and therapies have also led to improved understanding of TSC and improved clinical care. The development and improvements in magnetic resonance imaging (MRI) have allowed a better understanding of the neuroanatomic features of TSC, a way to monitor the development of subependymal giant cell tumors and even a way to diagnose TSC prenatally. Improved surgical and interventional techniques have made epilepsy surgery a better option and sometimes a cure for refractory epilepsy in TSC (Chapter 10) and have made embolization of renal angiomyolipoma a treatment option rather than nephrectomy (Chapter 15). Advances in laser therapies have also made treatment of facial angiofibroma and other dermatologic manifestations of TSC more effective (Chapter 14). Advances in the pharmaceutical industry have more than doubled the number of available anticonvulsant medications, making medical management of epilepsy in TSC more successful, and have also made the recent clinical trials evaluating the efficacy and tolerability of the drug rapamycin in TSC possible.

However, for the individuals with TSC and their family, the diagnosis continues to be overwhelming. Tuberous sclerosis complex is an extremely complicated disorder, affecting different organ systems at different times in an individual's life in different and often profound ways. Many individuals living with TSC have made the analogy to "walking in a mine field." It is often helpful for individuals with TSC, their families, and their health care providers to try and put different aspects and possible manifestations of the disorder into some type of framework, helping the patient and others know what to expect, and when.

The diagnosis of TSC continues to be a clinical diagnosis based on major and minor criteria (Chapter 3) [1]. DNA mutational analysis, although clinically available, is not included in the current diagnostic criteria, and it is recommended only when TSC is clinically diagnosed or highly suspected. The clinical features leading to diagnosis vary depending on age – in infants, either cardiac rhabdomyoma or onset of seizures usually leads to diagnosis; in adults, either dermatologic features, renal, or lung involvement leads to diagnosis, or diagnosis occurs only after a child is diagnosed with TSC [2]. And often when an individual is diagnosed, particularly in adulthood, "missed diagnoses" are found, that is, such aspects as seizures or hemorrhage of a renal angiomyolipoma in that person's medical history that could have and should have raised a suspicion of TSC much earlier than when the eventual diagnosis is made. This concept of "missed diagnosis" emphasizes the continued need for an increased awareness of TSC in both pediatric and internal medicine communities, particularly with current abilities to practice "preventive medicine" with several aspects of TSC as well as the anticipated emergence of more targeted therapies in the near future.

2.1 TSC: Multisystem Involvement

As mentioned above, TSC can affect most organ systems. The two most commonly affected organ systems are the brain and the skin, both of which are involved in 90–95% of individuals with TSC.

2.1.1 TSC and the Brain

The main neuropathologic features of TSC include cortical and subcortical tubers, subependymal nodules (SENs), and subependymal giant cell astrocytomas (SEGAs) (also referred to as subependymal giant cell tumors) (Chapters 9 and 11). Cortical tubers are found in the cerebral and cerebellar cortex and subcortical white matter and they vary widely in size and distribution. They often extend in a linear or wedge-shaped zone spanning the full thickness from the cortical surface to the ventricular wall. Histologically, tubers are characterized by a marked distortion of cortical lamination with dysplastic, hypomyelinated aggregates of abnormal glial and neural elements including giant cells. It is thought that they arise from progenitor cells in the subependymal matrix that give rise to abnormally migrating daughter cells, which in turn develop into individual tubers. However, the precise genetic (or other) mechanism of initiation and formation is not understood. It is thought that tubers develop during the same period as when normal brain development occurs, between 14 and

16 weeks gestation, the period of active neuronal proliferation and migration. Therefore, an individual's tuber "burden" or tuber "load" is essentially present at birth. However, visualization of tubers by neuroimaging improves with advancing postnatal age, related to ongoing myelination of the brain during early childhood. It is known that tubers can develop cyst-like changes or calcification postnatally, although the mechanisms involved in these changes and the possible clinical significance of these changes are uncertain [3, 4].

Subependymal nodules are located around the wall of the lateral ventricle, developing beneath the ependymal lining (Chapter 11). They most commonly occur in the region of the caudothalamic groove in the vicinity of the foramen of Monro, presumably since they arise from germinal matrix remnants in that region. Histologically, SENs consist of relatively large cells, somewhat similar to the giant cells seen in tubers, but are often somewhat spindle shaped [5]. Both tuber giant cells and SEN cells typically express astrocyte markers. In contrast to cortical tubers, SENs can grow over time and in approximately 10% of individuals, they develop into subependymal giant cell astrocytomas. However, for reasons that are unclear, they almost invariably stop growing and calcify by late adolescence.

SEGAs, as above, are thought to arise from SENs and are typically present in childhood to adolescence, although rarely they develop during infancy. Classically, the clinical symptoms of SEGA are those of increased intracranial pressure with headache, vomiting, and papilledema on fundoscopic examination. However, the presentation can be more subacute with changes in behavior or seizure activity. Current management of SEGA involves surgical resection although trials evaluating the efficacy of mTOR inhibitors in the treatment of SEGA are at present underway (Chapter 11) [6].

Other neuroanatomic features, such as radial glial lines, are also frequently seen on imaging studies of TSC patients. Although the significance of these is uncertain, their presence hints at a possible role of abnormal function of radial glia in the pathogenesis of tuber formation.

Concurrent with these neuroanatomic features, there are several neurologic clinical features of TSC, including epilepsy, cognitive impairment, autism spectrum disorders, and sleep disorders.

Epilepsy occurs in up to 90% of individuals with TSC, with close to 70% having seizure onset during the first year of life. One third of infants with TSC will also develop infantile spasm, which is considered a catastrophic epilepsy syndrome of childhood and is typically associated with subsequent profound neurocognitive impairment. Many individuals with TSC develop epilepsy that is refractory to medical therapy; in these patients, nonpharmacologic therapies, including dietary therapy and epilepsy surgery, may be effective (Chapter 10).

Approximately 50% of individuals with TSC have some degree of cognitive impairment, often profound, but 50% have normal cognition. Risk factors for cognitive impairment in TSC include a history of infantile spasms, refractory epilepsy, ongoing seizure activity, tuber burden, and mutation in the TSC2 gene [7, 8].

Autism spectrum disorders also occur commonly in TSC, affecting between 17 and 60% of individuals [9]. Autism appears to occur at similar rates in boys and girls with

TSC, as opposed to autism due to other and unknown causes, in which boys are affected four times more frequently than girls. Although a history of infantile spasms is often found in children with TSC and autism, the mechanisms leading to the development of autism in TSC are not understood.

Recently, there has been growing appreciation and identification of other mental health issues that affect both children and adults with TSC. Mental health disorders appear to be very common in individuals with TSC, particularly anxiety, and often significantly impact the individual's quality of life and ability to function (Chapter 12).

Sleep disorders are also common in TSC individuals, but have not been well characterized. For many, sleep disturbances are likely related to their epilepsy – both the seizures and the medications, although other factors may also contribute. This is often a major issue for both the individuals affected with TSC and their families.

2.1.2
TSC and the Skin

Dermatologic involvement occurs in approximately 90–95% of individuals with TSC, making the skin and the brain the two most frequently affected organs (Chapter 14). The main skin manifestations include hypopigmented macules, facial angiofibroma, shagreen patch, and periungual fibroma. Hypopigmented macules are regions where reduced pigmentation occurs in the skin, due to abnormalities in the production of melanin by melanocytes. They are often referred to as "ash leaf spots" due to their configuration. They sometimes are present at birth but are usually more easily identified as the child grows, because pigmentation increases overall and skin surface area increases. Usually they are easily identified in individuals with darker pigmented skin or with tanning by the sun; a Wood's lamp examination can facilitate their identification in fairer skinned individuals. Angiofibroma usually appear in early childhood, often between 2 and 5 years of age. They appear initially as red dots on the cheek surface when the child becomes excited, angry, or hot [10]. With time, they also become papular. There is a broad range of severity of angiofibroma in individuals, from very mild "freckle-like" appearance to prominent and disfiguring nodular and waxy appearance. Fortunately, evolving laser technologies can significantly minimize the appearance of angiofibroma, and hopefully topical drug therapies will become available in the future. Shagreen patches are collagenoma typically located in the lower lumbar sacral region of the back, although they can occur anywhere on the back and occasionally on the anterior abdomen. They can usually be identified in older infants and toddlers, although are easier to see as the child grows. Periungual fibroma are growths on the nails of the hands and the feet that can appear as subtle ridges in the nails to more obvious fleshy growths. Although they can be found occasionally in children, they are much more common and prominent with age and continue to appear in TSC individuals above 40.

The dermatologic manifestations of TSC do not typically cause medical complications, however, if prominent, they can be very disfiguring. This is very often a cause of significant psychological stress for individuals with TSC and their families.

2.1.3
TSC and the Heart

The heart manifestation of TSC, rhabdomyoma, is thought to occur in approximately 50% of individuals with TSC (Chapter 16). Rhabdomyoma are often identified in the fetal TSC heart during a late gestation ultrasound, and for many individuals with TSC, their identification represents the first clinical sign of TSC. Any infant found to have a cardiac rhabdomyoma should undergo complete evaluation for possible TSC; if the infant has multiple rhabdomyoma, it is extremely likely that the child will eventually meet criteria for the diagnosis of TSC. Although usually asymptomatic, rhabdomyoma can occasionally affect cardiac outflow and require surgical resection during infancy. They are also thought to be associated with cardiac dysrhythmias that can be seen in TSC, although it is possible that these are due to different mechanisms. For uncertain reasons, cardiac rhabdomyomas typically regress with age, although they can occasionally remain present or rarely grow [11]. Recently, we have seen several adults with TSC who have intraventricular masses that are thought to represent either lipoma or angiomyolipoma and that are of uncertain clinical significance.

2.1.4
TSC and the Kidney

The kidney is affected in approximately 80–85% of individuals with TSC, and involvement can be first detected at any age from infancy through adulthood (Chapter 15). Kidney manifestations include renal cysts and angiomyolipoma. Renal cysts are commonly seen in individuals with either TSC1 or TSC2 mutations and are typically simple cysts that do not significantly affect renal function, even though they may be numerous. About 3% of TSC individuals have large bilateral renal cysts, compatible with polycystic kidney disease. The majority of these patients have a genomic deletion involving both the *TSC2* gene on chromosome 16 and the adjacent polycystic kidney disease gene *PKD1*. This type of polycystic kidney disease typically has a major impact on renal function, leading to significant renal impairment and need for dialysis or renal transplantation by age 20.

Angiomyolipoma are benign tumors that develop in the kidney and contain varying amounts of smooth muscle, fat, and blood vessels. Individuals with TSC often have multiple AML, and they are often not clinically significant. However, AML can grow to be quite large, particularly in women. A major concern for AML is the risk of hemorrhage that appears to be proportional to size. Hemorrhage from AMLs is thought to be due to dysplastic blood vessels that form pseudoaneurysms. Hemorrhage from renal AML is a major source of morbidity and occasionally mortality in adults with TSC.

It is also thought that renal cell carcinoma is more common in TSC than the general population and can affect children as well as adults. However, it is still rare, occurring in no more than 3% of TSC individuals. Renal cell carcinoma is the only malignancy at present known to occur at an increased frequency in TSC.

Distinguishing a renal cell carcinoma from a lipid-poor or minimal fat AML, which occur very commonly, can be difficult, as both appear as solid tumors on imaging studies. Computed tomography (CT) scan is at present the most sensitive imaging modality as it can detect very small quantities of fat. However, radiologic guided biopsy of these lesions is often required to rule out malignancy.

2.1.5 TSC and the Lung

Lung involvement in TSC occurs only after puberty and shows a striking predilection for women in contrast to men (Chapter 17). Lymphangioleiomyomatosis can be detected by radiologic studies in about one third of women with TSC. Although the rate is much lower, there are several case reports of LAM affecting men with TSC. LAM is characterized by a proliferation of smooth muscle cells in nodules throughout the lung and the destruction of normal lung parenchyma by multiple thin-walled cysts that appear throughout the lungs, from the apices to the bases. Although most TSC women with LAM are asymptomatic, for some it is a progressive disease with diminishing pulmonary function mainly due to progressive cystic destruction. Chest pain, spontaneous pneumothorax, or insidious dyspnea are common clinical presentations of LAM. Although there are ongoing clinical trials for the treatment of LAM there is no effective medical treatment at present. Lung transplantation remains the only treatment option for women with severe and progressive disease. It is not understood why LAM affects women almost exclusively, and has been rarely found in men. However, due to the concern that estrogen may play a role, women with TSC at risk for LAM are often advised to limit exposure to estrogens such as in oral contraceptive agents. Women with LAM are often advised to not become pregnant due to the possibility of disease exacerbation from the high estrogen and progesterone levels that occur.

In addition to LAM, both men and women with TSC are often found to have micronodular multifocal pneumocyte hyperplasia (MMPH) on chest imaging. MMPH is thought not to be a clinically significant finding and not to affect respiratory function. However, it is important to realize that MMPH is commonly seen in TSC, as the possibility of another metastatic process is often raised, creating significant anxiety and occasionally morbidity.

2.1.6 TSC and the Eye

The most common ophthalmologic manifestation of TSC is retinal hamartoma (Chapter 13). These are seen in about 50% of individuals with TSC and are generally benign lesions that rarely affect vision. However, the natural history of retinal hamartoma is poorly understood, and it is unclear if they can develop after birth or like cortical tubers develop embryonically. The association between retinal hamartoma and subependymal nodules and subependymal giant cell astrocytomas has not been characterized, although pathologically they appear similar.

2.1.7
TSC and the Other Organ Systems

Many other organ systems can also be affected by TSC, although their involvement has not been well characterized, is not well understood, and is often thought not to be clinically significant. However, many of these features are recognized to occur commonly enough making up many of the current minor criteria for the clinical diagnosis of TSC.

The gastrointestinal tract is frequently involved in TSC, as at least 25% of individuals are found to have hepatic AML or lipoma and many have rectal polyps. AML are also often seen in the pancreas and other abdominal organs, and again these are typically thought not to be clinically significant although the natural history of these in large populations has not been studied.

The skeletal system is also frequently involved in TSC, although its involvement has also not been well characterized. Sclerotic bone lesions in the vertebrae and other bones and bone cysts are frequently noted on imaging reports. However, it is felt that these are in general not symptomatic and clinically significant, and whether their frequency is truly increased over that of the general population is unclear.

2.2
TSC: A Spectrum Across the Life Span

As described above, a variety of organ involvements in TSC can and do occur at various times during an individual's life. For every aspect of the disorder, there is a spectrum of severity, from those mildly affected to those severely affected. Historically, the concept of being "severely affected" referred to those with significant cognitive impairment and refractory epilepsy. However, with growing appreciation of the possible profound impact of many of the other manifestations of TSC, an individual "mildly affected" throughout childhood could potentially become "severely affected" in adulthood by developing progressive LAM or significant renal involvement. It is this continuing risk for development of new problems that leads patients and families to feel as though they are "walking in a mine field."

However, it is important to note that most individuals with TSC will have a normal life expectancy. Nonetheless, several aspects of TSC could impact a particular individual's life expectancy, including highly refractory epilepsy, polycystic kidney disease, AML with life-threatening hemorrhage, and progressive LAM. Therefore, it is very important that individuals with TSC and their health care providers be knowledgeable about the various manifestations of TSC and health risks at every age range and be aware of current treatment recommendations and options.

In order to minimize the potential complications of TSC, all children and adults with TSC should be followed on a regular basis. The most recent guidelines for ongoing surveillance were developed at the 1998 Tuberous Sclerosis Consensus Conference [12]. The recommendations were based on the known natural history of the various manifestations of TSC, particularly those that have the potential for significant morbidity and for which effective interventions exist. In children and adolescents with TSC, this includes an annual physical examination, including evaluation of dermatologic involvement, and an annual brain MRI to monitor the development of a subependymal giant cell astrocytoma. Due to the risk of cognitive and neurobehavioral difficulties, neuropsychological testing should be done at the time of diagnosis, and if diagnosis is made in infancy or early childhood, then it should be done again at the time of school entry, approximately 6 years of age. Subsequent testing should be determined on an individual basis in order to maximize the neurocognitive development and performance of every person with TSC. Abdominal imaging should be done every 1–3 years throughout life, and MRI is becoming the preferred technique due to better definition and resolution of AML. A reasonable strategy is to repeat MRI imaging every 2 years if there is no significant renal involvement; if there are significant AML or renal cysts or a significant change from most recent testing, imaging should be repeated at shorter intervals. It is very important that adults with TSC continue to have regular abdominal imaging, particularly because hemorrhage of AML is a major source of morbidity and occasional mortality. Regular cardiologic and ophthalmologic follow-up is indicated only if an individual has significant involvement. A high-resolution chest CT is recommended in late adolescence for women with TSC to evaluate the possible presence of LAM; subsequent testing should be considered if respiratory symptoms develop.

Although extremely important now, close monitoring of every individual with TSC will become even more important in the future as specific and effective treatments for the various manifestations of TSC will likely become available. And specific recommendations for ongoing surveillance will also change as our knowledge of the natural history of the various manifestations improves.

2.3
TSC: A "Model" System

As already described in this chapter, and elsewhere in this book in some detail, TSC is a disorder that affects almost every organ system. There is a spectrum of involvement for each manifestation, with some individuals being mildly affected and some severely affected. As a group, individuals with TSC have a high incidence of many significant medical problems that are also experienced by other people, including epilepsy, infantile spasms, autism, benign tumor growth, and LAM. Since individuals with TSC have the same "cause" for these conditions – that is, TSC – TSC can then be viewed as a "model" system for studying and better understanding these various disorders. For example, understanding factors leading to autism spectrum disorders in individuals with TSC may lead to better understanding of autism in the general population.

Further characterizing and understanding of many features of TSC will also likely lead to improved understanding of tumor biology in general. For example, why do subependymal giant cell tumors lose the propensity to grow in late adolescence? Why do rhabdomyomas regress?

Tuberous sclerosis complex is a disorder that often has a profound impact on the life of the affected individuals and their families. Our growing understanding of the clinical features of the disorder and their natural history is already leading to improved clinical care of both children and adults living with TSC. Our evolving understanding of the cellular and molecular mechanisms involved in the pathogenesis of TSC is already leading to concepts of specific treatment of disease rather than treatment of symptoms. These advances will undoubtedly improve the lives of individuals with TSC and will also likely lead to improved understanding of many aspects of human growth, nutrition, and disease.

References

1 Roach, E.S., Gomez, M.R., and Northrup, H. (1998) Tuberous sclerosis complex consensus conference: revised clinical diagnostic criteria. *J. Child Neurol.*, **13**, 624–628.

2 Jozwiak, S., Schwartz, R.A., Krysicka-Janniger, C., and Bielicka-Cymerman, J. (2000) Usefulness of diagnostic criteria of tuberous sclerosis complex in pediatric patients. *J. Child Neurol.*, **15**, 562–569.

3 Jurkiewicz, E., Jozwiak, S., Bekiesinska-Figatowska, M., Pakula-Kosciesza, I., and Walecki, J. (2006) Cyst-like cortical tubers in patients with tuberous sclerosis complex: MR imaging with the FLAIR sequence. *Pediatr. Radiol.*, **36**, 498–501.

4 Chu-Shore, C.J., Major, P., Montenegro, M., and Thiele, E. (2009) Cyst-like tubers are associated with TSC2 and epilepsy in tuberous sclerosis complex. *Neurology*, **72**, 1165–1169.

5 Mizuguchi, M. and Hino, O. (2003) Neuropathology, in *Tuberous Sclerosis Complex: From Basic Science to Clinical Phenotypes* (ed. Paolo Curatolo), Mac Keith Press, London, pp. 264–278.

6 Franz, D.N., Leonard, J., Tudor, C., Chuck, G., Care, M., Sethuraman, G., Dinopoulos, A., Thomas, G., and Crone, K.R. (2006) Rapamycin causes regression of astrocytomas in tuberous sclerosis complex. *Ann. Neurol.*, **59**, 490–498.

7 Jansen, F.E., Vincken, K.L., Algra, A., Anbeek, P., Braams, O., Nellist, M., Zonnenberg, B.A., Jennekens-Schinkel, A., van den Ouweland, A., Halley, D., van Huffelen, A.C., and van Nieuwenhuizen, O. (2008) Cognitive impairment in tuberous sclerosis complex is a multifactorial condition. *Neurology*, **70**, 916–923.

8 Winterkorn, E.B., Pulsifer, M.B., and Thiele, E.A. (2007) Cognitive prognosis of patients with tuberous sclerosis complex. *Neurology*, **68**, 62–64.

9 Curatolo, P., Porfirio, C., Manzi, B., and Seri, S. (2004) Autism in tuberous sclerosis. *Eur. J. Paediatr. Neurol.*, **8**, 327–332.

10 Jóźwiak, S., Schwartz, R.A., Janniger, C.K., Michałowicz, R., and Chmielik, J. (1998) Skin lesions in children with tuberous sclerosis complex: their prevalence, natural course, and diagnostic significance. *Int. J. Dermatol.*, **37**, 911–917.

11 Jóźwiak, S., Kotulska, K., Kasprzyk-Obara, J., Domańska-Pakieła, D., Tomyn-Drabik, M., Roberts, P., and Kwiatkowski, D. (2006) Clinical and genotype studies of cardiac tumors in 154 patients with tuberous sclerosis complex. *Pediatrics*, **118**, e1146–e1151.

12 Roach, E.S., DiMario, F.J., Kandt, R.S., and Northrup, H. (1999) Tuberous sclerosis consensus conference: recommendations for diagnostic evaluation. National Tuberous Sclerosis Association. *J. Child Neurol.*, **14**, 401–407.

3
Diagnostic Criteria for Tuberous Sclerosis Complex

E. Steve Roach and Steven P. Sparagana

Introduction

Well-designed clinical diagnostic criteria allow the clinician to confirm the diagnosis of tuberous sclerosis complex (TSC) quickly and confidently [1]. Clinical diagnostic criteria cost nothing to use, can be applied in any clinical setting, provide immediate answers, and are generally reliable. These clinical diagnostic criteria remain the "gold standard" for TSC diagnosis, when applied by an experienced clinician. Although there are TSC patients in whom no mutation in TSC1 or TSC2 can be identified by current means, this does not mean that those individuals do not have TSC. There are several possible explanations for this discrepancy between molecular findings and clinical diagnosis (see Chapters 4 and 5), and a clinical diagnosis of TSC using these criteria should lead to patient care and management, independent of molecular diagnostic findings.

The numerous clinical signs that occur with TSC, the well-known phenotypic variability of TSC, and the fact that many of its clinical manifestations are age related can make the diagnosis of TSC difficult, especially in young individuals or in those with subtle findings [2, 3]. Another challenge when devising clinical diagnostic criteria is that some clinical signs are more specific for TSC than others even though no single lesion may be pathognomonic. Multiple facial angiofibromas and subependymal giant cell tumors, for example, are unusual in individuals without TSC. Although a few cutaneous hypomelanotic macules and isolated areas of focal cortical dysplasia may be seen in the general population, they occur with increased frequency and numbers among individuals with TSC. To be useful, clinical diagnostic criteria must achieve the right balance of sensitivity and specificity lest we assign a diagnosis to individuals who do not have TSC or overlook those who do.

The diagnostic triad of epilepsy, mental retardation, and adenoma sebaceum (facial angiofibromas) that was proposed by Campbell in 1906 and Vogt in 1908 constituted the first rudimentary diagnostic criteria for TSC [4, 5]. As simple diagnostic criteria, however, the Vogt triad lacked sensitivity and would miss up to half of the individuals with TSC based on today's understanding of its clinical signs. Later TSC diagnostic criteria, such as those proposed by Gomez in 1979 and those offered by Roach *et al.* in 1992, included more of the lesions we now associate with

Tuberous Sclerosis Complex: Genes, Clinical Features, and Therapeutics
Edited by David J. Kwiatkowski, Vicky Holets Whittemore, and Elizabeth A. Thiele

ISBN: 978-3-527-32201-5

Table 3.1 Clinical diagnostic criteria for tuberous sclerosis complex.

Major features
Facial angiofibromas or forehead plaque
Nontraumatic ungual or periungual fibroma
Hypomelanotic macules (more than three)
Shagreen patch (connective tissue nevus)
Cortical tuber[a]
Subependymal nodule
Subependymal giant cell tumor (SGCT)
Multiple retinal nodular hamartomas
Cardiac rhabdomyoma, single or multiple
Lymphangioleiomyomatosis[b]
Renal angiomyolipomas (AML)[b]
Minor features
Multiple randomly distributed pits in dental enamel
Hamartomatous rectal polyps[c]
Bone cysts[d]
Prominent white-matter "migration tracts"[a,d]
Gingival fibromas
Nonrenal hamartoma[c]
Retinal achromic patch
"Confetti" skin lesions
Multiple renal cysts[d]

Adapted from Roach *et al.* [1]. Definite TSC: either two major features or one major feature plus two minor features; Probable TSC: one major feature plus one minor feature; and Suspected TSC: either one major feature or two or more minor features.

a) When cerebral cortical dysplasia and cerebral white-matter migration tracts occur together, they should be counted as one rather than two features of tuberous sclerosis complex.
b) When both lymphangioleiomyomatosis and renal angiomyolipomas are present, other features of tuberous sclerosis complex should be present before a definite diagnosis is assigned.
c) Histologic confirmation is suggested.
d) Radiographic confirmation is sufficient.
Definite TSC: Either two major features or one major feature plus two minor features.
Probable TSC: One major plus one minor feature.
Suspected TSC: Either one major or two or more minor features.

TSC, but both of these sets of criteria assumed that some lesions were pathognomonic when they are probably not [6, 7].

The consensus diagnostic criteria presented in this chapter were devised by a panel of international experts in 1998 at the Tuberous Sclerosis Complex Consensus Conference in Annapolis, Maryland [1, 8]. The 1998 criteria (Table 3.1) reflect a more current understanding of the clinical features of TSC and its genetic and molecular mechanisms. Essential to the revised criteria was the agreement among the experts that there may be no truly pathognomonic lesions for TSC because the signs that were at one time believed to be specific sometimes occur as isolated findings in individuals with no other clinical or genetic evidence of TSC. Therefore, the 1998 criteria require TSC-associated lesions of two or more organ systems, or at least two dissimilar lesions of the same organ, in order to confirm the diagnosis.

Table 3.2 Frequency of lesions in individuals with TSC versus other individuals.

Lesion	Tuberous sclerosis complex	Other individuals
Hypomelanotic macules	Occur in over 95% of TSC patients, often with many lesions [11]	Occur in up to 5% of the population (but usually fewer than three lesions per person) [12]
Facial angiofibromas	Eventually seen in 75% but less often in children [11]	Seen in individuals with multiple endocrine neoplasia type 1 and in a few sporadic families [13]
Shagreen patch	Up to 48% [11]	Occasional
Ungual fibromas	Seen in 15% but often not until develop in adulthood [11]	Occasionally sporadic or after nail trauma (but typically one lesion) [14]
Rhabdomyomas	One or more tumors seen in 47–65%, but much more common below 2 years of age. Up to 51% of patients with rhabdomyomas have TSC [15–17]	In 14–49% of rhabdomyoma patients, there are no other signs of TSC [16]
Renal AML	Often multiple AML occur in up to 80% of TSC patients by age 10 [18]	Sporadic AML occur but are typically solitary
Renal cysts	Polycystic kidneys occur in 3–5% of TSC patients. Smaller numbers of renal cysts are present in 15–20% [18]	There are both dominant and recessive polycystic kidney diseases. A few cysts are frequent sporadic findings
Cortical dysplasia/ tubers	90–95% and usually multiple lesions are present (MRI yields highest detection rate) [25]	Sporadic cortical dysplasia (typically one lesion) is common among individuals who have epilepsy not due to TSC
Subependymal nodules	83–93% [2, 25]	Rare, especially if calcified
Subependymal giant cell tumors	Up to 15% (using radiographic criteria) [19]	Rare in the absence of TSC
Cerebral migration tracks	Common (up to 40%) and sometimes prominent [20]	Common but often subtle
Lymphangioleio-myomatosis	Up to 40% of adult women with TSC (many asymptomatic)	About half of those affected with LAM do not have TSC
Retinal lesions	Occur in up to 87% of TSC patients when examined under ideal conditions [21]	Occasional sporadic retinal hamartomas
Dental pitting	Occur in up to 90–100% of adults and children [22, 23]	Found in 7% of adult controls and no childhood controls [22, 24]
Oral fibroma	Found in 69% of TSC patients [23]	Sporadic (usually single or fewer) [23]

Although the panel decided that no single lesion can be considered pathognomonic for TSC, some lesions are clearly more suggestive of the diagnosis and easier to confirm than others (Table 3.2). Facial angiofibromas, for example, are more likely to indicate TSC than hypomelanotic confetti lesions of the skin. The diagnostic criteria panel addressed this concern by creating two categories of signs, a group of "major" manifestations whose link to TSC is well accepted and believed to be more specific for TSC and a group of "minor" signs that are accepted findings whose specificity is lower or unknown. This tiered approach also created a way to estimate the level of certainty of a given individual's diagnosis. The diagnosis of TSC is considered definite if a person has two major manifestations or one major plus two minor manifestations, probable in an individual with one major feature plus one minor finding, and possible when there is one major finding or two or more minor findings (Table 3.1) [1].

The 1998 panel elected not to include epilepsy, autism, or mental retardation as either major or minor features in the TSC diagnostic criteria. The group reasoned that epilepsy and mental retardation were both so common in the general population, and their causes so numerous, that neither condition had enough specificity to be useful in the diagnosis [1]. In addition, most patients with epilepsy and mental retardation have cortical brain lesions on neuroimaging studies, and the number of such lesions tends to increase in rough proportion to the severity of the neurological problems [9, 10]. The panel was concerned that including both the symptoms and the lesions that caused them amounted to counting the same feature twice.

The consensus criteria were developed to create a reliable standard for the clinical diagnosis of TSC. These criteria generally compare favorably to the results of gene mutational analysis, and in fact the criteria sometimes identify individuals who are difficult to confirm with DNA testing due to mosaicism or other reasons. It is difficult, however, to devise criteria that will completely *exclude* TSC in affected individuals with few TSC signs or in young patients who have yet to develop the full array of manifestations. Nevertheless, the criteria are still useful even in very young patients or when the diagnostic evaluation is incomplete, because they can be used to assign a probable or suspected TSC diagnosis that can be periodically reassessed as additional findings manifest themselves [2].

The revised diagnostic criteria (Table 3.1) have been widely disseminated and have shown great utility in establishing the clinical diagnosis of TSC. Standardized criteria provide a quick, reliable, and economical method of establishing a diagnosis, and they help to ensure uniformity in clinical TSC research. Identification of a known disease-causing TSC gene mutation can also confirm the diagnosis, but mutational analysis is not always required. The addition of DNA-based testing complements clinical diagnosis and allows more precise genetic counseling and, in some cases, prenatal diagnosis.

References

1 Roach, E.S., Gomez, M.R., and Northrup, H. (1998) Tuberous sclerosis complex consensus conference: revised clinical diagnostic criteria. *J. Child Neurol.*, **13**, 624–628.

2 Jozwiak, S., Schwartz, R.A., Janniger, C.K., and Bielicka-Cymerman, J. (2000) Usefulness of diagnostic criteria of tuberous sclerosis complex in pediatric patients. *J. Child Neurol.*, **15**, 652–659.

3 Au, K.S., Williams, A.T., Roach, E.S. *et al.* (2007) Genotype/phenotype correlation in 325 individuals referred for a diagnosis of

tuberous sclerosis complex in the United States. *Genet. Med.*, **9**, 88–100.

4 Campbell, A.W. (1906) Cerebral sclerosis. *Brain*, **28**, 382–396.

5 Vogt, H. (1908) Diagnostik der tuberösen sklerose. *Zeitschrift für die Erforschung und Behandlung des jugendlichen Schwachsinns auf Wissenschaftlicher Grundlage*, **2**, 1–16.

6 Gomez, M.R.(ed.) (1979) *Tuberous Sclerosis*, 1st edn, Raven Press, New York.

7 Roach, E.S., Smith, M., Huttenlocher, P., Bhat, M., Alcorn, D., and Hawley, L. (1992) Diagnostic criteria: tuberous sclerosis complex. Report of the Diagnostic Criteria Committee of the National Tuberous Sclerosis Association. *J. Child Neurol.*, **7**, 221–224.

8 Hyman, M.H. and Whittemore, V.H. (2000) National institutes of health consensus conference: tuberous sclerosis complex. *Arch Neurol.*, **57**, 662–665.

9 Roach, E.S., Williams, D.P., and Laster, D.W. (1987) Magnetic resonance imaging in tuberous sclerosis. *Arch Neurol.*, **44**, 301–303.

10 Goodman, M., Lamm, S.H., Engel, A., Shepherd, C.W., and Gomez, M.R. (1997) Cortical tuber count: a biomarker indicating cerebral severity of tuberous sclerosis complex. *J. Child Neurol.*, **11**, 85–90.

11 Jozwiak, S., Schwartz, R.A., Janniger, C.K., Michalowicz, R., and Chmielik, J. (1998) Skin lesions in children with tuberous sclerosis complex: their prevalence, natural course, and diagnostic significance. *Int. J. Dermatol.*, **37**, 911–917.

12 Vanderhooft, S.L., Francis, J.S., Pagon, R.A., Smith, L.T., and Sybert, V.P. (1996) Prevalence of hypopigmented macules in a healthy population. *J. Pediatr.*, **129**, 355–361.

13 Darling, T.N., Skarulis, M.C., Steinberg, S.M., Marx, S.J., and Turner, M. (1997) Multiple facial angiofibromas and collagenomas in patients with multiple endocrine neoplasia type 1. *Arch Dermatol.*, **133**, 853–857.

14 Zeller, J., Friedmann, D., Clerici, T., and Revuz, J. (1995) The significance of a single periungual fibroma: report of seven cases. *Arch Dermatol.*, **131**, 1465–1466.

15 Gibbs, J.L. (1985) The heart and tuberous sclerosis. An echocardiographic and electrocardiographic study. *Br. Heart J.*, **54**, 596–599.

16 Harding, C.O. and Pagon, R.A. (1990) Incidence of tuberous sclerosis in patients with cardiac rhabdomyoma. *Am. J. Med. Genet.*, **37**, 443–446.

17 Jozwiak, S., Kawalec, W., Dluzewska, J., Daszkowska, J., Mirkowicz-Malek, M., and Michalowicz, R. (1994) Cardiac tumors in tuberous sclerosis: their incidence and course. *Eur. J. Pediatr.*, **153**, 155–157.

18 Ewalt, D.H., Sheffield, E., Sparagana, S.P., Delgado, M.R., and Roach, E.S. (1998) Renal lesion growth in children with tuberous sclerosis complex. *J. Urol.*, **160**, 141–145.

19 Torres, O.A., Roach, E.S., Delgado, M.R. *et al.* (1998) Early diagnosis of subependymal giant cell astrocytoma in patients with tuberous sclerosis. *J. Child Neurol.*, **13**, 173–177.

20 Iwasaki, S., Nakagawa, H., Kichikawa, K. *et al.* (1990) MR and CT of tuberous sclerosis: linear abnormalities in the cerebral white matter. *AJNR Am. J. Neuroradiol.*, **11**, 1029–1034.

21 Kiribuchi, K., Uchida, Y., Fukuyama, Y., and Maruyama, H. (1986) High incidence of fundus hamartomas and clinical significance of a fundus score in tuberous sclerosis. *Brain Dev.*, **8**, 509–517.

22 Mlynarczyk, G. (1991) Enamel pitting: a common symptom of tuberous sclerosis. *Oral Surg. Oral Med. Oral Pathol.*, **71**, 63–67.

23 Sparling, J.D., Hong, C.H., Brahim, J.S., Moss, J., and Darling, T.N. (2007) Oral findings in 58 adults with tuberous sclerosis complex. *J. Am. Acad. Dermatol.*, **56**, 786–790.

24 Russell, B.G., Russell, M.B., Praetorius, F., and Russell, C.A. (1996) Deciduous teeth in tuberous sclerosis. *Clin. Genet.*, **50**, 36–40.

25 Datta, A.N., Hahn, C.D., and Sahin, M. (2008) Clinical presentation and diagnosis of tuberous sclerosis complex in infancy. *J. Child Neurol.*, **23**, 268–273.

Part II
Genetics

Tuberous Sclerosis Complex: Genes, Clinical Features, and Therapeutics
Edited by David J. Kwiatkowski, Vicky Holets Whittemore, and Elizabeth A. Thiele

ISBN: 978-3-527-32201-5

4 Genetics of Tuberous Sclerosis Complex

David J. Kwiatkowski

4.1 Introduction

Over the past two decades, major progress has been made in our understanding of the human molecular genetics of tuberous sclerosis complex (TSC). In this chapter, these advances have been reviewed in detail, beginning with a historical review of the genetics, cloning, and initial characterization of the *TSC1* and *TSC2* genes. The genomic structure of each of these genes and their place in the surrounding genome are then reviewed. Structural features of the predicted proteins, alternative splice variants, and functional domains within the two proteins are then discussed. A comprehensive compilation and analysis of all reported *TSC1* and *TSC2* mutations is then presented. The potential role of these genes in malignancies beyond their role in TSC is then considered. Finally, consideration is given to special issues in TSC genetics, including the frequency and significance of mosaicism, patients in whom no mutation has yet been identified, and what the future is likely to hold in terms of the molecular diagnosis and genetic evaluation of TSC patients.

4.2 Historical Review of Linkage Analysis and Positional Cloning of the *TSC1* and *TSC2* Genes

4.2.1 Initial Linkage Studies

Family studies dating back to the early 1900s indicated that TSC was inherited as an autosomal dominant trait, with direct transmission from parent to child. Genetic linkage analysis in TSC families was initiated in the 1980s and led to the identification of linkage between TSC and markers on chromosome 9q34 in 1987 [1]. As this was the first TSC locus to be identified by linkage, it was named *TSC1*. Subsequent

Tuberous Sclerosis Complex: Genes, Clinical Features, and Therapeutics
Edited by David J. Kwiatkowski, Vicky Holets Whittemore, and Elizabeth A. Thiele

ISBN: 978-3-527-32201-5

linkage studies in other TSC families clearly indicated that there was locus heterogeneity in TSC (meaning that there was more than one gene involved) [2–6]. A multicenter linkage study on families not linked to *TSC1* then led to the identification of linkage to a second locus on chromosome 16p13.3, denoted *TSC2* [7]. Subsequently, in a multicenter study of families large enough to permit linkage analysis, approximately half showed linkage to 9q34 and half to 16p13, and there was no conclusive evidence for a third locus [8, 9]. Although many additional studies have been performed over the years, there is still no convincing evidence for a third TSC locus.

4.2.2
Positional Cloning of *TSC2* (1993)

Linkage studies, aided by the numerous genetic markers and genomic reagents developed in the effort to clone the gene (*PKD1*) for autosomal dominant polycystic kidney disease (ADPKD), were able to refine the *TSC2* candidate locus to a relatively small 1.5 Mb region [7, 10]. Subsequently, a family in which both TSC and ADPKD were seen was identified by an astute clinician and analyzed in detail [11]. Both the mother and the daughter each carried a balanced translocation involving 16p13.3 [46,XX, t(16;22)(p13.3;q11.21]. They had fairly typical ADPKD, but no evidence of TSC. A son had severe developmental delay, as well as seizures and autistic features. He had an unbalanced karyotype, 45, XY, −16, −22, +der(16)(16qter → 16 p13.3::22q11.21 → 22qter) and was therefore lacking one copy of the chromosomal regions 16p13.3 → 16pter and 22q11.21 → 22pter. Evaluation by an expert clinician demonstrated that he had both typical skin features of TSC, facial angiofibromas, and hypopigmented macules, as well as diagnostic brain features (subependymal calcification on CT scan).

The translocation breakpoint on chromosome 16 in this family was shown to disrupt the previously unidentified *PKD1* gene [12]. It was inferred that the son had TSC because the deleted region of his derivative chromosome 16 (16p13.3 → 16pter) contained the *TSC2* gene [11]. In addition, a *de novo* truncation of 16p was identified in a separate family with no clinical or radiological evidence of TSC [13]. The breakpoints in both families were mapped using a combination of fluorescence *in situ* hybridization (FISH) and Southern blotting using standard as well as pulsed field gel electrophoresis (PFGE). This reduced the region in which the *TSC2* gene was localized to a relatively small 300 kb interval. Facilitated by genomic resources gathered during the search for the *PKD1* gene, PFGE and Southern blotting were used to look for additional large genomic deletions. Five TSC patients were found to have large deletions involving the same 120 kb interval. cDNA clones were isolated corresponding to four genes from the interval, and one was found to be disrupted by all five deletions, making it a strong candidate for *TSC2*. Several additional smaller intragenic deletions were then found in this same gene in other TSC patients, including a *de novo* deletion. These findings confirmed the identity of the *TSC2* gene. The *TSC2* mRNA was predicted to encode a 198 kDa protein that was named tuberin [11].

4.2.3 Positional Cloning of *TSC1* (1997)

Though it was defined by linkage several years before *TSC2*, the positional cloning of *TSC1* took much longer. There were a number of reasons for this: mutations in the *TSC1* gene occur less frequently than in the *TSC2* gene (see further below); genomic resources for the *TSC1* region were less developed than for *TSC2*; and large genomic deletions at the *TSC1* locus are quite rare, so the attempts to refine the location of the gene by deletion mapping were not productive. Consequently, a detailed mutational analysis of each of many genes in the region had to be performed.

Since there were no chromosomal translocations or large genomic deletions, the *TSC1* candidate region on chromosome 9q34 was defined by the identification of key meiotic recombination events in large *TSC1* families [14, 15]. Two putative recombinants in unaffected individuals (which ultimately proved reliable) helped to narrow the candidate region to a 900 kb interval.

The *TSC1* region was unusually gene rich, with over 30 genes in the critical region. Several of these were assessed as candidates for *TSC1* without success [16]. Complete genomic sequencing of the region was initiated (as part of the steadily growing human genome project) and systematic mutation screening of exons using heteroduplex analysis was performed on a panel of unrelated familial TSC cases linked to 9q34. This comprehensive and laborious process finally encountered some good fortune, with evidence of mutation identified in an exon relatively early on [16]. This exon corresponded to a previously identified cDNA clone, which was extended by the usual techniques to obtain a full-length cDNA sequence. The *TSC1* mRNA was predicted to encode an 1164-amino acid/130 kDa protein that was named hamartin.

4.3 The *TSC1* and *TSC2* Genes: Genomic Structure, Splicing, Predicted Sequences, and Domains

4.3.1 Genomic Structure and Location of *TSC1* and *TSC2*

**TSC1* consists of 53 284 nucleotides (nt) from nt position 134 756 557 to 134 809 841 on chromosome 9q34 (March 2006, Human Genome Assembly, genome.ucsc.edu) (Figure 4.1). Approximately 30 kb upstream of the transcriptional start site of *TSC1* is the adjacent *GFI1B* gene, which encodes a zinc finger protein. Just 1 kb downstream of the 3′ untranslated region (UTR) of *TSC1* is the putative gene *C9orf9*. *TSC1* consists of 23 exons. The first two exons are relatively widely spaced, are a combined 154 nt in length, and comprise a portion of the 234 nt 5′ UTR of *TSC1*, upstream of the initiation codon (ATG) in exon 3. Exons 3 through 23 contain the coding sequence of the gene. Exons 3–14 and 16–22 are between 44 and 188 nt in size, while exon 15 is 559 nt in size. Exon 23 has 517 coding nt prior to the stop codon, followed by a 4885 nt 3′ UTR.

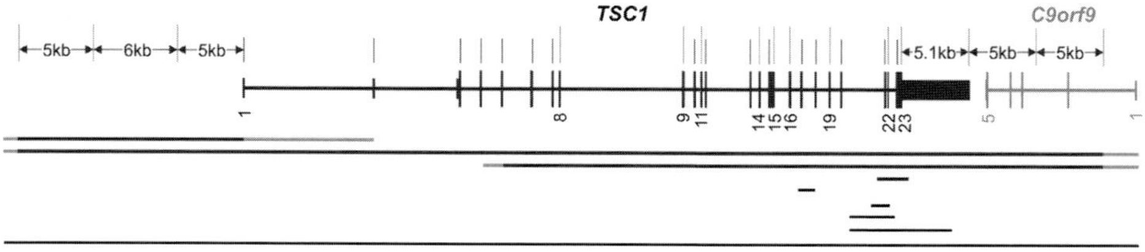

Figure 4.1 Map of *TSC1* and genomic deletion mutations. A map of *TSC1* and the adjacent flanking regions is shown at the top. Exons are indicated by black vertical lines with proportional spacing and width. Selected exons are numbered. Black and gray lines above indicate the position of MLPA probes. At bottom, each of nine deletions is represented by a black horizontal line, indicating the minimum deleted region. For those deletions not confirmed by breakpoint sequencing, gray lines indicate the maximum possible extent of the deletion. From Ref. [44]; with the bottom five lines from patients described in Refs [54, 55] and Dabora *et al.* unpublished observations.

TSC2 consists of 40 723 nt from nt position 2 037 991 to 2 078 714 on chromosome 16p13 (March 2006, Human Genome Assembly, genome.ucsc.edu) (Figure 4.2). It is in a highly gene-dense region. Just 123 nt upstream of the first exon of *TSC2*, in the opposite transcriptional orientation, lies the *NTLH1* gene. The 3′ UTR of *TSC2* abuts the 3′ UTR of the *PKD1* gene. *TSC2* is composed of 42 exons. Since the first exon was discovered several years after the identification of the gene, the first exon is usually called exon 1a, and the remaining exons are called exons 1 through 41. Exons 1–40 range from 49 to 213 nt in size, with the exception of exon 33 that consists of 488 nt. The 5′ UTR of *TSC2* consists of 106 nt in exons 1a and 1. The 3′ UTR consists of 101 nt following the stop (TGA) codon in exon 41.

4.3.2
Alternative Splicing of *TSC1* and *TSC2*

The only alternative splicing that is known for *TSC1* involves the absence of the noncoding exon 2 in some mRNAs. This has no effect on the encoded protein.

For *TSC2*, there is alternative splicing of moderate complexity. Some *TSC2* mRNA isoforms lack exon 25, the first 3bp of exon 26, and exon 31, individually or in combination [17–19]. Other possible splice variants have been suggested by the analysis of murine *TSC2* cDNAs [20]. Variation in the expression of the different *TSC2* mRNAs has been observed in various tissues and developmental stages, but without a clear pattern. The occurrence of the same alternative splice forms in all vertebrate organisms examined to date, including fish (*Fugu*), suggests that they have some functional and/or developmental significance. However, the absence of confirmed mutations in either exon 25 or 31 (see below) suggests that they do not have an important function. Many cell lines, including human lymphoblastoid and fibroblast lines, express large amounts of the isoform lacking exon 25. Functional differences among the isoforms are unknown at this time.

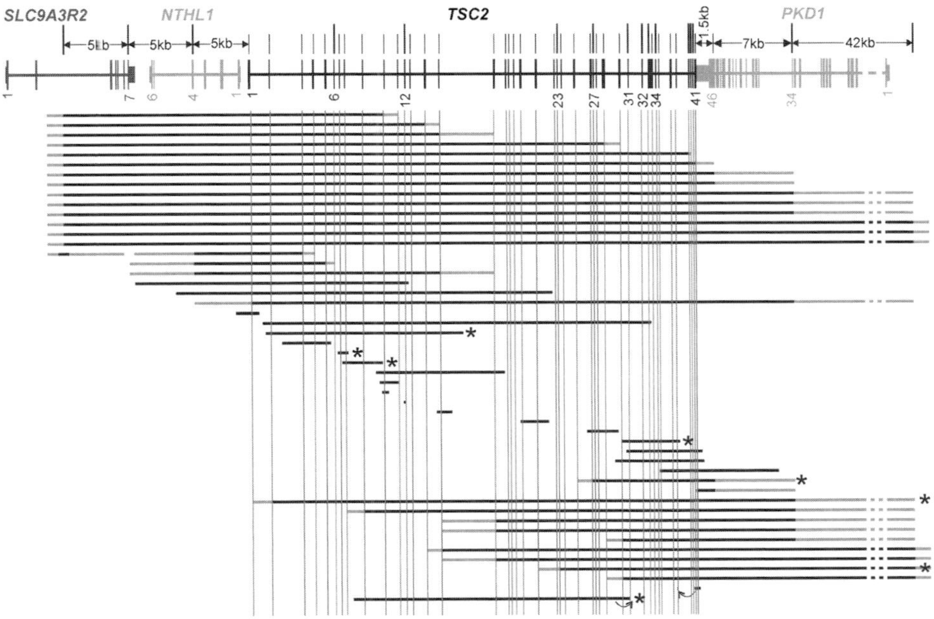

Figure 4.2 Map of *TSC2* and genomic deletion/duplication mutations. A map of *TSC2* and the adjacent flanking regions is shown at the top. Exons are indicated by black vertical lines with proportional spacing and width. Selected exons are numbered. Gray and black lines above indicate the position of MLPA probes. At bottom, each of 48 deletions is represented by a black horizontal line, indicating the minimum deleted region. Gray lines indicate the maximum possible extent of those deletions not confirmed by the breakpoint sequencing. Two duplications are shown at the bottom with black lines indicating the size of the duplicated region and an arrow showing the position of insertion. Mosaic mutations are indicated by asterisks. From Ref. [44].

4.3.3
Interspecies Comparisons of *TSC1* and *TSC2*

Thanks to ongoing efforts to sequence more and more genomes, there is considerable information on the orthologues of both *TSC1* and *TSC2* in other species. The three primate (orangutan, chimpanzee, and macaque) hamartin/TSC1 protein sequences that are available are 98–99% identical to that of human; while for tuberin/TSC2, this comparison shows 97–98% identity. Among other sequenced mammals (dog, cow, horse, mouse, and rat), this falls to 86–90% amino acid identity for TSC1 and 91–92% identity for TSC2. For chicken, platypus, and opossum, the rate of identity falls to 75–85% for TSC1 and 75–84% for TSC2. Two species of fish have been sequenced with 41–44% amino acid identity for TSC1 and 58–64% identity for TSC2. The fruit fly and honeybee sequences show 25–35% amino acid identity to each of TSC1 and TSC2, but in general over short regions of the protein, not over the entire length. The single region of TSC2 in which the amino acid sequence shows a higher level of conservation between species is near the COOH-terminus. This region consists of amino acids 1495–1756 (encoded by exons 34–40) of human TSC2 and is, for example, 47% identical to a similar region in *Drosophila melanogaster* (fly) TSC2. This region contains the GTPase activating protein (GAP) domain of TSC2 (see further below). Other shorter regions of sequence conservation are also seen in TSC2, and these may reflect regions of conserved function as well. However, this is uncertain. In contrast, most of the human TSC2 sequence shows no significant similarity to *Drosophila* TSC2.

In TSC1, regions of sequence conservation with *Drosophila* are seen near the N-terminus (residues 7–298; 31% identity) and C-terminus (residues 704–963; 25% identity) of the protein. However, the degree of amino acid conservation is much lower than for TSC2. Nonetheless, these observations suggest that these regions may have important functions.

4.3.4
Predicted Amino Acid Sequences of *TSC1* (Hamartin) and *TSC2* (Tuberin) and Their Functional Domains

Using the definition of a domain as a functional subunit of a protein that is stable when expressed by itself and has binding interactions and functions that are similar to that of the native full-length protein, the only true domain known for either TSC1 or TSC2 is the TSC2 GAP domain [21, 22]. However, there are several other regions in each protein that have been shown by truncation and mutation studies to have important binding activities [11]. The TSC2 GAP domain consists of ~160 amino acids and has homology to a region within the rap1GAP protein [23]. Although TSC2 was initially thought to function as a GAP for rap1, this was subsequently shown not to be the case. Instead, its GAP function is directed against the relatively unique ras family member, rheb (see Chapter 6) [11].

TSC1 and TSC2 interact to form a stable protein complex [24, 25] and amino acids 1–418 in TSC2 have been shown to bind to amino acids 302–430 in TSC1 [24]. However, the structure of the interface between these two regions is unknown. Both TSC1 and TSC2 are subject to multisite phosphorylation by a variety of kinases that regulate the intracellular localization and GAP activity of the TSC1–TSC2 complex (see Chapter 6). These sites are generally highly conserved among mammalian species and to a lesser extent among other species.

4.4
Mutational Spectrum of *TSC1* and *TSC2*

4.4.1
Introduction

Over 1500 mutations in TSC1 and TSC2 have been reported. For an earlier, detailed review of mutation analysis of these genes, see Ref. [26]. Here, all the reported mutations and sequence variants in these genes have been reviewed based upon the TSC1/TSC2 mutation database (http://chromium.liacs.nl/LOVD2/TSC/home.php), which has been generated, maintained, and updated through the continuing efforts of Rosemary Ekong and Sue Povey. This is an outstanding resource and is only limited by the fact that molecular diagnostic findings from some laboratories are not represented, or are outdated by several years. Nonetheless, it is unlikely that any major changes in the spectrum of mutations will occur, even if the database were doubled in size. Another source of information for this chapter was a review of the published literature on mutations in TSC1/TSC2 [27–46]. Major publications reporting mutation identification in these genes are summarized in Table 4.1.

A major consideration during the preparation of this chapter was how to interpret and classify all of the sequence variants that have been identified in TSC1 and TSC2. For many of the sequence variants identified, there is a clear and nearly certain adverse effect on the protein sequence, such that their status as mutation is quite clear. These include nonsense mutations and insertion and deletion mutations in which the reading frame is shifted (the number of bases added or lost is not divisible by 3). However, there are many other variants for which their status as disease causing is less certain. These include missense mutations, splice site mutations, and insertion and deletion mutations for which the reading frame is preserved (number of nucleotides gained or lost is divisible by 3). To assess each reported nucleotide change and to determine whether it was disease causing (pathogenic) or just a neutral sequence variant (benign polymorphism), a set of rules has been applied here. First, all sequence changes affecting the conserved splice set nucleotides were considered disease causing unless there was evidence to the contrary (nucleotide changes at the -1 and $+1$ through $+5$ positions relative to 5′ splice site, or at the -1 and -2 positions relative to the 3′ splice site) [47, 48]. Second, all in-frame insertion and deletion sequence changes were considered to be pathogenic, unless there was evidence to the contrary. Finally, missense changes were considered to be disease causing if (1) they were shown to be present in a sporadic TSC patient and not in that individual's parents; or (2) they showed segregation with disease in a family with TSC; and (3) the DNA sample had been screened without success for other *TSC1* and *TSC2* mutations. In the absence of definitive information on the screening of unaffected parents or other family members, missense mutations were considered pathogenic if they resulted in a significant alteration in the encoded amino acid (a negative score in the amino acid substitution matrix) [49] and there was no other mutation identified in the patient sample analyzed; or if the mutation had been shown to have reduced function in a cell-based assay [50, 51]. Using these criteria, there is little doubt that numerous miscalls have been made in both directions: judging some benign sequence variants as disease causing and judging some disease-causing mutations as benign sequence variants. Among this latter category are intronic and exonic variants that have a significant, adverse effect on splicing, but are not recognized as such. Nonetheless, although these miscalls are very important for the individual patient with the variation, they are relatively minor in viewing the overall spectrum of mutations in *TSC1* and *TSC2*. (See the glossary of terms to assist in understanding genetic variation and mutation, and the terminology used here.)

Table 4.1 Major published reports on mutation identification in *TSC1/TSC2*.

Mutation detection rate	*TSC1* : *TSC2* ratio	Methods	Population screened, origin	Mosaicism rate	Reference
25/27 (93%)	*TSC2/PKD1*	PFGE, Soublot, FISH	TSC + renal cystic disease, England	7/25	[27]
22/90 (24%)	*TSC2* only	SSCP TSC2	Texas	NR	[28]
27/79 (34%)	*TSC1* only	SSCP, SouBlot	England	1/27	[29]
23/38 (61%)	7 : 16	SSCP, seq	Japan	NR	[30]
28/48 (58%)	9 : 19	PTT	Germany	1/28	[31]
74/126 (59%)	16 : 58	SSCP	Boston	NR	[32]
120/150 (80%)	22 : 98	HD, SSCP, PFGE. Reg GE, LR-PCR	England	4/6 large deletions NR in others	[33]
29/225 (13%)	*TSC1* only	SSCP, SouBlot	The Netherlands	NR	[34]
10/27 (37%)	4 : 6	SSCP, seq	Japan	NR	[35]
186/224 (83%)	28 : 158	DHPLC, LRPCR, QPCR	Boston	NR	[36]
31/68 (46%)	2 : 29	SSCP	Germany	NR	[37]
6/33 (18%)	*TSC2* only	SSCP	Turkey	NR	[38]
51/65 (78%)	11 : 40	DGGE, MLPA, seq	Denmark	2/51	[39]
235/276 (85%)[a)]	53 : 182	DGGE, SSCP, seq, FISH, SouBlot	The Netherlands	8/235	[40]
12/24 (50%)	1 : 11	SSCP	India	NR	[41]
13/44 (30%)	2 : 11	DHPLC	Korea	NR	[42]
64/84 (76%)	9 : 55	DHPLC/seq	Taiwan	NR	[43]
54/261 (21%)	4 : 50	MLPA	Most with small mutation negative; Boston/ international	8/54	[44]
252/325 (78%)	61 : 182	Seq	Texas	NR	[45]
48/53 (91%)	17 : 31	SSCP, DGGE, seq, SouBlot, FISH, MLPA	The Netherlands	NR	[46]
Total	246 : 946 (21%:79%)				

PFGE, pulsed field gel electrophoresis; SouBlot, Southern blotting; FISH, fluorescence *in-situ* hybridization; seq, sequencing; PTT, protein truncation test; HD, heteroduplex analysis; LR-PCR, long-range PCR; DHPLC, denaturing high-pressure liquid chromatography; QPCR, quantitative PCR; DGGE, denaturing gradient gel electrophoresis; MLPA, multiplex ligation-dependent probe assay; NR, not reported.

a) Definite TSC cases only, in a mixed series.

Table 4.2 *TSC1* small mutation summary.

	Total		Unique	
Deletions	161	35.5%	111	40.7%
Deletions in-frame	6	1.3%	6	2.2%
Deletions–insertions	1	0.2%	1	0.4%
All deletions	168	37.1%	118	43.2%
Insertions	67	14.8%	51	18.7%
Missense	14	3.1%	14	5.1%
Nonsense	161	35.5%	62	22.7%
Splice	41	9.1%	26	9.5%
Splice deletions	1	0.2%	1	0.4%
Splice insertions	1	0.2%	1	0.4%
All splice	43	9.5%	28	10.3%
Total	453		273	

4.4.2
Overview of Types of Mutation and Mutation Frequencies for *TSC1* and *TSC2*

According to the TSC1/TSC2 mutation database, 453 small mutations have been identified in the *TSC1* gene, of which 273 are unique (Table 4.2). In the *TSC2* gene, 1162 small mutations have been identified, of which 714 are unique (Table 4.3). The term "small mutation" is used to denote mutations that affect just a single exon. Mutations affecting a larger portion of the *TSC1* or *TSC2* genes, most commonly deletions encompassing one or more exons, are termed "genomic mutations." These are very rare in *TSC1*, identified in about 0.5% of all TSC patients, but are relatively common in *TSC2*, occurring in about 6% of all TSC patients [44]. These types of mutations are discussed separately below.

Deletion and nonsense mutations occur at nearly equal frequency in *TSC1* (37.1 and 35.5%, respectively), while insertion and splice mutations are less common (14.8 and 9.5%, respectively), and missense mutations are rare (3.1%) (Table 4.2, Figure 4.3). Deletion, nonsense, and missense mutations all occur at nearly equal frequency in *TSC2* (22–27%), while splice and insertion mutations are less common (16.2 and 8.5%, respectively) (Table 4.3, Figure 4.3). Nearly half of all the mutations identified in *TSC1* (48%) and *TSC2* (49%) have been seen exactly once.

4.4.3
Distribution of Mutations Along the Length of *TSC1* and *TSC2*

The distribution of mutations within *TSC1* is highly nonuniform (Figures 4.4 and 4.5). Just over a quarter of all mutations are found in the relatively large exon 15,

Table 4.3 *TSC2* small mutation summary.

	Total		Unique	
Deletions	245	21.1%	195	27.3%
Deletions in-frame	60	5.2%	17	2.4%
Deletions –insertions	5	0.4%	5	0.7%
All deletions	310	26.7%	217	30.4%
Insertions	99	8.5%	93	13.0%
Insertions–deletions	3	0.3%	3	0.4%
Insertions in-frame	6	0.5%	6	0.8%
All insertions	108	9.3%	102	14.3%
Missense	293	25.2%	136	19.0%
Nonsense	263	22.6%	126	17.6%
Splice	173	14.9%	122	17.1%
Splice deletions	11	0.9%	7	1.0%
Splice insertions	4	0.3%	4	0.6%
All splice	188	16.2%	133	18.6%
Total	1162		714	

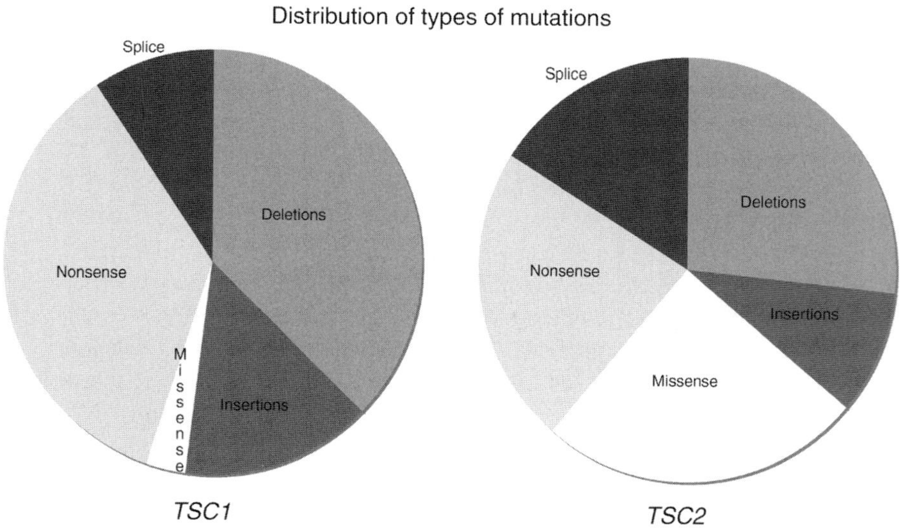

Figure 4.3 Distribution of types of small mutation in *TSC1* and *TSC2*. Pie charts are shown for each TSC gene, showing the relative frequency of five different kinds of small mutations.

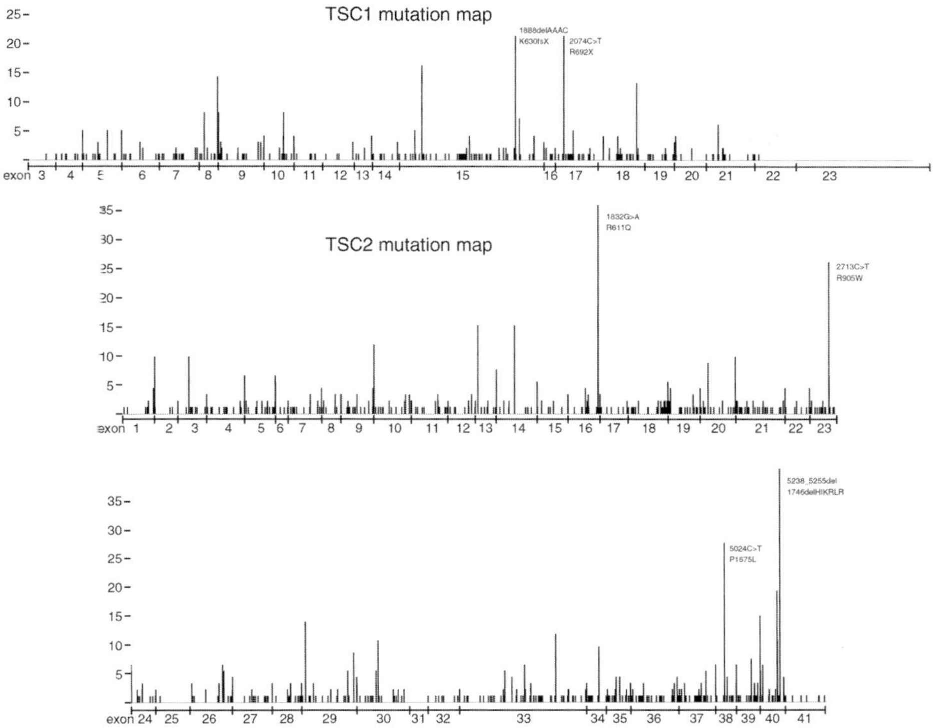

Figure 4.4 Mutation spectra of *TSC1* and *TSC2*. For each gene, each vertical line represents the number of mutations occurring at each individual nucleotide. Mutations that affect the same nucleotide but are distinct are binned together. The height of the vertical line reflects the number of mutations seen at each nucleotide position. Intronic mutations are indicated at the exon boundaries. The two most common mutations in *TSC1* and the four most common mutations in *TSC2* are labeled.

while exon 8 has the highest density of mutations per nucleotide. *TSC1* exons 17 and 18 are other relatively common sites of mutation. So far, only a single mutation has been identified in *TSC1* exon 3 and exon 22. Despite its relatively large size, no mutations have been identified in exon 23, the last exon of *TSC1*.

For *TSC2*, the distribution of mutations is also nonuniform, but not to the same extent. Exons 16, 33, and 40 have the highest numbers of mutations (each have 6–8% of the total), while exon 40 has the highest density of mutations per nucleotide. The alternatively spliced exons 25 and 31 have been found to contain only one unproven missense mutation, in exon 25. This suggests that these exons encode portions of TSC2 that have no important functional role in the structure and domains of the protein, or perhaps rather that they are not required for the function of TSC2 in the majority of tissues. Two other *TSC2* exons, exons 2 and 41, have mutation densities 20-fold less than that of exon 40.

Since the majority of *TSC1* and *TSC2* mutations (nonsense, most deletions and insertions, and splice site) lead to premature truncation of the protein product, with

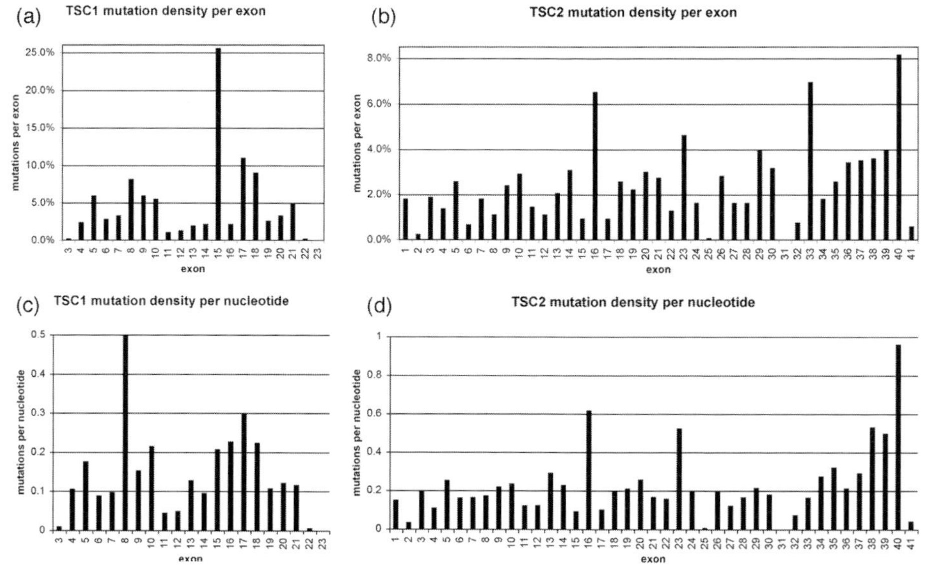

Figure 4.5 Distribution of small mutations in *TSC1* and *TSC2* by exon. The mutation density per exon (a and b) and the mutation density per nucleotide (c and d) are shown for *TSC1* (a and c) and *TSC2* (b and d). Mutations per exon are shown as a percentage of all mutations found in each gene. Note that each scale is different.

the production of a nonfunctional protein fragment, it is likely that the distribution of mutations in each gene reflects the intrinsic mutability of the underlying sequence rather than the targeting of certain specific encoded sequences or domains. This conclusion is also strongly supported by analyses of individual sites of mutation, as discussed below.

4.4.4
Single-Base Substitutions in *TSC1* and *TSC2*

In *TSC1*, nearly half of all mutations (216, 48%) are single base substitutions, and 75% (161 of 216) of these are nonsense mutations (Table 4.2). Only 62 of the 439 (14%) possible single base changes that would result in nonsense mutations in *TSC1* have been observed. The five most common nonsense mutations in *TSC1* (Table 4.4) were seen in 72 different index cases, and represent 45% of all the nonsense mutations identified so far in *TSC1* and 16% of all *TSC1* mutations. These five nonsense mutations are all C–T transitions at CGA codons (encoding arginine), which subsequently become the stop codon TGA. It is likely that these relatively high-frequency mutations in *TSC1* occur due to deamination of a methylated C residue at the CpG sequence in this codon, a common mechanism of genetic change [52]. The most common of these nonsense mutations (Arg692X, Table 4.4) was seen 21 times, accounting for 4.6% of all *TSC1* mutations reported (Figure 4.4).

Table 4.4 Relatively common mutations occurring in *TSC1*.

Gene	Exon	Mutation	aa Change	Type	Number of times seen	Percentage of all *TSC1* point mutations
TSC1	15	c.1888_1891del	Lys630GlnfsX22	del	21	4.6%
TSC1	17	c.2074C > T	Arg692X	non	21	4.6%
TSC1	15	c.1525C > T	Arg509X	non	16	3.5%
TSC1	8	c.733C > T	Arg245X	non	14	3.1%
TSC1	18	c.2356C > T	Arg786X	non	13	2.9%
TSC1	8	c.682C > T	Arg228X	non	8	1.8%
TSC1	15	c.1903_1904del	Thr635ArgfsX52	del	7	1.5%
TSC1	10	c.989dupT	Ser331GlufsX10	ins	7	1.5%
TSC1	21	c.2672dupA	Asn891LysfsX13	ins	6	1.3%

Fourteen missense mutations have been identified in *TSC1*. Each of these has been seen just once. In contrast to earlier views, at a time when there were no confirmed missense mutations in *TSC1*, functionally inactivating missense mutations in *TSC1* have been confirmed by functional studies in several cases [53] (Nellist *et al.* unpublished observations).

In the *TSC2* gene, 62.7% (729/1162) of the identified mutations are point mutations (Table 4.3). In contrast to *TSC1*, missense mutations make up a substantial fraction of the *TSC2* point mutations (293 of 729, or 40.2%). Nine point mutations (5 missense and 4 nonsense) were identified 11 or more times in *TSC2*, accounting for 14.1% (165 of 1162) of all small mutations identified in *TSC2* so far (Table 4.5).

Table 4.5 Relatively common mutations occurring in *TSC2*.

Gene	Exon	Mutation	aa Change	Type	Number of times seen	Percentage of all *TSC2* point mutations
TSC2	40	5238_5255del18	His1746_Arg1751 delinsGln	18nt In-frame deletion	38	3.3%
TSC2	15	1832G > A	Arg611Gln	Missense	33	2.8%
TSC2	38	5024C > T	Pro1675Leu	Missense	26	2.2%
TSC2	23	2713C > T	Arg905Trp	Missense	23	2.0%
TSC2	16	1831C > T	Arg611Trp	Missense	16	1.4%
TSC2	40	5227C > T	Arg1743Trp	Missense	15	1.3%
TSC2	13	1372C > T	Arg458X	Nonsense	14	1.2%
TSC2	14	1513C > T	Arg505X	Nonsense	14	1.2%
TSC2	29	3412C > T	Arg1138X	Nonsense	13	1.1%
TSC2	33	4375C > T	Arg1459X	Nonsense	11	0.9%
TSC2	30	3693_3696del	Ser1232ThrfsX92	4 nt Deletion	10	0.9%

All of these mutations occurred at CpG sites, and are therefore also likely to have occurred through deamination of a methylated C residue.

4.4.5
Insertions and Deletions in *TSC1* and *TSC2*

Two hundred and thirty-five insertion or deletion mutations have been identified in *TSC1*, of which 169 are unique (Table 4.2). Nearly all of these changes cause a frameshift in the *TSC1* coding sequence (232 of 235, 97.4%). Nearly all the small insertions occurred due to the duplication of an adjacent base or short sequence, while most deletions removed an element of a tandem repeat. The most common mutation in *TSC1*, identified 21 times and accounting for 4.6% of all *TSC1* mutations, is a four-nucleotide deletion in exon 15 (Table 4.4). Three other insertion or deletion mutations are also relatively common in *TSC1*. These include a two-nucleotide deletion and two single-nucleotide duplications (Table 4.4).

The spectrum of insertion/deletion mutations in the *TSC2* gene is also extensive, consisting of 319 unique insertions or deletions. Furthermore, similar to *TSC1*, nearly all insertion events are due to the duplication of single or multiple nucleotides, while most deletion events involve the removal of an element of a tandem repeat. The most common mutation in *TSC2* is a deletion of 18 nucleotides in exon 40, which leads to an in-frame deletion of six amino acids; this mutation has been identified in 38 patients/families and accounts for 3.3% of all *TSC2* mutations (Table 4.5). A second deletion, of four nucleotides in exon 30, was also relatively frequent.

4.4.6
Large Genomic Deletions/Rearrangements in *TSC1* and *TSC2*

Large genomic deletions and rearrangements in the *TSC2* gene are relatively common and were instrumental in the discovery of that gene [11]. In contrast, they are quite rare in *TSC1*. The total number of reported deletions and rearrangements in *TSC2* is over 130 [27, 44, 54]. Only nine large deletions have been reported in the *TSC1* gene thus far [44, 54, 55]. Of these, five are intragenic deletions, while four extend beyond *TSC1* (Figure 4.1). Interestingly, seven of the nine deletions, including three of size <5 kb, affect exons 21/22.

Figure 4.2 illustrates the size and location of 50 deletions or duplications that we recently identified and characterized in the *TSC2* gene [44]. This comprehensive analysis was done using multiple ligation-dependent probe assays (MLPA) with probes for every exon of *TSC1* and *TSC2*. Of the 50, 2 were duplications and 48 were deletions. Most *TSC2* deletions (73%, 35 of 48) extended beyond the gene [44]: 44% extended into the genomic region 5′ of *TSC2*; 50% extended 3′ of *TSC2* into the *PKD1* gene; and 21% extended into both the 5′ and 3′ regions. The majority of the deletions identified in the *TSC2* gene were large. Only 27% were less than 10 kb, and only 8% deleted a single exon. Sequence analysis of the junctions of the deletions did not reveal a consistent pattern in the sites or sequences involved in these genomic rearrangements, although for many deletions, short repeats of 2–7 nucleotides, which may have mediated the deletion event, were identified. Intron 29 of *TSC2* was somewhat enriched for breakpoints, but otherwise there was no particular region in *TSC2* in which breakpoints were particularly common [44]. *TSC2* deletions that extend into the adjacent *PKD1* gene are known to be associated with early-onset severe polycystic kidney disease (see Chapter 5) [27].

4.4.7 Polymorphisms

In *TSC1*, 33 different polymorphisms have been identified. Of these, 10 (30%) occur in intronic sequence, 13 (39%) are silent, and 10 (30%) cause missense changes, 4 of which are nonconservative according to the BLOSUM62 matrix [49]. Four of the *TSC1* polymorphisms are relatively common, with a rare allele frequency of 10% or greater.

In *TSC2*, 333 different polymorphisms have been reported. Of these, 227 (68%) have been seen exactly once; 42 have been seen 5 times or more; and 13 have been seen as heterozygotes in 5% or more of the screened population (though two of these have been reported in Asian populations only). Of the polymorphisms, 139 (42%) were identified in intronic sequence; 6 (2%) in the 3′ UTR; and 188 (56%) in the coding sequence. Among the polymorphisms seen in the coding sequence, 80 (42%) do not affect the amino acid sequence, while 108 (58%) change the amino acid sequence. As already discussed, it is likely that some of these rare polymorphisms are actually pathogenic mutations but are not recognized as such at this time.

It is interesting to note that despite roughly similar scrutiny for mutations, approximately 10-fold more polymorphisms have been identified in *TSC2* compared to *TSC1*.

4.4.8 Perspectives on Mutational Variation at the TSC Loci

The *TSC1* and *TSC2* mutation spectra are similar but not identical to those seen for human disease genes in general (Table 4.6; note that the comparison is being made

Table 4.5 Comparison between *TSC1* and *TSC2* small mutation frequencies.

	TSC1		*TSC2*		Genome average[a)]
Deletions	168	37.1%	310	26.7%	19.1%
Insertions	67	14.8%	108	9.3%	7.2%
Missense	14	3.1%	293	25.2%	61.5%
Nonsense	161	35.5%	263	22.6%	61.5%[b)]
Splice	43	9.5%	188	16.2%	10.5%[b)]
Regulatory					1.7%
Total	453		1162		

a) Genome average is taken from the Human Gene Mutation Database [56] (http://www.hgmd.cf.ac.uk/ac/index.php), which is based upon analysis of 79 077 mutations in 3000 human genes.

b) Note that missense and nonsense mutations are rooted together in this reference.

with a set of 3000 human disease genes, for which there are the same problems in mutation identification and analysis as discussed above for *TSC1* and *TSC2*. In addition, these genes have been scrutinized at highly variable levels, some much more than *TSC1* and *TSC2*, others less so) [56]. Point mutations at CpG sites are in general much more common than mutations involving other bases in the genome, with an estimated fivefold higher occurrence rate than the basal mutation rate [57]. Compared to other mutations, earlier calculations on *TSC1* and *TSC2* indicated that there was a 29-fold increase in CpG site mutations in *TSC1* and a 24-fold increase in *TSC2* [26]. Thus, this mechanism of mutation is particularly common in *TSC1* and *TSC2*.

The observed frequency of splice site point mutations in *TSC1* and *TSC2* matches the genome average fairly well, though it is somewhat higher in *TSC2* than average. Insertions occur somewhat less than half as frequently as deletions in both genes. This is consistent with mechanistic considerations and is also consistent with the genome average ratio between these two types of mutations (Table 4.6). However, in both *TSC1* and *TSC2*, deletions and insertions occur more often than the genome average. As seen elsewhere in the genome [58], slipped strand mispairing occurring at short runs of a simple sequence repeat (1–4 nucleotides) appears to be a common mechanism for the generation of deletions and insertions. There are clear hot spots for both insertion and deletion mutations in *TSC1* and *TSC2*, but it is not clear from the primary sequence why this might be the case.

Truncating mutations (nonsense, splicing, and frameshift) make up 97% of the reported mutations in *TSC1* and 71% of those in *TSC2*. This major difference is accounted for by the low frequency of missense mutations in *TSC1*.

From the TSC1/TSC2 mutation database (http://chromium.liacs.nl/LOVD2/TSC/home.php), *TSC1* mutations account for about 26% of all TSC mutations, while *TSC2* mutations account for about 74% (considering both small and genomic mutations). Published reports on the ratio of mutations between the two genes are shown in Table 4.1, where it can be seen that the summed ratio from all of these studies is 21% *TSC1* mutations and 79% *TSC2* mutations. Several differences between the two genes and their mutations may contribute to the difference in the observed mutation rates. First, missense mutations are rare in *TSC1*, while they are relatively common in *TSC2*. Second, large genomic deletions and duplications are also very rare in *TSC1*, while they are seen consistently in *TSC2*. Third, the *TSC2* coding region is about 1.5 times larger than the *TSC1* coding region, and the gene has nearly twice as many exons and splice sites that can be affected by mutation. However, apart from the difference in size of the two genes, differences in mutation frequency between *TSC1* and *TSC2* are largely due to the intrinsic mutability of the gene sequences themselves. This intrinsic difference in mutation rate is also revealed by the 10-fold difference in polymorphism frequency between the two genes. This difference in mutation rate is the real driver for the differences in mutation frequencies between *TSC1* and *TSC2*. Although in one sense surprising, it is consistent with the dramatic variability in mutation rates seen within each gene, considered from both the point of view of each individual exon and for each individual nucleotide (Figures 4.4 and 4.5).

4.5
Frequency and Significance of Mosaicism in TSC

Mosaicism comes about when a *TSC1* or *TSC2* mutation occurs during embryogenesis, but after the fertilized egg undergoes one or more divisions. The result is that not all the cells in the individual developing from that embryo carry the mutation. There are two forms of mosaicism: somatic (generalized) and germline (confined gonadal). Both forms have been described in TSC patients (see further below). In somatic mosaicism, the mutation is present at some frequency in nearly all tissues and cell types in the individual, consistent with the mutation occurring quite early during embryogenesis. In germline mosaicism, the mutation is present at a detectable frequency only in the gonadal germ cells (progenitors of sperm in males and eggs in females) of the individual, nowhere else, consistent with the mutation occurring in one of the progenitor cells of the gonad.

An important concept concerning mosaicism is that it occurs only once in a family. That is, the first generation in a TSC family may or may not have mosaicism, but the subsequent generations will either inherit the mutant allele of *TSC1* or *TSC2* and have it in all of their cells, or not inherit the mutant allele.

Our knowledge about mosaicism in TSC is strongly influenced by our ability to detect its occurrence. Somatic mosaicism is quite commonly seen in TSC patients with large genomic mutations in *TSC2*, being reported in 7 of 25 (28%) and 8 of 54 (15%), in two large series [27, 44]. As not all the parents were available for study in these reports, the actual frequency of mosaicism is probably even higher. Detection of mosaicism in these patients is relatively easy due to the quantitative nature of FISH, Southern blotting, and MLPA, the procedures used for both mutation and mosaicism detection.

Among patients with small mutations in *TSC1* and *TSC2*, somatic mosaicism has been reported at much lower frequencies. An early report [59] suggested a relatively high frequency (6/62, 10%) of somatic mosaicism in general in TSC patients. However, only three individuals in this study had both small mutations and evidence of somatic mosaicism in a parent, for an incidence of about 5% among small mutation patients. In addition, in subsequent studies from this group, the rate of somatic mosaicism for small mutations plummeted to an aggregate frequency of 4/342 (1%) [40]. The most likely explanation for this difference is the method of mutation detection used (denaturing gradient gel electrophoresis in combination with sequence analysis). Detection of mosaicism for small mutations can be very difficult, particularly when only sequencing is used for mutation analysis. It is more easily detected when denaturing high-pressure liquid chromatography (DHPLC), heteroduplex analysis, or single-strand conformation polymorphism (SSCP) analysis is performed [60, 61]. For example, we detected evidence of mosaicism in 9 of 317 (3%) TSC patients with small mutations using DHPLC for detection [62]. However, as the use of these techniques for mutation detection is declining, mosaicism detection rates are likely to fall.

We analyzed 30 sporadic TSC patients with small mutations and their parents, and found no evidence for even low levels (1%) of mosaicism among the 54 parents studied [62]. This suggests that low-level mosaicism is rare in this population. This finding clearly contrasts with the observation that among individuals with diagnosed TSC, the frequency of mosaicism is quite high (15–25% or more) for genomic deletions and in the range of 3% for small mutations. It suggests that individuals who are mosaic for a TSC gene mutation will in general have clinical features of TSC.

The severity of disease associated with somatic mosaicism might be predicted to correlate with the level of mosaicism present; that is, when fewer cells carry a TSC gene mutation, disease severity would be lower. Although this has been generally true in some reports [27, 59], some patients mosaic for a TSC gene mutation have a full and severe TSC phenotype [61]. This variability is likely to be due to both variable rates of mosaicism in different tissues and random chance in terms of frequency of second hit and other events during development that influence TSC pathogenesis.

Germline mosaicism is well known in TSC as well. About 2% of unselected TSC families have more than one member in one generation with TSC, with neither parent of that generation showing any evidence of the disease. This could be due to somatic mosaicism in one parent with no clinical manifestations of TSC. However, in several cases, mutation analysis has shown that a mutation is present in either *TSC1* or *TSC2* in the affected generation, but is not present in the blood cell DNA of either parent [63, 64]. These observations suggest that one of the parents has germline mosaicism, with the mutation present only in the gonadal cells of that parent. However, that has not been shown directly so far. Nonetheless, these observations provide further evidence that somatic mosaicism for a TSC gene mutation is generally, if not always, associated with a clinically detectable phenotype; so when parents are not affected after thorough screening, they are not likely to be somatic mosaics.

4.6
Considerations in Patients in Whom No Mutation Can Be Identified

Comprehensive efforts to identify mutations in the *TSC1* and *TSC2* genes among TSC patients have been pursued in several research and diagnostic laboratories. The results of these efforts are quite consistent in that mutations are detected in a maximum of 80–90% of patients, with only a single relatively small series achieving a 91% mutation detection rate (Table 4.1) [46]. In addition, three large series have shown that the group of TSC patients in whom no mutation has been identified (NMI) have similar, distinctive clinical features [36, 40, 45]. NMI patients have milder brain, neurological, and renal features than patients with *TSC2* mutations and somewhat milder brain features but more severe renal angiomyolipomas (AMLs) and LAM than *TSC1* patients (see Chapter 6 for further details.)

There are at least three possible reasons for NMI in TSC patients: (1) mutation detection failure; (2) the existence of a third, as yet undiscovered, TSC locus; and (3) mosaicism for mutations in the *TSC1* and *TSC2* genes. It is certain that mutation detection failure accounts for some NMI patients, as small mutations that affect transcription or mRNA processing may occur in locations remote from coding exons [65] or be unrecognized within coding exons [66]. Since current molecular diagnostic methods focus on exonic regions, intronic mutations will usually be missed. In addition, the analysis of the effects of putative intronic mutations is made difficult by the necessity for mRNA studies, which are not easily performed in a routine diagnostics laboratory. However, based upon experience with human disease genes in general, these mutations account for a small fraction (<5%) of all disease-causing mutations [56] (http://www.hgmd.cf.ac.uk/ac/index.php).

There is at present no substantial evidence for a third TSC gene locus beyond the occurrence of NMI TSC patients. The biochemical signaling pathway that the TSC1–TSC2 protein complex regulates does not suggest that there are other genes that might cause an identical syndrome. Mutations in other genes in the same signaling pathway cause related syndromes (e.g., *LKB1* mutations cause Peutz–Jeghers syndrome [67]; and *PTEN* mutations cause Cowden and a variety of hamartoma syndromes [68]), but these are clinically quite distinct from TSC. Nonetheless, a third TSC gene remains a possibility.

Mosaicism is very likely to contribute to a fraction of NMI patients. As discussed above, mosaicism is well documented in TSC. However, its frequency and precise contribution to NMI TSC patients is uncertain because, as already discussed, it is not easy to detect mosaicism for the predominant small mutations seen in *TSC1* and *TSC2*. The relatively mild phenotype of the NMI patient group is consistent with either mosaicism for a standard *TSC2* mutation or a partial function splicing mutation in *TSC2* as a molecular cause of TSC.

4.7 The Role of *TSC1* and *TSC2* in Tumor Development

4.7.1 The Role of *TSC1* and *TSC2* in Hamartoma Development in TSC Patients

The so-called two-hit model has been the working molecular model for tumor development in TSC for many years and has considerable support [69–72]. This model postulates that in TSC hamartomas and other proliferative lesions, loss of the second, normal *TSC1* or *TSC2* allele complements the constitutional inactivation of the first allele of that same gene (whichever the patient carries in the germline). This second hit loss might occur through different genetic and epigenetic mechanisms, but the most common mechanism is a large genomic deletion of the normal allele, which can be assessed by screening for loss of heterozygosity (LOH) using a genetic marker (e.g., the mutation present in the mutant allele, or a polymorphic genetic marker within or nearby the gene). Thus, the finding of LOH for markers within or near *TSC1* or *TSC2* in a TSC tumor or lesion provides evidence that the normal allele has been lost, and that there is no normal *TSC1* or *TSC2* transcribed in the cells from that lesion.

LOH for *TSC1* or *TSC2* markers has been demonstrated in 84 of 128 renal angiomyolipomas (66%) [70–72], and the smooth muscle, fat, and vascular cells from these lesions all show evidence of *TSC2* LOH [73]. In addition, in both TSC-associated and sporadic lymphangioleiomyomatosis (LAM), evidence of two-hit inactivation of the *TSC2* gene in the affected tissue has been demonstrated [74, 75].

The two-hit model of lesion pathogenesis is most disputed for the brain lesions of TSC (see also Chapter 8). We demonstrated that five of six subependymal giant cell astrocytomas (SEGAs) showed evidence for either LOH (four cases) or two point mutations (one case), consistent with the two-hit mechanism [76]. Reduced expression of TSC2 has also been seen in TSC-associated SEGAs [77–80]. In addition, two-hit inactivation of *TSC2* has been shown in a sporadic, non-TSC case of subependymal giant cell astrocytoma [81].

On the other hand, evidence for the two-hit model in TSC cortical tubers is much more limited [70, 72]. Detailed analysis of five cortical tubers, including the use of laser capture microdissection in two cases, failed to identify LOH [82]. In addition, phosphorylation of TSC2 by extracellular regulated kinase (ERK) and AKT has been seen in some TSC tubers, suggesting that the activation of specific kinases could cooperate with *TSC1* or *TSC2* haploinsufficiency to lead to tuber development [83, 84]. However, the mixture of cell types present in cortical tubers makes demonstration of LOH difficult, and this could explain the negative LOH studies. In addition, the giant cells of SEGAs and cortical tubers share many similarities, consistent with a common two-hit pathogenetic mechanism [76, 85].

Cardiac rhabdomyomas have not been studied in great detail, but show evidence of activation of the mammalian target of rapamycin complex 1 (mTORC1) signaling pathway (see Chapter 16) and reduced expression of TSC1 and TSC2, consistent with a two-hit mechanism [86]. However, ERK activation has also been demonstrated in rhabdomyomas, suggesting the possibility of a similar mechanism as to what has been proposed for tubers.

4.7.2
The Role of *TSC1* and *TSC2* Genes in Cancer Development in Non-TSC Patients

There are many factors that complicate the evaluation of a specific gene's role in cancer development. It has become clear that the common adult cancers sustain numerous genetic hits, and likely many other epigenetic hits, that are important in their development. Establishing a critical role for any individual event takes thorough analysis. However, it has become very clear in recent years that the highly conserved phosphatidyl inositol 3-kinase (PI3K)/3-phosphoinositide/PTEN/AKT pathway is perturbed and activated in the majority of human cancers [87]. The TSC1–TSC2 complex is immediately downstream of AKT in this pathway and also integrates signals from other pathways to regulate the state of activation of mTORC1. In this sense, there is involvement of TSC1 and TSC2 in the majority of human cancers. On the other hand, clear genetic evidence for the direct involvement of TSC1 or TSC2 in any of the common adult malignancies is quite limited. Here we review that evidence.

Chromosome 9q, site of the *TSC1* gene, appears to be a specific genomic target for single copy loss in several common cancers, including bladder carcinoma, ovarian carcinoma, gallbladder carcinoma, nasopharyngeal carcinoma, and nonsmall cell lung cancer [66, 88–90]. Bladder cancer has been studied in greatest detail. Fourteen percent of bladder cancers have a point or small mutation in *TSC1*, including nonsense, missense, frameshift, and splice mutations, which are inactivating with respect to TSC1 expression [66, 88]. Combined with 9q LOH, this means that TSC1 expression is completely ablated in a significant proportion of bladder cancers. Although this may be a coincidental bystander (or passenger) event, no other gene on 9q has yet been found to be mutated in bladder cancer. In addition, it is possible that biallelic inactivation of *TSC1* does not occur in all cases because *TSC1* haplo-loss confers a growth advantage to bladder epithelial cells [66], due to its effects on mTORC1 signaling, as has been noted in other contexts [91, 92]. Low TSC1 expression has also been associated with poor outcome in early-stage bladder cancers [93].

TSC1 LOH has also been seen in up to 53% of lung adenocarcinomas [89]. However, very few mutations in *TSC1* were identified in a subsequent study, with no cases of two-hit inactivation [94]. Thus, *TSC1* may possibly act as a haplo-tumor suppressor in lung adenocarcinoma. In breast cancer, TSC1 and TSC2 expression is markedly reduced in breast cancer cells compared to normal mammary epithelial cells, and in some cases the *TSC1* promoter region is methylated, suggesting that this could be a mechanism for reducing TSC1 expression [95]. Reduced expression also appears to be associated with poorer outcome. Reduced expression, LOH, and methylation of the *TSC1* and *TSC2* promoter regions are also reported in oral squamous cell carcinoma [96]. Inactivation of TSC2 via loss of expression or phosphorylation with activated mTORC1 signaling has also been seen in endometrial carcinoma [97].

Kidney cancer is recognized to occur somewhat more frequently and at a younger age in TSC patients than in the general population and to consist of both malignant AML and renal cell cancer [98]. However, LOH for *TSC1* or *TSC2* is not seen frequently in kidney cancers of various histologies from non-TSC patients, nor is there evidence of mutation in either gene [99].

Two other nervous system neoplasms, gangliogliomas [100] and medulloblastoma [101], have been screened for mutation in *TSC1* and *TSC2* with essentially negative results. On the other hand, reduced or absent TSC2 expression has been demonstrated in about 50% of both high-grade and low-grade astrocytomas from adults [102].

Pancreatic and other neuroendocrine neoplasms are recognized to occur at greater frequency in TSC patients than the general population [103]. LOH for the *TSC2* region was found in 36% of pancreatic neuroendocrine tumors [104].

Sacrococcygeal chordomas are very rare neoplasms that have been seen in several TSC patients. Molecular studies demonstrated that there was loss of the wild-type allele in two cases of TSC-associated chordomas [105].

PEComa is a pathologic term introduced in the past 10 years to refer to a family of related mesenchymal neoplasms, including renal AMLs, LAM, clear cell "sugar" tumor of the lung, and a group of rare tumors arising at a variety of visceral and soft tissue sites [106, 107]. Pathologically, PEComas are somewhat similar in appearance to AMLs and are composed of pleomorphic cells termed perivascular epithelioid cells (PECs) that express both smooth muscle and melanocytic markers, such as smooth muscle actin and the HMB-45 antigen. Similar to AML and LAM, extrarenal PEComas occur at increased frequency in TSC patients [107]; however, the great majority are reported in non-TSC patients [107]. Both non-TSC kidney AMLs and non-TSC extrarenal PEComas have been shown to have the same expression features seen in TSC-related AMLs, including increased expression of phosphorylated p70S6kinase (T389) and phospho-S6(S235–236) and reduced expression of phosphoAKT(S473) and TSC1 and TSC2 [108, 109]. In addition, 6 of 10 (60%) non-TSC PEComas studied showed evidence for *TSC2* LOH, and 10–36% of non-TSC AMLs showed evidence for either *TSC1* or *TSC2* LOH [109, 110]. Thus, there is strong evidence that this same pathway involving the TSC genes is critically involved in the pathogenesis of PEComas in non-TSC patients.

A link has also been established between inflammatory stimuli and inactivation of the TSC1–TSC2 complex. I kappa kinase β (IKKβ), a major downstream kinase in the tumor necrosis factor α (TNFα) signaling pathway, directly phosphorylates and inhibits the function of TSC1 [111]. Furthermore, in cell lines derived from Barrett's esophagus, a precursor lesion to esophageal adenocarcinoma due to gastroesophageal reflux disease, there was evidence of activation of IKKβ and phosphorylation of TSC1 and S6K1 in response to bile acid stimulation [112].

4.8
The Future of Molecular Diagnostics in TSC

DNA analytic technologies are developing at a rapid pace and it is anticipated that the goal of performing complete genome sequencing at a cost of $1000 per individual will be achieved in the next several years (http://grants2.nih.gov/grants/guide/rfa-files/RFA-HG-04-003.html). Although interpretation of the resulting sequence data is in fact a much more daunting task than generating the sequence information, it is possible to imagine a StarTrek-like situation in the not too distant future in which a doctor takes a saliva sample from a patient, puts it in a machine, and obtains full genome sequence data in a short period of time.

In the short term, currently available (or nearly available) high-throughput sequencing technologies have the potential to revolutionize genetic care in general, by providing the means for cheap and detailed sequencing information. In particular, it is likely to become possible to analyze the *TSC1* and *TSC2* genes in much greater detail, including all relevant upstream, downstream, and intronic regions, as well as performing multiple independent reads of both genes, to permit direct assessment for mosaicism. Interpretation of this sequence information will critically depend upon the development of appropriate bioinformatics tools and comparisons with multiple reference genomes. Even with such developments, many sequence changes will be found whose significance is uncertain, and this will continue to be an issue in interpretation.

Nonetheless, it is anticipated that these continuing developments in DNA analytic technologies will have a major impact on the clinical genetic care of TSC patients and will enable higher mutation detection rates than previously possible.

Glossary

Genetic code	Genes are transcribed into mRNA, which is then translated into a protein sequence by the ribosomes using a three-nucleotide (triplet) code, termed the genetic code. Each combination of three nucleotides is translated into an amino acid, apart from three three-nucleotide sequences that are called stop codons. Stop codons result in termination of protein translation at that point in the sequence.
Deletion mutation	In a deletion mutation, one or more of the nucleotides in a gene is deleted. Because the genetic code is a triplet code, whenever the number of nucleotides deleted is not divisible by 3, this has severe effects on the protein, typically resulting in premature truncation and functional inactivation. When the number of nucleotides lost is divisible by 3, this results in the loss of one or more amino acids from the protein, but does not disrupt the sequence of the protein following deletion. In this latter case, the protein may still be functional, but usually it is not.
Insertion mutation	In an insertion mutation, one or more nucleotides are inserted into a gene. Because the genetic code is a triplet code, whenever the number of nucleotides inserted is not divisible by 3, this has severe effects on the protein, usually resulting in premature truncation and functional inactivation. When the number of nucleotides inserted is divisible by 3, this results in the insertion of one or more amino acids into the protein, but does not disrupt the sequence of the protein following insertion. In this latter case, the protein may still be functional, but usually it is not.
Missense mutation	In a missense mutation, one of the nucleotides is changed to a different nucleotide such that the genetic code translation is changed from one amino acid to another. This single change in amino acid disrupts the normal structure and function of the protein in a significant way.
Mutation	A mutation is a change in the DNA sequence. Mutations occur at extremely low frequency in each individual nucleotide in the human genome, but due to the genome's large size (about 3 000 000 000 nucleotides), all of us have mutations that are not present in our parents' DNA. Most often mutations result in polymorphisms, which have no apparent effect (see below). Some mutations affect genes significantly, and this occurs in the TSC genes to cause TSC. Once a mutation occurs, it can then be passed on, or transmitted, from parent to child. Thus, a *TSC1* or *TSC2*

	gene mutation that causes TSC can be passed on from parent to child.
Nonsense mutation	In a nonsense mutation, one of the nucleotides is changed to a different nucleotide such that the genetic code translation is changed from an amino acid to a stop codon. This results in the premature truncation of the normal protein product of a gene, which usually results in a nonfunctional short protein fragment that is cleared rapidly from the cell.
Polymorphism (neutral variant)	There are many sites of sequence variation in the human genome (literally millions) that appear to have no significant effect on any gene, trait, or disease susceptibility. Some of these polymorphisms or neutral variants occur at high frequency, so many people have the variant. These are then relatively easily assessed as having no significant functional effect. Others are very rare, being seen in only a single individual out of thousands examined. The most common type of polymorphism in the human genome is a single-nucleotide polymorphism (SNP), in which one nucleotide is substituted for another, for example, a G nucleotide becomes a T nucleotide.
Sequence variant of uncertain significance (unclassified variant)	When DNA diagnostic studies are performed, examining the *TSC1* and *TSC2* genes, it often happens that a DNA sequence change is observed whose functional significance is unknown. Often these changes are very rare, having been seen only once or a just a few times. Such a DNA sequence change could be a mutation, or could be a polymorphism that has no significance. Thus, the terminology above is used.
Splice mutation	In a splice mutation, the nucleotides adjacent to an exon are either changed or deleted. Because splicing of the exons of a gene is critically important for the structure of the mRNA and therefore the protein that is encoded, these changes usually have severe effects on the function of the gene.

Acknowledgments

I wish to acknowledge the tremendous efforts of Rosemary Ekong and Sue Povey in developing and maintaining the Tuberous Sclerosis Complex Mutation Database, and in their provision of additional data extracted from that database, for the writing of this chapter. I would also like to thank the thousands of TSC patients and their families who have volunteered for genetic research on TSC over the past 25 years. I also thank Mark Nellist, Sandra Dabora, and Sue Povey for constructive critiques and editing of this chapter. I would like to acknowledge the financial support provided by NIH NINDS NS31535 and the Tuberous Sclerosis Alliance.

References

1 Fryer, A.E., Connor, J.M., Povey, S., Yates, J.R.W., Chalmers, A., Fraser, I., Yates, A.D., and Osborne, J.P. (1987) Evidence that the gene for tuberous sclerosis is on chromosome 9. *Lancet*, **1** (8534), 659–661.

2 Northrup, H., Beaudet, A.L., O'Brien, W.E., Herman, G.E., Lewis, R.A., and Pollack, M.S. (1987) Linkage of tuberous sclerosis to ABO blood group. *Lancet*, **2** (8562), 804–805.

3 Sampson, J.R., Yates, J.R.W., Pirrit, L.A., Fleury, P., Winship, I., Beighton, P., and Connor, J.M. (1989) Evidence for genetic heterogeneity in tuberous sclerosis. *J. Med. Genet.*, **26**, 511–516.

4 Janssen, L.A.J., Sandkuijl, L.A., Merkens, E.C., Maat-Kievit, J.A., Sampson, J.R., Fleury, P., Hennekam, R.C., Grosveld, G.C., Lindhout, D., and Halley, D.J. (1990) Genetic heterogeneity in tuberous sclerosis. *Genomics*, **8**, 237–242.

5 Northrup, H., Kwiatkowski, D., Roach, E., Dobyns, W., Lewis, R., Herman, G., Rodriguez, E., Daiger, S., and Blanton, S. (1992) Evidence for genetic heterogeneity in tuberous sclerosis: one locus on chromosome 9 and at least one locus elsewhere. *Am. J. Hum. Genet.*, **51**, 709–720.

6 Sampson, J., Janssen, L., and Sandkuijl, L. (1992) Linkage investigation of three putative tuberous sclerosis determining loci on chromosomes 9q, 11q, and 12q. *J. Med. Genet.*, **29**, 861–866.

7 Kandt, R., Haines, J., Smith, M., Northrup, H., Gardner, R., Short, M., Dumars, K., Roach, E., Steingold, S., Wall, S., Blanton, S., Flodman, P., Kwiatkowski, D., Jewell, A., Weber, J., Roses, A., and Pericak-Vance, M. (1992) Linkage of a major gene locus for tuberous sclerosis to a chromosome-16 marker for polycystic kidney disease. *Nat. Genet.*, **2**, 37–41.

8 Janssen, B., Sampson, J., van der Est, M., Deelen, W., Verhoef, S., Daniels, I., Hesseling, A., Brook, C.P., Nellist, M., Lindhout, D., Sandkuijl, L., and Halley, D. (1994) Refined localization of TSC1 by combined analysis of 9q34 and 16p13 data in 14 tuberous sclerosis families. *Hum. Genet.*, **94**, 437–440.

9 Povey, S., Burley, M.W., Attwood, J., Benham, F., Hunt, D., Jeremiah, S.J., Franklin, D., Gillett, G., Malas, S., Robson, E.B., Tippett, P., Edwards, J.H., Kwiatkowski, D.J., Super, M., Mueller, R., Fryer, A., Clarke, A., Webb, D., and Osborne, J. (1994) Two loci for tuberous sclerosis: one on 9q34 and one on 16p13. *Ann. Hum. Genet.*, **58**, 107–127.

10 Kwiatkowski, D.J., Armour, J., Bale, A.E., Fountain, J.W., Goudie, D., Haines, J.L., Knowles, M.A., Pilz, A., Slaugenhaupt, S., and Povey, S. (1993) Report and abstracts of the second international workshop on human chromosome 9 mapping 1993. *Cytogenet. Cell Genet.*, **64**, 93–121.

11 European Chromosome 16 Tuberous Sclerosis Consortium (1993) Identification and characterization of the tuberous sclerosis gene on chromosome 16. *Cell*, **75**, 1305–1315.

12 The European Polycystic Kidney Disease Consortium (1994) The polycystic kidney disease 1 gene encodes a 14 kb transcript and lies within a duplicated region on chromosome 16. *Cell*, **77**, 881–894.

13 Wilkie, A.O., Buckle, V.J., Harris, P.C., Lamb, J., Barton, N.J., Reeders, S.T., Lindenbaum, R.H., Nicholls, R.D., Barrow, M., Bethlenfalvay, N.C. *et al.* (1990) Clinical features and molecular analysis of the alpha thalassemia/mental retardation syndromes. I. Cases due to deletions involving chromosome band 16p13.3. *Am. J. Hum. Genet.*, **46**, 1112–1126.

14 Haines, J.L., Short, M.P., Kwiatkowski, D.J., Jewell, A., Andermann, E., Bejjani, B., Yang, C.H., Gusella, J.F., and Amos, J.A. (1991) Localization of one gene for tuberous sclerosis within 9q32–9q34, and further evidence for heterogeneity. *Am. J. Hum. Genet.*, **49**, 764–772.

15 Nellist, M., Brook-Carter, P.T., Connor, J.M., Kwiatkowski, D.J., Johnson, P., and Sampson, J.R. (1993) Identification of

markers flanking the tuberous sclerosis locus on chromosome 9 (TSC1). *J. Med. Genet.*, **30**, 224–227.

16 van Slegtenhorst, M., de Hoogt, R., Hermans, C., Nellist, M., Janssen, B., Verhoef, S., Lindhout, D., van den Ouweland, A., Halley, D., Young, J., Burley, M., Jeremiah, S., Woodward, K., Nahmias, J., Fox, M., Ekong, R., Osborne, J., Wolfe, J., Povey, S., Snell, R.G., Cheadle, J.P., Jones, A.C., Tachataki, M., Ravine, D., and Kwiatkowski, D.J. (1997) Identification of the tuberous sclerosis gene TSC1 on chromosome 9q34. *Science*, **277**, 805–808.

17 Kobayashi, T., Hirayama, Y., Kobayashi, E., Kubo, Y., and Hino, O. (1995) A germline insertion in the tuberous sclerosis (Tsc2) gene gives rise to the Eker rat model of dominantly inherited cancer. *Nat. Genet.*, **9**, 70–74.

18 Maheshwar, M.M., Sandford, R., Nellist, M., Cheadle, J.P., Sgotto, B., Vaudin, M., and Sampson, J.R. (1996) Comparative analysis and genomic structure of the tuberous sclerosis 2 (TSC2) gene in human and pufferfish. *Hum. Mol. Genet.*, **5**, 131–137.

19 Xu, L., Sterner, C., Maheshwar, M.M., Wilson, P.J., Nellist, M., Short, P.M., Haines, J.L., Sampson, J.R., and Ramesh, V. (1995) Alternative splicing of the tuberous sclerosis 2 (TSC2) gene in human and mouse tissues. *Genomics*, **27**, 475–480.

20 Kim, K.K., Pajak, L., Wang, H., and Field, L.J. (1995) Cloning, developmental expression, and evidence for alternative splicing of the murine tuberous sclerosis (TSC2) gene product. *Cell Mol. Biol. Res.*, **41**, 515–526.

21 Momose, S., Kobayashi, T., Mitani, H., Hirabayashi, M., Ito, K., Ueda, M., Nabeshima, Y., and Hino, O. (2002) Identification of the coding sequences responsible for Tsc2-mediated tumor suppression using a transgenic rat system. *Hum. Mol. Genet.*, **11**, 2997–3006.

22 Momose, S., Kobayashi, T., Tada, N., Itoyama, S., and Hino, O. (2007) N-terminal hamartin-binding and C-terminal GAP domain of tuberin can separate *in vivo*. *Biochem. Biophys. Res. Commun.*, **356**, 693–698.

23 Maheshwar, M.M., Cheadle, J.P., Jones, A.C., Myring, J., Fryer, A.E., Harris, P.C., and Sampson, J.R. (1997) The GAP-related domain of tuberin, the product of the TSC2 gene, is a target for missense mutations in tuberous sclerosis. *Hum. Mol. Genet.*, **6**, 1991–1996.

24 van Slegtenhorst, M., Nellist, M., Nagelkerken, B., Cheadle, J., Snell, R., van den Ouweland, A., Reuser, A., Sampson, J., Halley, D., and van der Sluijs, P. (1998) Interaction between hamartin and tuberin, the TSC1 and TSC2 gene products. *Hum. Mol. Genet.*, **7**, 1053–1058.

25 Plank, T.L., Yeung, R.S., and Henske, E.P. (1998) Hamartin, the product of the tuberous sclerosis 1 (TSC1) gene, interacts with tuberin and appears to be localized to cytoplasmic vesicles. *Cancer Res.*, **58**, 4766–4770.

26 Cheadle, J.P., Reeve, M.P., Sampson, J.R., and Kwiatkowski, D.J. (2000) Molecular genetic advances in tuberous sclerosis. *Hum. Genet.*, **107**, 97–114.

27 Sampson, J.R., Maheshwar, M.M., Aspinwall, R., Thompson, P., Cheadle, J.P., Ravine, D., Roy, S., Haan, E., Bernstein, J., and Harris, P.C. (1997) Renal cystic disease in tuberous sclerosis: role of the polycystic kidney disease 1 gene. *Am. J. Hum. Genet.*, **61**, 843–851.

28 Au, K.-S., Rodriguez, J.A., Finch, J.L., Volcik, K.A., Roach, E.S., Delgado, M.R., Rodriguez, E., and Northrup, H. (1998) Germ-line mutational analysis of the TSC2 gene in 90 tuberous-sclerosis patients. *Am. J. Hum. Genet.*, **62**, 286–294.

29 Young, J.M., Burley, M.W., Jeremiah, S.J., Jeganathan, D., Ekong, R., Osborne, J.P., and Povey, S. (1998) A mutation screen of the TSC1 gene reveals 26 protein truncating mutations and 1 splice site mutation in a panel of 79 tuberous sclerosis patients. *Ann. Hum. Genet.*, **62**, 203–213.

30 Zhang, H., Nanba, E., Yamamoto, T., Ninomiya, H., Ohno, K., Mizuguchi, M., and Takeshita, K. (1999) Mutational analysis of TSC1 and TSC2 genes in

Japanese patients with tuberous sclerosis complex. *J. Hum. Genet.*, **44**, 391–396.
31 Mayer, K., Ballhausen, W., and Rott, H.D. (1999) Mutation screening of the entire coding regions of the TSC1 and the TSC2 gene with the protein truncation test (PTT) identifies frequent splicing defects. *Hum. Mutat.*, **14**, 401–411.
32 Niida, Y., Lawrence-Smith, N., Banwell, A., Hammer, E., Lewis, J., Beauchamp, R.L., Sims, K., Ramesh, V., and Ozelius, L. (1999) Analysis of both TSC1 and TSC2 for germline mutations in 126 unrelated patients with tuberous sclerosis. *Hum. Mutat.*, **14**, 412–422.
33 Jones, A.C., Shyamsundar, M.M., Thomas, M.W., Maynard, J., Idziaszczyk, S., Tomkins, S., Sampson, J.R., and Cheadle, J.P. (1999) Comprehensive mutation analysis of TSC1 and TSC2-and phenotypic correlations in 150 families with tuberous sclerosis. *Am. J. Hum. Genet.*, **64**, 1305–1315.
34 van Slegtenhorst, M., Verhoef, S., Tempelaars, A., Bakker, L., Wang, Q., Wessels, M., Bakker, R., Nellist, M., Lindhout, D., Halley, D., and van den Ouweland, A. (1999) Mutational spectrum of the TSC1 gene in a cohort of 225 tuberous sclerosis complex patients: no evidence for genotype–phenotype correlation. *J. Med. Genet.*, **36**, 285–289.
35 Yamashita, Y., Ono, J., Okada, S., Wataya-Kaneda, M., Yoshikawa, K., Nishizawa, M. Hirayama, Y., Kobayashi, E., Seyama, K., and Hino, O. (2000) Analysis of all exons of TSC1 and TSC2 genes for germline mutations in Japanese patients with tuberous sclerosis: report of 10 mutations. *Am. J. Med. Genet.*, **90**, 123–126.
36 Dabora, S.L., Jozwiak, S., Franz, D.N., Roberts, P.S., Nieto, A., Chung, J., Choy, Y.S., Reeve, M.P., Thiele, E., Egelhoff, J.C., Kasprzyk-Obara, J., Domanska-Pakiela, D., and Kwiatkowski, D.J. (2001) Mutational analysis in a cohort of 224 tuberous sclerosis patients indicates increased severity of TSC2, compared with TSC1, disease in multiple organs. *Am. J. Hum. Genet.*, **68**, 64–80.
37 Langkau, N., Martin, N., Brandt, R., Zugge, K., Quast, S., Wiegele, G., Jauch, A., Rehm, M., Kuhl, A., Mack-Vetter, M., Zimmerhackl, B., and Janssen, B. (2002) TSC1 and TSC2 mutations in tuberous sclerosis, the associated phenotypes and a model to explain observed TSC1/TSC2 frequency ratios. *Eur. J. Pediatr.*, **161**, 393–402.
38 Apak, A., Haliloglu, G., Kose, G., Yilmaz, E., Anlar, B., and Aysun, S. (2003) Mutation analysis of TSC2 gene in 33 Turkish familial cases with tuberous sclerosis. *Turk. J. Pediatr.*, **45**, 1–5.
39 Rendtorff, N.D., Bjerregaard, B., Frodin, M., Kjaergaard, S., Hove, H., Skovby, F., Brondum-Nielsen, K., and Schwartz, M. (2005) Analysis of 65 tuberous sclerosis complex (TSC) patients by TSC2 DGGE, TSC1/TSC2 MLPA, and TSC1 long-range PCR sequencing, and report of 28 novel mutations. *Hum. Mutat.*, **26**, 374–383.
40 Sancak, O., Nellist, M., Goedbloed, M., Elfferich, P., Wouters, C., Maat-Kievit, A., Zonnenberg, B., Verhoef, S., Halley, D., and van den Ouweland, A. (2005) Mutational analysis of the TSC1 and TSC2 genes in a diagnostic setting: genotype–phenotype correlations and comparison of diagnostic DNA techniques in tuberous sclerosis complex. *Eur. J. Hum. Genet.*, **13**, 731–741.
41 Ali, M., Girimaji, S.C., Markandaya, M., Shukla, A.K., Sacchidanand, S., and Kumar, A. (2005) Mutation and polymorphism analysis of TSC1 and TSC2 genes in Indian patients with tuberous sclerosis complex. *Acta Neurol. Scand.*, **111**, 54–63.
42 Choi, J.E., Chae, J.H., Hwang, Y.S., and Kim, K.J. (2006) Mutational analysis of TSC1 and TSC2 in Korean patients with tuberous sclerosis complex. *Brain Dev.*, **28**, 440–446.
43 Hung, C.C., Su, Y.N., Chien, S.C., Liou, H.H., Chen, C.C., Chen, P.C., Hsieh, C.J., Chen, C.P., Lee, W.T., Lin, W.L., and Lee, C.N. (2006) Molecular and clinical analyses of 84 patients with tuberous sclerosis complex. *BMC Med. Genet.*, **7**, 72.
44 Kozlowski, P., Roberts, P., Dabora, S., Franz, D., Bissler, J., Northrup, H., Au,

K.S., Lazarus, R., Domanska-Pakiela, D., Kotulska, K., Jozwiak, S., and Kwiatkowski, D.J. (2007) Identification of 54 large deletions/duplications in TSC1 and TSC2 using MLPA, and genotype–phenotype correlations. *Hum. Genet.*, **121**, 389–400.

45 Au, K.S., Williams, A.T., Roach, E.S., Batchelor, L., Sparagana, S.P., Delgado, M.R., Wheless, J.W., Baumgartner, J.E., Roa, B.B., Wilson, C.M., Smith-Knuppel, T.K., Cheung, M.Y., Whittemore, V.H., King, T.M., and Northrup, H. (2007) Genotype/phenotype correlation in 325 individuals referred for a diagnosis of tuberous sclerosis complex in the United States. *Genet. Med.*, **9**, 88–100.

46 Jansen, F.E., Braams, O., Vincken, K.L., Algra, A., Anbeek, P., Jennekens-Schinkel, A., Halley, D., Zonnenberg, B.A., van den Ouweland, A., van Huffelen, A.C., van Nieuwenhuizen, O., and Nellist, M. (2008) Overlapping neurologic and cognitive phenotypes in patients with TSC1 or TSC2 mutations. *Neurology*, **70**, 908–915.

47 Zhang, M.Q. (1998) Statistical features of human exons and their flanking regions. *Hum. Mol. Genet.*, **7**, 919–932.

48 Schwarze, U., Goldstein, J.A., and Byers, P.H. (1997) Splicing defects in the COL3A1 gene: marked preference for 5′ (donor) spice-site mutations in patients with exon-skipping mutations and Ehlers–Danlos syndrome type IV. *Am. J. Hum. Genet.*, **61**, 1276–1286.

49 Henikoff, S. and Henikoff, J.G. (1992) Amino acid substitution matrices from protein blocks. *Proc. Natl. Acad. Sci. USA*, **89**, 10915–10919.

50 Nellist, M., Sancak, O., Goedbloed, M.A., Rohe, C., van Netten, D., Mayer, K., Tucker-Williams, A., van den Ouweland, A.M., and Halley, D.J. (2005) Distinct effects of single amino-acid changes to tuberin on the function of the tuberin–hamartin complex. *Eur. J. Hum. Genet.*, **13**, 59–68.

51 Nellist, M., Sancak, O., Goedbloed, M., Adriaans, A., Wessels, M., Maat-Kievit, A., Baars, M., Dommering, C., van den Ouweland, A., and Halley, D. (2008) Functional characterisation of the TSC1–TSC2 complex to assess multiple TSC2 variants identified in single families affected by tuberous sclerosis complex. *BMC Med. Genet.*, **9**, 10.

52 Cooper, D.N. and Youssoufian, H. (1988) The CpG dinucleotide and human genetic disease. *Hum. Genet.*, **78**, 151–155.

53 Jansen, F.E., Vincken, K.L., Algra, A., Anbeek, P., Braams, O., Nellist, M., Zonnenberg, B.A., Jennekens-Schinkel, A., van den Ouweland, A., Halley, D., van Huffelen, A.C., and van Nieuwenhuizen, O. (2008) Cognitive impairment in tuberous sclerosis complex is a multifactorial condition. *Neurology*, **70**, 916–923.

54 Longa, L., Saluto, A., Brusco, A., Polidoro, S., Padovan, S., Allavena, A., Carbonara, C., Grosso, E., and Migone, N. (2001) TSC1 and TSC2 deletions differ in size, preference for recombinatorial sequences, and location within the gene. *Hum. Genet.*, **108**, 156–166.

55 Nellist, M., Sancak, O., Goedbloed, M.A., van Veghel-Plandsoen, M., Maat-Kievit, A., Lindhout, D., Eussen, B.H., de Klein, A., Halley, D.J., and van den Ouweland, A.M. (2005) Large deletion at the TSC1 locus in a family with tuberous sclerosis complex. *Genet. Test.*, **9**, 226–230.

56 Stenson, P.D., Ball, E., Howells, K., Phillips, A., Mort, M., and Cooper, D.N. (2008) Human gene mutation database: towards a comprehensive central mutation database. *J. Med. Genet.*, **45**, 124–126.

57 Krawczak, M., Ball, E.V., and Cooper, D.N. (1998) Neighboring-nucleotide effects on the rates of germ-line single-base-pair substitution in human genes. *Am. J. Hum. Genet.*, **63**, 474–488.

58 Ball, E.V., Stenson, P.D., Abeysinghe, S.S., Krawczak, M., Cooper, D.N., and Chuzhanova, N.A. (2005) Microdeletions and microinsertions causing human genetic disease: common mechanisms of mutagenesis and the role of local DNA sequence complexity. *Hum. Mutat.*, **26**, 205–213.

59 Verhoef, S., Bakker, L., Tempelaars, A.M., Hesseling-Janssen, A.L., Mazurczak, T., Jozwiak, S., Fois, A., Bartalini, G., Zonnenberg, B.A., van Essen, A.J., Lindhout, D., Halley, D.J., and van den Ouweland, A.M. (1999) High rate of mosaicism in tuberous sclerosis complex. *Am. J. Hum. Genet.*, **64**, 1632–1637.

60 Emmerson, P., Maynard, J., Jones, S., Butler, R., Sampson, J.R., and Cheadle, J.P. (2003) Characterizing mutations in samples with low-level mosaicism by collection and analysis of DHPLC fractionated heteroduplexes. *Hum. Mutat.*, **21**, 112–115.

61 Kwiatkowska, J., Wigowska-Sowinska, J., Napierala, D., Slomski, R., and Kwiatkowski, D.J. (1999) Mosaicism in tuberous sclerosis as a potential cause of the failure of molecular diagnosis. *N. Engl. J. Med.*, **340**, 703–707.

62 Roberts, P.S., Dabora, S., Thiele, E.A., Franz, D.N., Jozwiak, S., and Kwiatkowski, D.J. (2004) Somatic mosaicism is rare in unaffected parents of patients with sporadic tuberous sclerosis. *J. Med. Genet.*, **41**, e69.

63 Yates, J.R., van Bakel, I., Sepp, T., Payne, S.J., Webb, D.W., Nevin, N.C., and Green, A.J. (1997) Female germline mosaicism in tuberous sclerosis confirmed by molecular genetic analysis. *Hum. Mol. Genet.*, **6**, 2265–2269.

64 Rose, V.M., Au, K.S., Pollom, G., Roach, E.S., Prashner, H.R., and Northrup, H. (1999) Germ-line mosaicism in tuberous sclerosis: how common? *Am. J. Hum. Genet.*, **64**, 986–992.

65 Mayer, K., Ballhausen, W., Leistner, W., and Rott, H. (2000) Three novel types of splicing aberrations in the tuberous sclerosis TSC2 gene caused by mutations apart from splice consensus sequences. *Biochim. Biophys. Acta*, **1502**, 495–507.

66 Pymar, L.S., Platt, F.M., Askham, J.M., Morrison, E.E., and Knowles, M.A. (2008) Bladder tumour-derived somatic TSC1 missense mutations cause loss of function via distinct mechanisms. *Hum. Mol. Genet.*, **17**, 2006–2017.

67 Launonen, V. (2005) Mutations in the human LKB1/STK11 gene. *Hum. Mutat.*, **26**, 291–297.

68 Eng, C. (2003) PTEN: one gene, many syndromes. *Hum. Mutat.*, **22**, 183–198.

69 Green, A.J., Johnson, P.H., and Yates, J.R. (1994) The tuberous sclerosis gene on chromosome 9q34 acts as a growth suppressor. *Hum. Mol. Genet.*, **3**, 1833–1834.

70 Henske, E.P., Scheithauer, B.W., Short, M.P., Wollmann, R., Nahmias, J., Hornigold, N., van Slegtenhorst, M., Welsh, C.T., and Kwiatkowski, D.J. (1996) Allelic loss is frequent in tuberous sclerosis kidney lesions but rare in brain lesions. *Am. J. Hum. Genet.*, **59**, 400–406.

71 Au, K.S., Hebert, A.A., Roach, E.S., and Northrup, H. (1999) Complete inactivation of the TSC2 gene leads to formation of hamartomas. *Am. J. Hum. Genet.*, **65**, 1790–1795.

72 Tucker, T. and Friedman, J.M. (2002) Pathogenesis of hereditary tumors: beyond the "two-hit" hypothesis. *Clin. Genet.*, **62**, 345–357.

73 Karbowniczek, M., Yu, J., and Henske, E.P. (2003) Renal angiomyolipomas from patients with sporadic lymphangiomyomatosis contain both neoplastic and non-neoplastic vascular structures. *Am. J. Pathol.*, **162**, 491–500.

74 Smolarek, T.A., Wessner, L.L., McCormack, F.X., Mylet, J.C., Menon, A.G., and Henske, E.P. (1998) Evidence that lymphangiomyomatosis is caused by TSC2 mutations: chromosome 16p13 loss of heterozygosity in angiomyolipomas and lymph nodes from women with lymphangiomyomatosis. *Am. J. Hum. Genet.*, **62**, 810–815.

75 Carsillo, T., Astrinidis, A., and Henske, E.P. (2000) Mutations in the tuberous sclerosis complex gene TSC2 are a cause of sporadic pulmonary lymphangioleiomyomatosis. *Proc. Natl. Acad. Sci. USA*, **97**, 6085–6090.

76 Chan, J.A., Zhang, H., Roberts, P.S., Jozwiak, S., Wieslawa, G., Lewin-Kowalik, J., Kotulska, K., and Kwiatkowski, D.J. (2004) Pathogenesis of tuberous sclerosis subependymal giant cell astrocytomas: biallelic inactivation of TSC1 or TSC2 leads to mTOR activation. *J. Neuropathol. Exp. Neurol.*, **63**, 1236–1242.

77 Henske, E.P., Wessner, L.L., Golden, J., Scheithauer, B.W., Vortmeyer, A.O., Zhang, Z., Klein-Szanto, A.J.P., Kwiatkowski, D.J., and Yeung, R.S. (1997) Loss of tuberin in both subependymal giant cell astrocytomas and angiomyolipomas supports a two-hit model for the pathogenesis of tuberous sclerosis tumors. *Am. J. Pathol.*, **151**, 1639–1647.

78 Jozwiak, S., Kwiatkowski, D., Kotulska, K., Larysz-Brysz, M., Lewin-Kowalik, J., Grajkowska, W., and Roszkowski, M. (2004) Tuberin and hamartin expression is reduced in the majority of subependymal giant cell astrocytomas in tuberous sclerosis complex consistent with a two-hit model of pathogenesis. *J. Child Neurol.*, **19**, 102–106.

79 Mizuguchi, M., Kato, M., Yamanouchi, H., Ikeda, K., and Takashima, S. (1996) Loss of tuberin from cerebral tissues with tuberous sclerosis and astrocytoma. *Ann. Neurol.*, **40**, 941–944 (see comments).

80 Arai, Y., Ackerley, C.A., and Becker, L.E. (1999) Loss of the TSC2 product tuberin in subependymal giant-cell tumors. *Acta Neuropathol. (Berl.)*, **98**, 233–239.

81 Ichikawa, T., Wakisaka, A., Daido, S., Takao, S., Tamiya, T., Date, I., Koizumi, S., and Niida, Y. (2005) A case of solitary subependymal giant cell astrocytoma: two somatic hits of TSC2 in the tumor, without evidence of somatic mosaicism. *J. Mol. Diagn.*, **7**, 544–549.

82 Niida, Y., Stemmer-Rachamimov, A.O., Logrip, M., Tapon, D., Perez, R., Kwiatkowski, D.J., Sims, K., MacCollin, M., Louis, D.N., and Ramesh, V. (2001) Survey of somatic mutations in tuberous sclerosis complex (TSC) hamartomas suggests different genetic mechanisms for pathogenesis of TSC lesions. *Am. J. Hum. Genet.*, **69**, 493–503.

83 Han, S., Santos, T.M., Puga, A., Roy, J., Thiele, E.A., McCollin, M., Stemmer-Rachamimov, A., and Ramesh, V. (2004) Phosphorylation of tuberin as a novel mechanism for somatic inactivation of the tuberous sclerosis complex proteins in brain lesions. *Cancer Res.*, **64**, 812–816.

84 Ma, L., Teruya-Feldstein, J., Bonner, P., Bernardi, R., Franz, D.N., Witte, D., Cordon-Cardo, C., and Pandolfi, P.P. (2007) Identification of S664 TSC2 phosphorylation as a marker for extracellular signal-regulated kinase mediated mTOR activation in tuberous sclerosis and human cancer. *Cancer Res.*, **67**, 7106–7112.

85 Jozwiak, J., Jozwiak, S., and Skopinski, P. (2005) Immunohistochemical and microscopic studies on giant cells in tuberous sclerosis. *Histol. Histopathol.*, **20**, 1321–1326.

86 Kotulska, K., Larysz-Brysz, M., Grajkowska, W., Jozwiak, J., Wlodarski, P., Sahin, M., Lewin-Kowalik, J., Domanska-Pakiela, D., and Jozwiak, S. (2009) Cardiac rhabdomyomas in tuberous sclerosis complex show apoptosis regulation and mTOR pathway abnormalities. *Pediatr. Dev. Pathol.*, **12** (2), 89–95.

87 Hay, N. (2005) The Akt-mTOR tango and its relevance to cancer. *Cancer Cell*, **8**, 179–183.

88 Adachi, H., Igawa, M., Shiina, H., Urakami, S., Shigeno, K., and Hino, O. (2003) Human bladder tumors with 2-hit mutations of tumor suppressor gene TSC1 and decreased expression of p27. *J. Urol.*, **170**, 601–604.

89 Takamochi, K., Ogura, T., Suzuki, K., Kawasaki, H., Kurashima, Y., Yokose, T., Ochiai, A., Nagai, K., Nishiwaki, Y., and Esumi, H. (2001) Loss of heterozygosity on chromosomes 9q and 16p in atypical adenomatous hyperplasia concomitant with adenocarcinoma of the lung. *Am. J. Pathol.*, **159**, 1941–1948.

90 Knowles, M.A., Hornigold, N., and Pitt, E. (2003) Tuberous sclerosis complex (TSC) gene involvement in sporadic tumours. *Biochem. Soc. Trans.*, **31**, 597–602.

91 Tavazoie, S.F., Alvarez, V.A., Ridenour, D.A., Kwiatkowski, D.J., and Sabatini, B.L. (2005) Regulation of neuronal morphology and function by the tumor suppressors Tsc1 and Tsc2. *Nat. Neurosci.*, **8**, 1727–1734.

92 Ehninger, D., Han, S., Shilyansky, C., Zhou, Y., Li, W., Kwiatkowski, D.J., Ramesh, V., and Silva, A.J. (2008)

Reversal of learning deficits in a Tsc2 (+/−) mouse model of tuberous sclerosis. *Nat. Med.*, **14**, 843–848.

93 Mhawech-Fauceglia, P., Alvarez, V., Fischer, G., Beck, A., and Herrmann, F.R. (2008) Association of TSC1/hamartin, 14-3-3sigma, and p27 expression with tumor outcomes in patients with pTa/pT1 urothelial bladder carcinoma. *Am. J. Clin. Pathol.*, **129**, 918–923.

94 Takamochi, K., Ogura, T., Yokose, T., Ochiai, A., Nagai, K., Nishiwaki, Y., Suzuki, K., and Esumi, H. (2004) Molecular analysis of the TSC1 gene in adenocarcinoma of the lung. *Lung Cancer*, **46**, 271–281.

95 Jiang, W.G., Sampson, J., Martin, T.A., Lee-Jones, L., Watkins, G., Douglas-Jones, A., Mokbel, K., and Mansel, R.E. (2005) Tuberin and hamartin are aberrantly expressed and linked to clinical outcome in human breast cancer: the role of promoter methylation of TSC genes. *Eur. J. Cancer*, **41**, 1628–1636.

96 Chakraborty, S., Mohiyuddin, S.M., Gopinath, K.S., and Kumar, A. (2008) Involvement of TSC genes and differential expression of other members of the mTOR signaling pathway in oral squamous cell carcinoma. *BMC Cancer*, **8**, 163.

97 Lu, K.H., Wu, W., Dave, B., Slomovitz, B.M., Burke, T.W., Munsell, M.F., Broaddus, R.R., and Walker, C.L. (2008) Loss of tuberous sclerosis complex-2 function and activation of mammalian target of rapamycin signaling in endometrial carcinoma. *Clin. Cancer Res.*, **14**, 2543–2550.

98 Al-Saleem, T., Wessner, L.L., Scheithauer, B.W., Patterson, K., Roach, E.S., Dreyer, S.J., Fujikawa, K., Bjornsson, J., Bernstein, J., and Henske, E.P. (1998) Malignant tumors of the kidney, brain, and soft tissues in children and young adults with the tuberous sclerosis complex. *Cancer*, **83**, 2208–2216.

99 Parry, L., Maynard, J.H., Patel, A., Clifford, S.C., Morrissey, C., Maher, E.R., Cheadle, J.P., and Sampson, J.R. (2001) Analysis of the TSC1 and TSC2 genes in sporadic renal cell carcinomas. *Br. J. Cancer*, **85**, 1226–1230.

100 Becker, A.J., Lobach, M., Klein, H., Normann, S., Nothen, M.M., von Deimling, A., Mizuguchi, M., Elger, C.E., Schramm, J., Wiestler, O.D., and Blumcke, I. (2001) Mutational analysis of TSC1 and TSC2 genes in gangliogliomas. *Neuropathol. Appl. Neurobiol.*, **27**, 105–114.

101 Przkora, R., Meyer-Puttlitz, B., Schmitt, O., Berthold, F., Nothen, M., Krauss, J., Tonn, J.C., von Deimling, A., Wiestler, O.D., and Pietsch, T. (2001) Analysis of the TSC2 gene in human medulloblastoma. *Acta Neuropathol. (Berl.)*, **102**, 380–384.

102 Wienecke, R., Guha, A., Maize, J.C., Jr., Heideman, R.L., DeClue, J.E., and Gutmann, D.H. (1997) Reduced TSC2 RNA and protein in sporadic astrocytomas and ependymomas. *Ann. Neurol.*, **42**, 230–235.

103 Verhoef, S., van Diemen-Steenvoorde, R., Akkersdijk, W.L., Bax, N.M., Ariyurek, Y., Hermans, C.J., van Nieuwenhuizen, O., Nikkels, P.G., Lindhout, D., Halley, D.J., Lips, K., and van den Ouweland, A.M. (1999) Malignant pancreatic tumour within the spectrum of tuberous sclerosis complex in childhood. *Eur. J. Pediatr.*, **158**, 284–287.

104 Chung, D.C., Brown, S.B., Graeme-Cook, F., Tillotson, L.G., Warshaw, A.L., Jensen, R.T., and Arnold, A. (1998) Localization of putative tumor suppressor loci by genome-wide allelotyping in human pancreatic endocrine tumors. *Cancer Res.*, **58**, 3706–3711.

105 Lee-Jones, L., Aligianis, I., Davies, P.A., Puga, A., Farndon, P.A., Stemmer-Rachamimov, A., Ramesh, V., and Sampson, J.R. (2004) Sacrococcygeal chordomas in patients with tuberous sclerosis complex show somatic loss of TSC1 or TSC2. *Genes Chromosomes Cancer*, **41**, 80–85.

106 Hornick, J.L. and Fletcher, C.D. (2006) PEComa: What do we know so far? *Histopathology*, **48**, 75–82.

107 Martignoni, G., Pea, M., Reghellin, D., Zamboni, G., and Bonetti, F. (2008) PEComas: the past, the present and the future. *Virchows Arch.*, **452**, 119–132.

108 Kenerson, H., Folpe, A.L., Takayama, T.K., and Yeung, R.S. (2007) Activation of the mTOR pathway in sporadic angiomyolipomas and other perivascular epithelioid cell neoplasms. *Hum. Pathol.*, **38**, 1361–1371.

109 Pan, C.C., Chung, M.Y., Ng, K.F., Liu, C.Y., Wang, J.S., Chai, C.Y., Huang, S.H., Chen, P.C., and Ho, D.M. (2008) Constant allelic alteration on chromosome 16p (TSC2 gene) in perivascular epithelioid cell tumour (PEComa): genetic evidence for the relationship of PEComa with angiomyolipoma. *J. Pathol.*, **214**, 387–393.

110 Henske, E.P., Neumann, H.P., Scheithauer, B.W., Herbst, E.W., Short, M.P., and Kwiatkowski, D.J. (1995) Loss of heterozygosity in the tuberous sclerosis (TSC2) region of chromosome band 16p13 occurs in sporadic as well as TSC-associated renal angiomyolipomas. *Genes Chromosomes Cancer*, **13**, 295–298.

111 Lee, D.F., Kuo, H.P., Chen, C.T., Hsu, J.M., Chou, C.K., Wei, Y., Sun, H.L., Li, L.Y., Ping, B., Huang, W.C., He, X., Hung, J.Y., Lai, C.C., Ding, Q., Su, J.L., Yang, J.Y., Sahin, A.A., Hortobagyi, G.N., Tsai, F.J., Tsai, C.H., and Hung, M.C. (2007) IKK beta suppression of TSC1 links inflammation and tumor angiogenesis via the mTOR pathway. *Cell*, **130**, 440–455.

112 Yen, C.J., Izzo, J.G., Lee, D.F., Guha, S., Wei, Y., Wu, T.T., Chen, C.T., Kuo, H.P., Hsu, J.M., Sun, H.L., Chou, C.K., Buttar, N.S., Wang, K.K., Huang, P., Ajani, J., and Hung, M.C. (2008) Bile acid exposure up-regulates tuberous sclerosis complex 1/mammalian target of rapamycin pathway in Barrett's-associated esophageal adenocarcinoma. *Cancer Res.*, **68**, 2632–2640.

5
Genotype–Phenotype Studies in TSC and Molecular Diagnostics

Kit S. Au and Hope Northrup

5.1
Introduction

Tuberous sclerosis complex (TSC) is an autosomal dominant disease affecting multiple organ systems (OMIM #191 100). The estimated incidence is 1 per 6000 live births with approximately two thirds of cases presenting *de novo* [1]. Diagnostic criteria have been published and revised to facilitate clinical diagnosis [2–4]. Mutations in two genes have been shown to contribute to the majority of TSC cases. The *TSC1* gene located at chromosome 9q34.3 was discovered a decade after the initial report of linkage [5, 6]. The existence of the *TSC2* gene was suggested through studies of families with no linkage to chromosome 9 [7] and was subsequently mapped and identified on chromosome 16p13.3 [8, 9]. Gene products of *TSC1* (hamartin) and *TSC2* (tuberin) have been demonstrated to downregulate small G-protein Ras-homologue enriched in brain (RHEB) activity via the GTPase-activating protein (GAP) domain in C-terminus of tuberin [10]. Once activated, RHEB in turn activates the mammalian target of rapamycin (mTOR) and the downstream signaling pathways to stimulate protein translation, cell growth, and proliferation. Cells with loss of function of tuberin or hamartin are no longer able to control cell growth and proliferation as discussed in Chapter 6. The release of this pathway within the cell leads to the phenotypes observed in patients who harbor *TSC1* or *TSC2* gene mutations. To date, approximately 75–85% of individuals who meet *definite* diagnostic criteria for TSC have a small identifiable *TSC1* or *TSC2* gene mutation [11–14]. Approximately 6% of TSC patients have large gene deletions (0.5% TSC1, 5.6% TSC2) [15]. The remaining 9–19% of patients with no mutation identified (NMI) are suspected to have somatic mosaicism for a mutation in one of the identified TSC genes, to possess a mutation in the unanalyzed noncoding regions of the TSC genes, or to have a mutation in an additional TSC gene locus yet to be identified. Detailed discussions regarding the genetics of TSC and mutations in the two causative genes are discussed in Chapter 4. As in many genetic diseases, the location and types of mutations present in affected individuals may correlate with the development of

Tuberous Sclerosis Complex: Genes, Clinical Features, and Therapeutics
Edited by David J. Kwiatkowski, Vicky Holets Whittemore, and Elizabeth A. Thiele

ISBN 978-3-527-32201-5

disease phenotypes and, in turn, can provide information to caregivers to formulate disease management strategies.

Worldwide, over 3600 individuals have undergone testing for mutations in their *TSC1* and *TSC2* genes with variants identified and recorded in a TSC variant database curated by Ekong and Povey at the Leiden Open Variation Database (LOVD) Web site [16]. They have reviewed and collected information on 1124 individuals with *TSC1* variants (413 unique variant entries and 97 nonpathogenic variant entries) and 2551 individuals with *TSC2* variants (1135 unique variant entries, 323 nonpathogenic variant entries, and 82 large gene deletion entries) from more than 30 published studies as well as unpublished information from TSC research groups worldwide. Almost all small mutations in the *TSC1* gene are predicted to cause premature protein termination; however, there are a few reported missense mutations. In contrast, approximately 70% of small mutations in *TSC2* are predicted to cause protein truncation and the remaining 30% are missense mutations confirmed to be *de novo* or shown to cause loss of function to suppress the RHEB/mTOR kinase activity. Many other missense variants in *TSC2* are reported, but their significance remains to be determined through functional assays and/or whether they are proven to have occurred *de novo*.

Sixteen years after the identification of the *TSC2* gene and 12 years after the identification of the *TSC1* gene, approximately 1000 unique disease-causing mutations have been reported in one of the two genes (Chapter 4) [16]. The high prevalence of early-onset infantile polycystic kidney disease was shown to associate with a contiguous gene deletion syndrome encompassing both the *TSC2* and *PKD1* genes soon after the identification of *TSC2* gene [17, 18]. Attempts to find correlation of TSC features with the underlying mutations on or before 2000 were not successful [19–28] with the exception of one study that found intellectual disability to be more frequent in sporadic patients with *TSC2* mutations [29]. Since 2000, several large comprehensive genotype–phenotype correlation studies have been completed that reported significant TSC features associated with the mutation genotypes [12–14]. These studies have aided in clarifying genotype–phenotype to help clinicians provide prognostic information to affected individuals and their families based on molecular testing.

5.2 Comprehensive Genotype–Phenotype Reports

There have been approximately 20 TSC mutation screening studies published attempting to address how the underlying mutations correlate to the clinical features (Table 5.1). The studies published on or before 2000 included smaller number of patients who had both detailed phenotypic features and TSC gene mutations identified for comprehensive genotype–phenotype correlation analysis [20–28]. These early studies reported no genotype–phenotype correlations, tended to have small sample sizes (<100), or were limited to one of the two *TSC* genes or even to the GAP domain exons of *TSC2*. Therefore, they had intrinsically limited statistical

Table 5.1 Published studies of genotype–phenotype correlations for TSC.

References	Study population	TSC gene	Features analyzed	Patients mutation/ total	Significant differences	Year
[12]	United States, Poland,	1 and 2	All	186/224	Yes	2001
[13]	Netherlands	1 and 2	All	234/490	Yes	2005
[14]	United States	1 and 2	All	236/325	Yes	2007
[15]	United States, Poland,	Large Del/ dup	All	54/261	Yes	2007
[29]	Cardiff/United Kingdom	1 and 2	All	101/150	Yes	1999
[30]	Cardiff/United Kingdom	1 and 2	Neurologic	92/157	Yes	2004
[31]	United States	1 and 2	Psychiatric	66–73/241	Yes	2007
[32]	United States	1 and 2	IQ/DQ	83/95	Yes	2007
[33]	United States	1 and 2	Renal	134/167	Yes	2006
[34]	Poland	1 and 2	Heart	108/127	Yes	2006
[20]	Cardiff/United Kingdom	2, exons 34–38	All	14/173	No	1997
[21]	United States	2 only	All	21/40	No	1998
[22]	United States	2 only	All	16/42	No	1998
[23]	Poland	1 only	All	21/148	No	1998
[25]	Netherlands	1 only	All	29/225	No	1999
[26]	United States	1 and 2	All	74/126	No	1999
[27]	Japan	1 and 2	All	23/38	No	1999
[28]	Japan	1 and 2	Five features	10/27	No	2000
[35]	Germany, Austria, Czech	1 and 2	All	31/68	No	2002
[36]	India	1 and 2	All	12/24	No	2005
[37]	Korea	1 and 2	All	11/44	No	2006
[38]	Taiwan	1 and 2	All	51/84	No	2006
[39]	Northern Ireland	1 and 2	All	35/73	No	2006

Del/dup, deletion/duplication.

power for comprehensive analyses. One of the early studies that included 58 TSC patients did report a significantly higher prevalence of mental retardation observed in individuals with *TSC2* mutations when compared to those with *TSC1* mutations, a finding validated in all subsequent and larger studies [29].

Studies reported since 2000 have been more comprehensive, including testing for mutations in both the *TSC1* and *TSC2* genes and also larger numbers of patients [12–15, 30–34]. Three of these studies [12–14] that included more comprehensive clinical and genetic information and much larger TSC patient cohorts will be the focus of this chapter. All three studies primarily included sporadic TSC patients and index patients from multigenerational TSC families in their analyses.

The first large comprehensive study published in 2001 [12] had 224 TSC patients enrolled primarily from pediatric neurology settings in the United States and Poland with a median age of 10 years and an average age of 11.5 years (range 1–51 years). The analyses included186 independently ascertained TSC patients with detailed records of major and many minor TSC features along with gene mutation information generated from several exhaustive mutation screening methods. Conclusions included a milder disease phenotype for patients with *TSC1* mutations with lower frequency for major features (cortical tubers, subependymal nodules (SENs), and kidney lesions) and one minor feature (retinal hamartoma). They also noted that the incidence of seizures and mental retardation were lower among patients with *TSC1* mutations although these two features due to their lack of specificity are not included in the diagnostic criteria for TSC.

The second study [13] enrolled 490 TSC patients from the Netherlands along with referrals from outside the Netherlands diagnosed by using a standardized clinical criteria of TSC features listed on an evaluation form prepared by the investigators. A total of 291 patients between 0 and 60 years (median age 13.0 years) had detailed clinical information for the analysis and 235 patients had an identified mutation. Similar to the first study, patients with *TSC1* mutations were found to have less severe disease and lower frequency of renal lesions, retinal phakoma, and mental retardation. However, frequencies for cortical tubers, subependymal nodules, and seizures were the same regardless of the *TSC* gene mutated and a higher frequency of shagreen patches was observed among the *TSC1* mutation group.

The third large study [14] enrolled 368 unrelated TSC patients with approximately one third from the University of Texas Medical School at Houston Genetics Clinic in Houston, Texas, one third from the Tuberous Sclerosis Clinic at Texas Scottish Rite Hospital in Dallas, Texas, and one third from throughout the United States referred by the Tuberous Sclerosis Alliance (TSA). All included patients were diagnosed utilizing standard diagnostic criteria [3, 4]. This study compared TSC features found in 59 patients with median age 10.7 years (average age 12.5 years) having *TSC1* mutations and 177 patients with median age 9.0 years (average age 11.2 years) having *TSC2* mutations. Consistent with the previous studies, subependymal nodules, mental retardation, renal lesions, and retinal phakoma were more frequently observed in individuals with *TSC2* mutations. In addition, other TSC findings (learning disability, hypomelanotic macules, facial angiofibromas, and forehead plaque) were also seen more often in those with *TSC2* mutations. A trend of a higher frequency of behavioral problems was noted in patients with *TSC1* mutations ($p = 0.057$).

All three large comprehensive studies [12–14] confirmed the finding of two smaller studies [29, 30] that mental retardation is more likely to occur among patients with *TSC2* mutations. In addition, the three studies also observed significantly higher prevalence of subependymal nodules, renal lesions, and retinal phakoma for patients with *TSC2* mutations. However, each of the three large studies also observed significant correlation of many TSC disease features not found by the other two

Table 5.2 Prevalence of TSC features in three comprehensive genotype–phenotype correlation studies.

Feature	All (% present)			*TSC1* (% present)			*TSC2* (% present)			NMI (% present)		
	[12]*	[13]*	[14]*	[12]*	[13]*	[14]*	[12]*	[13]*	[14]*	[12]*	[13]*	[14]*
Tub	88.5	87.0	84.0	83.3	83.3	84.4	93.0	89.4	89.6	79.2	77.8	71.0
SEN	91.7	89.8	83.4	80.0	91.9	74.5	97.8	90.6	93.2	76.5	81.8	67.2
SEGA	11.1	27.1	18.0	15.4	38.5	13.2	11.7	29.3	22.9	5.4	7.1	13.0
MR	65.8	71.9	46.5	50.0	48.7	21.4	70.8	82.7	57.3	54.5	36.4	38.6
Sz	90.6	88.6	74.8	85.7	91.3	72.2	96.8	90.9	82.6	67.6	68.0	59.6
HM	92.2	90.1	89.1	96.3	93.2	86.5	95.5	91.8	95.5	74.3	73.9	78.1
FA	74.5	77.9	60.1	64.3	71.1	39.6	74.8	82.2	70.6	58.3	63.6	52.6
SP	47.6	50.8	38.9	36.0	71.0	42.3	53.6	72.4	45.2	25.7	18.8	30.3
UF	18.1	35.6	24.2	20.0	45.2	25.5	20.0	33.7	30.8	8.6	26.7	13.6
FFP	34.1	37.4	25.6	12.0	27.3	13.3	41.3	44.9	39.1	22.9	13.3	10.8
CRM	51.3	40.6	46.2	42.9	42.8	38.2	51.4	41.6	56.1	55.9	33.3	33.8
RA	54.6	43.3	46.6	29.2	7.4	8.3	60.1	50.0	59.7	45.7	56.3	39.1
RC	25.1	24.6	24.8	12.5	10.7	2.5	28.9	27.6	30.0	17.6	31.3	29.9
RH	24.0	28.3	30.6	0	10.0	10.3	28.5	37.0	44.8	17.9	12.5	16.7
AP	/	22.6	8.3	/	0	10.5	/	31.4	10.1	/	10.0	5.4
GF	/	33.3	19.0	/	50.0	17.6	/	27.8	20.5	/	40.0	15.0
DEP	/	49.4	37.0	/	52.6	35.3	/	50.0	39.5	/	25.0	30.4

The first row indicates the groups of TSC patients enrolled in the study: all patients, those with mutations in *TSC1*, those with mutations in *TSC2*, and those who have no mutation identified (NMI). The percentage of positive TSC features is reported (% present). The number of TSC patients varied for each TSC feature reported due to differences in age of diagnosis and age-dependent presentation of some TSC features. *Represents reference number where data were extracted.

Abbreviations: Tub, Cortical tubers; SEN, subependymal nodules; SEGA, subependymal giant cell astrocytoma; MR, mental retardation; Sz, seizures; HM, hypomelanotic macules; FA, facial angiofibromas; SP, shagreen patches; UF, ungual fibromas; FFP, fibrous forehead plaques; CRM, cardiac rhabdomyomas; RA, renal angiomyolipomas; RC, renal cysts, RH, retinal phakoma or hamartomas; AP, retinal achromic patches; GF, gingival fibromas; DEP, dental enamel pits.

studies. Several possible factors (i.e., patient ascertainment, age, ethnic background, or geographic factors) could have resulted in these differences between studies affecting the genotype–phenotype correlation outcome (Table 5.2). Previous experience learned from underpowered TSC genotype–phenotype correlation studies [20–28] suggested that perhaps some of the studies did not find significant correlations because they had an insufficient number of patients reporting on presence or absence of the specific phenotypic features, thereby falling below the statistical threshold for analysis. One way to determine whether the observed differences represented true findings was to combine data of the studies, thus increasing the power through meta-analysis. This strategy allowed observations that had been identified in individual studies to be verified or refuted [14]. The strategy of using pooled data from several comparable comprehensive large studies increased the power of analysis for less prevalent phenotypes and boosted the numbers for

Table 5.3 Odds ratios of *TSC2* mutation versus *TSC1* mutation for phenotypic features observed in all TSC patients.

Feature	OR (CI) [12]*	OR (CI) [13]*	OR (CI) [14]*	OR (CI) Combined
Tub	2.64 (0.57,12.28)	1.69 (0.49,5.86)	1.48 (0.57,3.86)	1.72 (0.87,3.39)
SEN	**10.92 (2.42,49.26)**	0.85 (0.23,3.15)	**4.58 (1.93,10.9)**	**3.40 (1.85,6.27)**
MR	2.42 (0.92,6.37)	**5.03 (2.37,10.67)**	**4.38 (1.99,9.64)**	**3.96 (2.51,6.24)**
Sz	**5.10 (1.28,20.34)**	0.95 (0.30,2.98)	1.75 (0.86,3.56)	**1.91 (1.12,3.25)**
FA	1.65 (0.70,3.88)	1.88 (0.82,4.31)	**3.13 (1.66,5.89)**	**2.33 (1.53,3.53)**
SP	2.05 (0.85,4.93)	**0.40 (0.17,0.97)**	1.06 (0.57,1.99)	0.99 (0.65,1.51)
FFP	**5.16 (1.48,17.96)**	2.17 (0.77,6.13)	**3.37 (1.41,8.05)**	**3.29 (1.85,5.86)**
CRM	1.41 (0.56,3.56)	1.00 (0.40,2.49)	1.94 (0.91,4.11)	1.48 (0.91,2.41)
RA	**3.66 (1.43,9.38)**	**11.88 (2.69,52.44)**	**16.89 (4.92,57.93)**	**8.27 (4.36,15.7)**
RC	2.84 (0.81,10.02)	3.17 (0.88,11.36)	**19.55 (2.60,146.98)**	**5.24 (2.35,11.67)**
RH	16.44 (0.97,278.15)	**5.28 (1.49,18.71)**	**7.07 (2.04,24.53)**	**6.94 (2.94,16.39)**

*Represents reference number where data were extracted and analyzed. Combined – results generated from combining data in Refs [12–14]. Bolded OR (odds ratio) and CI (95% confidence interval) indicate significance with $p < 0.05$; total patients in combined study with $N = 656$. Only the TSC features that showed a significant difference are listed.

categories with small numbers in all three studies (i.e., the number of index patients from multigenerational TSC families and the number of patients determined to have *TSC1* mutations).

Meta-analysis studies using pooled data from several similar retrospective case-control studies commonly present the results as odds ratios (ORs). For analysis of the frequency of various clinical features among the TSC patient samples (Table 5.3), we compared the findings in patients with *TSC1* mutations (considering them as "controls") to those of patients with *TSC2* mutations (considering them as "cases"). Therefore, an OR greater than 1 for a specific TSC clinical feature indicates that a patient with a *TSC2* mutation is more likely to have that feature than a patient with a *TSC1* mutation. On the other hand, an OR less than 1 indicates that a patient with a *TSC1* mutation is more likely to have the feature than a patient with a *TSC2* mutation. An OR of about 1 indicates that patients with *TSC1* and *TSC2* mutations are about equally likely to have the disease feature. Statistical analysis is used to determine if each OR is really significantly different from 1, by determining the 95% confidence interval (CI) for each. Odds ratios are presented in each of Tables 5.3-5.6, for different subsets of TSC patients considered. In interpreting the data in Table 5.3, for example, for tubers (Tub), the combined odds ratio is 1.72 but the CI is 0.87–3.39. This means that tubers are seen somewhat more commonly in patients with TSC2 mutations than TSC1 mutations, but the observations are not quite statistically significant. On the other hand, for subependymal nodules (Table 5.2), the combined odds ratio is 3.40 (CI 1.85–6.27), meaning that SENs are seen significantly more often in patients with TSC2 mutations than TSC1 mutations.

5.3 Genotype–Phenotype Correlation

5.3.1 *TSC2* Versus *TSC1* Gene Mutations

By combining the data of the three large studies, information from a total of 656 patients with median age range of 9–13 (range 0–64 years) was used to perform odds ratio analyses [14]. Because all three studies were retrospective, odds ratios were calculated comparing the likelihood of observing a *TSC2* mutation compared to a *TSC1* mutation for a particular phenotypic feature. Results of the combined data analyses of observed TSC features from the three studies in relation to the *TSC* genotypes (*TSC2* versus *TSC1* mutation) are shown in Table 5.3. Analyses of the pooled data set reaffirmed that *TSC2* mutations are observed more commonly than *TSC1* mutations in patients who have subependymal nodules, mental retardation, renal angiomyolipomas, and retinal phakomas, as observed independently by all three studies. In addition, other features (seizures, renal cysts, forehead plaques, and facial angiofibromas) reported to be significantly associated with *TSC2* mutations by one or two large studies also showed significance in the combined data analyses. This exercise suggested meta-analysis is a powerful approach to perform genotype–phenotype correlation studies under the assumption that the pooled patient cohorts are diagnosed with comparable criteria by qualified professionals.

Higher prevalence of mental retardation has been consistently shown to be more likely in patients with mutations in *TSC2* rather than *TSC1* in many studies. Because mental retardation is common in the general population and it occurs secondary to a myriad of causes, it is not a diagnostic criterion for TSC [12–14, 29, 30]. Other related neuropsychiatric findings (autistic disorder, aggressive disruptive behavior, and poor cognitive outcome) have also been reported more often among patients with *TSC2* mutations than those with *TSC1* mutations [30, 32–40]. Intriguingly, genes that have been identified as resulting in mental retardation when disrupted are recognized to be central components in the protein networks regulating cytoskeleton dynamics and synthesis of new proteins critical for neural development [41]. One major function of tuberin and hamartin is to modulate protein synthesis by regulating RHEB/mTOR kinases. RHEB was initially identified as a Ras-family small GTP binding protein that is induced in brain neurons by synaptic activity and expressed at high levels during embryonic as well as pre- and postnatal brain development [42]. The RHEB GAP function appears to be an obvious explanation for the neural phenotypic features observed in TSC patients. In addition, functional hamartin is also required for tuberin to regulate RHEB activities, as discussed in Chapter 6. Loss of *Tsc1* or *Tsc2* in a rat model resulted in changes of neuronal morphology [43] that also affect neural development. How and why mutations of the *TSC1* gene appear to present less risk for development of mental retardation is yet to be determined. Future research studying the role each of these genes plays and their interactions in the signaling pathways should be helpful in shedding light on this and other questions now posed with the finding that *TSC2* mutations are more likely than *TSC1* mutations to be observed in TSC patients with neural phenotypes.

Renal angiomyolipomas, another feature consistently associated as occurring more often in patients who have *TSC2* mutations by many studies, consist of three cellular components (vascular, smooth muscle, and lipocytes) derived from progenitor cells via mesenchymal differentiation [33, 44]. Remarkable mesenchymal differentiation plasticity has been observed in cells lacking *TSC2* in these studies. Biochemical studies demonstrated that hamartin interacts with tuberin and facilitates tuberin GAP function. In the situation when *TSC1* is mutated, an unstable tuberin may still be functional, therefore contributing to a lower risk for developing TSC features. In addition to cases with contiguous deletion of the *TSC2* and *PKD1* genes, small *TSC2* mutations have also been associated with higher renal cysts occurrence. In a recent study, cells lacking tuberin or hamartin were found to have aberrant cilia length similar to cells lacking polycystins. The same study also showed that the number of renal cysts lesions found in mice heterozygous for *Tsc2* is approximately 15 times higher than mice heterozygous for *Tsc1* [45]. Studies comparing renal lesions that develop in the presence of a lack of *TSC1* function versus *TSC2* function should shed light on the differential prevalence of renal cysts observed among TSC patients.

Association of increased frequency of cardiac rhabdomyomas with *TSC2* mutation versus *TSC1* mutation has been reported in a study of 127 Polish patients [12, 34]. The study showed incidence of cardiac rhabdomyoma was higher for patients less than 2 years old (65.6%) and those 11–15 years old (54.5%) in comparison to patients 2–11 and 15–23 years old with frequencies of 26.1 and 33.3%, respectively. However, no association of a higher incidence of cardiac rhabdomyomas with *TSC2* mutations was found in the three larger genotype–phenotype correlation studies [12–14].

Affected individuals in multigenerational TSC families with *TSC2* mutations also have more severe TSC disease features than individuals from families with *TSC1* mutations as shown by meta-analyses of combined patient information between two large studies [13, 14]. The combined data provided sufficient sample size ($N = 131$) to conclude higher prevalence for mental retardation, renal angiomyolipomas, renal cysts, and retinal phakoma occurring in familial patients with *TSC2* mutations as compared to those with *TSC1* mutations (Table 5.4). From individual and combined data analyses, patients with *TSC2* mutations regardless of their mode of inheritance are shown to be at higher odds for having renal angiomyolipomas, retinal phakomas, renal cysts, and mental retardation. Further restricting the analyses to protein truncation mutations resulted in a similar conclusion [14]. In concert with this trend, it would not be surprising to find a higher prevalence for the remaining TSC features in those with *TSC2* mutations (both sporadic and familial patients) when the number of familial cases available for meta-analyses reaches a critical mass.

5.3.1.1 **NMI Patients**

Since almost all of the mutation screening methods used to detect TSC gene mutations can potentially miss mosaicism (mutation present in some cells but not all) for mutations, the TSC patient group with no mutation identified is likely to be

Table 5.4 Odds ratios of *TSC2* mutation versus *TSC1* mutation for phenotypic features observed in familial TSC patients.

Feature	OR (CI) [12]*	OR (CI) [13]*	OR (CI) [14]*	OR (CI) Combined
MR	1.56 (0.26,9.47)	1.43 (0.33,6.26)	6.14 (1.1,34.21)	**2.70 (1.17,6.23)**
AML	6.8 (0.66,69.64)	7.69 (0.37,157.91)	2.79 (0.26,30.27)	**8.92 (1.94,41.01)**
RC	9.29 (0.46,187.18)	1.33 (0.12,14.9)	1.88 (0.15,22.88)	**5.33 (1.15,24.77)**
RH	4.00 (0.19,84.20)	4.24 (0.2,89.34)	3.67 (0.37,35.98)	**8.64 (1.09,68.59)**

*Represents reference number where data were extracted and analyzed. Combined – results generated from combining data in Refs [12–14]. Bolded OR (odds ratio) and CI (95% confidence interval) indicate significance with $p < 0.05$. Total numbers of affected probands for Refs [12–14] and combined are 31, 47, 53, and 131, respectively. Actual number of probands having TSC features reported varied. Only the TSC features that showed a significant difference are listed.

enriched for patients having mosaic mutations. It has been shown for many single gene disorders that affected individuals having somatic mosaicism have a milder disease phenotype than those with germline mutations. Therefore, we expect the phenotypes observed in the NMI patients to be less severe than those observed in patients with germline *TSC2* mutations (Table 5.2). Observations from all three large studies and multiple other smaller studies showed data to support that the NMI patients generally have lower prevalence of TSC features than the mutation-identified group and the group with *TSC2* mutation regardless of mutation types as seen in Table 5.2 [12–14, 31–33]. Because the NMI patient group has TSC secondary to multiple different etiologies, some perhaps not even discovered, it is meaningless to discuss genotype–phenotype correlations for them. Genotype–phenotype correlations are not possible when we do not know the genotype.

5.3.1.2 Familial Versus Sporadic Cases

Individuals affected by TSC from families with multiple members having TSC (referred to from here onward as familial TSC patients) have been noted to have a milder TSC phenotype when compared to TSC patients who are the only affected family member (sporadic TSC patients; individuals with a *de novo* mutation). A survey [46] of 404 sporadic and 69 familial TSC patients in the United Kingdom found significantly more mental retardation reported in sporadic versus familial patients (65 versus 46%, chi square = 10.14, $p = 0.0015$, and OR for MR in sporadic 2.44 (1.98, 0.81). In a study comparing findings of 38 familial and 183 sporadic TSC cases, a higher prevalence of hypomelanotic macules was found for sporadic patients along with a trend for higher seizure prevalence [12]. Another study of 63 familial and 137 sporadic patients found the sporadic TSC patients had a higher prevalence for subependymal nodules, hypomelanotic macules, and retinal phakomas than the familial patients [13]. They also found that seizure prevalence for patients with *de novo* *TSC2* mutation was higher when compared to individuals with inherited *TSC2* mutations. The Au *et al.* study [14] analyzed TSC features for 70 familial index patients and 297 sporadic patients according to modes of inheritance and effects of

Table 5.5 Odds ratios of sporadic patients versus familial patients for frequencies of phenotypic features.

Feature	OR (CI) [12]*	OR (CI) [13]*	OR (CI) [14]*	OR (CI) Combined
MR	1.66 (0.71,3.89)	1.57 (0.64,3.86)	**2.17 (1.00,4.72)**	1.55 (0.98,2.43)
SZ	**5.89 (1.49,23.19)**	3.00 (0.80,11.23)	1.35 (0.62,2.93)	**1.88 (1.07,3.29)**
SEN	**6.76 (2.1,21.78)**	**18.58 (2.18,158.42)**	**4.14 (1.63,10.48)**	**4.73 (2.55,8.75)**
RA	0.98 (0.45,2.12)	0.8 (0.31,2.04)	**4.49 (1.71,11.79)**	**1.64 (1.01,2.64)**
RH	0.77 (0.32,1.85)	**3.44 (1.04,11.4)**	**2.80 (1.06,7.40)**	**1.83 (1.05,3.21)**

*Represents reference number where data were extracted and analyzed. Combined – results generated from combining data in Refs [12–14]. Bolded OR (odds ratio) and CI (95% confidence interval) indicate significance with $p < 0.05$. Only the TSC features that showed a significant difference are listed.

gene mutated. The ages at time of comparison for features between the familial index patients (median age 10.5 years) and sporadic patients (median age 9 years) was not significant. Significantly less frequent mental retardation, subependymal nodules, renal angiomyolipomas, and retinal phakomas were observed among the familial index patients. The number of familial index patients studied in all three studies limits a comparison of phenotypic features between familial cases and sporadic cases. Meta-analyses were performed on the combined data of the three larger studies [12–14] and supported the finding that subependymal nodules, renal angiomyolipomas, retinal phakoma, and seizures are less frequently observed among familial TSC cases compared to sporadic TSC cases (Table 5.5).

5.3.2
Protein Truncation Versus Missense Mutations

Almost all mutations identified in *TSC1* are predicted to cause premature termination of hamartin. In contrast, about 30% of all *TSC2* mutations are missense and cause change of a single amino acid. The most recurrent missense or in-frame deletion mutations identified are p.1746delHIKRLR, p.R611Q, p.R611W, p.P1675L, p.R905W, and p.R905Q. In most cases, the mutation results in loss of all activity needed for suppression of RHEB/mTOR kinase [16]. The type of mutations in many single gene disorders affects disease severity or associates with specific disease phenotype. For example, in von Hippel–Lindau (VHL) disease, missense mutations are associated with a high risk of pheochromocytoma, while pheochromocytoma is rarely seen in VHL patients with inactivating germline mutations of the *VHL* gene [47]. The effect of type of mutations on the presence of TSC features by grouping patients into two main categories, mutations predicted to cause protein truncation (including nonsense, out-of-frame insertion/deletion, and cryptic splice sites) and mutations predicted to cause amino acid substitution (missense mutation and in-frame insertion/deletion), has been evaluated [14]. It was concluded that missense mutations in the *TSC2* gene cause similar disease phenotypes as protein truncation mutations in the *TSC2* gene. Almost all features were equally represented between the groups of patients having *TSC2* missense mutations versus *TSC2* protein truncation mutations. Interestingly, the missense mutation group showed less frequent ungual fibromas and nonrenal hamartomas [14]. At the time of the analysis, patients with *TSC2* missense mutations were younger with median age of 7 years while 10.7 years was the median age for patients with *TSC2* protein truncation mutations. This difference in age was not significant ($p = 0.9416$). In a similar study [13], *TSC2* nonsense and frameshift mutations presented with more frequent renal angiomyolipomas than the missense mutations (64 and 38%, $p = 0.050$). Yet another study showed a higher prevalence of shagreen patches in patients who had protein termination mutations (59 and 38%, $p = 0.04$) [12]. It appears that missense mutations in *TSC2*, in general, are as pathogenic as the *TSC2* protein truncating mutations. The Au study [14] also found that missense mutations in the *TSC2* gene resulted in more severe disease than protein truncation mutations in the *TSC1* gene. The finding remained when comparing all patients with *TSC1* mutations to all patients with *TSC2* mutations.

Despite the general finding that *TSC2* missense mutations are equally pathogenic as *TSC2* protein truncation mutations, several *TSC2* missense mutations have been found to associate with very mild TSC disease phenotypes. Two large Utah TSC families with the *TSC2* c.4508A > C (p.Q1503P) mutation present displayed very mild findings, including mood disorder, anxiety disorder, and autism [48]. O'Connor *et al.* [49] reported a family with a *TSC2* c.3106T > C (p.S1036P) mutation as having only seizures and white matter abnormalities. Two individuals in this family had no clinical features but were found to harbor the mutation. Mayer *et al.* [50] identified a c.4684G > A (p.G1556S) *TSC2* mutation in multiple members of a family but only the index patient fulfilled diagnostic criteria for TSC. Other family members with the mutation did not have sufficient findings for diagnosis. The p.G1556S tuberin mutant was hypophosphorylated, had reduced hamartin interacting ability, and lacked RHEB inactivation activity to inhibit phosphorylation of ribosomal protein S6. Jansen *et al.* [51] in a study of 19 familial TSC cases reported the relatively common c.2713C > T (p.R905Q) tuberin mutation to be associated with mild TSC findings. Some mutation carriers in the families did not meet diagnostic criteria, a phenomenon consistent with slightly reduced protein function. Certain *TSC2* missense mutations of tuberin retain partial GAP activity, thus resulting in a milder phenotype [52].

5.3.3 Whole Gene/Large Deletion Versus Small Mutation

5.3.3.1 *TSC1* Large Deletions

Multiple exon or whole gene deletion of the *TSC1* gene is rare in TSC patients and estimated to account for 0.5% of all mutations [15]. One report [53] described a familial case with the entire *TSC1* gene and two flanking genes deleted in six affected individuals meeting diagnostic criteria (with facial angiofibromas, forehead plaques, ungual fibromas, shagreen patches, and subependymal nodules). In this report, the second-generation affected individuals had normal mental capabilities, but the third-generation affected individuals were all mentally disabled. We are aware of a similar case with deletion of the entire *TSC1* gene and findings, including hypomelanotic macules, facial angiofibromas, forehead plaques, shagreen patches, confetti skin lesions, cortical tubers, cardiac rhabdomyomas, and seizures (unpublished data). Coincidentally, no renal findings were observed among all the affected individuals in both studies. A large study [54] of 202 TSC patients using Southern blot analyses utilizing two restriction enzymes found large deletions in 19 patients, including 3 with *TSC1* mutations and 16 with *TSC2* mutations. All *TSC1* deletions were within the 3′-end of the transcriptional unit.

5.3.3.2 *TSC2* Large Deletions

Large deletion/duplication mutations in the *TSC2* gene are estimated to account for approximately 5.6% of mutations identified in TSC patients [15]. Consistently, young TSC patients with early-onset infantile polycystic kidney phenotypes have frequently been found to have a contiguous gene deletion syndrome involving the *TSC2* gene and the *PKD1* gene [15, 17, 18]. Patients who have the 5′-end of the *TSC2* gene deleted do not have the polycystic kidney phenotype [15]. In a study [54] of 202 patients, 16 (7.9%) large *TSC2* deletions were identified. Of the 16 *TSC2* deletions, 12 had at least one end of the deletion mapping beyond the *TSC2* gene, with 6 ($\sim$3%) extending into the *PKD1* gene (all these patients had polycystic kidney disease). Of the 16 *TSC2* large deletions, 2 were mosaic. Several other reports described adult patients with confirmed deletions involved *TSC2* and *PKD1* who were not found to have polycystic kidneys until the time of nephrectomies [55–58]. These reports provide additional evidence for variable expressivity in TSC patients and highlight that the relationship of TSC clinical expression is influenced by the stochastic "second hits" that disable the remaining functional copy of the gene in a particular cell as well as by other modifying factors. In the UT Medical School at Houston (UTMSH) cohort of 325 TSC patients [14], 20 patients had large *TSC2* deletion mutations. Of the 20, 8 had part of or the entire *TSC2* and *PKD1* genes deleted. Of the 20, 14 had information on renal findings available, including 6 who had involvement of both the *TSC2* and *PKD1* genes. Five patients who had *TSC2* and *PKD1* gene deletions also had polycystic kidneys, with three (aged 0–2) harboring germline mutations and two (aged 4.4 and 33.5, respectively) with mutations in a mosaic state. While one patient who only had the entire *TSC2* gene deleted had polycystic kidneys at 6 years of age, another patient with both *TSC2* and *PKD1* deleted was not reported to have cystic kidneys until 4 years of age. Overall, patients who have a contiguous deletion of *TSC2* and *PKD1* are mostly likely to develop early-onset infantile polycystic kidney disease.

5.3.4 Mutations in *TSC2* GAP Domain

5.3.4.1 *TSC2* GAP Domain Mutations

No significant genotype–phenotype correlation was found in a study of mutations of the *TSC2* gene GAP domain (exons 34–38) [20]. However, another study [13] showed lower prevalence of renal angiomyolipomas and renal cysts to be associated with *TSC2* mutations in exons 35–39 when compared to nonsense or frameshift mutations in other part of *TSC2* (39–64% and 10–37%, $p = 0.032$ and 0.018, respectively). In analyzing 48 affected TSC patients (17 with *TSC1* mutations and 31 with *TSC2* mutations), Jansen *et al.* [40] found that mutations predicted to cause loss of tuberin GAP activity were associated with more tubers and a higher proportion of total brain volume occupied by tubers. However, it is not known if truncated tuberin or hamartin present in the cells of TSC patients affect function of the normal tuberin/hamartin. Nor do we know if missense mutations outside the GAP domain affect GAP activity. It is not prudent to assume that loss of GAP activity is only related to missense mutations located within the GAP domain of tuberin. Nellist *et al.* [52] has demonstrated missense mutations (R611Q, R611W, A614D, C696Y, V769E) far from the GAP domain (amino acids 1499–1720) to cause loss of GAP activity, inability to interact with hamartin, and failure to inhibit RHEB to activate S6K via phosphorylation of T389. Peptides containing amino acid 1–1240 of tuberin have been shown to lack suppression of S6 phosphorylation, while the peptides consisting of amino acids 1125–1784 of tuberin are capable of partially suppressing S6 phosphorylation. Therefore, intact GAP activity of tuberin requires intact tuberin and interaction with intact functional hamartin. There also exist mutations (i.e., p.609insS, F615S) that result in loss of hamartin interaction but retain weak RHEB GAP activity and are able to suppress S6 phosphorylation. Other mutations (i.e., R905Q) retain hamartin interaction but lose RHEB GAP activity. Until all missense mutations are functionally characterized, genotype–phenotype correlation of missense mutations by location associated with the GAP domain will be premature.

5.3.4.2 *TSC2* Gene Amino-Termini Mutants Versus Carboxy-Termini Mutants

At the UTMSH, we have compared groups of patients with *TSC2* missense mutations by stratifying and evaluating differences between patients having missense mutations in the amino (N)-terminal exons (1–33, $n = 37$) of *TSC2* versus those with missense mutations on the carboxy (C)-terminal exons (34–41, including the GAP domain, $n = 21$). We found that significantly less renal cysts and retinal phakoma were associated with the N-terminal missense mutations (3/27, 11.1%; 6/25, 24%, respectively) versus the C-terminal missense mutations (6/14, 42.9%; 7/12, 58.3%, respectively) (Fisher's exact $p = 0.0287$, $p = 0.0475$, respectively). No significant differences were found for other features. Caution is needed to interpret phenotypes correlated with missense mutations simply by location since several missense mutations in the N-terminus of tuberin have been found to affect the RHEB-GAP function located in the C-terminus [52]. Unlike DNA and mRNA, secondary and tertiary folding of proteins brings amino acids from different locations together to form domains that carry on the functions of the protein.

Similar analyses of the UTMSH cohort with protein truncation mutations (exon 1–33, $n = 100$; exons 34–41, $n = 22$) did not yield significant differences between the two groups. Nonsense-mediated mRNA decay could be the dominating factor for these mutation types where truncated mRNAs are rapidly recycled. Evidence for the presence of C-termini-truncated tuberin or hamartin in TSC patients is lacking; thus, there is no support for the hypothesis that truncated proteins act in a dominant negative fashion.

5.3.5
Mosaicism

As discussed in Chapter 4, different frequencies of somatic mosaicism among TSC patients have been reported by different studies and patients with mosaic mutations are generally reported to have milder disease phenotypes [13, 17]. In Sancak's study of sporadic cases, they found less than 1% of the 235 patients with defined small mutations were mosaic [13]. Much higher rates of somatic mosaicism (26%, 7/27) have been reported in TSC patients with contiguous deletions of *TSC2-PKD1* [17], a finding that is consistent with speculation that somatic mosaicism is a mechanism leading to better survival rate and less severe disease phenotypes. Incidentally, 6 of 20 (30%) large *TSC2* deletion patients in the UTMSH cohort of 325 TSC patients have been found to be mosaic for their mutation. The average age for the nonmosaic group is 5.7 years (median age 4 years), while for the mosaic group, the average age is 16.5 years (median age 4.4 years) with $p = 0.034$ by *t*-test. The only significant difference observed between the two groups is a much lower incidence of epilepsy for the mosaic group (2/5) versus the nonmosaic group (12/13) with an OR of 0.056 (CI 0.004–0.838; $p = 0.017$ by chi-square test) consistent with findings concluded in a larger study [15]. General consensus regarding the proportion of individuals who have TSC secondary to mosaic mutations cannot be determined from current reports. An accurate, comprehensive study to find the proportion of somatic mosaicism in TSC patients is extremely challenging due to lack of effective and efficient technologies to screen for mosaic mutations. In addition, mosaic TSC cases generally have milder symptoms, therefore, leading to a bias of ascertainment in these individuals coming to medical attention.

5.3.6
Male Versus Female Sex

With the important roles of tuberin and hamartin in cell signaling, genes products such as hormones that are active in the cell signaling networks, are potential modifiers of tuberin and hamartin functions. There is evidence that sex plays a role in the prevalence of some TSC features. Examples where gender has been shown to be important include mental retardation and retinal phakoma [13] as well as lymphangioleiomyomatosis (LAM) [59]. In TSC, LAM affects female patient almost exclusively most likely secondary to hormonal differences between females and males [44, 60, 61]. In comparing male and female TSC patients, Au *et al.* [14] found male patients to have higher prevalence of many TSC features than female patients. However, in this study the specific types of features (cortical tuber, SEN, mental retardation, and seizures) differed from the study of Sancak *et al.* who also reported a higher prevalence of some TSC findings in males [13]. The proportion of mutations for *TSC1* and *TSC2* does not differ between male and female TSC patients. A meta-analysis of two large

Table 5.6 Odds ratios of male versus female TSC patients for frequencies of phenotypic features.

Feature	OR (CI) [13]*	OR (CI) [14]*	OR (CI) Combined
Tub	0.86 (0.27–2.71)	**2.32 (1.13–4.77)**	**1.80 (1.00–3.28)**
SEN	1.15 (0.43–3.05)	**2.21 (1.12–4.35)**	**1.85 (1.07–3.20)**
MR	**2.23 (1.15–4.31)**	**1.68 (1.00–2.81)**	**1.96 (1.33–2.89)**
Sz	1.56 (0.67–3.59)	**2.34 (1.38–3.99)**	**2.17 (1.40–3.37)**
UF	1.92 (0.91–4.06)	1.50 (0.87–2.58)	**1.67 (1.08–2.58)**
RC	2.04 (0.91–4.58)	1.53 (0.85–2.76)	**1.70 (1.06–2.73)**
RH	**2.42 (1.06–5.50)**	1.77 (0.97–3.21)	**1.97 (1.22–3.19)**
AP	**9.78 (1.76–54.26)**	0.50 (0.17–1.49)	1.43 (0.65–3.13)
GF	**5.14 (1.29–20.52)**	0.83 (0.26–2.62)	1.73 (0.75–3.99)

Bolded OR (odds ratio) and CI (95% confidence interval) indicate significance with $p < 0.05$. Only the TSC features that showed a significance difference are listed.

studies reporting these male/female differences has been undertaken [13, 14]. Of particular interest, males were more likely to exhibit more neurological features (cortical tuber, SEN, mental retardation, and seizures) in the combined data sets, leading to increased morbidity over that observed in females. The odds for male patients to have these neurological features were approximately twice that of female patients (Table 5.6). A higher odds ratio for other features, including retinal phakomas, ungual fibromas, and renal cysts, among male patients was also observed in the meta-analysis. The reason for these observed differences is not known. One possible explanation could be modifier genes coded for on the X chromosome or involvement of X-inactivation among modifier genes. Another possibility would be effect of age-dependent hormonal influences between sexes [60–62]. Future research will be needed to sort out these differences for both provision of prognostic information as well as therapeutic interventions.

5.4 Molecular Diagnostic Methods

With the identification of the disease-causing genes (*TSC1* and *TSC2*) for TSC more than a decade ago, TSC research groups began utilizing different mutation screening techniques to rapidly identify DNA sequence variants in patient DNAs in a cost-effective and efficient manner. There are now a wide variety of mutation detection methods available, including automated sequencing by capillary array electrophoresis, microarrays, and methods designed to differentiate wild-type and variant DNA fragments by electrophoresis or high-performance liquid chromatography. The pros and cons of these methods from the clinical diagnostic application perspective have been discussed in the literature [63].

To comprehensively screen the TSC genes for small mutations requires examining 3.5 kb of coding nucleotides in 21 coding exons for *TSC1* and 5.5 kb of coding nucleotides in 41 exons for *TSC2*. Screening mutation analysis for the two breast cancer-causing genes (*BRCA1* and *BRCA2*) by sequencing is a model, but the cost of this methodology can be prohibitive. There are less-expensive techniques available that exploit electrophoretic mobility differences of DNA conformers containing wild-type or variant sequences. Among these less-expensive methods, single-stranded conformation polymorphism (SSCP) and heteroduplex analysis (HA) each achieve detection rates from 50–75% and have been used most commonly for TSC testing [11–14, 19–28, 64–67]. The denaturing gradient gel electrophoresis method that separates homoduplexes from heteroduplexes in the presence of a gradient of urea or formamide has also been used by a few groups [13, 68–70]. In many cases, the mutation detection rate improves as the laboratory personnel gain more experience with the specific detection method used.

Denaturing high-performance liquid chromatography (DHPLC) separates wild-type and variant homoduplexes and heteroduplexes based on their differential retention to the column matrix with no electrophoresis involved. TSC mutation detection rate using DHPLC is suggested to be superior (70–80%) to SSCP and HA [12, 26, 37, 38, 68].

Mutation screening by direct sequencing of exons bypasses the process of screening for differentially migrated heteroduplex variants before sequencing the exon carrying the variant. Sequencing represents the "gold standard" for detection of small mutations (representing the majority of mutations in TSC cases). At present, commercially available testing for TSC gene mutations includes direct sequencing of all the exons of *TSC1* and *TSC2*. It is, unfortunately, cost prohibitive for some patients. The cost of direct sequencing is expected to reduce dramatically in the future. The National Human Genome Research Institute (NHGRI) launched the National Nanotechnology Initiative with the aim of facilitating efficient and cost-effective DNA sequencing technology development. A goal of the initiative is to sequence full mammalian genomes for $100, 000 by 2009 and then to reduce the cost to $1000 by 2014.

Large gene deletion is traditionally detected by Southern blot analysis of aberrantly migrated *TSC* gene DNA restriction fragments separated by normal agarose gel electrophoresis or by pulse field gel electrophoresis [9, 11, 13, 17, 18, 23–25, 35, 54, 55, 67, 69–71]. Due to labor intensiveness and the need for large quantities of intact genomic DNA necessary for successful Southern analyses, other improved methods such as qPCR, LR-PCR were explored [12, 29, 31, 54, 55]. Fluorescent probe *in situ* hybridization has also been used to identify large *TSC2* gene deletions on patient's DNA and cells [13, 30]. The development of multiplex ligation probes amplification (MLPA) techniques coupled with statistical programs for computing changes in quantity of exon-specific probes compared to internal reference probes has proven to be an effective and less labor intense method to detect large gene deletions and duplications [15, 70].

A few groups used RT-PCR and protein truncation testing to examine *TSC* mutations that produce aberrantly spliced *TSC1* or *TSC2* mRNA in patients' cells [21, 22, 70–72]. The assay is labor intensive, low throughput, and remains in use only in a research laboratory setting.

When *de novo* missense mutations of unknown pathogenic status have been identified, assays detecting the loss of functional activity of hamartin/tuberin have been used to differentiate polymorphic variants from pathogenic mutations [52, 71, 73]. Protein functional assays are labor intensive. A recent development of higher throughput and higher efficiency *in vitro* cell-based immunoassay has greatly improved the process to demonstrate loss of function of missense mutations with unknown significance [74].

Like all other genetic diseases, the least studied mutation type for TSC patients are mutations within the noncoding sequences of the *TSC1* and *TSC2* genes. Many research groups have identified multiple noncoding variants within *TSC1* and *TSC2* genes of NMI patients, but the potential disease-causing roles of these variants remain to be verified (unpublished data). The biggest hurdle to define mutations in noncoding regions of genes is to verify their pathogenic nature through robust *in vitro* assays at the least and, further, to demonstrate their pathogenic nature in patients' cells or tissues.

Technologies for detecting TSC somatic mosaic mutation generally involve cloning an individual's exons and examining tens to hundred of clones to identify a mutation that could be present in only a few clones for low-level mosaicism. This process has to be applied to every exon of the disease-causing gene until a pathogenic mutation is identified. A report identified subtle abnormal DHPLC profiles of TSC exons from lymphocyte DNA of three TSC cases and then identified mosaic mutations from 7 to 18 clones out of over a hundred of clones sequenced [75]. One of the three cases is the father of a patient with a known single-base substitution mutation (TSC1 c.2724–1 g > c; 6.5% mosaic). The second patient had a 42 bp deletion (TSC2 p.1462–1428del; 17.5% mosaic) and the third had a 4 bp deletion (TSC2 p.1772–1774del, 7.5% mosaic). The author suggested that some of the mosaic TSC cases can be identified using DHPLC method. However, because a "false positive" rate was not discussed, it is difficult to estimate the extent of effort needed to determine every minute profile change observed to determine the true causative mutation. This dilemma holds true for users of any mutation screening method.

5.5 Conclusion

Tuberin and hamartin function, in heterodimer form, as tumor suppressors to regulate RHEB/mTOR activity and subsequently regulate protein synthesis and cell proliferation in response to cell energy states and extracellular factors affecting cell growth and proliferation. Loss of function of either protein through germline mutation (first hit) with subsequent somatic mutation (second hit) leads to disease features we observe in the TSC patients. Variable disease expressivity of TSC is well documented and can be attributed to the chance second hit as well as to genetic/epigenetic modifiers; therefore, using genotype to predict TSC disease phenotype is not straightforward.

From the genotype–phenotype correlation studies of TSC patients, general predictions can be made for an individual with a *TSC1* or *TSC2* mutation regarding the odds for developing particular TSC features. So far, at least two published genotype–phenotype correlation studies have suggested that patients with *TSC2* mutations have higher odds for SEN (91–98% versus 75–92%, OR = 3.4, CI 1.85–6.27), mental retardation (57–83% versus 21–49%, OR = 3.96, CI 2.51–6.24), forehead plaques (23–45% versus 12–27%, OR = 3.29, CI 1.85–5.86), renal angiomyolipomas (50–60% versus 7–29%, OR = 8.27, CI 4.36–15.7), and retinal phakomas (29–45% versus 0–10%, OR = 6.94, CI 2.94–16.39). Individuals with contiguous deletion of the *TSC2* and *PKD1* genes have much higher odds of developing early-onset infantile polycystic kidney disease. This is a well-established phenotype that the physician needs to discuss with the family in anticipation of disease management options despite the few cases that have been reported as exceptions. Missense mutations in *TSC2* are equally pathogenic as protein-truncating mutations and do not predict lower disease severity unless functional tests have proven otherwise. A few known specific *TSC2* missense mutations (p.R905Q, p.S1036P, p.Q1503P, and p.G1556S) are associated with very mild TSC disease phenotypes and in these cases we can make cautious predictions. However, we cannot predict how much disease modifiers (genetic and epigenetic) influence the phenotype in a particular patient. Other genetic factors differ between genders and appear to contribute to a higher prevalence of neurologic features for male patients (OR = 2.2, CI 1.4–3.4), and a higher prevalence of LAM associated with *TSC2* mutations almost exclusively in female patients. Understanding of how gender-specific factors regulate tuberin/hamartin function may lead to better disease management specific to gender.

Advances in mutation detection methodologies have greatly facilitated the finding of nearly 1000 unique pathogenic mutations since the identification of two TSC causative genes. The most robust mutation screening methods for TSC appears to be direct sequencing and DHPLC analysis of heteroduplexes, both having a small mutation detection rate of 75–85% for TSC patients with *Definite* diagnosis. Sequencing protocol and sequence data interpretation is very straightforward. Cost is the major prohibiting factor to sequencing all 64 exons for both TSC genes, but future high-throughput targeted sequencing platforms should make it much more cost-effective. DHPLC has been a less-expensive and robust tool for screening mutation for TSC patients. Success in DHPLC testing requires reproducible day-to-day sample preparation process to make the same heteroduplexes to be tested and may vary depending on the operator. Introduction of MLPA (now available in commercial testing) has made detection of large gene deletion/duplication mutations more efficient and less labor intensive. Many *de novo* missense variants have been identified from TSC patients and their pathogenic status remains to be tested by assessing ability to suppress RHEB/mTOR kinase. A robust and cost-effective assay to test missense variants suitable for clinical diagnostic laboratories is difficult to develop, but has begun to emerge [74]. There are also noncoding variants identified in TSC patients potentially affecting normal splicing but their pathogenic status needs to be verified. One can isolate and examine abnormally spliced TSC gene transcript by

RT-PCR to make cDNA for protein truncation testing but the process is technically challenging.

Unfortunately, 10–15% of TSC patients do not have a mutation identified within the coding sequences or the splice donor/acceptor. These patients may be mosaic for a mutation or have a pathogenic mutation in the noncoding regions of the *TSC1* or *TSC2* gene. In general, there is no simple efficient method to detect unknown low-level mosaic mutation. The chance of detecting a low-level unknown mosaic mutation depends on both the level of mosaicism and whether the tissue being tested harbors the mutation. In our experience, testing the DNA isolated from certain lesions may be advantageous (i.e., renal angiomyolipomas), but that from other lesions (i.e., facial angiofibromas, ungual fibromas, cortical tubers, and cardiac rhabdomyomas) appears to be less successful. New targeted sequencing platforms capable of high-speed, high-coverage sequencing of millions of bases per run may potentially identify low amounts of disease variant among a background of normal DNA sequence. It remains to be determined whether future targeted sequencing platforms are capable of detecting low-level mosaic mutations and if using these platforms will be cost-effective for disease mutation screening.

References

1 Osborne, J.P., Fryer, A., and Webb, D. (1991) Epidemiology of tuberous sclerosis. *Ann. N. Y. Acad. Sci.*, **615**, 125–127.

2 Gomez, M.R. (1988) Varieties of expression of tuberous sclerosis. *Neurofibromatosis*, **1**, 330–338.

3 Roach, E.S., Gomez, M.R., and Northrup, H. (1998) Tuberous sclerosis complex consensus conference: revised clinical diagnostic criteria. *J. Child. Neurol.*, **13**, 624–628.

4 Roach, E.S. and Sparagana, S.P. (2004) Diagnosis of tuberous sclerosis complex. *J. Child. Neurol.*, **19**, 643–649.

5 Fryer, A.E., Chalmers, A., Connor, J.M., Fraser, I., Povey, S., Yates, A.D., Yates, J.R., and Osborne, J.P. (1987) Evidence that the gene for tuberous sclerosis is on chromosome 9. *Lancet*, **1**, 659–661.

6 van Slegtenhorst, M., de Hoogt, R., Hermans, C., Nellist, M., Janssen, B., Verhoef, S., Lindhout, D., van den Ouweland, A., Halley, D., Young, J. *et al.* (1997) Identification of the tuberous sclerosis gene TSC1 on chromosome 9q34. *Science*, **277**, 805–808.

7 Northrup, H., Kwiatkowski, D.J., Roach, E.S., Dobyns, W.B., Lewis, R.A., Herman, G.E., Rodriguez, E., Jr., Daiger, S.P., and Blanton, S.H. (1992) Evidence for genetic heterogeneity in tuberous sclerosis: one locus on chromosome 9 and at least one locus elsewhere. *Am. J. Hum. Genet.*, **51**, 709–720.

8 Kandt, R.S., Haines, J.L., Smith, M., Northrup, H., Gardner, R.J., Short, M.P., Dumars, K., Roach, E.S., Steingold, S., Wall, S. *et al.* (1992) Linkage of an important gene locus for tuberous sclerosis to a chromosome 16 marker for polycystic kidney disease. *Nat. Genet.*, **2**, 37–41.

9 European Chromosome 16 Tuberous Sclerosis Consortium (1993) Identification and characterization of the tuberous sclerosis gene on chromosome 16. *Cell*, **31**, 1305–1315.

10 Inoki, K., Li, Y., Xu, T., and Guan, K.L. (2003) Rheb GTPase is a direct target of TSC2 GAP activity and regulates mTOR signaling. *Genes Dev.*, **17** (15), 1829–1834.

11 Jones, A.C., Daniells, C.E., Snell, R.G., Tachataki, M., Idziaszczyk, S.A.,

Krawczak, M., Sampson, J.R., and Cheadle, J.P. (1997) Molecular genetic and phenotypic analysis reveals differences between TSC1 and TSC2 associated familial and de novo tuberous sclerosis. *Hum. Mol. Genet.*, **6**, 2155–2161.

12 Dabora, S.l., Jozwiak, S., Franz, D.N., Roberts, P.S., Nieto, A., Chung, J., Choy, Y.S., Reeve, M.P., Thiele, E., Egelhoff, J.C., Kasprzyk-Obara, J., Domanska-Pakiela, D., and Kwiatkowski, D.J. (2001) Mutational analysis in a cohort of 224 tuberous sclerosis patients indicates increased severity of TSC2, compared with TSC1, disease in multiple organs. *Am. J. Hum. Genet.*, **68**, 64–80.

13 Sancak, O., Nellist, M., Goedbloed, M., Elfferich, P., Wouters, C., Maat-Kievit, A., Zonnenberg, B., Verhoef, S., Halley, D., and van den Ouweland, A. (2005) Mutational analysis of the TSC1 and TSC2 genes in a diagnostic setting: genotype-phenotype correlations and comparison of diagnostic DNA techniques in Tuberous Sclerosis Complex. *Eur. J. Hum. Genet.*, **13**, 731–741.

14 Au, K.S., Williams, A.T., Roach, E.S., Batchelor, L., Sparagana, S.P., Delgado, M.R., Wheless, J.W., Baumgartner, J.E., Roa, B.B., Wilson, C.M., Smith-Knuppel, T.K., Cheung, M.-Y.C., Whittemore, V.H., King, T.M., and Northrup, H. (2007) Genotype/phenotype correlation in 325 individuals referred for a diagnosis of tuberous sclerosis complex in the United States. *Genet. Med.*, **9**, 88–100.

15 Kozlowski, P., Roberts, P., Dabora, S., Franz, D., Bissler, J., Northrup, H., Au, K.S., Lazarus, R., Domanska-Pakiela, D., Kotulska, K., Jozwiak, S., and Kwiatkowski, D.J. (2007) Identification of 54 large deletions/duplications in TSC1 and TSC2 using MLPA, and genotype-phenotype correlations. *Hum. Genet.*, **121**, 389–400.

16 Ekong, R., and Povey, S. (2008) Tuberous Sclerosis database: In Leiden Open Variatnions Database at Leiden University Medical Center, Leiden, Netherland. http://chromium.liacs.nl/LOVD2/TSC/home.php?action=switch_db.

17 Brook-Carter, P.T., Peral, B., Ward, C.J., Thompson, P., Hughes, J., Maheshwar, M.M., Nellist, M., Gamble, V., Harris, P.C., and Sampson, J.R. (1994) Deletion of the TSC2 and PKD1 genes associated with severe infantile polycystic kidney disease – a contiguous gene syndrome. *Nat. Genet.*, **8**, 328–332.

18 Sampson, J.R., Maheshwar, M.M., Aspinwall, R., Thompson, P., Cheadle, J.P., Ravine, D., Roy, S., Haan, E., Bernstein, J., and Harris, P.C. (1997) Renal cystic disease in tuberous sclerosis: role of the polycystic kidney disease 1 gene. *Am. J. Hum. Genet.*, **61**, 843–851.

19 Wilson, P.J., Ramesh, V., Kristiansen, A., Bove, C., Jozwiak, S., Kwiatkowski, D.J., Short, M.P., and Haines, J.L. (1996) Novel mutations detected in the TSC2 gene from both sporadic and familial TSC patients. *Hum. Mol. Genet.*, **5**, 249–256.

20 Maheshwar, M.M., Cheadle, J.P., Jones, A.C., Myring, J., Fryer, A.E., Harris, P.C., and Sampson, J.R. (1997) The GAP-related domain of tuberin, the product of the TSC2 gene, is a target for missense mutations in tuberous sclerosis. *Hum. Mol. Genet.*, **6**, 1991–1996.

21 Beauchamp, R.l., Banwell, A., McNamara, P., Jacobsen, M., Higgins, E., Northrup, H., Short, P., Sims, K., Ozelius, L., and Ramesh, V. (1998) Exon scanning of the entire TSC2 gene for germline mutations in 40 unrelated patients with tuberous sclerosis. *Hum. Mutat.*, **12**, 408–416.

22 Gilbert, J.R., Guy, V., Kumar, A., Wolpert, C., Kandt, R., Aylesworth, A., Roses, A.D., and Pericak-Vance, M.A. (1998) Mutation and polymorphism analysis in the tuberous sclerosis 2 (TSC2) gene. *Neurogenetics.*, **1**, 267–272.

23 Kwiatkowska, J., Jozwiak, S., Hall, F., Henske, E.P., Haines, J.L., McNamara, P., Braiser, J., Wigowska-Sowinska, J., Kasprzyk-Obara, J., Short, M.P., and Kwiatkowski, D.J. (1998) Comprehensive mutational analysis of the TSC1 gene: observations on frequency of mutation, associated features, and nonpenetrance. *Ann. Hum. Genet.*, **62**, 277–285.

24 Young, J.M., Burley, M.W., Jeremiah, S.J., Jeganathan, D., Ekong, R., Osborne, J.P., and Povey, S. (1998) A mutation screen of the TSC1 gene reveals 26 protein truncating mutations and 1 splice site

mutation in a panel of 79 tuberosis patients. *Ann. Hum. Genet.*, **62**, 203–213.

25 van Slegtenhorst, M., Verhoef, S., Tempelaars, A., Bakker, L., Wang, Q., Wessels M., Bakker, R., Nellist, M., Lindhout, D., Halley, D., and van den Ouweland, A. (1999) Mutational spectrum of the TSC1 gene in a cohort of 225 tuberous sclerosis complex patients: no evidence for genotype-phenotype correlation. *J. Med. Genet.*, **36**, 285–289.

26 Niida, Y., Lawrence-Smith, N., Banwell, A., Hammer, E., Lewis, J., Beauchamp, R.L., Sims, K., Ramesh, V., and Ozelius, L. (1999) Analysis of both TSC1 and TSC2 for germline mutations in 126 unrelated patients with tuberous sclerosis. *Hum. Mutat.*, **14**, 412–422.

27 Zhang H., Nanba, E., Yamamoto, T., Ninomiya, H., Ohno, K., Mizuguchi, M., and Takeshita, K. (1999) Mutational analysis of TSC1 and TSC2 genes in Japanese patients with tuberous sclerosis. *J. Hum. Genet.*, **44**, 391–396.

28 Yamashita, Y., Ono, J., Okada, S., Wataya-Kaneda, M., Yoshikawa, K., Nishizawa, M., Hirayama, Y., Kobayashi, E., Seyama, K., and Hino, O. (2000) Analysis of all exons of TSC1 and TSC2 genes for germline mutations in Japanese patients with tuberous sclerosis: report of 10 mutations. *Am. J. Med. Genet.*, **90**, 123–126.

29 Jones, A.C., Shyamsundar, M.M., Thomas, M.W., Maynard, J., Idziaszczyk, S., Tomkins, S., Sampson, J.R., and Cheadle, J.P. (1999) Comprehensive mutation analysis of TSC1 and TSC2-and phenotypic correlations in 150 families with tuberous sclerosis. *Am. J. Hum. Genet.*, **64**, 1305–1315.

30 Lewis, J.C., Thomas, H.V., Murphy, K.C., and Sampson, J.R. (2004) Genotype and psychological phenotype in tuberous sclerosis. *J. Med. Genet.*, **41**, 203–207.

31 Muzykewicz, D.A., Newberry, P., Danforth, N., Halpern, E.F., and Thiele, E.A. (2007) Psychiatric comorbid conditions in a clinic population of 241 patients with tuberous sclerosis complex. *Epilepsy Behav.*, **11**, 506–513.

32 Winterkorn, E.B., Pulsifer, M.B., and Thiele, E.A. (2007) Cognitive prognosis of patients with tuberous sclerosis complex. *Neurology.*, **68**, 62–64.

33 Rakowski, S.K., Winterkorn, E.B., Paul, E., Steele, D.J.R., Halpern, E.F., and Thiele, E.A. (2006) Renal manifestations of tuberous sclerosis complex: Incidence, prognosis, and predictive factors. *Kidney Intl.*, **70**, 1777–1782.

34 Jozwiak, S., Kotulska, K., Kasprzyk-Obara, J., Domanska-Pakiela, D., Tomyn-Drabik, M., Roberts, P., and Kwiatkowski, D. (2006) Clinical and genotype studies of cardiac tumors in 154 patients with tuberous sclerosis complex. *Pediatrics.*, **118**, e1146–e1151.

35 Langkau, N., Martin, N., Brandt, R., Zugge, K., Quast, S., Wiegele, G., Jauch, A., Rehm, M., Kuhl, A., Mack-Vetter, M., Zimmerhackl, L.B., and Janssen, B. (2002) TSC1 and TSC2 mutations in tuberous sclerosis, the associated phenotypes and a model to explain observed TSC1/ TSC2 frequency ratios. *Eur. J. Pediatr.*, **161**, 393–402.

36 Ali, M., Girimaji, S.C., Markandaya, M., Shukla, A.K., Sacchidanand, S., and Kumar, A. (2005) Mutation and polymorphism analysis of TSC1 and TSC2 genes in Indian patients with tuberous sclerosis complex. *Acta Neurol. Scand.*, **111**, 54–63.

37 Choi, J.-E., Chae, J.-H., Hwang, Y.-S., and Kim, K.-J. (2006) Mutational analysis of TSC1 and TSC2 in Korean patients with tuberous sclerosis complex. *Brain Dev.*, **28**, 440–446.

38 Hung, C.C., Su, Y.-N., Chien, S.-C., Liou, H.-H., Chen, C.-C., Chen, P.-C., Hsieh, C.-J., Chen, C.-P., Lee, W.-T., Lin, W.-L., and Lee, C.-N. (2006) Molecular and clinical analyses of 84 patients with tuberous sclerosis complex. *BMC Med. Genet.*, **7**, 72.

39 Devlin, L.A., Shepherd, C.H., Crawford, H., and Morrison, P.J. (2006) Tuberous sclerosis complex: clinical features, diagnosis, and prevalence within Northern Ireland. *Dev. Med. & Child Neurol.*, **48**, 495–499.

40 Jansen, F.E., Braams, O., Vincken, K.L., Algra, A., Anbeek, P., Jennekens-Schinkel, A., Halley, D., Zonnenberg, B.A., van den Ouweland, A., van Huffelen, A.C., van Nieuwenhuizen, O., and Nellist,

M. (2007) Overlapping neurologic and cognitive phenotypes in patients with TSC1 or TSC2 mutations. *Neurology.*, **70**, 908–915.

41 Vaillend, C., Poirier, R., and Laroche, S. (2008) Genes, plasticity and mental retardation. *Behav. Brain Res.*, 2008 Jan 31 [Epub ahead of print].

42 Yamagata, K., Sanders, L.K., Kaufmann, W.E., Yee, W., Barnes, C.A., Nathans, D., and Worley, P.F. (1995) rheb, a growth factor- and synaptic activity-regulated gene, encodes a novel Ras-related protein. *J Biol Chem.*, **269**, 16333–9.

43 Tavazoie, S.F., Alvarez, V.A., Ridenour, D.A., Kwiatkowski, D.J., and Sabatini, B.L. (2005) Regulation of neuronal morphology and function by the tumor suppressors Tsc1 and Tsc2. *Nat. Neurosci.*, **8**, 1727–1734.

44 Henske, E.P. (2005) Tuberous sclerosis and the kidney: from mesenchyme to epithelium, and beyond. *Pediatr. Nephrol.*, **20**, 854–857.

45 Bonnet, C.S., Aldred, M., von Ruhland, C., Harris, R., Sandford, R., and Cheadle, J.P. (2009) Defects in cell polarity underlie TSC and ADPKD-associated cystogenesis. *Hum Mol Genet.*, **18** (12), 2166–76.

46 De Vries, P.J., and Bolton, P.F. (2000) Genotype-phenotype correlations in tuberous sclerosis. *J. Med. Genet.*, **37**, E3.

47 Crossey, P.A., Richards, F.M., Foster, K., Green, J.S., Prowse, A., Latif, F., Lerman, M.I., Zbar, B., Affara, N.A., Ferguson-Smith, M.A., and Maher, E.R. (1994) Identification of intragenic mutations in the Von Hippel Lindau disease tumor suppressor gene and correlation with disease phenotype. *Hum. Mol. Genet.*, **3**, 1303–1308.

48 Khare, L., Strizheva, G.D., Bailey, J.N., Au, K.S., Northrup, H., Smith, M., Smalley, S.L., and Henske, E.P. (2001) A novel missense mutation in the GTPase activating protein homology region of TSC2 in two large families with tuberous sclerosis complex. *J. Med. Genet.*, **38**, 347–349.

49 O'Connor, S.E., Kwiatkowski, D.J., Roberts, P.S., Wollmann, R.L., and Huttenlocher, P.R. (2003) A family with seizures and minor features of tuberous sclerosis and a novel TSC2 mutation. *Neurology.*, **61**, 409–412.

50 Mayer, K., Goedbloed, M., van Zijl, K., Nellist, M., and Rott, H.D. (2004) Characterisation of a novel TSC2 missense mutation in the GAP related domain. *J. Med. Genet.*, **41**, e64.

51 Jansen, A.A., Sancak, O., D'Agostino, M.D., Badhwar, A., Roberts, P., Gobbi, G., Wilkinson, R., Melanson, D., Tampieri, D., Koenekoop, R., Gans, M., Maat-Kievit, A., Goedbloed, M., van den Ouweland, A.M.W., Nellist, M., Pandolfo, M., McQueen, M., Sims, K., Thiele, E.A., Dubeau, F., Andermann, F., Kwiatkowski, D.J., Halley, D.J.J., and Andermann, E. (2006) Unusually mild tuberous sclerosis phenotype is associated with TSC2 R905Q mutation. *Ann. Neurol.*, **60**, 528–539.

52 Nellist, M., Sancak, O., Goedbloed, M.A., Rohe, C., van Netten, D., Mayer, K., Tucker-Williams, A., van den Ouweland, A.M.W., and Halley, D.J. (2005) Distinct effects of single amino-acid changes to tuberin on the function of the tuberin-hamartin complex. *Eur. J. Hum. Genet.*, **13**, 59–68.

53 Nellist, M., Sancak, O., Goedbloed, M.A., van Veghel-Plandsoen, M., Maat-Kievit, A., Lindhout, D., Eussen, B.H., de Klein, A., Halley, D.J.J., and van den Ouweland, A.M. (2005) Large deletion at the TSC1 locus in a family with tuberous sclerosis complex. *Genet. Test.*, **9**, 226–230.

54 Longa, L., Saluto, A., Brusco, A., Polidoro, S., Padovan, S., Allavena, A., Carbonara, C., Grosso, E., and Migone, N. (2001) TSC1 and TSC2 deletions differ in size, preference for recombinatorial sequences and location within the gene. *Hum. Genet.*, **108**, 156–166.

55 Martignoni, G., Bonetti, F., Pea, M., Tardanico, R., Brunelli, M., and Eble, J.N. (2002) Renal disease in adults with TSC2/PKD1 contiguous gene syndrome. *Am. J. Surg. Pathol.*, **26**, 198–205.

56 Smulders, Y.M., Eussen, B.H.J., Verhoef, S., and Wouters, C.H. (2003) Large deletion causing the TSC2-PKD1 contiguous gene syndrome without infantile polycystic kidney disease. *J. Med. Genet.*, **40**, E17.

57 Boehm, D., Bacher, J., and Neumann, H.P.H. (2007) Gross genomic rearrangement involving the TSC2-PKD1 contiguous deletion syndrome: characterization of the deletion event by quantitative polymerase chain reaction deletion assay. *Am. J. Kidney Dis.*, **49**, e11–21.

58 Yadaden, T., Molinie, V., Ples, R., Lazure, T., Benoit, G., Yonneau, L., and Ferlicot, S. (2007) [TSC2/PKD1 contiguous gene syndrome. Report of two cases.] *Ann. Pathol*, **27**, 136–140.

59 Strizheva, G.D., Carsillo, T., Kruger, W.D., Sullivan, E.J., Ryu, J.H., and Henske, E.P. (2001) The spectrum of mutations in TSC1 and TSC2 in women with tuberous sclerosis and lymphangiomyomatosis. *Am. J. Respir. Crit. Care Med.*, **163**, 253–258.

60 Kwiatkowski, D.J., Zhang, H., Bandura, J.L., Heiberger, K.M., Glogauer, M., el-Hashemite, N., and Onda, H. (2002) A mouse model of TSC1 reveals sex-dependent lethality from liver hemangiomas, and up-regulation of p70S6 kinase activity in TSC1 null cells. *Hum. Mol. Genet.*, **11**, 525–534.

61 Yu, J., Astrinidis, A., Howard, S., and Henske, E.P. (2004) Estradiol and tamoxifen stimulate LAM-associated angiomyolipoma cell growth and activate both genomic and non-genomic signaling pathways. *Am. J. Physiol. Lung Cell Mol. Physiol.*, **286**, L694–700.

62 Smalley, S.l., Tanguay, P.E., Smith, M., and Gutierrez, G. (1992) Autism and tuberous sclerosis. *J. Autism Dev. Disord.*, **22**, 339–355.

63 Hestekin, C.N., and Barron, A.E. (2006) The potential of electrophoretic mobility shift assays for clinical mutation detection. *Electrophoresis.*, **27**, 3805–3815.

64 Au, K.S., Rodriguez, J.A., Rodriguez, E., Jr., Dobyns, W.B., Delgado, M.R., and Northrup, H. (1997) Mutations and polymorphism in the tuberous sclerosis complex gene on chromosome 16. *Hum. Mutat.*, **9**, 23–29.

65 Au, K.S., Rodriguez, J.A., Finch, J.L., Volcik, K.A. et al. (1998) Germ-line mutational analysis of the TSC2 gene in 90 tuberous sclerosis patients. *Am. J. Hum. Genet.*, **62**, 286–294.

66 Ali, J.B.M., Sepp, T., Ward, S., Green, A.J., and Yates, J.R.w. (1998) Mutations in the TSC1 gene account for a minority of patients with tuberous sclerosis. *J. Med. Genet.*, **35**, 969–972.

67 Hodges, A.K., Li, S., Maynard, J., Parry, L., Braverman, R., Cheadle, J.P., DeClue, J.E., and Sampson, J.R. (2001) Pathological mutations in TSC1 and TSC2 disrupt the interaction between hamartin and tuberin. *Hum. Mol. Genet.*, **10**, 2899–2905.

68 Choy, Y.S., Dabora, S.L., Hall, F., Ramesh, V., Niida, Y., Franz, D., Kasprzyk-Obara, J., Reeve, M.P., and Kwiatkowski, D.J. (1999) Superiority of denaturing high performance liquid chromatography over single-stranded conformation and conformation-sensitive gel electrophoresis for mutation detection in TSC2. *Ann. Hum. Genet.*, **63**, 383–391.

69 Dabora, S.L., Sigalas, I., Hall, F., Eng, C., Vijg, J., and Kwiatkowski, D.J. (1998) Comprehensive mutation analysis of TSC1 using two-dimensional DNA electrophoresis with DGGE. *Ann. Hum. Genet.*, **62**, 491–504.

70 Rendtorff, N.D., Bjerregaard, B., Frodin, M., Kjaergaard, S., Hove, H., Skovby, F., Brondum-Nielsen, K., and Schwartz, M. (2005) Analysis of 65 tuberous sclerosis complex (TSC) patients by TSC2 DGGE, TSC1/TSC2 MLPA, and TSC1 Long-Range PCR sequencing, and report of 28 novel mutations. *Hum. Mutat.*, **26**, 374–383.

71 Mayer, K., Ballhausen, W., and Rott, H.D. (1999) Mutation screening of the entire coding regions of the TSC1 and the TSC2 gene. *Hum. Mutat.*, **14**, 401–411.

72 van Bakel, I., Sepp, T., Ward, S., Yates, J.R., and Green, A.J. (1997) Mutations in the TSC2 gene: analysis of the complete coding sequence using the protein truncation test (PTT). *Hum. Mol. Genet.*, **6**, 1409–1414.

73 Nellist, M., Sancak, O., Goedbloed, M.A., Adriaans, A., Wessels, M., Maat-Kievit, A., Baars, M., Dommering, C., van den Ouweland, A., and Halley, D. (2008) Functional characterization of the TSC1-TSC2 complex to assess multiple TSC2

variants identified in single families affected by tuberous sclerosis complex. *BMC Med. Genet.*, **9**, 10 [Epub ahead of print].

74 Coevoets, R., Arican, S., Hoogeveen-Westerveld, M., Simons, E., van den Ouweland, A., Halley, D., and Nellist, M. (2009) A reliable cell-based assay for testing unclassified TSC2 gene variants. *Eur J Hum Genet.*, **17**, 301–310.

75 Jones, A.C., Sampson, J.R., and Cheadle, J.P. (2001) Low level mosaicism detectable by DHPLC but not by direct sequencing. *Hum. Mutat.*, **17**, 233–234.

Part III
Basic Science

Tuberous Sclerosis Complex: Genes, Clinical Features, and Therapeutics
Edited by David J. Kwiatkowski, Vicky Holets Whittemore, and Elizabeth A. Thiele

ISBN: 978-3-527-32201-5

6
The Role of Target of Rapamycin Signaling in Tuberous Sclerosis Complex

Brendan D. Manning

6.1
The Target of Rapamycin: An Evolutionarily Conserved Regulator of Cell Growth and Proliferation

6.1.1
Rapamycin and the Discovery of TOR Proteins

In 1975, a strain of bacteria called *Streptomyces hygroscopicus* was isolated from a soil sample taken from Rapa Nui (Easter Island) and was found to produce an antifungal antibiotic dubbed rapamycin [1]. Like a number of other naturally occurring antibiotics, rapamycin (or sirolimus) is classified as a macrolide. Part of the large rapamycin molecule is identical to the macrolide FK506 (or tacrolimus), and both compounds have immunosuppressive activities that make them useful in the prevention of transplant rejection [2]. The chemical moiety shared between rapamycin and FK506 associates with the FK506-binding protein of 12 kD (FKBP12), a peptidyl-prolyl isomerase of the immunophilin family [3]. Genetic screens for rapamycin-resistant mutants in the budding yeast *Saccharomyces cerevisiae* revealed a homologue of FKBP12, demonstrating that this binding is conserved and important for rapamycin's mode of action [4, 5]. However, inhibition of FKBP12 itself could not account for the immunosuppressant effects of rapamycin on T lymphocytes or the G1 cell cycle arrest caused by rapamycin in yeast [3, 4]. Two other mutants were identified that confer rapamycin resistance in yeast, and these affected genes assigned the name target of rapamycin 1 and 2 (*TOR1* and *TOR2*) [4]. More important, deletion analyses revealed that the products encoded by these genes were the likely targets through which rapamycin causes cell cycle arrest in yeast [6, 7]. Independent biochemical studies with mammalian cells identified a single mammalian homologue of the yeast TOR proteins as a direct binder of the FKBP12–rapamycin complex, designated FKBP–rapamycin-associated protein (FRAP) [8], and rapamycin and FKBP12 target 1 (RAFT) [9]. This protein, universally referred to as mammalian TOR (mTOR), is now recognized as the primary target of rapamycin in cells, and mTOR is directly inhibited by rapamycin only when the compound is bound to the FKBP12 protein. Due to their high degree of specificity for TOR proteins, rapamycin and its synthetic

Tuberous Sclerosis Complex: Genes, Clinical Features, and Therapeutics
Edited by David J. Kwiatkowski, Vicky Holets Whittemore, and Elizabeth A. Thiele

ISBN: 978-3-527-32201-5

analogues (e.g., CCI-779/temsirolimus and RAD001/everolimus) have been instrumental in delineating TOR-dependent cellular functions. More important, these compounds are also now being tested for efficacy in the treatment of a variety of human tumor syndromes and cancers, which often exhibit elevated mTOR activity due to a variety of oncogenic lesions (detailed below).

6.1.2
Molecular Characteristics of mTOR and Its Complexes

The domain structure of TOR is conserved from yeast to humans (Figure 6.1a), and the large TOR proteins are 40–50% identical in their primary sequences (2549-amino acids for mTOR). The highest degree of conservation lies within a C-terminal kinase domain that resembles the lipid kinase domain of phosphatidylinositol 3-kinases (PI3K). However, TOR proteins are ser/thr protein kinases and are not believed to

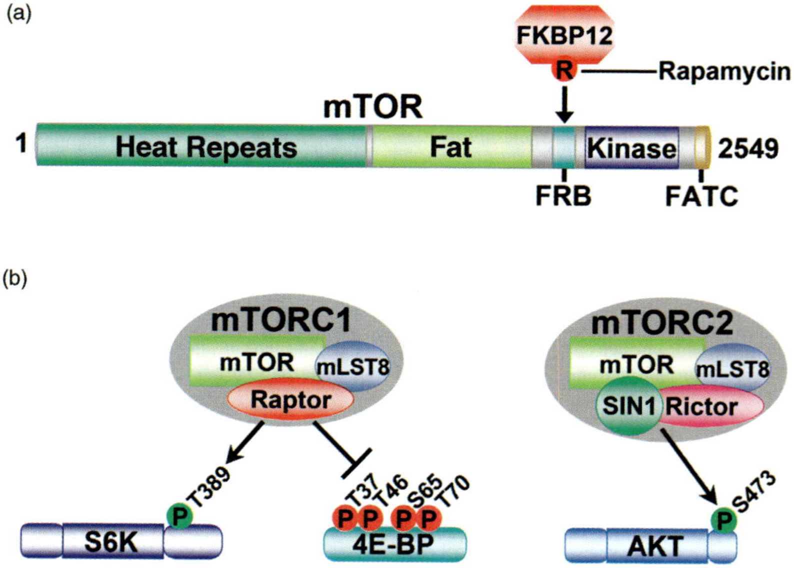

Figure 6.1 Schematic of the mammalian target of rapamycin protein, complexes, and substrates. (a) TOR proteins from all eukaryotes contain the domain structure pictured for mTOR. The N-terminal half is comprised of 20 HEAT repeats, named for proteins first found to contain these repeats (*H*untington, *e*IF1A, PP2*A*, *T*or), followed by the conserved FAT domain, named for PIKK family members (*F*RAP/mTOR, *A*TM, *T*RAPP2), all of which contain this region. The FKBP12-rapamycin-binding (FRB) domain precedes the highly conserved protein kinase domain. Finally, another short stretch of amino acids highly conserved in PIKK family members is found at the extreme C-terminus of mTOR, referred to as the FATC domain. (b) mTOR assembles into two functionally distinct complexes. In addition to mTOR, mTORC1 contains mLST8 and RAPTOR, while mTORC2 contains mLST8, SIN1, and RICTOR. On the amino acid residues pictured, mTORC1 can directly phosphorylate S6K1, leading to its activation, and 4E-BP1, leading to its inhibition, while mTORC2 phosphorylates Akt on S473, thereby contributing to its activation. See text for details.

possess significant kinase activity toward phosphatidylinositols. TOR was the first identified member of the PI3K-related kinase (PIKK) family of protein kinases, which also includes ATM, ATR, DNA-PK, and SMG-1. Interestingly, the high degree of similarity between the TOR kinase domain and that of PI3K renders TOR sensitive to a variety of PI3K-inhibiting compounds (e.g., Ref. [10]). The molecular mechanism of rapamycin's remarkable specificity for TOR proteins became obvious when it was found that the FKBP12–rapamycin complex does not bind to the kinase domain, like most kinase inhibitors, but rather to an adjacent domain found exclusively in TOR proteins [11, 12], called the FKBP12–rapamcyin binding domain (FRB). Mutations in this highly conserved region were responsible for the identification of *TOR1* and *TOR2* in the original yeast screen for rapamycin-resistant mutants [4, 7]. TOR proteins share two other domains of unknown function, the FAT and FATC domains, which are found in other PIKK family members. Finally, all TOR proteins possess a large region at their N-terminus consisting of tandem HEAT repeats. In general, the evolutionary pressure underlying the conservation of this domain structure in TOR proteins from all eukaryotes can be explained by the fact that TOR proteins form larger macromolecular complexes that are very similar in yeast and humans [13].

In the early studies on TOR proteins, yeast genetic experiments demonstrated that a subset of Tor2p functions were distinct from Tor1p, and those functions were resistant to inhibition by rapamycin [7, 14]. A molecular mechanism for these separate functions came from the finding that TOR proteins in yeast and humans could be found in two distinct complexes (Figure 6.1b), only one of which is acutely sensitive to rapamycin. Mammalian TOR complex 1 (mTORC1) is robustly inhibited by rapamycin and consists of mTOR, raptor, and mLST8 (also referred to as GβL), all with orthologues in *S. cerevisiae* TORC1 [13, 15–17]. An additional mTORC1 component, without a known yeast orthologue, is PRAS40 [18, 19]. However, there is some controversy as to whether this is a regulatory subunit or downstream substrate of mTORC1. Mammalian TOR complex 2 (mTORC2) consists of the conserved subunits mTOR, Rictor, SIN1, and mLST8 [13, 20–23], and the novel component Protor/PRR5 [24, 25]. Aside from the mTOR kinase, the molecular functions of the various subunits of these two complexes are poorly understood. However, Raptor appears to direct substrate binding for mTORC1 [26–28]. While only mTORC1 is acutely sensitive to inhibition by rapamycin, mTORC2 stability is affected by prolonged exposure to rapamycin, and in some cell types (e.g., endothelial cells), this can inhibit mTORC2 function [29].

6.1.3
Downstream of mTOR

Even before the discovery of TOR proteins, studies on the cellular effects of rapamycin revealed a conserved role for its target in promoting cell growth and proliferation. We now know that these studies were primarily focused on TORC1. Rapamycin is a potent inhibitor of proliferation in yeast, lymphocytes, and cancer cell lines, resulting in a G1-phase cell cycle arrest (e.g., Refs [4, 30–32]). Rapamycin treatment is known to elevate levels of the cyclin-dependent kinase inhibitor p27 [33], but this is an indirect effect and is not sufficient for the cell cycle arrest induced by rapamcyin [34]. More important, rapamycin was found to inhibit the growth factor and cytokine-stimulated activation of the 70 kD ribosomal S6 kinase (S6K1) [31, 35] and phosphorylation of the eukaryotic translation initiation factor 4E (eIF4E) binding protein 1 (4E-BP1) [36, 37], two proteins involved in the control of mRNA translation. Rapamycin was found to inhibit cap-dependent translation, and mRNAs with 5′-terminal oligopyrimidine (5′-TOP) tracts, including those encoding translation factors and ribosomal proteins, are particularly sensitive [37–39]. While the downstream mechanisms appear to vary, the inhibitory effects of rapamycin on mRNA translation and ribosome biogenesis are conserved in yeast and are linked to the ability of this compound to block cell growth and proliferation (e.g., Ref. [40]).

The effects of rapamycin on S6K1 and 4E-BP1 depend on its ability to inhibit mTOR [41, 42], and mTOR can directly phosphorylate S6K1 and 4E-BP1 *in vitro* [43]. There are many rapamycin-sensitive phosphorylation sites on these two proteins [44], but not all of them are the direct result of mTOR kinase activity. MTOR directly phosphorylates T389 in a hydrophobic motif C-terminal to the S6K1 kinase domain (Figure 6.1b), which is a critical regulatory site conserved in the majority of AGC family kinases [43, 45]. This phosphorylation event triggers a number of other sites to be phosphorylated leading to full activation of S6K1. Once active, S6K1 and its homologue S6K2, which is also regulated by mTOR, phosphorylate downstream targets, including the ribosomal S6 protein and eIF4B, to promote protein synthesis and cell growth [46–48]. On 4E-BP1, rapamycin inhibits the phosphorylation of multiple proline-directed (S/T-P) sites (S37, T46, S65, and T70), all of which can be phosphorylated directly by mTOR (Figure 6.1b) [44]. However, S65 and T70 appear to be the sites that are acutely regulated by mTOR activation, with S37 and T46 acting more as priming sites [49]. MTOR-dependent phosphorylation of 4E-BP1, and its homologues (4E-BP2 and 4E-BP3), triggers its release from the 5′-cap binding protein eIF4E, thereby allowing eIF4G association and subsequent ribosome recruitment preceding translation initiation [50]. The specificity of these targets to phosphorylation by mTOR within mTORC1 comes from the fact that the mTORC1 component raptor binds directly to a TOR signaling (TOS) motif shared between these substrates [26–28]. This substrate recognition mechanism could explain the complete lack of sequence similarity between the mTOR-mediated phosphorylation sites on the S6Ks and those on the 4EBPs. While mTORC1 is very likely to have other direct targets important for its cellular functions, cell culture experiments suggest that the inhibition of mTOR signaling to S6K and 4E-BP accounts for the primary inhibitory effects of rapamycin on cell growth and proliferation [51, 52].

Due to its resistance to rapamcyin, knowledge of direct targets downstream of mTORC2 has lagged behind that of mTORC1. While mTORC2 appears to play some role in regulation of the actin cytoskeleton [22, 53], which may be conserved in yeast [13], the direct targets involved are currently unknown. In fact, the only confirmed direct target of mTORC2 identified to date is the ser/thr kinase Akt/PKB [54], an AGC family kinase whose catalytic domain is very similar to that of S6K. Interestingly, mTOR within mTORC2 phosphorylates a residue within the conserved hydrophobic motif on Akt (S473 on human Akt1) (Figure 6.1b) that is strikingly similar to that surrounding T389 on S6K1, which is phosphorylated by mTORC1. There is also genetic evidence that mTORC2 phosphorylates this same motif on other AGC kinases, such as PKC isoforms [55]. However, there is at present no biochemical evidence of a direct connection between mTORC2 and other protein kinases.

6.1.4 Upstream of mTOR

Although very little is known regarding upstream regulation of mTORC2, the rapamycin-sensitive functions of mTOR (i.e., those primarily mediated by mTORC1) have been known for some time to be under the control of a large variety of upstream inputs. Protein synthesis is the most energy-consuming process in the cell, and the majority of its regulation occurs at the translation initiation steps controlled by mTORC1. In addition, ribosome biogenesis, which is regulated by mTORC1 in all eukaryotes, has been estimated to consume up to 80% of the total energy output of a cell [56]. Therefore, it is not surprising that mTORC1 is under tight control and has evolved mechanisms to sense a myriad of signals indicating cellular growth conditions. These include the presence of nutrients (amino acids, glucose, and oxygen), availability of cellular energy (ATP), conditions of cellular stress, and, in higher eukaryotes, signals emanating from growth factors and cytokines (Figure 6.2). However, the molecular mechanisms by which mTORC1 senses any of these signals were unknown until the seminal placement of a critical negative regulatory complex upstream of mTORC1, comprised of the tuberous sclerosis complex gene products, TSC1 (or hamartin) and TSC2 (or tuberin).

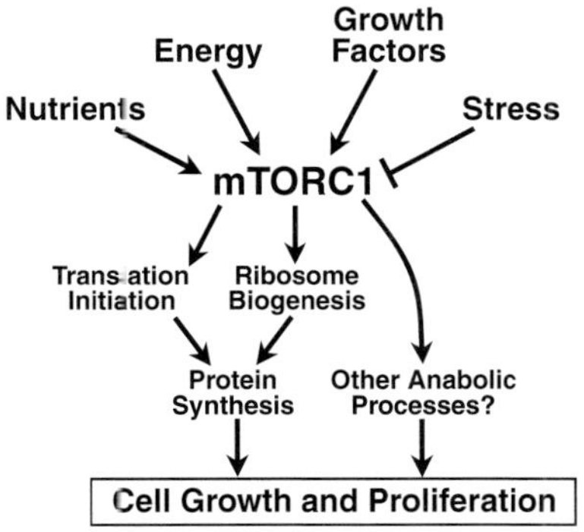

Figure 6.2 Cellular inputs into mTORC1 activation and mTORC1-dependent outputs promoting cell growth and proliferation. mTORC1, as a critical regulator of anabolic processes promoting cell growth, is exquisitely sensitive to cellular growth conditions. mTORC1 can sense the presence of nutrients, cellular energy (ATP), and growth factors, which are all required for full activation of mTORC1, while conditions of cellular stress generally block mTORC1 function. The best-characterized functions of mTORC1 are to promote translation initiation and ribosome biogenesis, thereby increasing the protein synthetic capacity of the cell. However, it is very likely that mTORC1 drives other anabolic processes such as lipid/membrane biosynthesis. See text for more details.

6.2
Genetic and Biochemical Studies Link the TSC1–TSC2 Complex to Cell Growth Control Through mTORC1

6.2.1
Drosophila Genetics Lays the Groundwork

The TSC gene products were put on the map of a well-known signal transduction pathway by three nearly identical genetic screens for mutants leading to an overgrowth phenotype in the *Drosophila* eye [57–59]. The *Drosophila* orthologue of *TSC1* (*dTsc1*) was found in all of these screens, and a variety of genetic epistasis experiments suggested that both *dTsc1* and *dTsc2* negatively regulate cell growth and proliferation in a pathway downstream of, or parallel to, the insulin/insulin-like growth factor (IGF)-PI3K–Akt pathway. Furthermore, the *dTsc* genes appeared to function upstream of the *Drosophila* S6K orthologue (dS6k) to block the growth-promoting activities of dS6K [58–61]. Strikingly, the larval lethality of *dTsc1* mutants could be rescued by mutations reducing dS6K activity [61], demonstrating that inhibition of S6K is an essential function of TSC1 in the fly (see Chapter 13 for details on these genetic studies).

6.2.2
Biochemical Studies Fill in the Gaps

The *Drosophila* genetic studies paved the way for a number of biochemical studies aimed at understanding the molecular regulation and function of the TSC1–TSC2 complex as it relates to this evolutionarily conserved pathway controlling cell growth. In a study independent of the *Drosophila* findings, the mammalian TSC2 protein was identified in an unbiased screen for new *in vivo* substrates of Akt [62]. TSC2 was subsequently shown to be a direct target of Akt in cells, and Akt phosphorylates TSC2 on two sites (S939 and T1462) conserved in dTsc2. Previous studies had demonstrated that activation of the PI3K–Akt pathway is the major mechanism by which growth factors stimulate S6K activity in mammalian cells [63–65], but the mechanism was unknown. Taken together with the *Drosophila* studies described above [57–59], the finding that Akt directly phosphorylates TSC2 suggested that this might be the long sought link between PI3K–Akt signaling and activation of S6K. Indeed, expression of a TSC2 mutant lacking the two major Akt phosphorylation sites blocked growth factor-mediated activation of S6K1 [62]. Based on the *Drosophila* studies, two other groups also examined Akt-mediated phosphorylation of TSC2. One group fused peptide sequences representing every Akt consensus site (R-X-R-X-X-S/T) found in TSC2 to the GST protein and showed that Akt was capable of phosphorylating all of these sites *in vitro* [66]. A second group used two-dimensional phosphopeptide mapping to identify Akt-mediated phosphorylation sites on TSC2 *in vivo* [67]. While it remains unclear whether all of the Akt consensus sites identified in the *in vitro* study are indeed phosphorylated by Akt *in vivo*, the second study identified the two conserved sites (S939 and T1462) and an additional site (S1130 and/or S1132) on TSC2 as bona fide phosphorylation sites in cells. Finally, the sites corresponding to S939 and T1462 in dTsc2 were found to be phosphorylated by Akt in *Drosophila* cells [68]. It is now known that Akt activates S6K, at least in part, through multisite phosphorylation of TSC2, which somehow relieves the ability of the TSC1–TSC2 complex to inhibit S6K.

More important, the TSC1–TSC2 complex inhibits S6K through the upstream inhibition of mTORC1. Like *dS6k*, loss of function mutations in the gene encoding the single *Drosophila* TOR orthologue (*dTor*) were found to suppress the larval lethality of *dTsc1* or *dTsc2* mutants [60, 61]. More important, cooverexpression of *TSC1* and *TSC2* inhibits, while loss of either gene stimulates, the mTORC1-dependent phosphorylation of both S6K1 and 4E-BP1 [60, 62, 67, 69–72]. In addition to distinct branches downstream of mTORC1 being affected by the TSC1–TSC2 complex, the use of rapamycin-resistant (i.e., mTORC1-independent) versions of S6K1 further suggested that the TSC1–TSC2 complex acts upstream of mTOR to inhibit S6K1 [67, 71]. Strikingly, S6K1 and 4E-BP1 were found to be phosphorylated independent of growth factors and nutrients in cells lacking a functional TSC1–TSC2 complex [60, 69, 70, 72], suggesting for the first time that constitutive mTORC1 signaling might contribute to TSC pathology (see below).

6.2.3
Rheb: A Direct Target of the TSC1–TSC2 Complex That Regulates mTORC1

The mechanism by which the TSC1–TSC2 complex negatively regulates mTORC1 was unknown until parallel genetic and biochemical studies collectively identified a poorly understood member of the Ras superfamily as the key molecular link [73]. Rheb (Ras-homologue enriched in brain) is a small GTPase with a high degree of sequence and structural similarity to isoforms of Rap and Ras [74]. Despite its name, Rheb, like mTOR, is ubiquitously expressed. A major breakthrough came when, like *dTsc1*, the gene encoding the *Drosophila* orthologue of Rheb (dRheb) was found in three independent screens for regulators of cell and organ growth [75–77]. However, contrary to the *dTsc* genes, *dRheb* was found to promote rather than suppress cell growth. Epistasis analyses placed *dRheb* downstream of *dAkt* and the *dTsc* genes and upstream of *dTOR* and *dS6K* [75, 76, 78]. This genetic evidence of a small G protein acting between the TSC1–TSC2 complex and TOR brought to light a long-standing question among TSC researchers: What is the true molecular target of the GTPase-activating protein (GAP) domain found at the C-terminus of TSC2? A number of missense mutations affecting this domain had been identified in TSC patients [79], and important biochemical studies had demonstrated that the domain did indeed possess GAP activity [80, 81]. However, evidence that the small G proteins identified, Rap1 and Rab5, were bona fide *in vivo* targets of the TSC2 GAP domain was lacking. Therefore, even before the breakthrough in *Drosophila*, an intense search was on for a GTPase that is targeted by this domain. Several independent groups identified Rheb as a small G protein whose *in vitro* GTPase activity is strongly activated in the presence of the TSC1–TSC2 complex [78, 82, 83]. Furthermore, cell biological studies demonstrated that Rheb potently activates mTORC1

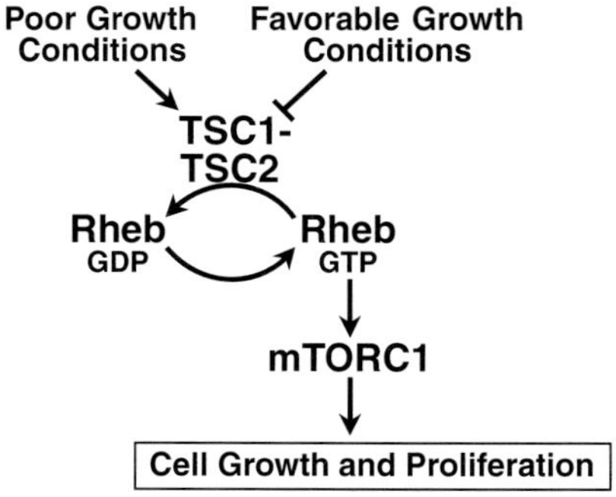

Figure 6.3 The TSC1-TSC2 complex acts as a molecular switch to turn mTORC1 off in response to perturbation in cellular growth conditions. The TSC1-TSC2 complex acts as a GTPase-activating protein (GAP) for the Rheb GTPase, thereby stimulating the conversion of GTP-bound active Rheb, which turns on mTORC1, to GDP-bound inactive Rheb, which is incapable of activating mTORC1. This function of the TSC1-TSC2 complex is regulated by cellular growth conditions, such that poor growth conditions activate the complex, while growth-promoting conditions inhibit it. Therefore, the TSC1-TSC2 complex senses the growth environment and puts a break on cell growth and proliferation when unfavorable growth conditions exist.

signaling and that, through its GAP activity, the TSC1–TSC2 complex blocks this activation [78, 82–85]. Therefore, when the TSC1–TSC2 complex is active (i.e., under poor growth conditions), TSC2 within this complex stimulates the intrinsic GTPase activity of Rheb, thereby facilitating the conversion of Rheb from its GTP-bound active state to its GDP-bound inactive state. As Rheb-GTP is required for mTORC1 activation, the GAP activity of the TSC1–TSC2 complex effectively inhibits mTORC1 signaling (Figure 6.3).

6.2.4 The TSC–Rheb–mTORC1 Circuit: Important Remaining Questions

While there has been great progress in understanding how the TSC1–TSC2 complex regulates Rheb and mTORC1, some critical mechanistic questions remain. For instance, is there a guanine–nucleotide exchange factor (GEF) for Rheb that counters the activity of the TSC1–TSC2 complex and promotes Rheb-GTP accumulation? In the literature, a chaperone-like protein called TCTP (translationally controlled tumor protein) has been proposed to possess such activity [86]. However, follow-up studies by other groups have provided compelling evidence against TCTP being a GEF for Rheb or regulating mTORC1 in any way [152, 153]. A true Rheb-GEF could be an ideal therapeutic target to block the molecular effects of TSC1–TSC2 complex disruption. The molecular mechanism by which Rheb-GTP activates mTORC1 is also not fully understood. Evidence exists that overexpressed Rheb can associate with mTOR [87, 88] and that Rheb-GTP can stimulate mTORC1 activity in a dose-dependent manner *in vitro* [18]. Rheb has also been proposed to activate mTORC1 through binding to an FKBP12 homologue called FKBP38 (also referred to as

FKBP8), which was found in one study to bind to the FRB domain of mTOR and inhibit its function in a manner similar to rapamycin complexed with FKBP12 [89]. However, a number of independent studies, including one published work [153], have failed to reproduce these findings, and the majority of evidence is against FKBP38 being an effector of Rheb in the regulation of mTORC1. More biochemical studies are clearly needed to determine the precise molecular mechanism of mTORC1 activation by GTP-bound Rheb. Finally, the subcellular setting in which this small G protein circuit functions is poorly understood, and it remains possible that independent pools of the TSC1–TSC2 complex, Rheb, and mTORC1 exist at distinct locations within the cell to respond to different stimuli. Interestingly, one study has suggested that Rheb activates mTORC1 at a late endosomal compartment, which also contains the Rab7 protein, upon stimulation with amino acids [154].

6.3
The TSC1–TSC2 Complex as a Critical Sensor of Cellular Growth Conditions

Since its placement in 2001 and 2002 as a key molecular link between insulin/IGF1 signaling and mTORC1 activation, the TSC1–TSC2 complex has emerged as a central signaling hub that monitors a variety of cellular growth conditions to properly regulate mTORC1. A myriad of signaling pathways merge at the TSC1–TSC2 complex with many stimulus-specific kinases phosphorylating the TSC1 and TSC2 proteins to modulate the function of the complex and affect downstream signaling to mTORC1 (Figure 6.4). The ser/thr kinases Akt, Erk, RSK, IKKβ, AMPK, GSK3, MAPKAP-K2, and CDK1 have all been demonstrated to phosphorylate TSC1 or TSC2 on distinct residues and have been suggested to affect the ability of the TSC1–TSC2 complex to act as a GAP for Rheb [62, 67, 90–97]. Some of the prominent pathways regulating mTORC1 through the TSC1–TSC2 complex are discussed briefly below. For a more thorough discussion of this subject, see Ref. [98].

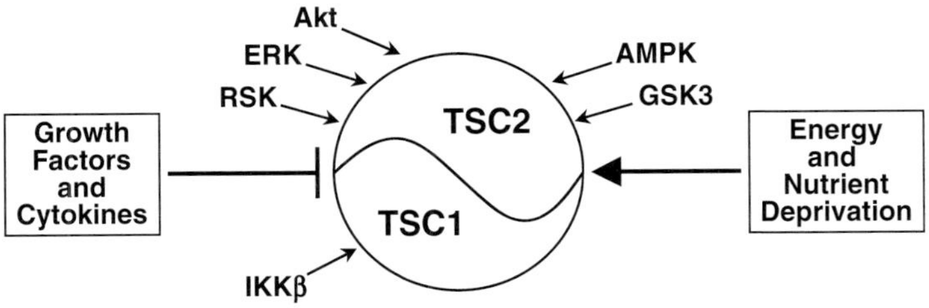

Figure 6.4 Cellular growth conditions signal to the TSC1-TSC2 complex through a variety of protein kinases. In response to a variety of growth factors and cytokines, Akt, ERK, and RSK can phosphorylate TSC2 and IKKb can phosphorylate TSC1, all of which have been shown to exert inhibitory effects ont the TSC1-TSC2 complex. In response to energy/nutrient stress, AMPK and GSK3 can phosphorylate TSC2, which is thought to promote the activity of the TSC1-TSC2 complex.

6.3.1
Growth Factors and Cytokines

The majority of growth factors and cytokines activate mTORC1. A predominant mechanism of mTORC1 activation by these extracellular factors is a common receptor-mediated activation of the PI3K–Akt pathway, leading to subsequent phosphorylation and inhibition of the TSC1–TSC2 complex. This mechanism is identical to that of insulin signaling to mTORC1 and involves direct multisite phosphorylation of TSC2 by Akt [62, 67]. It appears that phosphorylation of some of the Akt sites on TSC2 (e.g., S939, S981, and T1462) creates binding sites for 14-3-3 proteins [99, 100], which might block the ability of TSC2 to act as a GAP for Rheb within the cell.

In addition to Akt, another major growth factor-stimulated pathway, the Erk–RSK pathway, also activates mTORC1 through modification of TSC2. Erk1 and Erk2 are ubiquitous mitogen-activated protein kinases (MAPK) that phosphorylate and regulate a number of downstream targets, including the RSK subfamily of AGC kinases. Interestingly, both Erk and RSK have been found to directly phosphorylate TSC2 and contribute to mTORC1 activation. RSK has overlapping substrate specificity with Akt [101] and can phosphorylate the major Akt sites on TSC2 (S939 and T1462), as well as an additional site (S1798). This mechanism of TSC2 inhibition is particularly prominent under conditions when the PI3K–Akt pathway is not active [96, 102]. Finally, Erk has been found to directly phosphorylate TSC2 on S540 and S664, and this appears to disrupt the TSC1–TSC2 complex [103]. Therefore, through a combination of Akt and Erk signaling, most, if not all, major growth factors and cytokines affect the function of the TSC1–TSC2 complex.

6.3.2
Energy and Nutrients

A number of other fundamental cellular properties important for cell growth are also sensed by mTORC1 in a manner dependent on the TSC1–TSC2 complex. The best understood among these are cellular energy levels, sensed through AMP-dependent protein kinase (AMPK), which is activated under conditions of low intracellular ATP. Under conditions of cellular energy depletion, AMPK is activated and through phosphorylation of downstream targets stimulates catabolic processes and inhibits anabolic processes [104]. AMPK has been found to directly phosphorylate TSC2 on S1271 and S1387, and this appears to promote inhibition of mTORC1 through activation of the TSC1–TSC2 complex [93, 97]. However, like many of the phosphorylation sites on TSC2, the molecular mechanism of TSC2 regulation by modification of these sites is unknown. Interestingly, AMPK has also been found to directly phosphorylate raptor, thereby inhibiting mTORC1 activity through an additional mechanism independent of the TSC1–TSC2 complex [105].

Cellular oxygen levels also affect the TSC1–TSC2 complex, and this appears to occur through both AMPK-dependent and AMPK-independent mechanisms. As oxygen depletion, or hypoxia, blocks oxidative phosphorylation in the mitochondria, it can lead to rapid decreases in intracellular ATP levels, thereby activating AMPK. Therefore, the acute effects of hypoxia on the TSC1–TSC2 complex and mTORC1 can be mediated by AMPK [106]. However, an additional mechanism involves a regulator of the TSC1–TSC2 complex that was originally identified in yet another *Drosophila* genetic screen for regulators of cell and organ growth. *Scylla* and *charybdis* encode two very similar *Drosophila* proteins and were found to inhibit cell growth at the level of the *dTsc* genes in epistasis experiments [107]. Interestingly, the single mammalian orthologue, called REDD1 (also referred to as RTP801, DDIT4, or Dig2), is a transcriptional target of hypoxia-inducible factor α (HIFα), which is a critical mediator of adaptation to hypoxia. More important, hypoxia-mediated downregulation of mTORC1 requires both the TSC1–TSC2 complex and REDD1 [108]. REDD1 is not a kinase and its molecular function is unknown. However, REDD1 has been shown to bind 14-3-3 proteins, and this has been suggested to hinder their binding to residues on TSC2 phosphorylated by Akt, thereby alleviating the inhibition of the TSC1–TSC2 complex [100].

mTORC1 signaling is acutely sensitive to amino acid availability [109]. However, whether the TSC1–TSC2 complex and/or Rheb are involved in sensing amino acids is not clear from the data published to date. Amino acid withdrawal leads to acute inhibition of mTORC1, but this effect is significantly blunted in the absence of the TSC1–TSC2 complex. In *Tsc* null *Drosophila* or mammalian cells, amino acid starvation leads to a partial decrease in mTORC1-mediated S6K1 and 4E-BP1 phosphorylation [60, 75, 110], demonstrating that, unlike growth factor signaling, some aspects of amino acid sensing by mTORC1 can occur in the absence of the TSC1–TSC2 complex. The partial response in these cells relative to wild-type cells might be due to a role for Rheb downstream of the TSC–TSC2 complex, as Rheb overexpression can clearly activate mTORC1 in the absence of amino acids [75, 76, 82, 83, 85, 87, 110, 111]. Therefore, elevated levels of Rheb-GTP in cells lacking the TSC1–TSC2 complex might override the effects of amino acid withdrawal on mTORC1. Furthermore, siRNA-mediated knockdown of Rheb has demonstrated that it is required for mTORC1 activation upon amino acid refeeding [75, 112], but conflicting data exist as to whether Rheb-GTP levels are affected by the presence or absence of amino acids [78, 110, 112, 113].

An important mTORC1-proximal component of amino acid sensing is the Rag family of GTPases [154, 155]. The Rag proteins function as heterodimers and appear to bind directly to mTORC1 in response to amino acids [154]. While the mechanism is at present unknown, this binding appears to stimulate the trafficking of mTORC1 to a late endosomal/Rab7-containing vesicular compartment where Rheb is also localized. This mechanism would support a model in which amino acids do not regulate Rheb-GTP levels *per se*, but Rheb is still required for amino acids to activate mTORC1. The molecular sensor(s) of amino acids and its downstream signaling pathway leading to Rag protein activation is unknown and should be an exciting avenue of research in the coming years. It will be particularly interesting to see how previously proposed mechanisms of amino acid signaling to mTORC1, such as VPS34 [156, 157] and MAP4K3 [158], relate to the Rag proteins and Rheb.

6.4
Primary mTOR-Related Signaling Defects Triggered by Disruption of the TSC1–TSC2 Complex

6.4.1
Constitutive and Elevated mTORC1 Signaling

Due to the numerous regulatory inputs that control mTORC1 through modulating the function of the TSC1–TSC2 complex, mTORC1 activity is constitutively high in cells lacking this complex. In other words, mTORC1 can no longer sense fluctuations in cellular growth conditions and remains maximally activated in TSC1- or TSC2-deficient cells. This is reflected in elevated and unregulated phosphorylation of S6K on T389 (Figure 6.5) and of 4E-BP1 on its four regulatory sites. Therefore, S6K is constitutively activated and able to phosphorylate its downstream targets involved in promoting mRNA translation, including ribosomal S6 and eIF4B. On the other hand, 4E-BP1 is constitutively inhibited and unable to bind to its target eIF4E at the 5′-7-methyl-GTP cap of mRNAs, thereby allowing translation initiation complexes to form at the cap, even under poor growth conditions that would normally inhibit

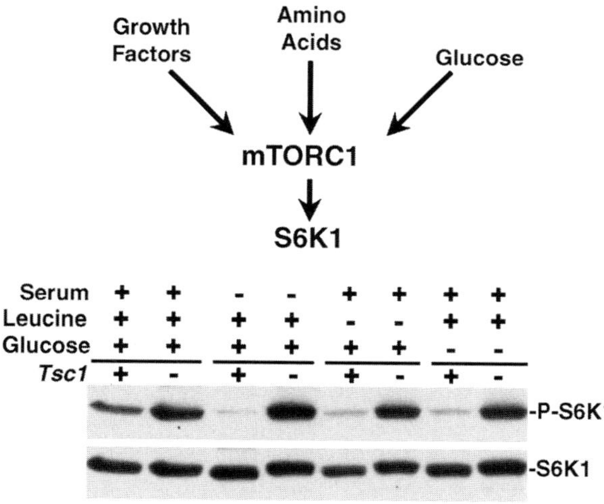

Figure 6.5 In the absence of the TSC1-TSC2 complex, mTORC1 cannot sense perturbations in cellular growth conditions. Growth-promoting signals emanating from growth factors, amino acids, and glucose are all essential to fully activate mTORC1 signaling to its downstream target S6K1. While mTORC1-dependent phosphorylation and activation of S6K1 is blocked upon removal of any one of these factors in wild-type cells, mTORC1 signaling is unresponsive and constitutive in cells lacking the TSC1-TSC2 complex. This is illustrated by the pictured immunoblot of S6K phosphorylation on T389 in cell lysates from littermate-derived $Tsc1^{+/+}$ and $Tsc1^{-/-}$ mouse embryo fibroblasts (derived in the laboratory of D.J. Kwiatkowski) under different growth conditions. Cells were grown in medium containing all essential nutrients in the presence or absence of 10% serum (dialyzed to remove amino acids and glucose), 100 mg/ml L-leucine, or 4.5 mg/ml D-glucose, as indicated. Serum withdrawal was for 16 hours, while leucine and glucose were removed for two hours prior to cell lysis.

cap-dependent translation (e.g., growth factor withdrawal, energy stress, and so on). Through these targets, and likely other unknown substrates, mTORC1 signaling is particularly potent at stimulating the translation of mRNAs with 5′-TOP tracts, including those encoding translation factors and ribosomal proteins. In fact, this class of mRNAs is enriched in actively translating polysome fractions from TSC-deficient cells, even under serum starvation conditions [114]. Therefore, it is predicted that in all cells lacking the TSC genes, unregulated mTORC1 signaling will trigger aberrant mRNA translation and, through increased ribosomal biogenesis, global protein synthesis. However, this has yet to be definitively demonstrated in any TSC-deficient system. In addition to promoting protein synthesis, mTORC1 inhibits protein degradation by inhibiting autophagy, albeit through an unknown mechanism. Again, a clear defect in the regulation of autophagy has yet to be demonstrated in TSC-deficient cells. Therefore, while the normal signaling events downstream of mTORC1 are clearly misregulated in the absence of the TSC genes, the cellular consequences have yet to be fully established.

In addition to constitutive and elevated signaling to well-known substrates of mTORC1, a number of other downstream consequences of aberrantly high mTORC1 activity have been uncovered in cells lacking the TSC1–TSC2 complex (Figure 6.6). mTORC1 signaling is known to stimulate translation of the mRNA encoding the HIFα transcription factor through a 5′-TOP sequence [115]. However, more direct posttranslational mechanisms of HIFα activation by mTORC1 are also possible [116]. The HIFα protein is rapidly degraded in the presence of oxygen (normoxia) and is only stabilized under conditions of hypoxia. However, in cells lacking the TSC genes, elevated mTORC1 signaling leads to significant accumulation of HIFα and its transcriptional targets under normoxic conditions [117]. Elevated mTORC1 signaling has also been found to cause endoplasmic reticulum (ER) stress in TSC1- and TSC2-deficient cells [118]. ER stress is caused by the

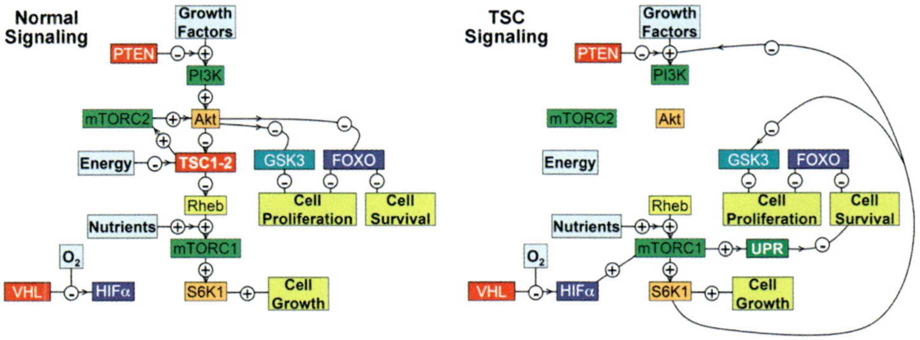

Figure 6.6 Molecular rewiring of signaling pathways upon loss of the TSC1-TSC2 complex. A model of growth factor signaling to mTORC1 in normal cells (left panel) versus those lacking a functional TSC1-TSC2 complex (right panel) is shown. Positive and negative regulatory inputs are marked with '+' and '-', respectively, and the direction of signaling is indicated by arrowheads. All abbreviations are protein names, except UPR (unfolded protein response). See text for the details regarding these individual connections and their rewiring in TSC-deficient cells.

inability of the protein-folding capacity of the ER to keep up with its protein load, thereby leading to accumulation of unfolded proteins in the ER lumen. ER stress triggers the unfolded protein response (UPR), which through a series of signaling pathways reduces cap-dependent translation, increases ER-associated protein degradation, and enhances ER protein folding. Cells lacking a functional TSC1–TSC2 complex display basally activated UPR, which can be reversed by prolonged inhibition of mTORC1 with rapamycin, inhibition of protein synthesis with cyclohexamide, or increasing ER-folding capacity with chemical chaperones [118]. Therefore, likely through elevated rates of protein synthesis, aberrantly high mTORC1 activity in TSC null cells causes ER stress and activates the UPR. Constitutive mTORC1-mediated activation of S6K1 also triggers the phosphorylation of new downstream targets normally regulated by other kinases. In response to growth factors, GSK3α and GSK3β are normally inhibited by Akt through direct phosphorylation of S21 and S9, respectively. Interestingly, in TSC-deficient cells, S6K1, rather than Akt, phosphorylates these same residues on GSK3, but does so in an unregulated manner, thereby leading to its constitutive inhibition [119]. However, it appears that this aberrant regulation of GSK3 does not occur in all settings of TSC gene disruption (e.g., Ref. [159]). The potential pathological consequences of these aberrant mTORC1-dependent signaling events in TSC are discussed below.

6.4.2 mTORC1-Dependent Feedback Inhibition of PI3K Signaling

In addition to constitutive mTORC1 signaling, another signature of loss of TSC1–TSC2 complex function is a severe decrease in Akt phosphorylation and activation. This attenuation of Akt signaling has been found to occur through multiple mechanisms, including both mTORC1-dependent and mTORC1-independent processes.

The insulin and IGF-1 receptors activate PI3K through tyrosine phosphorylation of insulin receptor substrate (IRS) proteins, which bind directly to the p85 regulatory subunit of PI3K. Numerous studies of insulin resistance in adipocyte and myocyte cultures have demonstrated that mTORC1 activation negatively influences insulin signaling and IRS protein function [120]. Due to their robust constitutive activation of mTORC1, cells lacking a functional TSC1–TSC2 complex are an excellent model for this so-called negative feedback mechanism [121, 122]. TSC-deficient cells are unresponsive to insulin and IGF-1 for phosphorylation of Akt, and treatment with rapamycin can significantly rescue this defect. There appears to be multiple mTORC1-dependent mechanisms leading to downregulation of IRS-1, and perhaps IRS-2, in these cells. Upon mTORC1 activation, IRS-1 is hyperphosphorylated on serine residues by both mTORC1 and S6K1, and this triggers IRS-1 protein degradation [121–125]. Through an unknown mechanism, IRS-1 mRNA levels are also decreased in *Tsc2*$^{-/-}$ MEFs [121]. While rapamycin treatment leads to rapid dephosphorylation of these serines on IRS-1, prolonged exposure to rapamycin (>12 h) is required to restore IRS-1 mRNA and protein levels and insulin/IGF-1 signaling to Akt. Interestingly, previous studies on mechanisms of insulin resistance found that ER stress and activation of the UPR can lead to downregulation of IRS-1 [126]. This mechanism also appears to contribute to the decrease in IRS-1 protein levels in *Tsc1*$^{-/-}$ and *Tsc2*$^{-/-}$ MEFs and likely accounts for the need for extended rapamycin treatment to fully restore IRS-1 levels and signaling [118].

A less well-defined feedback mechanism in TSC-deficient MEFs affects the PDGF receptors [127, 128]. These cells are substantially less responsive to PDGF for phosphorylation of both Akt and Erk, and this coincides with decreased expression of both PDGFRα and PDGFRβ. Like IRS-1, PDGFRβ protein levels are increased by prolonged exposure of TSC null MEFs to rapamycin, and both PDGFRα and PDGFRβ mRNA levels are somewhat increased by rapamcyin treatment [128]. Therefore, the elevated mTORC1 signaling in cells lacking the TSC1–TSC2 complex leads to downregulation of both insulin/IGF-1- and PDGF-mediated activation of PI3K through distinct mechanisms.

6.4.3 Loss of mTORC2 Activity

Although there are clearly mTORC1-dependent mechanisms to block the activation of PI3K signaling in response to specific growth factors in cells lacking the TSC1–TSC2 complex, loss of Akt phosphorylation appears to be a more general phenomenon in this setting. In fact, decreases in TSC2 levels can lead to a decrease in Akt phosphorylation prior to detectable increases in mTORC1 signaling [129], suggesting that an mTORC1-independent mechanism is also involved. Interestingly, the kinase activity of mTORC2, which is responsible for Akt phosphorylation on S473, was found to be severely attenuated in a variety of cell lines with loss of TSC1 or TSC2. Furthermore, mTORC2 activity was not restored by blocking mTORC1-dependent feedback mechanisms in these cells. Very little is currently known regarding how mTORC2 is regulated, but these data suggest that the TSC1–TSC2 complex is required for its normal activation. More important, this effect is independent of the GAP activity of TSC2 and its regulation of Rheb, suggesting a very distinct mechanism of regulation from that of mTORC1. Finally, the TSC1–TSC2 complex was found to physically associate with mTORC2 but not with mTORC1, indicating that this mode of regulation is likely to be rather direct [129]. Therefore, there are both indirect mTORC1-dependent feedback mechanisms and more direct effects on mTORC2 accounting for the defect in Akt signaling in cells lacking the TSC1–TSC2 complex.

6.5 Pathological Consequences of mTOR Dysregulation in TSC

All of the cellular effects of TSC gene disruption already described, including elevated phosphorylation events downstream of mTORC1, attenuation of Akt phosphorylation and downstream signaling, elevated phosphorylation of GSK3, and activation of the UPR have also been detected in at least a subset of tumors from TSC mouse models and human TSC or LAM lesions [69, 72, 118, 119, 130–135]. While certainly not the entire picture underlying the complex clinical features of TSC, these signaling defects triggered by aberrant mTORC1 activation and mTORC2 inactivation are likely to explain many of the unique pathological consequences resulting from the loss of TSC1–TSC2 complex function.

6.5.1 Neoplastic Lesions

Much of the aberrant proliferation of TSC tumors is likely explained by constitutive activation of mTORC1 in the initiating cell type. In addition to its well-characterized role in promoting increases in cell size, mTORC1 plays an essential role in cell proliferation. This is best illustrated by the fact that most cell types exhibit a G1 phase cell cycle arrest upon treatment with rapamycin. This is thought to be due, at least in part, to mTORC1 activity being required for the translation of proteins that drive cell cycle entry, such as cyclin D1 and c-myc [136–138]. Due to activation of mTORC1, these translational targets would be predicted to be elevated in cells lacking a functional TSC1–TSC2 complex. Furthermore, GSK3, which phosphorylates many of these proteins and targets them for degradation, is constitutively inhibited by S6K1 in TSC-deficient cells, and this contributes to the abnormal proliferation properties of these cells [119]. Therefore, upon disruption of the TSC genes, at least in cell culture models, mTORC1 is activated and GSK3 is inhibited, thereby leading to the accumulation of proteins that promote cell cycle progression. Cell culture studies have demonstrated that *Tsc2* null MEFs can proliferate, in an mTORC1-dependent (i.e., rapamycin-sensitive) manner, under conditions in which their wild-type counterparts undergo cell cycle arrest, such as serum withdrawal [127, 119] or hypoxia [108]. Thus, within a TSC patient, a cell undergoing *TSC1* or *TSC2* loss of heterozygosity will likely gain a proliferative advantage under otherwise suboptimal conditions due to the immediate and robust activation of mTORC1 signaling.

6.5.2 Benign Tumors

Unlike other tumor syndromes in which mTORC1 activity is elevated, TSC is characterized predominately by the development of benign tumors. These other diseases, which often arise due to loss of tumor suppressors acting upstream of the TSC1–TSC2 complex, such as Cowden disease (PTEN mutations) and Peutz–Jeghers syndrome (LKB1 mutations), are true cancer predisposition syndromes, whereas malignancies are quite rare in TSC. Although more genetic studies are required, recent work has suggested that this feature of TSC might be due to the severe loss of Akt signaling in TSC tumors [135]. Upon loss of function of the TSC1–TSC2 complex, the activation of Akt is blocked due to a combination of mTORC1-dependent feedback mechanisms [121, 122, 128] and loss of mTORC2 activity [129], as already described. Akt activation is a common feature in human cancers and, through phosphorylation of numerous downstream targets, is known to promote many oncogenic cellular events [139]. Therefore, the attenuation of Akt signaling in the context of TSC gene mutations is likely to limit the malignant potential of TSC tumors.

6.5.3 Specific Clinical Features

Uncontrolled mTORC1 activity is also likely to contribute to some of the unique clinical manifestations of TSC. Although these are discussed in detail elsewhere in this book, it is worth speculating here on the potential role of mTORC1-dependent signaling defects in the development and characteristic properties of certain TSC lesions. As a central player in the control of cell growth, dysregulated mTORC1 activity in neuronal precursor cells within the TSC brain is predicted to be a major source of the giant cells and dysplastic neurons common to cortical tubers [130]. In addition, proper spatial regulation of GSK3 is required for the maintenance of neuronal polarity [140, 141], and GSK3 has been shown to be phosphorylated, and therefore inhibited, in tuber giant cells [119]. However, a mouse model with neuron-specific deletion of *Tsc1* showed reduced, rather than elevated, GSK3 phosphorylation, demonstrating that defects in GSK3 regulation vary between cell types and tissues with loss of TSC gene function [159]. Studies have found that mTORC1 activity, likely through a combination of S6K activation [142] and 4E-BP inhibition [143, 144], promotes localized translation within synapses, the proper regulation of which is essential for synaptic plasticity [145]. This suggests that unregulated mTORC1 signaling in certain neuronal populations within the TSC brain could have profound regional effects on synaptic wiring. Future work in this area will be of great importance to our understanding of the complex neuropsychiatric features of TSC [146].

Given that GSK3 is constitutively phosphorylated and inhibited by S6K1 in some TSC cells and tumors [119] and is dephosphorylated and activated due to Akt attenuation in others [159], it will be important to determine the status of GSK3 regulation in specific TSC lesions. In addition to phosphorylating a large number of neuronal targets, GSK3 normally phosphorylates and regulates other substrates that, if misregulated, could contribute to TSC pathology [147]. For instance, GSK3 normally inhibits its namesake substrate glycogen synthase (GS), suggesting that GS activity and, hence, glycogen synthesis might be elevated in TSC lesions. This could explain the massive accumulation of glycogen globules in cardiac rhabdomyomas, a heart tumor common in newborns with TSC but not the general population. Interestingly, GSK3 has recently been found to mediate the proper localization of polycystin-2, encoded by the polycystic kidney disease 2 (PKD2) gene [148]. It will be interesting to determine whether polycystin-2 localization is normal in renal epithelial cells lacking the TSC1–TSC2 complex and whether this is due to constitutive inhibition of GSK3, thereby explaining the development of kidney cysts in TSC. Finally, while the responsible downstream substrate is unknown, inhibition of GSK3 has been shown to promote smooth muscle cell survival under hypoxic conditions [149], which could contribute to the development of angiomyolipomas and LAM nodules.

While the risk of renal cell carcinoma (RCC) is only slightly higher in TSC patients, it occurs at much younger ages than in the normal population [150]. Genetic lesions leading to normoxic accumulation of HIFα transcription factors, such as mutations in the von Hippel–Lindau tumor suppressor [151], are extremely common in RCC. Therefore, it seems likely that the mTORC1-dependent elevation in HIFα levels and transcriptional targets observed in TSC-deficient cells [117] might account for the increased susceptibility and earlier onset of RCC in TSC patients. A recent study in our laboratory has found that HIFα levels are greatly increased in a variety of *in vitro* and *in vivo* settings where the function of the TSC1–TSC2 complex has been compromised. Furthermore, transcripts regulated by HIFα comprise the primary transcriptional response to mTORC1 activation in cells lacking TSC1 or TSC2 (K. Düvel, S. Menon, and B.D. Manning, unpublished). Therefore, aberrant HIFα upregulation appears to be a major event downstream of mTORC1 activation with the potential to contribute substantially to TSC pathology. For instance, the highly vascular nature of some TSC lesions, such as angiomyolipomas and angiofibromas, might be the result of strong angiogenic signals generated downstream of HIFα activation, such as the production of vascular endothelial growth factor (VEGF). Future genetic studies will help determine the pathological consequences of mTORC1-dependent HIFα activation in TSC.

6.6
Therapeutic Opportunities: Rapamycin and Beyond

Given that constitutive activation of mTORC1 is the primary molecular defect triggered by genetic disruption of the TSC genes, mTORC1 inhibitors should be the ideal targeted therapeutics for the treatment of TSC. Due to the fact that rapamycin and its analogues are allosteric inhibitors, they are extremely specific to mTOR and do not target similar kinases, thereby limiting concerns regarding off-target effects. However, in most cell types, rapamycin has cytostatic effects rather than the cytotoxic effects required to induce tumor cell apoptosis resulting in ultimate elimination of the tumor. Furthermore, by blocking mTORC1-dependent feedback mechanisms, inhibition of mTORC1 resensitizes the PI3K–Akt pathway to growth factors, such as IGF1 and PDGF [121, 122, 128]. Therefore, while mTORC1 inhibitors will clearly block the primary tumor-promoting event downstream of the TSC genes, they have the potential to reactivate this critical cell survival pathway, which could prevent tumor cell apoptosis. This dual effect could necessitate chronic, perhaps life-long, treatment with such drugs to maintain the beneficial inhibitory effects on tumor growth.

One exciting possibility to bypass the effects of mTORC1 inhibitors on tumor-suppressive feedback loops is to use them in combination with inhibitors of receptor tyrosine kinases (e.g., IGF1-R or PDGFR) or PI3K. Interestingly, as mTOR is a PI3K-related kinase, many PI3K inhibitors are also catalytic domain inhibitors of mTOR and, therefore, can both block mTORC1 activity and prevent the reactivation of Akt. While these compounds are likely to be more toxic, in general, than rapamycin-like compounds, they could be more effective at causing permanent tumor regression in TSC. Such inhibitors would have the added benefit of also inhibiting mTORC2, which normally phosphorylates and activates Akt. However, given the finding that the TSC1–TSC2 complex promotes mTORC2 activity [129], it is unclear whether mTORC2 is ever active in TSC tumors. If rapamycin analogues ultimately fail in the clinic, combination therapies or dual PI3K/mTOR inhibitors offer promising alternatives for future clinical trials. However, the preclinical trials done so far on *Tsc2*$^{+/-}$ mice have demonstrated incredible effectiveness of rapamycin analogues as single therapeutic agents (e.g., Ref. [160]), and a recent study showed that a dual PI3K/mTOR inhibitor was equivalent, but not superior, to a rapamycin analogue (D.J. Kwiatkowski, unpublished). A major issue with testing these therapeutic approaches is the current lack of preclinical animal models that recapitulate the manifestations and therapeutic responses of TSC patients.

An alternative approach to targeting tumors with aberrantly high mTORC1 activity is to take advantage of resulting defects in critical cell survival and death pathways. For instance, rather than targeting mTORC1 activation *per se*, it is worth thinking of strategies to selectively kill cells with defects in Akt activation. Akt phosphorylates and inhibits many proapoptotic proteins to promote cell survival [139]. Therefore, cells lacking a functional TSC1–TSC2 complex, which have defects in Akt survival signaling due to both mTORC1-dependent feedback mechanisms and defects in mTORC2 activation (see above), should theoretically be more prone to some apoptotic stimuli. An increased susceptibility to such stimuli would offer a therapeutic window to induce apoptosis specifically in TSC null tumor cells. A similar therapeutic opportunity is offered by activation of the UPR downstream of mTORC1-driven ER stress in TSC null cells and tumors [118]. Initially, the UPR is an adaptive response aimed at increasing the folding capacity of the ER, but upon elevated or chronic levels of ER stress, the UPR triggers an apoptotic response. Due to higher basal levels of UPR pathway activation, TSC-deficient cells are hypersensitive to ER stress-inducing agents (e.g., thapsigargin and tunicamycin) and conditions (e.g., glucose starvation) and undergo apoptosis at doses that do not significantly affect wild-type cells [118]. There are many available classes of compounds that cause ER stress and activate the UPR. While many more *in vitro* and *in vivo* studies are needed, low doses of such compounds might offer potent anti-TSC therapeutics in the future. As the UPR is an adaptive response to elevated mTORC1 signaling, it is also possible that in some settings of TSC gene disruption, basal activation of the UPR might be a prosurvival mechanism. Therefore, agents that block the UPR might also have therapeutic value in treating TSC tumors.

In conclusion, targeting some of the signaling defects and cellular responses triggered by aberrant mTORC1 activation, rather than mTORC1 itself, presents an additional and perhaps more cytotoxic approach to TSC therapeutics. Finally, it is worth noting that this chapter has focused on mTOR as the critical downstream target of the TSC1–TSC2 complex. However, it is likely that both the TSC1–TSC2 complex and Rheb have other downstream effectors that, once identified, will provide further therapeutic opportunities for the treatment of TSC and LAM.

Acknowledgments

I thank David J. Kwiatkowski for critical comments on the chapter. I apologize to those colleagues in the TSC and TOR fields whose work was not discussed here due to the relatively narrow focus of this chapter. Research in my laboratory on the regulation and function of the TSC1–TSC2 complex and mTOR was supported by grants from the Tuberous Sclerosis Alliance, LAM Foundation, American Diabetes Association, and National Institutes of Health (R01-CA122617 and P01-CA120964).

References

1 Vezina, C., Kudelski, A., and Sehgal, S.N. (1975) Rapamycin (AY-22,989), a new antifungal antibiotic. I. Taxonomy of the producing streptomycete and isolation of the active principle. *J. Antibiot. (Tokyo)*, **28**, 721–726.

2 Collier, S.J. (1989) Immunosuppressive drugs. *Curr. Opin. Immunol.*, **2**, 854–858.

3 Bierer, B.E., Somers, P.K., Wandless, T.J., Burakoff, S.J., and Schreiber, S.L. (1990) Probing immunosuppressant action with a nonnatural immunophilin ligand. *Science*, **250**, 556–559.

4 Heitman, J., Movva, N.R., and Hall, M.N. (1991) Targets for cell cycle arrest by the immunosuppressant rapamycin in yeast. *Science*, **253**, 905–909.

5 Koltin, Y., Faucette, L., Bergsma, D.J., Levy, M.A., Cafferkey, R., Koser, P.L., Johnson, R.K., and Livi, G.P. (1991) Rapamycin sensitivity in *Saccharomyces cerevisiae* is mediated by a peptidyl–prolyl *cis–trans* isomerase related to human FK506-binding protein. *Mol. Cell. Biol.*, **11**, 1718–1723.

6 Kunz, J., Henriquez, R., Schneider, U., Deuter-Reinhard, M., Movva, N.R., and Hall, M.N. (1993) Target of rapamycin in yeast, TOR2, is an essential phosphatidylinositol kinase homolog required for G1 progression. *Cell*, **73**, 585–596.

7 Helliwell, S.B., Wagner, P., Kunz, J., Deuter-Reinhard, M., Henriquez, R., and Hall, M.N. (1994) TOR1 and TOR2 are structurally and functionally similar but not identical phosphatidylinositol kinase homologues in yeast. *Mol. Biol. Cell*, **5**, 105–118.

8 Brown, E.J., Albers, M.W., Shin, T.B., Ichikawa, K., Keith, C.T., Lane, W.S., and Schreiber, S.L. (1994) A mammalian protein targeted by G1-arresting rapamycin–receptor complex. *Nature*, **369**, 756–758.

9 Sabatini, D.M., Erdjument-Bromage, H., Lui, M., Tempst, P., and Snyder, S.H. (1994) RAFT1: a mammalian protein that binds to FKBP12 in a rapamycin-dependent fashion and is homologous to yeast TORs. *Cell*, **78**, 35–43.

10 Brunn, G.J., Williams, J., Sabers, C., Wiederrecht, G., Lawrence, J.C., Jr., and Abraham, R.T. (1996) Direct inhibition of the signaling functions of the mammalian target of rapamycin by the phosphoinositide 3-kinase inhibitors, wortmannin and LY294002. *Embo J.*, **15**, 5256–5267.

11 Chiu, M.I., Katz, H., and Berlin, V. (1994) RAPT1, a mammalian homolog of yeast Tor, interacts with the FKBP12/rapamycin complex. *Proc. Natl. Acad. Sci. USA*, **91**, 12574–12578.

12 Stan, R., McLaughlin, M.M., Cafferkey, R., Johnson, R.K., Rosenberg, M., and Livi, G.P. (1994)

Interaction between FKBP12-rapamycin and TOR involves a conserved serine residue. *J. Biol. Chem.*, **269**, 32027–32030.

13 Loewith, R., Jacinto, E., Wullschleger, S., Lorberg, A., Crespo, J.L., Bonenfant, D., Oppliger, W., Jenoe, P., and Hall, M.N. (2002) Two TOR complexes, only one of which is rapamycin sensitive, have distinct roles in cell growth control. *Mol. Cell*, **10**, 457–468.

14 Zheng, X.F., Florentino, D., Chen, J., Crabtree, G.R., and Schreiber, S.L. (1995) TOR kinase domains are required for two distinct functions, only one of which is inhibited by rapamycin. *Cell*, **82**, 121–130.

15 Hara, K., Maruki, Y., Long, X., Yoshino, K., Oshiro, N., Hidayat, S., Tokunaga, C., Avruch, J., and Yonezawa, K. (2002) Raptor, a binding partner of target of rapamycin (TOR), mediates TOR action. *Cell*, **110**, 177–189.

16 Kim, D.H., Sarbassov, D.D., Ali, S.M., King, J.E., Latek, R.R., Erdjument-Bromage, H., Tempst, P., and Sabatini, D.M. (2002) mTOR interacts with raptor to form a nutrient-sensitive complex that signals to the cell growth machinery. *Cell*, **110**, 163–175.

17 Kim, D.H., Sarbassov, D.D., Ali, S.M., Latek, R.R., Guntur, K.V., Erdjument-Bromage, H., Tempst, P., and Sabatini, D.M. (2003) GbetaL, a positive regulator of the rapamycin-sensitive pathway required for the nutrient-sensitive interaction between raptor and mTOR. *Mol. Cell*, **11**, 895–904.

18 Sancak, Y., Thoreen, C.C., Peterson, T.R., Lindquist, R.A., Kang, S.A., Spooner, E., Carr, S.A., and Sabatini, D.M. (2007) PRAS40 is an insulin-regulated inhibitor of the mTORC1 protein kinase. *Mol. Cell*, **25**, 903–915.

19 Vander Haar, E., Lee, S.I., Bandhakavi, S., Griffin, T.J., and Kim, D.H. (2007) Insulin signalling to mTOR mediated by the Akt/PKB substrate PRAS40. *Nat. Cell Biol.*, **9**, 316–323.

20 Frias, M.A., Thoreen, C.C., Jaffe, J.D., Schroder, W., Sculley, T., Carr, S.A., and Sabatini, D.M. (2006) mSin1 is necessary for Akt/PKB phosphorylation, and its isoforms define three distinct mTORC2s. *Curr. Biol.*, **16**, 1865–1870.

21 Jacinto, E., Facchinetti, V., Liu, D., Soto, N., Wei, S., Jung, S.Y., Huang, Q., Qin, J., and Su, B. (2006) SIN1/MIP1 maintains rictor-mTOR complex integrity and regulates Akt phosphorylation and substrate specificity. *Cell*, **127**, 125–137.

22 Sarbassov, D.D., Ali, S.M., Kim, D.H., Guertin, D.A., Latek, R.R., Erdjument-Bromage, H., Tempst, P., and Sabatini, D.M. (2004) Rictor, a novel binding partner of mTOR, defines a rapamycin-insensitive and raptor-independent pathway that regulates the cytoskeleton. *Curr. Biol.*, **14**, 1296–1302.

23 Yang, Q., Inoki, K., Ikenoue, T., and Guan, K.L. (2006) Identification of Sin1 as an essential TORC2 component required for complex formation and kinase activity. *Genes Dev.*, **20**, 2820–2832.

24 Pearce, L.R., Huang, X., Boudeau, J., Pawlowski, R., Wullschleger, S., Deak, M., Ibrahim, A.F., Gourlay, R., Magnuson, M.A., and Alessi, D.R. (2007) Identification of Protor as a novel Rictor-binding component of mTOR complex-2. *Biochem. J.*, **405**, 513–522.

25 Woo, S.Y., Kim, D.H., Jun, C.B., Kim, Y.M., Haar, E.V., Lee, S.I., Hegg, J.W., Bandhakavi, S., Griffin, T.J., and Kim, D.H. (2007) PRR5, a novel component of mTOR complex 2, regulates platelet-derived growth factor receptor beta expression and signaling. *J. Biol. Chem.*, **282**, 25604–25612.

26 Schalm, S.S. and Blenis, J. (2002) Identification of a conserved motif required for mTOR signaling. *Curr. Biol.*, **12**, 632–639.

27 Nojima, H., Tokunaga, C., Eguchi, S., Oshiro, N., Hidayat, S., Yoshino, K., Hara, K., Tanaka, N., Avruch, J., and Yonezawa, K. (2003) The mammalian target of rapamycin (mTOR) partner, raptor, binds the mTOR substrates p70 S6 kinase and 4E-BP1 through their TOR signaling (TOS) motif. *J. Biol. Chem.*, **278**, 15461–15464.

28 Schalm, S.S., Fingar, D.C., Sabatini, D.M., and Blenis, J. (2003) TOS

motif-mediated raptor binding regulates 4E-BP1 multisite phosphorylation and function. *Curr. Biol.*, **13**, 797–806.

29 Sarbassov, D.D., Ali, S.M., Sengupta, S., Sheen, J.H., Hsu, P.P., Bagley, A.F., Markhard, A.L., and Sabatini, D.M. (2006) Prolonged rapamycin treatment inhibits mTORC2 assembly and Akt/PKB. *Mol. Cell*, **22**, 159–168.

30 Kay, J.E., Kromwel, L., Doe, S.E., and Denyer, M. (1991) Inhibition of T and B lymphocyte proliferation by rapamycin. *Immunology*, **72**, 544–549.

31 Price, D.J., Grove, J.R., Calvo, V., Avruch, J., and Bierer, B.E. (1992) Rapamycin-induced inhibition of the 70-kilodalton S6 protein kinase. *Science*, **257**, 973–977.

32 Albers, M.W., Williams, R.T., Brown, E.J., Tanaka, A., Hall, F.L., and Schreiber, S.L. (1993) FKBP-rapamycin inhibits a cyclin-dependent kinase activity and a cyclin D1-Cdk association in early G1 of an osteosarcoma cell line. *J. Biol. Chem.*, **268**, 22825–22829.

33 Nourse, J., Firpo, E., Flanagan, W.M., Coats, S., Polyak, K., Lee, M.H., Massague, J., Crabtree, G.R., and Roberts, J.M. (1994) Interleukin-2-mediated elimination of the p27Kip1 cyclin-dependent kinase inhibitor prevented by rapamycin. *Nature*, **372**, 570–573.

34 Nakayama, K., Ishida, N., Shirane, M., Inomata, A., Inoue, T., Shishido, N., Horii, I., Loh, D.Y., and Nakayama, K. (1996) Mice lacking p27(Kip1) display increased body size, multiple organ hyperplasia, retinal dysplasia, and pituitary tumors. *Cell*, **85**, 707–720.

35 Chung, J., Kuo, C.J., Crabtree, G.R., and Blenis, J. (1992) Rapamycin-FKBP specifically blocks growth-dependent activation of and signaling by the 70 KD S6 protein kinases. *Cell*, **69**, 1227–1237.

36 Graves, L.M., Bornfeldt, K.E., Argast, G.M., Krebs, E.G., Kong, X., Lin, T.A., and Lawrence, J.C., Jr. (1995) cAMP- and rapamycin-sensitive regulation of the association of eukaryotic initiation factor 4E and the translational regulator PHAS-I in aortic smooth muscle cells. *Proc. Natl. Acad. Sci. USA*, **92**, 7222–7226.

37 Beretta, L., Gingras, A.C., Svitkin, Y.V., Hall, M.N., and Sonenberg, N. (1996) Rapamycin blocks the phosphorylation of 4E-BP1 and inhibits cap-dependent initiation of translation. *Embo J.*, **15**, 658–664.

38 Jefferies, H.B., Reinhard, C., Kozma, S.C., and Thomas, G. (1994) Rapamycin selectively represses translation of the "polypyrimidine tract" mRNA family. *Proc. Natl. Acad. Sci. USA*, **91**, 4441–4445.

39 Terada, N., Patel, H.R., Takase, K., Kohno, K., Nairn, A.C., and Gelfand, E.W. (1994) Rapamycin selectively inhibits translation of mRNAs encoding elongation factors and ribosomal proteins. *Proc. Natl. Acad. Sci. USA*, **91**, 11477–11481.

40 Barbet, N.C., Schneider, U., Helliwell, S.B., Stansfield, I., Tuite, M.F., and Hall, M.N. (1996) TOR controls translation initiation and early G1 progression in yeast. *Mol. Biol. Cell*, **7**, 25–42.

41 Brown, E.J., Beal, P.A., Keith, C.T., Chen, J., Shin, T.B., and Schreiber, S.L. (1995) Control of p70 s6 kinase by kinase activity of FRAP *in vivo*. *Nature*, **377**, 441–446.

42 Brunn, G.J., Hudson, C.C., Sekulic, A., Williams, J.M., Hosoi, H., Houghton, P.J., Lawrence, J.C., Jr., and Abraham, R.T. (1997) Phosphorylation of the translational repressor PHAS-I by the mammalian target of rapamycin. *Science*, **277**, 99–101.

43 Burnett, P.E., Barrow, R.K., Cohen, N.A., Snyder, S.H., and Sabatini, D.M. (1998) RAFT1 phosphorylation of the translational regulators p70 S6 kinase and 4E-BP1. *Proc. Natl. Acad. Sci. USA*, **95**, 1432–1437.

44 Harris, T.E. and Lawrence, J.C., Jr. (2003) TOR signaling. *Sci. STKE*, **2003**, re15.

45 Pearson, R.B., Dennis, P.B., Han, J.W., Williamson, N.A., Kozma, S.C., Wettenhall, R.E., and Thomas, G. (1995) The principal target of rapamycin-induced p70s6k inactivation is a novel phosphorylation site within a conserved hydrophobic domain. *Embo. J.*, **14**, 5279–5287.

46 Holz, M.K., Ballif, B.A., Gygi, S.P., and Blenis, J. (2005) mTOR and S6K1

mediate assembly of the translation preinitiation complex through dynamic protein interchange and ordered phosphorylation events. *Cell*, **123**, 569–580.

47 Raught, B., Peiretti, F., Gingras, A.C., Livingstone, M., Shahbazian, D., Mayeur, G.L., Polakiewicz, R.D., Sonenberg, N., and Hershey, J.W. (2004) Phosphorylation of eucaryotic translation initiation factor 4B Ser422 is modulated by S6 kinases. *Embo J.*, **23**, 1761–1769.

48 Ruvinsky, I. and Meyuhas, O. (2006) Ribosomal protein S6 phosphorylation: from protein synthesis to cell size. *Trends. Biochem. Sci.*, **31**, 342–348.

49 Gingras, A.C., Raught, B., Gygi, S.P., Niedzwiecka, A., Miron, M., Burley, S.K., Polakiewicz, R.D., Wyslouch-Cieszynska, A., Aebersold, R., and Sonenberg, N. (2001) Hierarchical phosphorylation of the translation inhibitor 4E-BP1. *Genes Dev.*, **15**, 2852–2864.

50 Richter, J.D. and Sonenberg, N. (2005) Regulation of cap-dependent translation by eIF4E inhibitory proteins. *Nature*, **433**, 477–480.

51 Fingar, D.C., Salama, S., Tsou, C., Harlow, E., and Blenis, J. (2002) Mammalian cell size is controlled by mTOR and its downstream targets S6K1 and 4EBP1/eIF4E. *Genes Dev.*, **16**, 1472–1487.

52 Fingar, D.C., Richardson, C.J., Tee, A.R., Cheatham, L., Tsou, C., and Blenis, J. (2004) mTOR controls cell cycle progression through its cell growth effectors S6K1 and 4E-BP1/eukaryotic translation initiation factor 4E. *Mol. Cell. Biol.*, **24**, 200–216.

53 Jacinto, E., Loewith, R., Schmidt, A., Lin, S., Ruegg, M.A., Hall, A., and Hall, M.N. (2004) Mammalian TOR complex 2 controls the actin cytoskeleton and is rapamycin insensitive. *Nat. Cell Biol.*, **6**, 1122–1128.

54 Sarbassov, D.D., Guertin, D.A., Ali, S.M., and Sabatini, D.M. (2005) Phosphorylation and regulation of Akt/PKB by the rictor-mTOR complex. *Science*, **307**, 1098–1101.

55 Guertin, D.A., Stevens, D.M., Thoreen, C.C., Burds, A.A., Kalaany, N.Y., Moffat, J., Brown, M., Fitzgerald, K.J., and Sabatini, D.M. (2006) Ablation in mice of the mTORC components raptor, rictor, or mLST8 reveals that mTORC2 is required for signaling to Akt-FOXO and PKCalpha, but not S6K1. *Dev. Cell*, **11**, 859–871.

56 Sollner-Webb, B. and Tower, J. (1986) Transcription of cloned eukaryotic ribosomal RNA genes. *Annu. Rev. Biochem.*, **55**, 801–830.

57 Gao, X. and Pan, D. (2001) *TSC1* and *TSC2* tumor suppressors antagonize insulin signaling in cell growth. *Genes Dev.*, **15**, 1383–1392.

58 Potter, C.J., Huang, H., and Xu, T. (2001) *Drosophila Tsc1* functions with *Tsc2* to antagonize insulin signaling in regulating cell growth, cell proliferation, and organ size. *Cell*, **105**, 357–368.

59 Tapon, N., Ito, N., Dickson, B.J., Treisman, J.E., and Hariharan, I.K. (2001) The *Drosophila* tuberous sclerosis complex gene homologs restrict cell growth and cell proliferation. *Cell*, **105**, 345–355.

60 Gao, X., Zhang, Y., Arrazola, P., Hino, O., Kobayashi, T., Yeung, R.S., Ru, B., and Pan, D. (2002) Tsc tumour suppressor proteins antagonize amino acid-TOR signalling. *Nat. Cell Biol.*, **4**, 699–704.

61 Radimerski, T., Montagne, J., Hemmings-Mieszczak, M., and Thomas, G. (2002) Lethality of *Drosophila* lacking TSC tumor suppressor function rescued by reducing dS6K signaling. *Genes Dev.*, **16**, 2627–2632.

62 Manning, B.D., Tee, A.R., Logsdon, M.N., Blenis, J., and Cantley, L.C. (2002) Identification of the tuberous sclerosis complex-2 tumor suppressor gene product tuberin as a target of the phosphoinositide 3-kinase/Akt pathway. *Mol. Cell*, **10**, 151–162.

63 Chung, J., Grammer, T.C., Lemon, K.P., Kazlauskas, A., and Blenis, J. (1994) PDGF- and insulin-dependent pp70S6k activation mediated by phosphatidylinositol-3-OH kinase. *Nature*, **370**, 71–75.

64 Weng, Q., Andrabi, K., Kozlowski, M.T., Grove, J.R., and Avruch, J. (1995)

Multiple independent inputs are required for activation of the p70 S6 kinase. *Mol. Cell. Biol.*, **15**, 2333–2340.

65 Burgering, B.M.T. and Coffer, P.J. (1995) Protein kinase B (c-Akt) in phosphatidylinositol-3-OH kinase signal transduction. *Nature*, **376**, 599–602.

66 Dan, H.C., Sun, M., Yang, L., Sui, R.M., Yeung, R.S., Halley, D.J.J., Nicosia, S.V., Pledger, W.J., and Cheng, J.Q. (2002) PI3K/Akt pathway regulates TSC tumor suppressor complex by phosphorylation of tuberin. *J. Biol. Chem.*, **277**, 35364–35370.

67 Inoki, K., Li, Y., Zhu, T., Wu, J., and Guan, K.L. (2002) TSC2 is phosphorylated and inhibited by Akt and suppresses mTOR signalling. *Nat. Cell Biol.*, **4**, 648–657.

68 Potter, C.J., Pedraza, L.G., and Xu, T. (2002) Akt regulates growth by directly phosphorylating Tsc2. *Nat. Cell Biol.*, **4**, 658–665.

69 Goncharova, E.A., Goncharov, D.A., Eszterhas, A., Hunter, D.S., Glassberg, M.K., Yeung, R.S., Walker, C.L., Noonan, D., Kwiatkowski, D.J., Chou, M.M., Panettieri, R.A., Jr., and Krymskaya, V.P. (2002) Tuberin regulates p70 S6 kinase activation and ribosomal protein S6 phosphorylation: a role for the *TSC2* tumor suppressor gene in pulmonary lymphangioleiomyomatosis (LAM). *J. Biol. Chem.*, **277**, 30958–30967.

70 Jaeschke, A., Hartkamp, J., Saitoh, M., Roworth, W., Nobukuni, T., Hodges, A., Sampson, J., Thomas, G., and Lamb, R. (2002) Tuberous sclerosis complex tumor suppressor-mediated S6 kinase inhibition by phosphatidylinositide-3-OH kinase is mTOR independent. *J. Cell Biol.*, **159**, 217–224.

71 Tee, A.R., Fingar, D.C., Manning, B.D., Kwiatkowski, D.J., Cantley, L.C., and Blenis, J. (2002) Tuberous sclerosis complex-1 and -2 gene products function together to inhibit mammalian target of rapamycin (mTOR)-mediated downstream signaling. *Proc. Natl. Acad. Sci. USA*, **99**, 13571–13576.

72 Kwiatkowski, D.J., Zhang, H., Bandura, J.L., Heiberger, K.M., Glogauer, M., el-Hashemite, N., and Onda, H. (2002) A mouse model of TSC1 reveals sex-dependent lethality from liver hemangiomas, and up-regulation of p70S6 kinase activity in TSC1 null cells. *Hum. Mol. Genet.*, **11**, 525–534.

73 Manning, B.D. and Cantley, L.C. (2003) Rheb fills a GAP between TSC and TOR. *Trends Biochem. Sci.*, **28**, 573–576.

74 Yu, Y., Li, S., Xu, X., Li, Y., Guan, K., Arnold, E., and Ding, J. (2005) Structural basis for the unique biological function of small GTPase RHEB. *J. Biol. Chem.*, **280**, 17093–17100.

75 Saucedo, L.J., Gao, X., Chiarelli, D.A., Li, L., Pan, D., and Edgar, B.A. (2003) Rheb promotes cell growth as a component of the insulin/TOR signaling network. *Nat. Cell Biol.*, **5**, 566–571.

76 Stocker, H., Radimerski, T., Schindelholz, B., Wittwer, F., Belawat, P., Daram, P., Breuer, S., Thomas, G., and Hafen, E. (2003) Rheb is an essential regulator of S6K in controlling cell growth in *Drosophila*. *Nat. Cell Biol.*, **5**, 559–566.

77 Patel, P.H., Thapar, N., Guo, L., Martinez, M., Maris, J., Gau, C.L., Lengyel, J.A., and Tamanoi, F. (2003) Drosophila Rheb GTPase is required for cell cycle progression and cell growth. *J. Cell Sci.*, **116**, 3601–3610.

78 Zhang, Y., Gao, X., Saucedo, L.J., Ru, B., Edgar, B.A., and Pan, D. (2003) Rheb is a direct target of the tuberous sclerosis tumour suppressor proteins. *Nat. Cell Biol.*, **5**, 578–581.

79 Maheshwar, M.M., Cheadle, J.P., Jones, A.C., Myring, J., Fryer, A.E., Harris, P.C., and Sampson, J.R. (1997) The GAP-related domain of tuberin, the product of the TSC2 gene, is a target for missense mutations in tuberous sclerosis. *Hum. Mol. Genet.*, **6**, 1991–1996.

80 Wienecke, R., Konig, A., and DeClue, J.E. (1995) Identification of tuberin, the tuberous sclerosis-2 product. Tuberin possesses specific Rap1GAP activity. *J. Biol. Chem.*, **270**, 16409–16414.

81 Xiao, G.H., Shoarinejad, F., Jin, F., Golemis, E.A., and Yeung, R.S. (1997) The tuberous sclerosis 2 gene product,

tuberin, functions as a Rab5 GTPase activating protein (GAP) in modulating endocytosis. *J. Biol. Chem.*, **272**, 6097–6100.

82 Inoki, K., Li, Y., Xu, T., and Guan, K.-L. (2003) Rheb GTPase is a direct target of TSC2 GAP activity and regulates mTOR signaling. *Genes Dev.*, **17**, 1829–1834.

83 Tee, A.R., Manning, B.D., Roux, P.P., Cantely, L.C., and Blenis, J. (2003) Tuberous sclerosis complex gene products, tuberin and hamartin, control mTOR signaling by acting as a GTPase-activating protein complex toward Rheb. *Curr. Biol.*, **13**, 1259–1268.

84 Castro, A.F., Rebhun, J.F., Clark, G.J., and Quilliam, L.A. (2003) Rheb binds tuberous sclerosis complex 2 (TSC2) and promotes S6 kinase activation in a rapamycin- and farnesylation-dependent manner. *J. Biol. Chem.*, **278**, 32493–32496.

85 Garami, A., Zwartkruis, F.J.T., Nobukuni, T., Joaquin, M., Roccio, M., Stocker, H., Kozma, S.C., Hafen, E., Bos, J.L., and Thomas, G. (2003) Insulin activation of Rheb, a mediator of mTOR/S6K/4E-BP signaling, is inhibited by TSC1 and 2. *Mol. Cell*, **11**, 1457–1466.

86 Hsu, Y.C., Chern, J.J., Cai, Y., Liu, M., and Choi, K.W. (2007) Drosophila TCTP is essential for growth and proliferation through regulation of dRheb GTPase. *Nature*, **445**, 785–788.

87 Long, X., Lin, Y., Ortiz-Vega, S., Yonezawa, K., and Avruch, J. (2005) Rheb binds and regulates the mTOR kinase. *Curr. Biol.*, **15**, 702–713.

88 Tee, A.R., Blenis, J., and Proud, C.G. (2005) Analysis of mTOR signaling by the small G-proteins, Rheb and RhebL1. *FEBS Lett.*, **579**, 4763–4768.

89 Bai, X., Ma, D., Liu, A., Shen, X., Wang, Q.J., Liu, Y., and Jiang, Y. (2007) Rheb activates mTOR by antagonizing its endogenous inhibitor, FKBP38. *Science*, **318**, 977–980.

90 Astrinidis, A., Senapedis, W., Coleman, T.R., and Henske, E.P. (2003) Cell cycle-regulated phosphorylation of hamartin, the product of the tuberous sclerosis complex 1 gene, by cyclin-dependent kinase 1/cyclin B. *J. Biol. Chem.*, **278**, 51372–51379.

91 Ballif, B.A., Roux, P.P., Gerber, S.A., MacKeigan, J.P., Blenis, J., and Gygi, S.P. (2005) Quantitative phosphorylation profiling of the ERK/p90 ribosomal S6 kinase-signaling cassette and its targets, the tuberous sclerosis tumor suppressors. *Proc. Natl. Acad. Sci. USA*, **102**, 667–672.

92 Inoki, K., Ouyang, H., Zhu, T., Lindvall, C., Wang, Y., Zhang, X., Yang, Q., Bennett, C., Harada, Y., Stankunas, K., Wang, C.Y., He, X., MacDougald, O.A., You, M., Williams, B.O., and Guan, K.L. (2006) TSC2 integrates Wnt and energy signals via a coordinated phosphorylation by AMPK and GSK3 to regulate cell growth. *Cell*, **126**, 955–968.

93 Inoki, K., Zhu, T., and Guan, K.L. (2003) TSC2 mediates cellular energy response to control cell growth and survival. *Cell*, **115**, 577–590.

94 Lee, D.F., Kuo, H.P., Chen, C.T., Hsu, J.M., Chou, C.K., Wei, Y., Sun, H.L., Li, L.Y., Ping, B., Huang, W.C., He, X., Hung, J.Y., Lai, C.C., Ding, Q., Su, J.L., Yang, J.Y., Sahin, A.A., Hortobagyi, G.N., Tsai, F.J., Tsai, C.H., and Hung, M.C. (2007) IKK beta suppression of TSC1 links inflammation and tumor angiogenesis via the mTOR pathway. *Cell*, **130**, 440–455.

95 Li, Y., Inoki, K., Vacratsis, P., and Guan, K.L. (2003) The p38 and MK2 kinase cascade phosphorylates tuberin, the tuberous sclerosis 2 gene product, and enhances its interaction with 14-3-3. *J. Biol. Chem.*, **278**, 13663–13671.

96 Roux, P.P., Ballif, B.A., Anjum, R., Gygi, S.P., and Blenis, J. (2004) Tumor-promoting phorbol esters and activated Ras inactivate the tuberous sclerosis tumor suppressor complex via p90 ribosomal S6 kinase. *Proc. Natl. Acad. Sci. USA*, **101**, 13489–13494.

97 Shaw, R.J., Bardeesy, N., Manning, B.D., Lopez, L., Kosmatka, M., DePinho, R.A., and Cantley, L.C. (2004) The LKB1 tumor suppressor negatively regulates mTOR signaling. *Cancer Cell*, **6**, 91–99.

98 Huang, J. and Manning, B.D. (2008) The TSC1–TSC2 complex: a molecular

switchboard controlling cell growth. *Biochem. J.*, **412**, 179–190.
99 Cai, S.L., Tee, A.R., Short, J.D., Bergeron, J.M., Kim, J., Shen, J., Guo, R., Johnson, C.L., Kiguchi, K., and Walker, C.L. (2006) Activity of TSC2 is inhibited by AKT-mediated phosphorylation and membrane partitioning. *J. Cell Biol.*, **173**, 279–289.
100 Deyoung, M.P., Horak, P., Sofer, A., Sgroi, D., and Ellisen, L.W. (2008) Hypoxia regulates TSC1/2 mTOR signaling and tumor suppression through REDD1-mediated 14–3–3 shuttling. *Genes Dev.*, **22**, 239–251.
101 Alessi, D.R., Caudwell, F.B., Andejelkovic, M., Hemmings, B.A., and Cohen, P. (1996) Molecular basis for the substrate specificity of protein kinase B; comparison with MAPKAP kinase-1 and p70 S6 kinase. *FEBS Lett.*, **399**, 333–338.
102 Tee, A.R., Anjum, R., and Blenis, J. (2003) Inactivation of the tuberous sclerosis complex-1 and -2 gene products occurs by phosphoinositide 3-kinase/Akt-dependent and -independent phosphorylation of tuberin. *J. Biol. Chem.*, **278**, 37288–37296.
103 Ma, L., Chen, Z., Erdjument-Bromage, H., Tempst, P., and Pandolfi, P.P. (2005) Phosphorylation and functional inactivation of TSC2 by Erk implications for tuberous sclerosis and cancer pathogenesis. *Cell*, **121**, 179–193.
104 Hardie, D.G. (2005) New roles for the LKB1 → AMPK pathway. *Curr. Opin. Cell Biol.*, **17**, 167–173.
105 Gwinn, D.M., Shackelford, D.B., Egan, D.F., Mihaylova, M.M., Mery, A., Vasquez, D.S., Turk, B.E., and Shaw, R.J. (2008) AMPK phosphorylation of raptor mediates a metabolic checkpoint. *Mol. Cell*, **30**, 214–226.
106 Liu, L., Cash, T.P., Jones, R.G., Keith, B., Thompson, C.B., and Simon, M.C. (2006) Hypoxia-induced energy stress regulates mRNA translation and cell growth. *Mol. Cell*, **21**, 521–531.
107 Reiling, J.H. and Hafen, E. (2004) The hypoxia-induced paralogs Scylla and Charybdis inhibit growth by down-regulating S6K activity upstream of TSC in Drosophila. *Genes Dev.*, **18**, 2879–2892.
108 Brugarolas, J., Lei, K., Hurley, R.L., Manning, B.D., Reiling, J.H., Hafen, E., Witters, L.A., Ellisen, L.W., and Kaelin, W.G., Jr. (2004) Regulation of mTOR function in response to hypoxia by REDD1 and the TSC1/TSC2 tumor suppressor complex. *Genes Dev.*, **18**, 2893–2904.
109 Hara, K., Yonezawa, K., Weng, Q.P., Kozlowski, M.T., Belham, C., and Avruch, J. (1998) Amino acid sufficiency and mTOR regulate p70 S6 kinase and eIF-4E BP1 through a common effector mechanism. *J. Biol. Chem.*, **273**, 14484–14494.
110 Smith, E.M., Finn, S.G., Tee, A.R., Browne, G.J., and Proud, C.G. (2005) The tuberous sclerosis protein TSC2 is not required for the regulation of the mammalian target of rapamycin by amino acids and certain cellular stresses. *J. Biol. Chem.*, **280**, 18717–18727.
111 Long, X., Ortiz-Vega, S., Lin, Y., and Avruch, J. (2005) Rheb binding to mTOR is regulated by amino acid sufficiency. *J. Biol. Chem.*, **280**, 23433–23436.
112 Nobukuni, T., Joaquin, M., Roccio, M., Dann, S.G., Kim, S.Y., Gulati, P., Byfield, M.P., Backer, J.M., Natt, F., Bos, J.L., Zwartkruis, F.J., and Thomas, G. (2005) Amino acids mediate mTOR/raptor signaling through activation of class 3 phosphatidylinositol 3OH-kinase. *Proc. Natl. Acad. Sci. USA*, **102**, 14238–14243.
113 Roccio, M., Bos, J.L., and Zwartkruis, F.J. (2006) Regulation of the small GTPase Rheb by amino acids. *Oncogene*, **25**, 657–664.
114 Bilanges, B., Argonza-Barrett, R., Kolesnichenko, M., Skinner, C., Nair, M., Chen, M., and Stokoe, D. (2007) Tuberous sclerosis complex proteins 1 and 2 control serum-dependent translation in a TOP-dependent and -independent manner. *Mol. Cell. Biol.*, **27**, 5746–5764.
115 Thomas, G.V., Tran, C., Mellinghoff, I.K., Welsbie, D.S., Chan, E., Fueger, B., Czernin, J., and Sawyers, C.L. (2006) Hypoxia-inducible factor determines sensitivity to inhibitors of mTOR in kidney cancer. *Nat. Med.*, **12**, 122–127.
116 Land, S.C. and Tee, A.R. (2007) Hypoxia-inducible factor 1alpha is regulated by the

mammalian target of rapamycin (mTOR) via an mTOR signaling motif. *J. Biol. Chem.*, **282**, 20534–20543.

117 Brugarolas, J.B., Vazquez, F., Reddy, A., Sellers, W.R., and Kaelin, W.G., Jr. (2003) TSC2 regulates VEGF through mTOR-dependent and -independent pathways. *Cancer Cell*, **4**, 147–158.

118 Ozcan, U., Ozcan, L., Yilmaz, E., Duvel, K., Sahin, M., Manning, B.D., and Hotamisligil, G.S. (2008) Loss of the tuberous sclerosis complex tumor suppressors triggers the unfolded protein response to regulate insulin signaling and apoptosis. *Mol. Cell*, **29**, 541–551.

119 Zhang, H.H., Lipovsky, A.I., Dibble, C.C., Sahin, M., and Manning, B.D. (2006) S6K1 regulates GSK3 under conditions of mTOR-dependent feedback inhibition of Akt. *Mol. Cell*, **24**, 185–197.

120 Manning, B.D. (2004) Balancing Akt with S6K: implications for both metabolic diseases and tumorigenesis. *J. Cell Biol.*, **167**, 399–403.

121 Harrington, L.S., Findlay, G.M., Gray, A., Tolkacheva, T., Wigfield, S., Rebholz, H., Barnett, J., Leslie, N.R., Cheng, S., Shepherd, P.R., Gout, I., Downes, C.P., and Lamb, R.F. (2004) The TSC1-2 tumor suppressor controls insulin-PI3K signaling via regulation of IRS proteins. *J. Cell Biol.*, **166**, 213–223.

122 Shah, O.J., Wang, Z., and Hunter, T. (2004) Inappropriate activation of the TSC/Rheb/mTOR/S6K cassette induces IRS1/2 depletion, insulin resistance, and cell survival defects. *Curr. Biol.*, **14**, 1650–1656.

123 Shah, O.J. and Hunter, T. (2006) Turnover of the active fraction of IRS1 involves raptor-mTOR- and S6K1-dependent serine phosphorylation in cell culture models of tuberous sclerosis. *Mol. Cell. Biol.*, **26**, 6425–6434.

124 Tzatsos, A. and Kandror, K.V. (2006) Nutrients suppress phosphatidylinositol 3-kinase/Akt signaling via raptor-dependent mTOR-mediated insulin receptor substrate 1 phosphorylation. *Mol. Cell. Biol.*, **26**, 63–76.

125 Tremblay, F., Brule, S., Um, S.H., Li, Y., Masuda, K., Roden, M., Sun, X.J., Krebs, M., Polakiewicz, R.D., Thomas, G., and Marette, A. (2007) Identification of IRS-1 Ser-1101 as a target of S6K1 in nutrient- and obesity-induced insulin resistance. *Proc. Natl. Acad. Sci. USA*, **104**, 14056–14061.

126 Ozcan, U., Cao, Q., Yilmaz, E., Lee, A.H., Iwakoshi, N.N., Ozdelen, E., Tuncman, G., Gorgun, C., Glimcher, L.H., and Hotamisligil, G.S. (2004) Endoplasmic reticulum stress links obesity, insulin action, and type 2 diabetes. *Science*, **306**, 457–461.

127 Zhang, H., Cicchetti, G., Onda, H., Koon, H.B., Asrican, K., Bajraszewski, N., Vazquez, F., Carpenter, C.L., and Kwiatkowski, D.J. (2003) Loss of Tsc1/Tsc2 activates mTOR and disrupts PI3K–Akt signaling through downregulation of PDGFR. *J. Clin. Invest.*, **112**, 1223–1233.

128 Zhang, H., Bajraszewski, N., Wu, E., Wang, H., Moseman, A.P., Dabora, S.L., Griffin, J.D., and Kwiatkowski, D.J. (2007) PDGFRs are critical for PI3K/Akt activation and negatively regulated by mTOR. *J. Clin. Invest.*, **117**, 730–738.

129 Huang, J., Dibble, C.C., Matsuzaki, M., and Manning, B.D. (2008) The TSC1–TSC2 complex is required for proper activation of mTOR complex 2. *Mol. Cell. Biol.*, **28**, 4104–4115.

130 Crino, P.B. (2004) Molecular pathogenesis of tuber formation in tuberous sclerosis complex. *J. Child Neurol.*, **19**, 716–725.

131 El-Hashemite, N., Walker, V., Zhang, H., and Kwiatkowski, D.J. (2003) Loss of Tsc1 or Tsc2 induces vascular endothelial growth factor production through mammalian target of rapamycin. *Cancer Res.*, **63**, 5173–5177.

132 El-Hashemite, N., Zhang, H., Henske, E.P., and Kwiatkowski, D.J. (2003) Mutation in TSC2 and activation of mammalian target of rapamycin signalling pathway in renal angiomyolipoma. *Lancet*, **361**, 1348–1349.

133 Kenerson, H., Folpe, A.L., Takayama, T.K., and Yeung, R.S. (2007) Activation of the mTOR pathway in

sporadic angiomyolipomas and other perivascular epithelioid cell neoplasms. *Hum. Pathol.*, **38**, 1361–1371.

134 Kenerson, H.L., Aicher, L.D., True, L.D., and Yeung, R.S. (2002) Activated mammalian target of rapamycin pathway in the pathogenesis of tuberous sclerosis complex renal tumors. *Cancer Res.*, **62**, 5645–5650.

135 Manning, B.D., Logsdon, M.N., Lipovsky, A.I., Abbott, D., Kwiatkowski, D.J., and Cantley, L.C. (2005) Feedback inhibition of Akt signaling limits the growth of tumors lacking Tsc2. *Genes Dev.*, **19**, 1773–1778.

136 Gera, J.F., Mellinghoff, I.K., Shi, Y., Rettig, M.B., Tran, C., Hsu, J.H., Sawyers, C.L., and Lichtenstein, A.K. (2004) AKT activity determines sensitivity to mammalian target of rapamycin (mTOR) inhibitors by regulating cyclin D1 and c-myc expression. *J. Biol. Chem.*, **279**, 2737–2746.

137 Muise-Helmericks, R.C., Grimes, H.L., Bellacosa, A., Malstrom, S.E., Tsichlis, P.N., and Rosen, N. (1998) Cyclin D expression is controlled post-transcriptionally via a phosphatidylinositol 3-kinase/Akt-dependent pathway. *J. Biol. Chem.*, **273**, 29864–29872.

138 Nelsen, C.J., Rickheim, D.G., Tucker, M.M., Hansen, L.K., and Albrecht, J.H. (2003) Evidence that cyclin D1 mediates both growth and proliferation downstream of TOR in hepatocytes. *J. Biol. Chem.*, **278**, 3656–3663.

139 Manning, B.D. and Cantley, L.C. (2007) AKT/PKB signaling: navigating downstream. *Cell*, **129**, 1261–1274.

140 Jiang, H., Guo, W., Liang, X., and Rao, Y. (2005) Both the establishment and the maintenance of neuronal polarity require active mechanisms: critical roles of GSK-3beta and its upstream regulators. *Cell*, **120**, 123–135.

141 Yoshimura, T., Kawano, Y., Arimura, N., Kawabata, S., Kikuchi, A., and Kaibuchi, K. (2005) GSK-3beta regulates phosphorylation of CRMP-2 and neuronal polarity. *Cell*, **120**, 137–149.

142 Khan, A., Pepio, A.M., and Sossin, W.S. (2001) Serotonin activates S6 kinase in a rapamycin-sensitive manner in Aplysia synaptosomes. *J. Neurosci.*, **21**, 382–391.

143 Banko, J.L., Hou, L., Poulin, F., Sonenberg, N., and Klann, E. (2006) Regulation of eukaryotic initiation factor 4E by converging signaling pathways during metabotropic glutamate receptor-dependent long-term depression. *J. Neurosci.*, **26**, 2167–2173.

144 Banko, J.L., Poulin, F., Hou, L., DeMaria, C.T., Sonenberg, N., and Klann, E. (2005) The translation repressor 4E-BP2 is critical for eIF4F complex formation, synaptic plasticity, and memory in the hippocampus. *J. Neurosci.*, **25**, 9581–9590.

145 Sutton, M.A. and Schuman, E.M. (2006) Dendritic protein synthesis, synaptic plasticity, and memory. *Cell*, **127**, 49–58.

146 Prather, P. and de Vries, P.J. (2004) Behavioral and cognitive aspects of tuberous sclerosis complex. *J. Child Neurol.*, **19**, 666–674.

147 Jope, R.S. and Johnson, G.V. (2004) The glamour and gloom of glycogen synthase kinase-3. *Trends Biochem. Sci.*, **29**, 95–102.

148 Streets, A.J., Moon, D.J., Kane, M.E., Obara, T., and Ong, A.C. (2006) Identification of an N-terminal glycogen synthase kinase 3 phosphorylation site which regulates the functional localization of polycystin-2 *in vivo* and *in vitro*. *Hum. Mol. Genet.*, **15**, 1465–1473.

149 Loberg, R.D., Vesely, E., and Brosius, F.C., 3rd. (2002) Enhanced glycogen synthase kinase-3beta activity mediates hypoxia-induced apoptosis of vascular smooth muscle cells and is prevented by glucose transport and metabolism. *J. Biol. Chem.*, **277**, 41667–41673.

150 Bjornsson, J., Short, M.P., Kwiatkowski, D.J., and Henske, E.P. (1996) Tuberous sclerosis-associated renal cell carcinoma. Clinical, pathological, and genetic features. *Am. J. Pathol.*, **149**, 1201–1208.

151 Kim, W.Y. and Kaelin, W.G. (2004) Role of VHL gene mutation in human cancer. *J. Clin. Oncol.*, **22**, 4991–5004.

152 Rehmann, H., Bruning, M., Berghaus, C., Schwarten, M., Kohler, K., Stocker, H., Stoll, R., Zwartkruis, F.J., and

Wittinghofer, A. (2008) Biochemical characterisation of TCTP questions its function as a guanine nucleotide exchange factor for Rheb. *FEBS Lett.*, **582**, 3005–3010.

153 Wang, X., Fonseca, B.D., Tang, H., Liu, R., Elia, A., Clemens, M.J., Bommer, U.A., and Proud, C.G. (2008) Re-evaluating the roles of proposed modulators of mammalian target of rapamycin complex 1 (mTORC1) signaling. *J. Biol. Chem.*, **283**, 30482–30492.

154 Sancak, Y., Peterson, T.R., Shaul, Y.D., Lindquist, R.A., Thoreen, C.C., Bar-Peled, L., and Sabatini, D.M. (2008) The Rag GTPases bind raptor and mediate amino acid signaling to mTORC1. *Science*, **320**, 1496–1501.

155 Kim, E., Goraksha-Hicks, P., Li, L., Neufeld, T.P., and Guan, K.L. (2008) Regulation of TORC1 by Rag GTPases in nutrient response. *Nat. Cell Biol.*, **10**, 935–945.

156 Nobukuni, T., Joaquin, M., Roccio, M., Dann, S.G., Kim, S.Y., Gulati, P., Byfield, M.P., Backer, J.M., Natt, F., Bos, J.L., Zwartkruis, F.J., and Thomas, G. (2005) Amino acids mediate mTOR/raptor signaling through activation of class 3 phosphatidylinositol 3OH-kinase. *Proc. Natl. Acad. Sci. USA*, **102**, 14238–14243.

157 Gulati, P., Gaspers, L.D., Dann, S.G., Joaquin, M., Nobukuni, T., Natt, F., Kozma, S.C., Thomas, A.P., and Thomas, G. (2008) Amino acids activate mTOR complex 1 via Ca2 + /CaM signaling to hVps34. *Cell Metab.*, **7**, 456–465.

158 Findlay, G.M., Yan, L., Procter, J., Mieulet, V., and Lamb, R.F. (2007) A MAP4 kinase related to Ste20 is a nutrient-sensitive regulator of mTOR signalling. *Biochem. J.*, **403**, 13–20.

159 Meikle, L., Pollizzi, K., Egnor, A., Kramvis, I., Lane, H., Sahin, M., and Kwiatkowski, D.J. (2008) Response of a neuronal model of tuberous sclerosis to mammalian target of rapamycin (mTOR) inhibitors: effects on mTORC1 and Akt signaling lead to improved survival and function. *J. Neurosci.*, **28**, 5422–5432.

160 Lee, L., Sudentas, P., Donohue, B., Asrican, K., Worku, A., Walker, V., Sun, Y., Schmidt, K., Albert, M.S., El-Hashemite, N., Lader, A.S., Onda, H., Zhang, H., Kwiatkowski, D.J., and Dabora, S.L. (2005) Efficacy of a rapamycin analog (CCI-779) and IFN-gamma in tuberous sclerosis mouse models. *Genes Chromosomes Cancer*, **42**, 213–227.

7
Rat and Mouse Models of Tuberous Sclerosis

David J. Kwiatkowski

7.1
Introduction

Animal models are well known as invaluable tools in the understanding of human disease. They provide the ability to investigate disease development in ways that are not possible in humans. For example, one can use an animal model to very precisely evaluate the multiple molecular steps that occur in tumor development, in a way that is difficult or impossible to do in patients. Animals can be evaluated by both noninvasive and humane means, as well as at specified dates for autopsy, in great detail. Tumors and other lesions can be harvested and prepared for pathologic studies in pristine condition. Animal models have particular benefit in the study of brain diseases, since in humans very limited direct sampling of the brain can be performed prior to natural death. Mice of the appropriate genotype may be sacrificed at the time chosen by the scientist, with rapid and complete collection of all important tissues and fluids. Finally, animal models are ideal (and in many cases mandatory) for the testing of possible therapeutic interventions, of all kinds, the so-called preclinical testing.

Animal model studies in TSC have been pursued with considerable vigor. Early studies were accelerated by the fortuitous discovery of a spontaneous rat model of *Tsc2* mutation (the Eker rat, see below) and were focused on analysis of tumors and their pathology and development. More recent work, over the past 10 years, has focused on the development of mouse models of TSC, using genetic engineering. Mouse models have the advantage that there is a large and ever-expanding collection of other mutant alleles available, which can be used for genetic interaction analysis *in vivo*. In addition, conditional alleles in the mouse permit targeted loss of the gene in tissues of interest.

This chapter reviews the various rodent models of TSC that have been developed (Tables 7.1, 7.2). For each model considered, we attempt to take a critical perspective in assessing the value of the model: (1) Does the rodent model reproduce the pathology and pathogenesis that occurs in human TSC patients? (2) What have we learned from the model? (3) What are the limitations of the model, and how can it be

Tuberous Sclerosis Complex: Genes, Clinical Features, and Therapeutics
Edited by David J. Kwiatkowski, Vicky Holets Whittemore, and Elizabeth A. Thiele

ISBN: 978-3-527-32201-5

improved? This chapter requires some consideration of the molecular and cellular functions of the TSC1 and TSC2 protein products and the pathogenesis of TSC in patients. However, since these topics are the primary subject of other chapters in this book, they will be discussed here in limited detail.

7.2 The Eker Rat

7.2.1 Historical Review: The Eker Rat: A Unique Spontaneous Mutation in Rat *Tsc2*

The Eker rat was first described as an autosomal dominant, hereditary model of predisposition to renal adenoma and carcinoma by Eker [1]. It is truly amazing to imagine how Eker managed to both discover and maintain these rats, since carrier status for the trait could be determined only at the time of sacrifice or by performing unilateral nephrectomy. Decades later, these animals came to the attention of Alfred Knudson, one of the premier cancer biologists of the twentieth century, who recognized their value and brought them to the Fox Chase Cancer Center in Philadelphia in the 1980s. He then led a series of pathologic and genetic studies, which culminated in the recognition that the underlying genetic defect was a spontaneous mutation in the *Tsc2* gene.

7.2.2 The Eker Rat *Tsc2* Model

In the Eker rat, genotype $Tsc2^{Ek/+}$, cystadenoma lesions in the kidney are the predominant pathologic lesion and vary in morphology from pure cysts to cysts with papillary projections, to solid adenomas [2], which can be seen as early as 4 months of age (Figure 7.1). A small minority of these tumors become malignant, with nuclear atypia, and expand to replace the entire kidney with occasional metastasis to the lungs, pancreas, and liver. Although there is strain variance in severity (see below), there is 100% penetrance of kidney involvement in Eker rats. The kidney tumors develop in the outer cortex and have variable staining characteristics even within a single cystadenoma [2]. Histological studies have identified proximal tubular and collecting duct epithelial cells as the cells of origin of the Eker solid and cystic adenomas, respectively [3]. Three-dimensional reconstruction studies demonstrate that these lesions grow through the kidney by extending along the tubule of origin with localized regions of cystic expansion with or without papillary growth [2]. Although these tumors occur in an organ that is commonly involved by angiomyolipoma in TSC patients, these cystadenomas are epithelial neoplasms and there is no pathological relationship between angiomyolipomas and cystadenomas. Renal cysts are present in a substantial fraction of TSC patients, though are most severe in those with combined deletion of both *TSC2* and *PKD1*. However, in patients without that

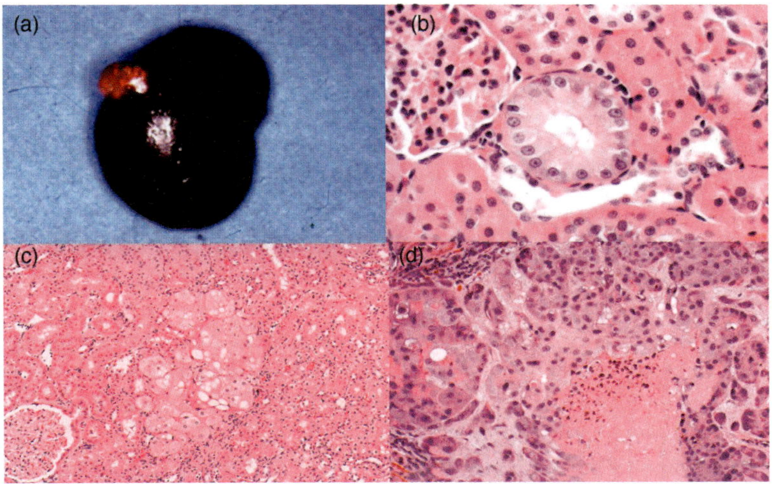

Figure 7.1 Eker rat kidney tumors. (a) Gross view of an Eker rat kidney with an exophytic renal tumor. (b–d) H&E stained pictures of kidney lesions from Eker rats. (b) Earliest stage of a tubule showing atypical hyperplasia. (c) Renal adenoma. (d) Renal carcinoma. Courtesy of Okio Hino.

combined deletion syndrome, they typically remain small and asymptomatic and do not progress with papillary growth. Thus, renal cysts in TSC patients have a distinct biologic behavior that contrasts with what is seen in the Eker rat.

Homozygous *Tsc2*$^{Ek/Ek}$ pups die at about embryonic day 11 (E11) to 12 and are smaller and less developed than littermates beginning at E9 [4]. Brain developmental defects are consistently seen in *Tsc2*$^{Ek/Ek}$ pups in the Long–Evans strain, and consist of dysraphia and forebrain papillary overgrowth of the neuroepithelium. However, these neurodevelopmental abnormalities are not seen in *Tsc2*$^{Ek/Ek}$ pups in the Fisher 344 strain, despite embryonic death in that strain at E11–E12 [4]. Thus, the cause of death in these embryos is uncertain.

Eker rats also develop pituitary adenomas (55% at 2 years), uterine leiomyomas and leiomyosarcomas (47–62% of females at 14 months–2 years), and splenic hemangiosarcomas (23–68% at 14 months–2 years) [5, 6]. They also develop brain hamartomas resembling human TSC subependymal nodules at very low frequency (see below) [7].

Genetic linkage studies in the Eker rat, followed by identification of *TSC2* in the syntenic region of linkage, led to the identification of the Eker mutation in *Tsc2* [8, 9]. It is an insertion of a 6.3 kb intracisternal A particle into *Tsc2* that likely occurred as a spontaneous transposon insertional mutagenesis event. The insertion disrupts codon 1272 of rat *Tsc2* and is predicted to lead to production of an aberrant, larger protein, which has never been detected [10], likely due to its poor stability and rapid clearance.

The Eker rat has provided strong evidence that the *Tsc2* gene functions in a tumor suppressor gene fashion, fitting the classic Knudson model. In that model, germline inactivation of *Tsc2* in the Eker allele is complemented by second-hit loss of the other, wild-type allele in tumors. This second-hit event is postulated to be the critical initiating event leading to tumor formation. This model is supported by analysis of lesions developing in the Eker rat, which have shown consistent loss of the wild-type (*Tsc2*$^{+}$) allele. Loss of heterozygosity (LOH), consistent with a second-hit deletion of *Tsc2*$^{+}$, is seen in 40–60% of renal adenomas, 36% of uterine leiomyomas, and 35% of pituitary adenomas [11, 12]. In contrast, 0% of splenic hemangiomas show this finding, possibly related to admixture of normal cell types [11]. Further analysis has shown that about one third of both spontaneous and ENU-induced renal adenomas in the Eker rat are due to small mutations in the *Tsc2* gene [13, 14].

Several carcinogens are capable of inducing renal tumors in rats not bearing the *Tsc2*^{Ek} allele. Analysis of such tumors in non-Eker rats has shown that inactivating point mutations in *Tsc2* occur, and in some cases mutations inactivating both *Tsc2* alleles were seen [15].

The ultimate proof of the two-hit model of pathogenesis was obtained by Hino and coworkers [16]. They generated a transgenic rat bearing an additional copy of the wild-type rat *Tsc2* gene and its upstream promoter element. The transgene completely compensated for the *Tsc2*^{Ek} allele in a dosage-dependent manner. These investigators then extended this study by analyzing the effects of expression of various truncates of Tsc2 on renal tumor development and embryonic survival. High-level expression of the GTPase activating protein (GAP) domain of Tsc2 alone caused a significant reduction in the both the size and number of renal tumors that developed, but did not rescue the embryonic lethality of *Tsc2*$^{Ek/Ek}$ pups. A minor C-terminal truncate (55-amino acids removed) Tsc2 transgene completely rescued renal tumor development and partially rescued embryonic lethality. Interestingly, they went on to show that transgenes separately encoding the N-terminus of Tsc2 (residues 1–1424) and the C-terminus GAP domain (residues 1425–1755) could synergistically completely rescue embryonic lethality of *Tsc2*$^{Ek/Ek}$ pups [17]. This suggests that these two Tsc2 truncates/domains interact with each other and with Tsc1 *in vivo* to achieve normal function and regulation.

Early-onset severe polycystic kidney disease has been seen in occasional Eker rats, and shown to be due to widespread loss of the wild-type *Tsc2* allele [18]. Analysis of the tissues and a derivative cell line suggests that this occurred through a mechanism of chromosomal nondisjunction, occurring early in development. This observation provides one potential mechanism by which relatively severe TSC manifestations might occur in a TSC patient.

Eker rats develop uterine leiomyomas at high frequency, and have been studied in detail by Walker *et al.* [19]. LOH for *Tsc2* and loss of Tsc2 expression are seen in most tumors, oophorectomy is highly effective at leiomyoma suppression, and pregnancy also reduces tumor incidence [20]. However, TSC patients do not appear to be at increased risk of uterine leiomyomas, though this should be studied in greater detail, and human uterine tumors show no direct evidence of involvement of the TSC genes [19]. Thus, although there is some relevance and interest to the human situation, it is modest.

7.2.3
Genetic Modifiers in the Eker Rat

The size of kidney tumors that develop in the Eker rat varies as a function of strain, even though the number of tumors does not vary significantly [21]. Using a backcross analysis between two strains that showed a substantial difference in tumor size, a quantitative trait locus, *Mot1*, influencing tumor size was localized to rat chromosome 3q [21]. In Eker rats treated with ENU to enhance the rate of tumor development, a major strain difference was seen in the number of lesions that developed [22]. Through backcross analysis, a modifier locus for this effect was mapped to rat chromosome 5. However, whether this locus influences tumor development *per se*, or *in vivo* pharmacokinetics of ENU uptake, metabolism, and clearance, is not known.

7.2.4
Pathway Studies in the Eker Rat and Rapamycin Treatment

All stages of the renal tumors developing in the Eker rat have been shown to express markers of mTORC1 activation, including phospho-S6(S235–236) and phospho-p70S6K(T389) [23], consistent with the critical role of Tsc2 in the regulation of rheb-GTP levels and mTORC1 activation, and in accordance with the two-hit model of tumor development. In addition, these expression features are eliminated and proliferation, as assessed by PCNA staining, is markedly reduced in response to short-term treatment with rapamycin [23]. Serial ultrasound imaging has also been used in the Eker rat model to track the size of kidney cystadenomas during treatment [24]. The majority of lesions showed a dramatic decrease in estimated tumor volume (46–98%), with histologic confirmation of "tumor scars," fibrotic lesions with few viable tumor cells, in the treated rats. However, there were also other tumors that seemed unaffected by rapamycin therapy [24]. Note as well that in this study, pituitary tumors were found to be the major cause of death. Rapamycin was also found to be highly effective in reversing morbidity thought to be due to the pituitary tumors. In addition, there was evidence of reduced mTORC1 activity and increased apoptosis in the pituitary tumors [24].

Blockade of TGF-beta signaling with a TGF-beta receptor kinase inhibitor, SB-525334, has also been tested in the Eker rat [25]. The compound was found to reduce uterine leiomyoma incidence and size. However, it also appeared to increase the size of the renal cystadenomas.

7.2.5
Brain and Neurologic Features of the Eker Rat

Although neurologic features are not apparent in Eker rats, 27 of 43 (63%) $Tsc2^{Ek/+}$ rats at 1.5–2 years of age had brain lesions (all <2 mm in diameter), consisting of mixtures of large and elongated cells that were found in both subependymal and subcortical regions [7, 26]. Intralesional calcification was also seen, similar to human subependymal nodules. The large cells stained with glial fibrillary acidic protein (GFAP) and other glial and astrocytic markers. However, the cells in these lesions did not show evidence of LOH of the wild-type *Tsc2* allele [7]. A single cortical tuber has also been identified in an Eker rat, overlying a subcortical hamartoma, which after laser capture microdissection also did not show evidence for loss of the wild-type *Tsc2* allele [27, 28].

Detailed studies of the cortex in the aged Eker rat have revealed the presence of rare giant cells [26]. These include large dysmorphic pyramid-like cells expressing NeuN, tuberin, and EAAC-1 in layers IV–VI and abnormal cytomegalic cells expressing GFAP, vimentin, and nestin in deep cortical layers or along the white matter.

Prenatal and neonatal irradiation (E18-P6, 400–750cGy whole body) of Eker rats has no obvious neurologic consequences, including seizure activity, as assessed with electrode monitoring [29, 30]. However, irradiation led to the appearance of many dysmorphic large neurons in the neocortex and giant GFAP + astrocyte-like cells in cortical and subcortical regions. The numbers of these cells increased significantly with age, being about three times higher at age 3 months versus 1 month, with little relationship to the timing or dose of radiation [30]. In addition, these abnormal cells expressed both Tsc2 and phospho-S6 [29, 30]. Despite these findings, evoked seizure thresholds did not differ as assessed in electrophysiological slice experiments [30].

The molecular mechanism by which the rare spontaneous giant cells, and those induced by radiation treatment, occur in the Eker rat is uncertain and of considerable interest. This has obvious importance in considering mechanisms of development of the giant cells that are seen in cortical tubers (see Chapter 8). The enhanced incidence with radiation strongly suggests a genetic mechanism, while the persistent expression of Tsc2 argues against this being a simple second-hit ablation of the wild-type allele of *Tsc2*.

Synaptic plasticity is altered in Eker rats [31]. Paired-pulse plasticity is enhanced in $Tsc2^{Ek/+}$ rats at the Schaffer collateral-CA1 synapse, while both long-term potentiation (LTP) and long-term depression are decreased at this same synapse [31]. In addition, young adult $Tsc2^{Ek/+}$ rats exhibit enhanced episodic-like memory and enhanced responses to chemically induced kindling, with higher expression of phospho-p42-MAPK in the hippocampus than in controls [32]. These studies were the first to provide *in vivo* evidence that haploinsufficiency due to a *Tsc2* gene mutation might have physiologic consequences in brain function, and possibly contribute to some of the clinical neurologic and developmental phenotypes seen in TSC patients [31].

7.3
TSC Models in the Mouse

7.3.1
Tsc2 Knockout Mice

Two different knockout (null, −) alleles of *Tsc2* were independently derived and reported in 1999 (Table 7.1) [33, 34]. The two different alleles were each made by targeting the second coding exon of *Tsc2*, though the precise methods somewhat differ. Both of these $Tsc2^{+/-}$ mice have essentially identical phenotypes. More recently, a third null allele of *Tsc2* has been reported, in which exons 2–4 are deleted, and which appears to have similar findings, though it has been studied in much less detail [35].

$Tsc2^{+/-}$ mice develop kidney tumors that are very similar to those of the Eker rat (Figure 7.3). Kidney tumors develop by 6–12 months of age, and grow progressively throughout the life of the mouse [33, 34]. The tumors are cystadenomas consisting of a spectrum from pure cysts to cysts with papillary projections, to solid adenomas. Major strain differences are seen in the number and size of these tumors (see further below). Renal carcinoma, characterized by nuclear atypia, massive growth, and metastatic disease, develops in 5–10% of mice by 18 months, indicating a very low rate (~1 in 1000) of malignant progression given the overall numbers of these tumors, suggesting that additional genetic or epigenetic events are required for malignant progression [34].

$Tsc2^{+/-}$ mice also develop liver hemangiomas, characterized mainly by endothelial and some smooth muscle proliferation with large vascular spaces (Figure 7.2) [33, 34]. These lesions are seen in about half of $Tsc2^{+/-}$ mice by 18 months of age and can rupture with intraperitoneal hemorrhage leading to death. Hemangiosarcomas develop on the tail, paws, or mouth region in about 5% of $Tsc2^{+/-}$ mice by age

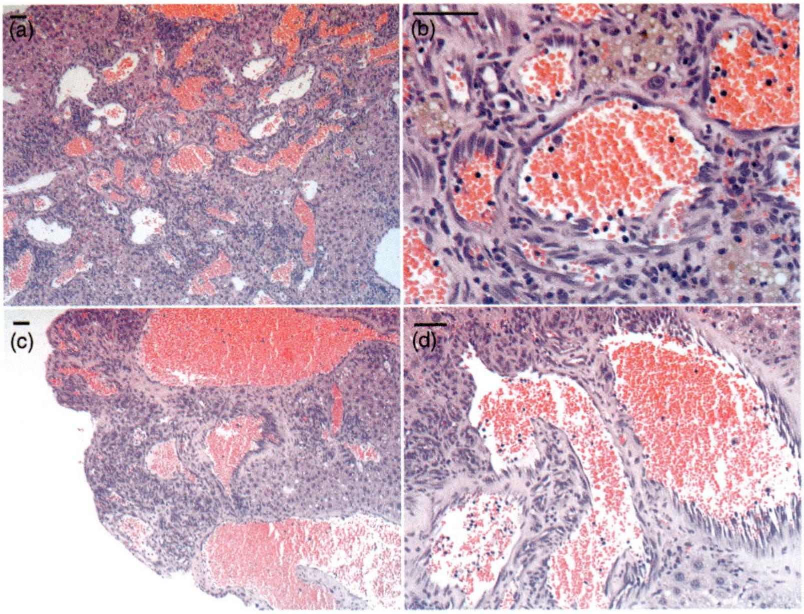

Figure 7.2 Tsc mouse liver hemangiomas. H&E stained pictures of liver hemangiomas from the $Tsc1^{+/-}$ (a and b) (14 months old) and $Tsc2^{+/-}$ mice (c and d) (12 months old) are shown. Note the dilated aberrant, disorganized vascular channels with prominent hyperplastic endothelium, containing red blood cells. In (c), some vessel lumens approach 1 mm in diameter, very large for the mouse. Scale bars are 50 μm.

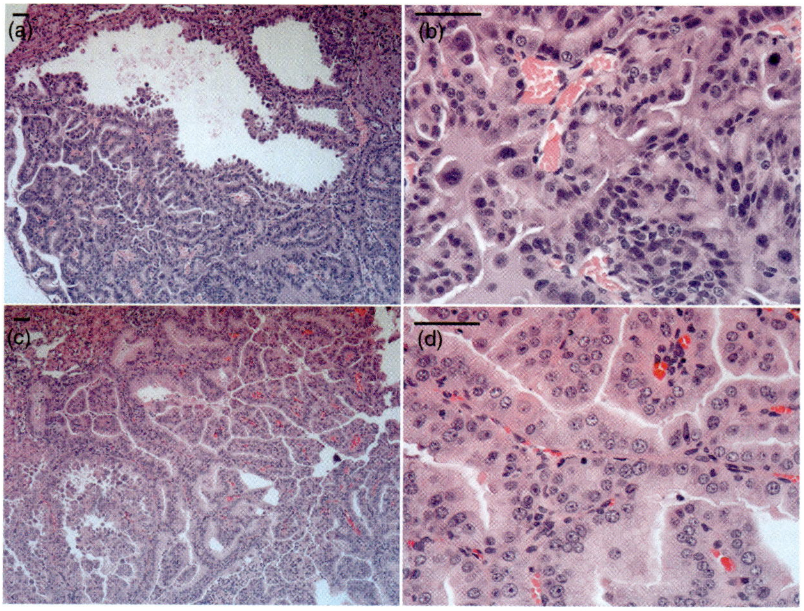

Figure 7.3 Tsc mouse kidney cystadenomas. H&E stained pictures of kidney cystadenomas from the *Tsc1*$^{+/-}$ (a and b) (14 months old) and *Tsc2*$^{+/-}$ mice (c and d) (12 months old) are shown. Note the prominent hyperplastic epithelial cells making up these tumors, with cell fragments in the lumen of some cystic regions. In (b), note the lack of organization and nuclear and cytoplasmic pleomorphism, consistent with malignancy. In all images, note the prominent vascularization. Scale bars are 50 μm.

of 12 months. These lesions, consisting of proliferative spindle cells and aberrant vascular channels, do not metastasize but are malignant by cytologic criteria and exhibit bone invasion. No brain lesions have been identified in *Tsc2*$^{+/-}$ mice.

Similar to findings in the Eker rat, major strain-dependent differences in tumor development are seen in *Tsc2*$^{+/-}$ mice [34] (Kwiatkowski *et al.*, unpublished observations). The most dramatic difference is seen in comparison of the A/J and 129Sv/Jae strains. Age-matched *Tsc2*$^{+/-}$ A/J strain mice have about 10 times as many kidney tumors and 20 times the extent of kidney tumor as *Tsc2*$^{+/-}$ 129/SvJae mice. In contrast, liver hemangiomas are much more prevalent in *Tsc2*$^{+/-}$ 129/SvJae mice than in any other six strains tested (Kwiatkowski *et al.* unpublished observations).

Histologic and immunohistochemical studies indicate that the cystadenomas arise in the cortical region of the kidney and have an expression profile consistent with origin from the interstitial cell of the cortical collecting duct [34]. The actin binding protein gelsolin, which is normally expressed in interstitial cells of the kidney, was found to be overexpressed in all forms of cystadenoma and is a sensitive marker of early tumor development.

Similar to the Eker rate, LOH is seen fairly consistently (24–50%) in both renal cystadenomas and liver hemangiomas, consistent with second-hit genomic loss of the wild-type *Tsc2* allele [33, 34].

Tsc2$^{-/-}$ embryos die at midgestation (E10.5–12.5) and are less developed than *Tsc2*$^{+/-}$ and wild-type littermates [33, 34]. Consistent with the generalized developmental delay, a proportion of E8–E9 embryos had open neural tubes. *Tsc2*$^{-/-}$ embryos are pale, edematous, and had pericardial effusions. Histologic sections demonstrated liver hypoplasia that was quite striking and appeared to be the primary cause of fetal demise, with secondary growth retardation and circulatory failure from anemia [34]. Cardiac hypertrophy was seen in some cases [33, 34].

VEGF levels are increased in both the kidney tumors and in the serum of *Tsc2*$^{+/-}$ mice and *Tsc2*$^{Eker/+}$ rats [36, 37]. VEGF levels are increased due to both mTOR-dependent and mTOR-independent pathways, and were associated with higher levels of each of HIF1α and HIF2α [36, 38].

7.3.2
Hypomorphic Alleles of Tsc2

Two hypomorphic alleles of *Tsc2* have been discribed. The first (*Tsc2*^{neo6}) was made in the course of construction of a conditional allele of *Tsc2*, in which a neocassette was inserted into the first intron of *Tsc2* [35]. *Tsc2*$^{neo/neo}$ embryos survive out to E17 in some cases. Only renal cysts, and no renal tumors, were seen in *Tsc2*$^{+/neo}$ mice at up to 20 months of age. Surprisingly, immunoblotting showed near-absence of Tsc2 protein in a *Tsc2*$^{neo/neo}$ embryo [35].

The second hypomorphic allele of *Tsc2*, in which there is deletion of exon 3, has been called "*del3*" [71]. *Tsc2*$^{del3/del3}$ embryos survive out to E13.5, longer than completely null Tsc2 embryos, and show liver hypoplasia, hemorrhage, and aberrant vascular structures. In a study carefully matched for strain, *Tsc2*$^{+/del3}$ mice develop kidney tumors at a rate 10–15-fold lower than that seen in *Tsc2*$^{+/-}$ mice. Analysis of both MEF cell lines and embryos from these mice revealed that there was both activation of mTORC1 with phosphorylation of S6 as well as reduced phosphorylation of Akt at the S473 site.

7.3.3
Tsc1 Knockout Mice

Two different knockout (null, −) alleles in *Tsc1* were developed and reported in 2001–2002 (Table 7.1) [39, 40]. One null allele was derived by insertion of a neocassette and deletion of exons 6–8 of *Tsc1* [39], while the other was generated by a loxP strategy that generated a conditional allele of *Tsc1* simultaneously and had deletion of exons 17 and 18 of *Tsc1* [40]. They have very similar phenotypes.

The tumors that occur in *Tsc1*$^{+/-}$ mice are entirely similar to those seen in *Tsc2*$^{+/-}$ mice (Figures 7.2 and 7.3). However, renal tumor development was somewhat less in *Tsc1*$^{+/-}$ mice than in age-matched *Tsc2*$^{+/-}$ mice [39, 40]. In one *Tsc1*$^{+/-}$ mouse line developed, there was a clear sex-dependent difference in the frequency and severity of liver hemangiomas (higher in females than males), and this resulted in reduced survival of female compared with male mice [40]. In addition, this sex-related difference in liver hemangiomas was confirmed in studies that examined the effects of estrogen and tamoxifen treatment on these mice. Estrogen enhanced liver hemangioma development in both male and female mice, while tamoxifen suppressed liver hemangioma development in females [41].

Analysis of one of these mouse models also provided the first *in vivo* evidence in a model system of the important role of the TSC1/TSC2 complex in regulating the state of mTORC1. Phospho-S6 levels were increased and phospho-AKT levels were reduced in mouse embryo fibroblasts that were *Tsc1*$^{-/-}$, and phospho-S6 levels were shown to be increased in kidney tumor lysates from the *Tsc1*$^{+/-}$ mice [40].

Similar to *Tsc2*, embryos that are *Tsc1*$^{-/-}$ die in the range of E10.5–13.5. For one of the *Tsc1* null alleles, survival seemed entirely equivalent to *Tsc2*$^{-/-}$ pups [39], while for the other, survival appeared to be extended by about one embryonic day [40].

A third *Tsc1* null allele was generated more recently by deletion of part of exon 6 and all of exons 7 and 8 [42]. The phenotype of these *Tsc1*$^{+/-}$ mice appeared somewhat different than the earlier two *Tsc1*$^{+/-}$ mice. First, in the C57BL/6 strain, but not the C3H or Balb/c strains, about one fourth of *Tsc1*$^{+/-}$ pups died between birth and weaning of uncertain causes. Second, there was a strikingly high incidence of renal carcinoma in *Tsc1*$^{+/-}$ BALB/c mice in comparison to two other strains, including some with sarcomatoid features and lung metastases (Figure 7.4) [42]. A comprehensive analysis of second-hit events in the kidney lesions of these *Tsc1*$^{+/-}$ mice demonstrated that half to two thirds of cystadenomas and renal cell carcinomas

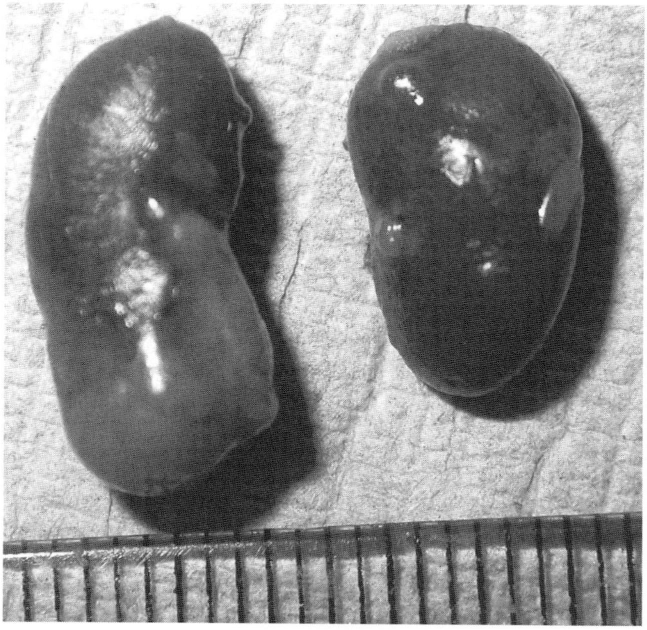

Figure 7.4 Tsc1$^{+/-}$ BALB/c mouse kidney tumor. A gross picture of kidneys removed from a *Tsc1*$^{+/-}$ mouse at 11 months of age [42]. Note large renal tumor in lower pole of kidney on left. Courtesy of Catherine Wilson.

showed LOH for the wild-type allele, and about half of the remainder have a point mutation in the wild-type *Tsc1* allele [43]. However, very few cysts showed LOH or a small mutation in the wild-type allele of *Tsc1*. The authors concluded that this indicated that cysts could develop from a haploinsufficiency mechanism [43]. However, it seems possible that the cysts with a wild-type *Tsc1* allele were sampled up- or downstream from a tubule that had been blocked/distorted by a cystadenoma [2].

In all, the relatively small differences seen in these three different *Tsc1*$^{+/-}$ mice suggest that the alleles have the same genetic and physiologic effects. The reported differences are likely due to either strain effects, or reflect the more detailed studies of certain aspects of phenotype in individual laboratories.

7.3.4
Mouse Studies: Interbreeding with Other Alleles

Both *Tsc1*$^{+/-}$ and *Tsc2*$^{+/-}$ mice have been interbred with other mutant alleles to explore signaling pathway interactions and generate additional interesting models.

Tsc2$^{+/-}$*Pten*$^{+/-}$ mice display reduced survival in comparison to *Pten*$^{+/-}$ mice, which in one lab was due to several causes: (1) enhanced inflammation including development of pelvic abscesses in 35% females, (2) cutaneous squamous cell carcinomas in 10%, and (3) carcinoma of the prostate in nearly all males by age 9 months(Figure 7.5) [44]. All of these pathologic features were markedly enhanced in the double hets, compared to *Pten*$^{+/-}$ mice. Surprisingly, the prostate carcinomas in the *Tsc2*$^{+/-}$*Pten*$^{+/-}$ mice showed retention of the wild-type allele of each of *Pten* and *Tsc2*, suggesting that they developed through a haploinsufficiency synergistic mechanism involving each gene. On the other hand, renal cystadenomas were not

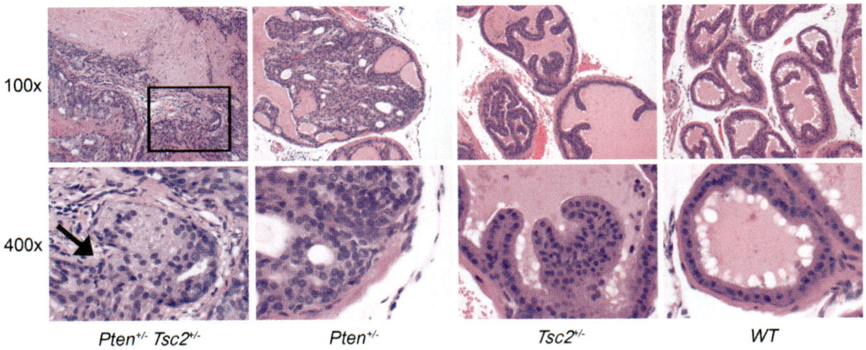

Figure 7.5 Prostate cancer development in *Tsc2*$^{+/-}$*Pten*$^{+/-}$ mice. Prostate cancer development in the antral lobe of a 9-month-old *Tsc2*$^{+/-}$*Pten*$^{+/-}$ mouse is shown, along with comparable sections from littermate mice of the same age with other genotypes. 100× and 400× indicate magnification. The boxed region is an area of prostate cancer invasion in the *Tsc2*$^{+/-}$*Pten*$^{+/-}$ mouse, which is indicated by an arrow. Both the wild-type (WT) and the *Tsc2*$^{+/-}$ mice show normal prostate epithelium, while the *Pten*$^{+/-}$ mouse shows prostate intraepithelial neoplasia (PIN). From Ref. [44], courtesy of Li Ma.

enhanced in number or growth in the *Tsc2*$^{+/-}$*Pten*$^{+/-}$ mice, compared to *Tsc2*$^{+/-}$ mice, and did show LOH for the wild-type *Tsc2* allele. This suggests that the heterozygosity for *Pten* had no effect on growth of the renal tumors. Signaling studies on MEFs from these mice indicated that there was a synergistic effect of haplo-loss of *Tsc2* and *Pten* on mTORC1 activation [44].

In a second study, strikingly different results were seen [45] for uncertain reasons. Survival was reduced in the *Tsc2*$^{+/-}$*Pten*$^{+/-}$ mice compared to *Pten*$^{+/-}$ mice. However, this was due to progressive lymphadenopathy in the *Tsc2*$^{+/-}$*Pten*$^{+/-}$ mice, more severe than in the *Pten*$^{+/-}$ mice. However, abscesses, squamous cell carcinoma, and prostate cancer were not seen at appreciable frequency in the double hets. *Tsc2*$^{+/-}$*Pten*$^{+/-}$ mice had much larger and progressive liver hemangioma than control *Tsc2*$^{+/-}$ mice, and this was shown to be due to reversal of the Akt inactivation that usually occurs in cells lacking Tsc2, due to haploinsufficiency of Pten expression (Figure 7.6) [45]

Interbreeding of the *Tsc1*$^{-}$ allele with a hypomorphic allele of *Blm*, the Bloom syndrome gene that encodes a helicase required for double-stranded DNA break repair, yielded *Tsc1*$^{+/-}$*Blm*$^{m3/m3}$ mice with an interesting phenotype [46]. These mice had 4–10-fold increased numbers of gross and microscopic kidney tumors, due to an enhanced rate of somatic recombination leading to LOH, but no other difference in phenotype compared to *Tsc1*$^{+/-}$ mice.

The *Tsc2*$^{-}$ allele has also been interbred with a null allele of p53 (*TP53*$^{-}$) (Kwiatkowski *et al.*, unpublished observations). Interestingly, neither absence of p53 (in *Tsc2*$^{+/-}$*TP53*$^{-/-}$ mice) nor haplo-loss of p53 (in *Tsc2*$^{+/-}$*TP53*$^{+/-}$ mice) appeared to have a major effect on the rate of development of either renal or liver tumors.

The *Tsc2*$^{-}$ allele has also been interbred with a null allele of *p27*. p27 is a member of the Cip/Kip family of cyclin-dependent kinase inhibitors (CKIs). Neither kidney tumor incidence nor multiplicity was enhanced in either *Tsc2*$^{+/-}$*p27*$^{+/-}$ or *Tsc2*$^{+/-}$*p27*$^{-/-}$ mice in comparison to *Tsc2*$^{+/-}$*p27*$^{+/+}$ mice [47]. However, the apoptotic index was significantly higher in tumors that developed in p27-deficient animals compared to p27-intact mice.

7.3.5
Mouse Models: Results from Tissue-Restricted Knockout of Tsc1 or Tsc2

Several studies have combined conditional alleles of *Tsc1* or *Tsc2* with tissue-specific Cre alleles to attempt to model human TSC phenotypes more accurately and/or to explore physiological functions of Tsc1/Tsc2 (Table 7.2).

To develop a mouse model of cardiac rhabdomyomas, we used a conditional, floxed allele of Tsc1 and a myosin light chain 2v allele driving expression of Cre recombinase in ventricular myocytes [48]. Mice with ventricular loss of Tsc1 had a median survival of 6 months and developed a dilated cardiomyopathy with the occurrence of scattered foci of enlarged ventricular myocytes, whose staining characteristics were similar to those of TSC rhabdomyoma cells. However, the mice showed no evidence of fetal/neonatal demise, and there was no evidence of proliferation in the lesions. We

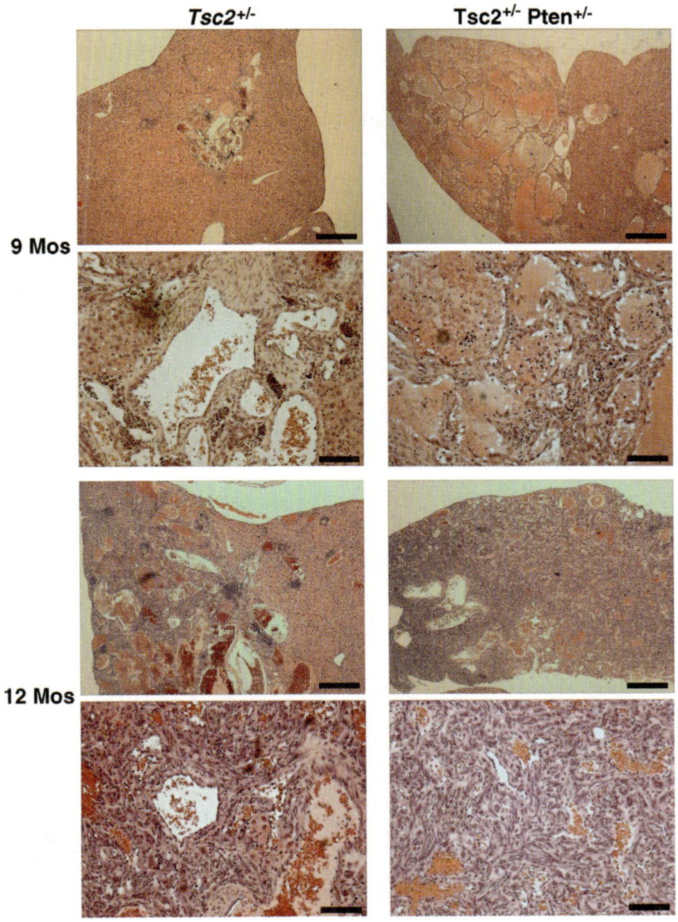

Figure 7.6 Liver hemangioma development in the *Tsc2*$^{+/-}$*Pten*$^{+/-}$ mice. Liver hemangiomas are shown from *Tsc2*$^{+/-}$*Pten*$^{+/-}$ and *Tsc2*$^{+/-}$ mice at two ages. Note the more extensive liver hemangioma development in the *Tsc2*$^{+/-}$*Pten*$^{+/-}$ mice. Scale bars are 0.5 mm (low power) and 100 μm (high power). From Ref. [45], courtesy of Brendan Manning.

suspect that these differences were due to the timing of loss of Tsc1 in the ventricular myocytes and/or the truncated gestational period in the mouse compared with humans, during which progestational hormones may accentuate the growth of patient rhabdomyomas [48].

Overexpression of Tsc1 in mouse skeletal muscle leads to increased expression of both Tsc1 and Tsc2, reduced muscle and body weight, but normal signaling in response to insulin, with no effect on survival [49].

Conditional deletion of *Tsc2* in pancreatic β cells results in lower glucose levels, hyperinsulinemia, and improved glucose tolerance, as well as expansion of the β cell mass by increased proliferation and cell size [50, 51]. Rapamycin treatment reversed these metabolic changes by induction of insulin resistance and reduction of β cell mass [50]. However, at the age of about 40 weeks, there was a dramatic reversal with hyperglycemia and hypoinsulinemia in the mice with loss of *Tsc2* in islets [51]. This phenomenon is thought to be due to the counterbalancing effects of mTORC1 activation and AKTand Foxo inactivation in the β cells [51]. The observations confirm that TSC1/TSC2/mTOR signaling is very important in the pancreas in normal physiology and suggest that this pathway could represent a target for treatment in patients with diabetes.

7.3.6 Mouse Models of TSC Brain Disease

Modeling TSC brain disease has considerable interest due to the importance of neurologic and neurocognitive effects in TSC patients. Both null and conditional alleles of *Tsc1* and *Tsc2* have proven valuable in the study of the effect of haploinsufficiency and complete loss of these genes during brain development and in neurologic function. However, initial detailed neuropathological studies of both $Tsc1^{+/-}$ and $Tsc2^{+/-}$ mice failed to demonstrate evidence of neuropathology similar to that seen in TSC patients [34, 52, 53].

$Tsc2^{-/-}$neuroepithelial progenitor (NEP) cells have enhanced growth characteristics, overexpress GFAP, show evidence of mTORC1 activation, and underexpress early neuronal markers in comparison to control NEP cells [52]. In addition, they could be differentiated into giant cells that resembled the giant cells of TSC cortical tubers. These observations provided circumstantial evidence that tuber giant cells are likely to be generated through a two-hit, complete loss of Tsc2 mechanism. Evidence of mTORC1 activation was also seen in Tsc1 null astrocytes [54].

An astrocyte-specific knockout of Tsc1 was achieved using a GFAP-Cre allele and a conditional allele of *Tsc1* [55]. These mice have been labeled $Tsc1^{GFAP}$CKO and have been studied in multiple respects by Gutmann and coworkers (Figure 7.7) [55]. There may be some neuronal populations in this $Tsc1^{GFAP}$CKO model that also sustain loss of *Tsc1*, but most of the effects seen in this model are due to loss in astrocytes. The mice develop seizures at about 2 months of age, and have a median survival of 12 weeks and maximum survival of 22 weeks. Histological studies demonstrate progressive astrogliosis with proliferation, as well as abnormal neuronal organization in the dentate region of the hippocampus by 1 month of age [55]. The observations highlight the importance of astrocyte expression of TSC1/TSC2 and suggest that complete loss of TSC1/TSC2 in astrocytes may contribute in a significant manner to the common seizure disorders seen in TSC patients.

Several mechanisms have been explored to explain the seizure tendency of the $Tsc1^{GFAP}$CKO mice. The expression of the glutamate transporters, GLT-1 and GLAST, is reduced in the astrocytes of these mice [56]. Moreover, there is a functional decrease in glutamate transport currents of Tsc1 null astrocytes in hippocampal slice and astrocyte cultures [56]. These changes appear to be responsible for an approximately 50% increase in glutamate levels in hippocampi of

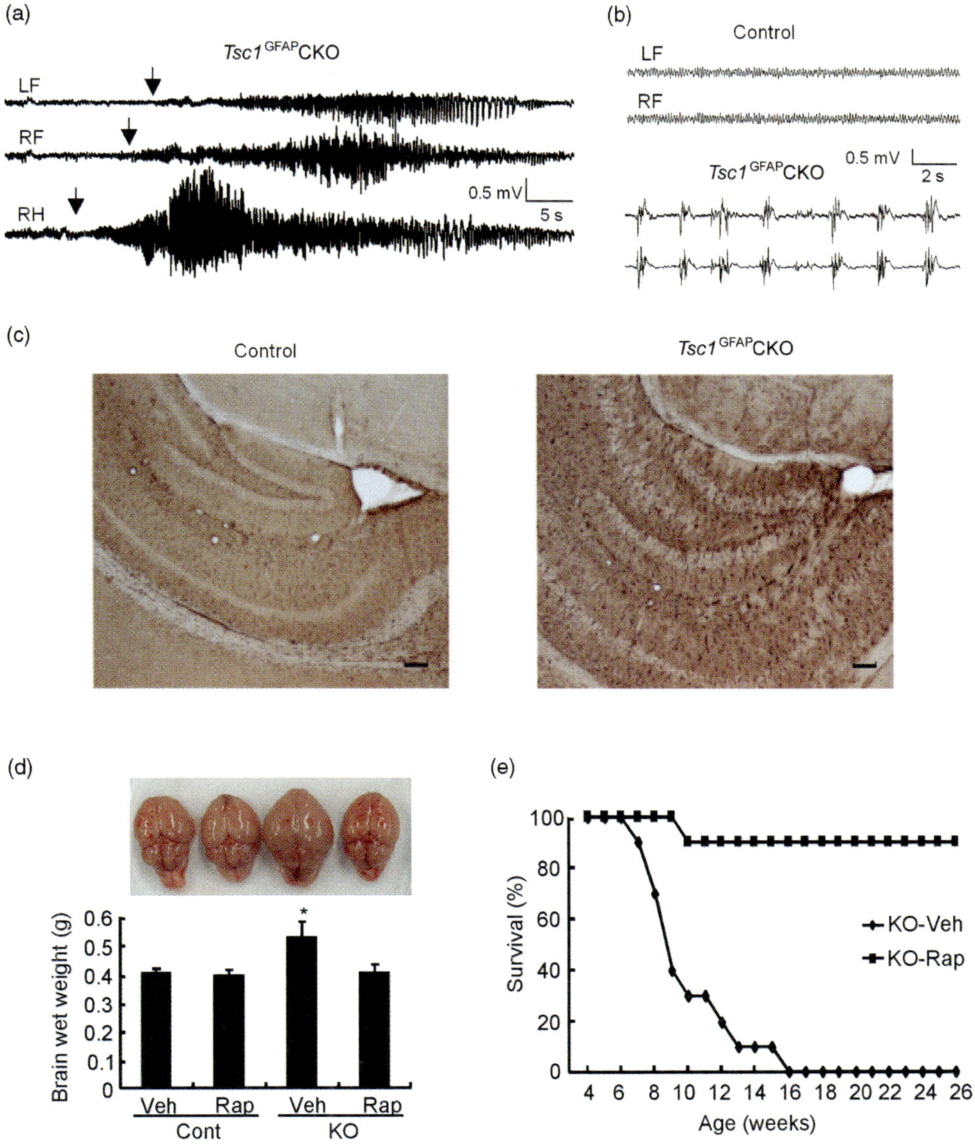

Figure 7.7 Epilepsy and glial abnormalities in *Tsc1*^{GFAP}CKO mice. (a) EEG recordings from a *Tsc1*^{GFAP}CKO mouse document a seizure that originates in the hippocampus. LF, left frontal electrode; RF, right frontal electrode; RH, right hippocampal depth electrode. (b) EEG recordings from a control mouse (top), and the *Tsc1*^{GFAP}CKO mice (bottom) show that the *Tsc1*^{GFAP}CKO mice have a severely abnormal interictal EEG, with periodic bursts of spikes. (c) GFAP immunohistochemistry staining of hippocampus from 6-week-old control (left) and *Tsc1*^{GFAP}CKO mice (right) shows diffuse glial proliferation, which contributes to megencephaly in the *Tsc1*^{GFAP}CKO mouse. Scale bar is 200 μm. (d and e) Rapamycin treatment, starting at 2 weeks of age, inhibits the glial proliferation and megencephaly (d) and prolongs survival of the *Tsc1*^{GFAP}CKO mice (e). Veh, vehicle treated; Rap, rapamycin treated. Adapted from Refs [69, 70], courtesy of Michael Wong.

Tsc1^{GFAP}CKO mice compared to controls [57]. This elevation in glutamate levels appears to explain widespread cortical and hippocampal neuronal death in these mice, prior to onset of seizure activity [57]. Inward rectifier potassium (Kir) channel expression is also reduced in the Tsc1 null astrocytes and is not responsive to rapamycin treatment, suggesting it is independent of mTORC1 activation [58]. Hippocampal slices from *Tsc1*^{GFAP}CKO mice also show decreased astrocytic Kir currents, as well as increased susceptibility to potassium-induced epileptiform activity [58]. Impairment of long-term potentiation with tetanic stimulation was observed in hippocampal slices from *Tsc1*^{GFAP}CKO mice and was reversed by low concentrations of NMDA antagonist, indicating that excessive synaptic glutamate directly inhibited LTP [57]. Thus, there are several mechanisms by which astrocyte abnormalities may contribute to epileptogenesis in TSC.

Tsc1^{GFAP}CKO mice have increased expression of vimentin, BLBP, cdk5, reelin, and others proteins typically associated with immature astroglia [59]. *Tsc1*^{GFAP}CKO mice also show impairment in two different tasks that require hippocampal function, a Pavlovian-conditioned fear task and Morris water navigation task that tests spatial learning [57]. They did not have global deficits in sensorimotor tests.

In vitro hippocampal slice cultures have also been used in an elegant study to examine the effects of loss of either Tsc1 or Tsc2 in postmitotic, hippocampal pyramidal neurons in both mice and rats [60]. Loss of either protein by genetic strategy or RNAi knockdown led to the enlargement of somas and dendritic spines and altered the properties of glutamatergic synapses. In addition, loss of a single copy of *Tsc1* was sufficient to perturb dendritic spine structure, indicating a haploinsufficiency effect [60]. However, similar detailed studies performed on *Tsc1*$^{+/-}$ mice showed no evidence of a difference in neuronal size, spine density, branching, or number of primary or secondary dendrites, as assessed by Golgi-Cox staining [53]. It was shown that expression of a mutant, nonphosphorylatable form of cofilin, S3A-cofilin, blocked the increase in cell size and dendritic spine changes, suggesting that effects on the actin cytoskeleton were critical in mediating these changes consequent to loss of Tsc1 or Tsc2 [60].

A neuron-specific knockout of *Tsc1* has also been studied in some detail [61, 62]. The *Tsc1*$^{null\text{-}neuron}$ mice display tremor, clasping, Straub tail, and kyphosis beginning at P5 with subsequent failure to thrive and median survival of 35 days (Figure 7.8) [61]. Of these mice, 10–20% also display clinical and electrographic seizures both spontaneously and with physical stimulation, and some seizures end in a fatal tonic phase. Cortical and hippocampal neurons are enlarged by about 50% and most are dysplastic with aberrant orientation of the apical dendrite (Figure 7.9). *Tsc1*$^{null\text{-}neuron}$ mouse neurons also strongly express pS6(S235–236), and are ectopic in multiple sites in the cortex and hippocampus (Figure 7.8). There is a striking delay in myelination in the mutant mice, which appears to be caused by an inductive neuronal defect (Figure 7.9) [61].

A third brain model of TSC has also been reported, in which the αCaMKII-Cre allele was combined with the *Tsc1* conditional allele, to generate mice that survive a few weeks for the most part, develop macrocephaly, and have both neuronal hypertrophy and astrogliosis [63].

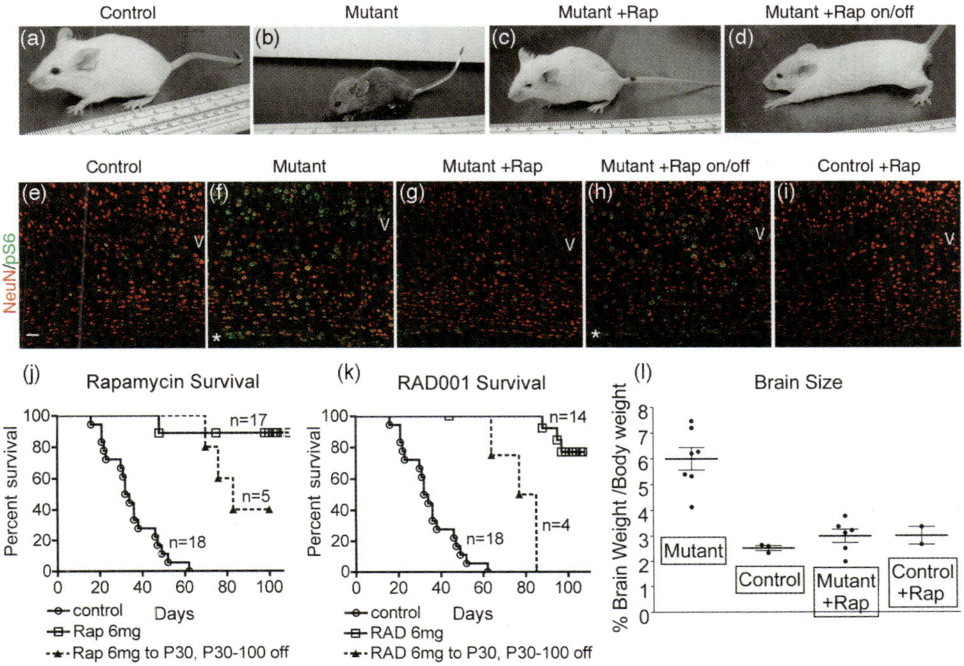

Figure 7.8 Clinical features in *Tsc1*$^{null\text{-}neuron}$ mice and improvement in response to rapamycin/RAD001. (a–d) Pictures of mice. Control and mutant (*Tsc1*$^{null\text{-}neuron}$) mice with or without rapamycin treatment are shown. Rapamycin was given IP 6 mg/kg every other day beginning at P7-9 and was discontinued at P30 in the on/off mouse. (a, c, and d) are all P45 and (b) is P30. Note abnormal posture of mutant and mutant on/off treatment mice. (e–i) Combined NeuN (red)–pS6(S235/236) (green) stains of base of cortex over the anterior hippocampus. P45 control, P30 mutant (*Tsc1*$^{null\text{-}neuron}$), P45 rapamycin-treated mutant, P45 rapamycin-treated mutant until P30 then taken off drug (on/off), and P45 rapamycin-treated control mouse sections are shown. Scale bar is 50 μm. Note the enlarged pS6+ cells in the mutant, their absence in the treated mutant, and presence in the on/off mutant. * indicates base of cortex where enlarged, pS6+ neurons are prominent. (j and k) Survival curves of *Tsc1*$^{null\text{-}neuron}$ mice treated with rapamycin (j) or RAD001 (k). Doses of each drug, periods of treatment, and numbers of treated mice are indicated. Treatment began at P7–9 for all mice, and was given every other day. Note that dashed lines reflect cohorts in which treatment was discontinued after P30. (l) Brain/body weight ratio in these same groups of mice, all at P30. Each point is a different mouse; the line indicates the mean. $P < 0.01$ for each group in comparison to untreated mutant. Note the dramatic reduction in brain/body weight ratio in the treated mice. From Refs [61, 68].

7.3.7
Neurocognitive Studies in *Tsc1*$^{+/-}$ and *Tsc2*$^{+/-}$ Mice

Two recent studies have examined in some detail the neurocognitive phenotype of *Tsc1*$^{+/-}$ and *Tsc2*$^{+/-}$ mice. *Tsc1*$^{+/-}$ mice showed impaired hippocampus-dependent spatial learning in the Morris water maze task, as well as a hippocampal learning deficit in a contextual fear conditioning test [53]. In addition, *Tsc1*$^{+/-}$ mice showed

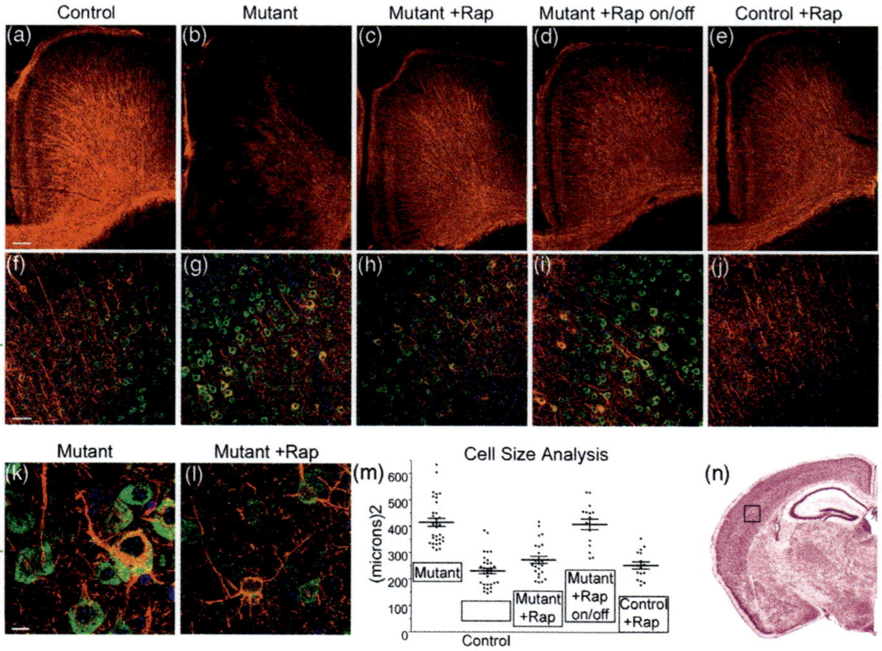

Figure 7.9 Effects on myelination, neuronal size, and orientation in *Tsc1*$^{null\text{-}neuron}$ mice treated with rapamycin. (a–j) Each of the five columns of two images is taken from P45 control, P30 mutant (*Tsc1*$^{null\text{-}neuron}$), P45 rapamycin-treated mutant, P45 rapamycin-treated mutant until P30 then taken off drug ("on/off") mutant, and P45 rapamycin-treated control, respectively. (a–e) Myelin stain of the retrosplenial granular region of cortex demonstrates that there is a marked reduction in myelination in untreated *Tsc1*$^{null\text{-}neuron}$ mice, which improves with treatment, and that myelination in the on/off mice also begins to look somewhat more patchy than in other groups. Scale bar is 100 μm. (f–j) Nonphosphorylated neurofilament (SMI311) expression (red) with pS6(S235/236) expression (green) and DAPI (blue) stains of lateral cortex. Note the abnormal SMI311+ cells that are enlarged and dysplastic in the untreated mutant mice. These cells have reduced cell size but not the degree of dysplasia in response to rapamycin. In mice taken off rapamycin, there is recurrence of cell enlargement. (k and l) Higher power view of cells from images (g) and (h) highlight the aberrant polarity of the apical dendrite of these neurons, despite the smaller size of the treated mutant neuron (l). Scale bar is 10 μm. (m) Cell size measurements in these same five categories of mice, from SMI311-stained lateral cortex. Cell size is indicated in μm^2. Each point is the size of one SMI311-stained neuron, with eight neurons counted per mouse section. The average size and SEM are shown. Comparisons between each of the mutants and the on/off mutants and each of the controls, treated mutants, and treated controls are significant at $P = 0.02$ to <0.0001. All other comparisons have $P > 0.05$. (n) Cortical section indicates region of cortex and orientation shown in (f–j). From Ref. [68].

significantly less social interaction and reduced nest building behavior [53]. However, although these findings met statistical significance, they were not profound [53].

Similarly, Ehninger *et al.* found that *Tsc2*$^{+/-}$ mice demonstrated impaired spatial learning in the Morris water maze task and impaired working memory in the eight-arm radial maze test and in context conditioning [63]. However, they also

Table 7.1 *Tsc1* and *Tsc2* alleles in mice and rats.

Species	Gene	Allele name	Exon targeted/ mutation	Major features of heterozygote animals	Major references
Rat	*Tsc2*	Eker	IAP element insertion into codon 1272	Cystadenomas–carcinomas of kidney Splenic hemangiomas Uterine leiomyomas Pituitary adenomas Subependymal and subcortical hamartomas	[2, 5–7]
Mouse	*Tsc2*	–, Kwiatkowski	Neomycin cassette insertion into exon 2	Cystadenomas of kidney Liver hemangiomas Extremity angiosarcomas	[34]
Mouse	*Tsc2*	–, Hino	Neomycin cassette insertion into exon 2, deletion of exons 2–5	Cystadenomas of kidney Liver hemangiomas	[33]
Mouse	*Tsc1*	–, Hino	Neomycin cassette insertion and deletion of exons 6–8	Cystadenomas of kidney Liver hemangiomas	[39]
Mouse	*Tsc1*	–, Kwiatkowski	Deletion of exons 17 and 18	Cystadenomas of kidney Liver hemangiomas	[40]
Mouse	*Tsc1*	–, Cheadle	Deletion of exons 6–8 with insertion of neomycin cassette	Cystadenomas of kidney	[42]

(*Continued*)

Table 7.1 (*Continued*)

Species	Gene	Allele name	Exon targeted/ mutation	Major features of heterozygote animals	Major references
				Liver hemangiomas Reduced survival of *Tsc1*$^{+/-}$ when in C57BL/6 strain *Tsc1*$^{+/-}$ kidney cancer in BALB/c strain	
Mouse	*Tsc2*	*neo*, Gambello	Neomycin cassette insertion into exon 1	Hypomorphic allele: renal cysts only at 20 months of age, and *Tsc2*$^{neo/neo}$ embryos survive to E17 in some cases	[35]
Mouse	*Tsc2*	*del3*, Kwiatkowski	Deletion of exon 3	Hypomorphic allele: reduced severity of renal tumors, 1–2 day longer survival of *Tsc2*$^{del3/del3}$ embryos	[71]
Mouse	*Tsc2*	*KO*, Gambello	Deletion of exons 2–4	Cystadenomas of kidney Little published data	[35]
Mouse	*Tsc2*	–, Kobayashi	Deletion of exons 3–4	Little published data	[51]

demonstrated that Tsc2 protein levels were reduced about 25% and pS6(S235–236) levels were increased in brain lysates from the *Tsc2*$^{+/-}$ mice in comparison to controls [63]. *Tsc2*$^{+/-}$ mice also showed a lowered threshold for late-phase long-term potentiation in brain slice experiments than controls [63].

Table 7.2 Conditional alleles of *Tsc1* and *Tsc2*.

Species	Gene	Allele name	Exons targeted	Major references
Mouse	*Tsc1*	c, Kwiatkowski	17–18	[40, 48]
Mouse	*Tsc2*	*Flox*, Gambello	2–4	[35, 50]
Mouse	*Tsc2*	*Flox*, Kobayashi	3–4	[51]
Mouse	*Tsc2*	*c-del3*, Kwiatkowski	3	Kwiatkowski *et al.*,

These studies, although not entirely parallel in their findings, strongly indicate that there is a neurocognitive phenotype due to single allele loss (haploinsufficiency) of either Tsc1 or Tsc2. These observations are extremely important in addressing the clinical question as to whether all neurocognitive effects in TSC patients are due to individual cortical tubers that perturb local cortical function, aggregate cortical tuber burden, or seizures. The answer is no; there are physiologic consequences of haploinsufficiency for each of these genes.

7.3.8 Treatment Studies in the Mouse Models of TSC

Similar to results seen in the Eker rat, rapamycin and related mTORC1 inhibitors are very effective in blocking tumor development in the *Tsc2*$^{+/-}$ mouse model and in subcutaneous tumor models using Tsc2 null MEF lines [64]. Based upon an association between milder TSC renal disease and a high expressing allele of interferon gamma [65], treatment with interferon gamma has also been tested in the subcutaneous tumor model. Interferon gamma had some benefit but was not as effective as the rapamycin analogue CCI-779 [66]. Combination therapy with both did not appear to provide significant additional benefit over CCI-779 alone in the nude mouse model [67].

Treatment with rapamycin has been shown to have dramatic therapeutic benefit in several mouse brain/neurologic models of TSC (Figures 7.7–7.9). In *Tsc2*$^{+/-}$ mice, rapamycin treatment improved context discrimination, spatial learning in the Morris water maze task, and normalized late-phase long-term potentiation [63]. Rapamycin improved the median survival of *Tsc1*$^{null\text{-}neuron}$ mice by over threefold, markedly improved both neurologic phenotype and weight gain, and reduced the brain–body weight ratio (Figures 7.8 and 7.9) [68]. It also corrected signaling effects of mTORC1 activation in the brain and reduced neuronal enlargement, though neuronal dysplasia persisted, and there were only modest effects on dendritic spine density and length [68]. Rapamycin had similar dramatic benefit on the αCaMKII-Cre *Tsc1* mice [63].

Rapamycin treatment of the *Tsc1*$^{null\text{-}astrocyte}$ mice results in reduced astrocyte proliferation, normalized brain weight, and increased Glt-1 and GLAST expression (Figure 7.7) [69]. When given beginning at P14, nearly all treated mice survived over 6 months, in comparison to a median survival of 9 weeks in vehicle-treated controls. When given beginning at 6 weeks, after seizure onset, rapamycin led to marked reduction in seizures and interictal EEG abnormalities and markedly improved survival [69].

7.4 Concluding Remarks

Although there was initial disappointment due to the lack of similarity between the Tsc mouse and rat models in comparison to human TSC, more detailed studies on these models over the years have provided more and more pathophysiological insights, which have clear relevance to the human disease. The models have also provided the opportunity for treatment trials. Here the tangible insights achieved are summarized and both limitations and avenues for further progress are highlighted.

First, the positive aspects of the existing standard *Tsc1*$^{+/-}$ and *Tsc2*$^{+/-}$ mice can be summarized. Although they do not develop tumors that match those seen in TSC patients to any major extent, their tumors follow the two-hit mechanism of TSC tumor development and demonstrate the same pathway effects that are seen in more limited detail in TSC patient tumors and lesions. They have also provided the opportunity to explore the interaction of *Tsc1/Tsc2 in vivo* with other genes (e.g., *Pten* and *Blm*), and will continue to be used in this way to explore pathways and physiology. They have been major tools in the assessment of potential therapies for TSC. Three different mTORC1 inhibitors (rapamycin, CCI-779, and RAD001) have been shown to have major therapeutic benefit in one TSC model or another. These studies have been critical in enabling and stimulating the current set of clinical trials with these compounds in TSC patients.

The recent findings [53, 63] that there are neurocognitive and social defects in the *Tsc1*$^{+/-}$ and *Tsc2*$^{+/-}$ mice are of considerable interest, as elaborated above, in highlighting haploinsufficiency as a probable contributor to neurocognitive and developmental issues in TSC patients, independent of tuber burden and seizure history. These studies also highlight the importance of detailed and refined studies to assess phenotype with sophisticated methodology in strain-purified mice, studies that may not always capture the excitement of a funding agency or grant review panel. The responsiveness of one of these models with neurocognitive deficits to rapamycin treatment provides hope that such treatment may also be of benefit for neurocognitive issues in TSC patients.

The conditional alleles of *Tsc1* and *Tsc2* provide the opportunity for a variety of studies that explore how loss of these genes in specific organ systems or cell types affect physiology and health. Although several examples are cited above, and many more are in the pipeline, arguably the most important have been the use of Cre alleles that target specific brain cell populations to achieve loss of *Tsc1* or *Tsc2* in different cell types and regions and at different times during development. The *Tsc1*^{GFAP}CKO mice demonstrate the critical importance of expression of these proteins in astrocytes. Although this model does not replicate the precise neuropathology of human TSC, it has provided considerable insights into the important functions of TSC1/TSC2 in the astrocyte and how its loss can lead to epileptogenesis. The *Tsc1*$^{null\text{-}neuron}$ mice demonstrate the critical importance of TSC1/TSC2 in neuronal function and underscore the likelihood that all brain cell types need a normal dose of TSC1/TSC2 complex for normal function.

It is disappointing that none of the mouse models yet created generates the specific pathology of the cortical tuber. This may be difficult to achieve in the mouse, given the small size of the mouse neocortex, in comparison to the human neocortex, and even in comparison to a single human tuber (!). However, it is still critical to achieve such a model to be able to explore in detail the genesis and physiologic effects of tubers and their giant cells on cortical function. There are continuing efforts in many labs to get closer to the goal of an authentic mouse model of cortical tubers, and some degree of success is anticipated in the next few years.

Another major TSC-related pathology for which a mouse model is desired is the angiomyolipoma–lymphangioleiomyomatosis spectrum of disorders. As discussed in Chapters 15 and 17, these two disorders are closely related from a genetic and pathophysiological point of view, and each represents a major contributor to morbidity and mortality in TSC adults. Thus, an animal model is highly desirable. Although the liver and spleen hemangiomas of Tsc mice and rats bear some resemblance to these tumors, they are far from adequate. A major problem in developing such a model is insight into the cell type that gives rise to these lesions. Nonetheless, again multiple approaches are being taken in multiple laboratories, and major progress is expected within the next several years.

Acknowledgments

I thank Okio Hino, Catherine Wilson, Li Ma, Brendan Manning, and Mike Wong for contributing figures to this chapter. Supported by NIH NINDS NS031535 and P01 NS24279, NIH NCI P01 CA120964, and the Tuberous Sclerosis Alliance.

References

1 Eker, R. (1954) Familial renal adenomas in Wistar rats; a preliminary report. *Acta Pathol. Microbiol. Scand.*, **34**, 554–562.

2 Eker, R., Mossige, J., Johannessen, J.V., and Aars, H. (1981) Hereditary renal adenomas and adenocarcinomas in rats. *Diagn. Histopathol.*, **4**, 99–110.

3 Wolf, D.C., Whiteley, H.E., and Everitt, J.I. (1995) Preneoplastic and neoplastic lesions of rat hereditary renal cell tumors express markers of proximal and distal nephron. *Vet. Pathol.*, **32**, 379–386.

4 Rennebeck, G., Kleymenova, E.V., Anderson, R., Yeung, R.S., Artzt, K., and Walker, C.L. (1998) Loss of function of the tuberous sclerosis 2 tumor suppressor gene results in embryonic lethality characterized by disrupted neuroepithelial growth and development. *Proc. Natl. Acad. Sci. USA*, **95**, 15629–15634.

5 Everitt, J., Goldsworthy, T., Wolf, D., and Walker, C. (1992) Hereditary renal cell carcinoma in the Eker rat: a rodent familial cancer syndrome. *J. Urol.*, **148**, 1932–1936.

6 Hino, O., Klein-Szanto, A.J.P., Freed, J.J., Testa, J.R., Brown, D.Q., Vilensky, M., Yeung, R.S., Tartof, K.D., and Knudson, A.G. (1993) Spontaneous and radiation-induced renal tumors in the Eker rat model of dominantly inherited cancer. *Proc. Natl. Acad. Sci. USA*, **90**, 327–331.

7 Yeung, R.S., Katsetos, C.D., and Klein-Szanto, A. (1997) Subependymal astrocytic hamartomas in the Eker rat model of tuberous sclerosis. *Am. J. Pathol.*, **151**, 1477–1486.

8 Yeung, R.S., Xiao, G.H., Jin, F., Lee, W.C., Testa, J.R., and Knudson, A.G. (1994) Predisposition to renal carcinoma in the Eker rat is determined by germ-line mutation of the tuberous sclerosis 2 (TSC2) gene. *Proc. Natl. Acad. Sci. USA*, **91**, 11413–11416.

9 Kobayashi, T., Hirayama, Y., Kobayashi, E., Kubo, Y., and Hino, O. (1995) A germline insertion in the tuberous sclerosis (Tsc2) gene gives rise to the Eker rat model of dominantly inherited cancer. *Nat. Genet.*, **9**, 70–74.

10 Yeung, R.S. (2004) Lessons from the Eker rat model: from cage to bedside. *Curr. Mol. Med.*, **4**, 799–806.

11 Kubo, Y., Kikuchi, Y., Mitani, H., Kobayashi, E., Kobayashi, T., and Hino, O. (1995) Allelic loss at the tuberous sclerosis (Tsc2) gene locus in spontaneous uterine leiomyosarcomas and pituitary adenomas in the Eker rat model. *Jpn. J. Cancer Res.*, **86**, 828–832.

12 Yeung, R.S., Xiao, G.H., Everitt, J.I., Jin, F., and Walker, C.L. (1995) Allelic loss at the tuberous sclerosis 2 locus in spontaneous tumors in the Eker rat. *Mol. Carcinog.*, **14**, 28–36.

13 Kobayashi, T., Urakami, S., Hirayama, Y., Yamamoto, T., Nishizawa, M., Takahara, T., Kubo, Y., and Hino, O. (1997) Intragenic Tsc2 somatic mutations as Knudson's second hit in spontaneous and chemically induced renal carcinomas in the Eker rat model. *Jpn. J. Cancer Res.*, **88**, 254–261.

14 Kubo, Y., Mitani, H., and Hino, O. (1994) Allelic loss at the predisposing gene locus in spontaneous and chemically induced renal cell carcinomas in the Eker rat. *Cancer Res.*, **54**, 2633–2635.

15 Satake, N., Urakami, S., Hirayama, Y., Izumi, K., and Hino, O. (1998) Biallelic mutations of the Tsc2 gene in chemically induced rat renal cell carcinoma. *Int. J. Cancer*, **77**, 895–900.

16 Kobayashi, T., Mitani, H., Takahashi, R., Hirabayashi, M., Ueda, M., Tamura, H., and Hino, O. (1997) Transgenic rescue from embryonic lethality and renal carcinogenesis in the Eker rat model by introduction of a wild-type Tsc2 gene. *Proc. Natl. Acad. Sci. USA*, **94**, 3990–3993.

17 Momose, S., Kobayashi, T., Tada, N., Itoyama, S., and Hino, O. (2007) N-terminal hamartin-binding and C-terminal GAP domain of tuberin can separate *in vivo*. *Biochem. Biophys. Res. Commun.*, **356**, 693–698.

18 Cai, S., Everitt, J.I., Kugo, H., Cook, J., Kleymenova, E., and Walker, C.L. (2003) Polycystic kidney disease as a result of loss of the tuberous sclerosis 2 tumor suppressor gene during development. *Am. J. Pathol.*, **162**, 457–468.

19 Walker, C.L., Hunter, D., and Everitt, J.I. (2003) Uterine leiomyoma in the Eker rat: a unique model for important diseases of women. *Genes Chromosomes Cancer*, **38**, 349–356.

20 Walker, C.L., Cesen-Cummings, K., Houle, C., Baird, D., Barrett, J.C., and Davis, B. (2001) Protective effect of pregnancy for development of uterine leiomyoma. *Carcinogenesis*, **22**, 2049–2052.

21 Yeung, R.S., Gu, H., Lee, M., and Dundon, T.A. (2001) Genetic identification of a locus, Mot1, that affects renal tumor size in the rat. *Genomics*, **78**, 108–112.

22 Kikuchi, Y., Sudo, A., Mitani, H., and Hino, O. (2004) Presence of a modifier gene(s) affecting early renal carcinogenesis in the Tsc2 mutant (Eker) rat model. *Int. J. Oncol.*, **24**, 75–80.

23 Kenerson, H.L., Aicher, L.D., True, L.D., and Yeung, R.S. (2002) Activated mammalian target of rapamycin pathway in the pathogenesis of tuberous sclerosis complex renal tumors. *Cancer Res.*, **62**, 5645–5650.

24 Kenerson, H., Dundon, T.A., and Yeung, R.S. (2005) Effects of rapamycin in the Eker rat model of tuberous sclerosis complex. *Pediatr. Res.*, **57**, 67–75.

25 Laping, N.J., Everitt, J.I., Frazier, K.S., Burgert, M., Portis, M.J., Cadacio, C., Gold, L.I., and Walker, C.L. (2007) Tumor-specific efficacy of transforming growth factor-beta RI inhibition in Eker rats. *Clin. Cancer Res.*, **13**, 3087–3099.

26 Takahashi, D.K., Dinday, M.T., Barbaro, N.M., and Baraban, S.C. (2004) Abnormal cortical cells and astrocytomas in the Eker rat model of tuberous sclerosis complex. *Epilepsia*, **45**, 1525–1530.

27 Mizuguchi, M., Takashima, S., Yamanouchi, H., Nakazato, Y., Mitani, H., and Hino, O. (2000) Novel cerebral lesions in the Eker rat model of tuberous sclerosis: cortical tuber and anaplastic ganglioglioma. *J. Neuropathol. Exp. Neurol.*, **59**, 188–196.

28 Mizuguchi, M., Mori, M., Nozaki, Y., Momoi, M.Y., Itoh, M., Takashima, S., and Hino, O. (2004) Absence of allelic loss in cytomegalic neurons of cortical tuber in

the Eker rat model of tuberous sclerosis. *Acta Neuropathol.*, **107**, 47–52.

29 Wenzel, H.J., Patel, L.S., Robbins, C.A., Emmi, A., Yeung, R.S., and Schwartzkroin, P.A. (2004) Morphology of cerebral lesions in the Eker rat model of tuberous sclerosis. *Acta Neuropathol.*, **108**, 97–108.

30 Tschuluun, N., Wenzel, H.J., and Schwartzkroin, P.A. (2007) Irradiation exacerbates cortical cytopathology in the Eker rat model of tuberous sclerosis complex, but does not induce hyperexcitability. *Epilepsy Res.*, **73**, 53–64.

31 von der Brelie, C., Waltereit, R., Zhang, L., Beck, H., and Kirschstein, T. (2006) Impaired synaptic plasticity in a rat model of tuberous sclerosis. *Eur. J. Neurosci.*, **23**, 686–692.

32 Waltereit, R., Welzl, H., Dichgans, J., Lipp, H.P., Schmidt, W.J., and Weller, M. (2006) Enhanced episodic-like memory and kindling epilepsy in a rat model of tuberous sclerosis. *J. Neurochem.*, **96**, 407–413.

33 Kobayashi, T., Minowa, O., Kuno, J., Mitani, H., Hino, O., and Noda, T. (1999) Renal carcinogenesis, hepatic hemangiomatosis, and embryonic lethality caused by a germ-line Tsc2 mutation in mice. *Cancer Res.*, **59**, 1206–1211.

34 Onda, H., Lueck, A., Marks, P.W., Warren, H.B., and Kwiatkowski, D.J. (1999) Tsc2 (+/−) mice develop tumors in multiple sites that express gelsolin and are influenced by genetic background. *J. Clin. Invest.*, **104**, 687–695.

35 Hernandez, O., Way, S., McKenna, J., 3rd, and Gambello, M.J. (2007) Generation of a conditional disruption of the Tsc2 gene. *Genesis*, **45**, 101–106.

36 Liu, M.Y., Poellinger, L., and Walker, C.L. (2003) Up-regulation of hypoxia-inducible factor 2alpha in renal cell carcinoma associated with loss of Tsc-2 tumor suppressor gene. *Cancer Res.*, **63**, 2675–2680.

37 El-Hashemite, N., Walker, V., Zhang, H., and Kwiatkowski, D.J. (2003) Loss of Tsc1 or Tsc2 induces vascular endothelial growth factor production through mammalian target of rapamycin. *Cancer Res.*, **63**, 5173–5177.

38 Brugarolas, J.B., Vazquez, F., Reddy, A., Sellers, W.R., and Kaelin, W.G., Jr. (2003) TSC2 regulates VEGF through mTOR-dependent and -independent pathways. *Cancer Cell*, **4**, 147–158.

39 Kobayashi, T., Minowa, O., Sugitani, Y., Takai, S., Mitani, H., Kobayashi, E., Noda, T., and Hino, O. (2001) A germ-line Tsc1 mutation causes tumor development and embryonic lethality that are similar, but not identical to, those caused by Tsc2 mutation in mice. *Proc. Natl. Acad. Sci. USA*, **98**, 8762–8767.

40 Kwiatkowski, D.J., Zhang, H., Bandura, J.L., Heiberger, K.M., Glogauer, M., el-Hashemite, N., and Onda, H. (2002) A mouse model of TSC1 reveals sex-dependent lethality from liver hemangiomas, and up-regulation of p70S6 kinase activity in Tsc1 null cells. *Hum. Mol. Genet.*, **11**, 525–534.

41 El-Hashemite, N., Walker, V., and Kwiatkowski, D.J. (2005) Estrogen enhances whereas tamoxifen retards development of Tsc mouse liver hemangioma: a tumor related to renal angiomyolipoma and pulmonary lymphangioleiomyomatosis. *Cancer Res.*, **65**, 2474–2481.

42 Wilson, C., Idziaszczyk, S., Parry, L., Guy, C., Griffiths, D.F., Lazda, E., Bayne, R.A., Smith, A.J., Sampson, J.R., and Cheadle, J.P. (2005) A mouse model of tuberous sclerosis 1 showing background specific early post-natal mortality and metastatic renal cell carcinoma. *Hum. Mol. Genet.*, **14**, 1839–1850.

43 Wilson, C., Bonnet, C., Guy, C., Idziaszczyk, S., Colley, J., Humphreys, V., Maynard, J., Sampson, J.R., and Cheadle, J.P. (2006) Tsc1 haploinsufficiency without mammalian target of rapamycin activation is sufficient for renal cyst formation in Tsc1 +/− mice. *Cancer Res.*, **66**, 7934–7938.

44 Ma, L., Teruya-Feldstein, J., Behrendt, N., Chen, Z., Noda, T., Hino, O., Cordon-Cardo, C., and Pandolfi, P.P. (2005) Genetic analysis of Pten and Tsc2 functional interactions in the mouse reveals asymmetrical haploinsufficiency

in tumor suppression. *Genes Dev.*, **19**, 1779–1786.
45 Manning, B.D., Logsdon, M.N., Lipovsky, A.I., Abbott, D., Kwiatkowski, D.J., and Cantley, L.C. (2005) Feedback inhibition of Akt signaling limits the growth of tumors lacking Tsc2. *Genes Dev.*, **19**, 1773–1778.
46 Wilson, C., Idziaszczyk, S., Colley, J., Humphreys, V., Guy, C., Maynard, J., Sampson, J.R., and Cheadle, J.P. (2005) Induction of renal tumorigenesis with elevated levels of somatic loss of heterozygosity in Tsc1 + /− mice on a Blm-deficient background. *Cancer Res.*, **65**, 10179–10182.
47 Short, J.D., Houston, K.D., Dere, R., Cai, S.L., Kim, J., Johnson, C.L., Broaddus, R.R., Shen, J., Miyamoto, S., Tamanoi, F., Kwiatkowski, D.J., Mills, G.B., and Walker, C.L. (2008) AMP-activated protein kinase signaling results in cytoplasmic sequestration of p27. *Cancer Res.*, **68**, 6496–6506.
48 Meikle, L., McMullen, J.R., Sherwood, M.C., Lader, A.S., Walker, V., Chan, J.A., and Kwiatkowski, D.J. (2005) A mouse model of cardiac rhabdomyoma generated by loss of Tsc1 in ventricular myocytes. *Hum. Mol. Genet.*, **14**, 429–435.
49 Wan, M., Wu, X., Guan, K.L., Han, M., Zhuang, Y., and Xu, T. (2006) Muscle atrophy in transgenic mice expressing a human TSC1 transgene. *FEBS Lett.*, **580**, 5621–5627.
50 Rachdi, L., Balcazar, N., Osorio-Duque, F., Elghazi, L., Weiss, A., Gould, A., Chang-Chen, K.J., Gambello, M.J., and Bernal-Mizrachi, E. (2008) Disruption of Tsc2 in pancreatic {beta} cells induces {beta} cell mass expansion and improved glucose tolerance in a TORC1-dependent manner. *Proc. Natl. Acad. Sci. USA*, **105**, 9250–9255.
51 Shigeyama, Y., Kobayashi, T., Kido, Y., Hashimoto, N., Asahara, S., Matsuda, T., Takeda, A., Inoue, T., Shibutani, Y., Koyanagi, M., Uchida, T., Inoue, M., Hino, O., Kasuga, M., and Noda, T. (2008) Biphasic response of pancreatic beta-cell mass to ablation of tuberous sclerosis complex 2 in mice. *Mol. Cell Biol.*, **28**, 2971–2979.
52 Onda, H., Crino, P.B., Zhang, H., Murphey, R.D., Rastelli, L., Gould Rothberg, B.E., and Kwiatkowski, D.J. (2002) Tsc2 null murine neuroepithelial cells are a model for human tuber giant cells, and show activation of an mTOR pathway. *Mol. Cell. Neurosci.*, **21**, 561–574.
53 Goorden, S.M., van Woerden, G.M., van der Weerd, L., Cheadle, J.P., and Elgersma, Y. (2007) Cognitive deficits in Tsc1 + /− mice in the absence of cerebral lesions and seizures. *Ann. Neurol.*, **62**, 648–655.
54 Uhlmann, E.J., Li, W., Scheidenhelm, D.K., Gau, C.L., Tamanoi, F., and Gutmann, D.H. (2004) Loss of tuberous sclerosis complex 1 (Tsc1) expression results in increased Rheb/S6K pathway signaling important for astrocyte cell size regulation. *Glia*, **47**, 180–188.
55 Uhlmann, E.J., Wong, M., Baldwin, R.L., Bajenaru, M.L., Onda, H., Kwiatkowski, D.J., Yamada, K., and Gutmann, D.H. (2002) Astrocyte-specific TSC1 conditional knockout mice exhibit abnormal neuronal organization and seizures. *Ann. Neurol.*, **52**, 285–296.
56 Wong, M., Ess, K.C., Uhlmann, E.J., Jansen, L.A., Li, W., Crino, P.B., Mennerick, S., Yamada, K.A., and Gutmann, D.H. (2003) Impaired glial glutamate transport in a mouse tuberous sclerosis epilepsy model. *Ann. Neurol.*, **54**, 251–256.
57 Zeng, L.H., Ouyang, Y., Gazit, V., Cirrito, J.R., Jansen, L.A., Ess, K.C., Yamada, K.A., Wozniak, D.F., Holtzman, D.M., Gutmann, D.H., and Wong, M. (2007) Abnormal glutamate homeostasis and impaired synaptic plasticity and learning in a mouse model of tuberous sclerosis complex. *Neurobiol. Dis.*, **28**, 184–196.
58 Jansen, L.A., Uhlmann, E.J., Crino, P.B., Gutmann, D.H., and Wong, M. (2005) Epileptogenesis and reduced inward rectifier potassium current in tuberous sclerosis complex-1-deficient astrocytes. *Epilepsia*, **46**, 1871–1880.
59 Ess, K.C., Uhlmann, E.J., Li, W., Li, H., Declue, J.E., Crino, P.B., and Gutmann, D.H. (2004) Expression profiling in tuberous sclerosis complex (TSC) knockout mouse astrocytes to characterize

human TSC brain pathology. *Glia*, **46**, 28–40.

60 Tavazoie, S.F., Alvarez, V.A., Ridenour, D.A., Kwiatkowski, D.J., and Sabatini, B.L. (2005) Regulation of neuronal morphology and function by the tumor suppressors Tsc1 and Tsc2. *Nat. Neurosci.*, **8**, 1727–1734.

61 Meikle, L., Talos, D.M., Onda, H., Pollizzi, K., Rotenberg, A., Sahin, M., Jensen, F.E., and Kwiatkowski, D.J. (2007) A mouse model of tuberous sclerosis: neuronal loss of Tsc1 causes dysplastic and ectopic neurons, reduced myelination, seizure activity, and limited survival. *J. Neurosci.*, **27**, 5546–5558.

62 Wang, Y., Greenwood, J.S., Calcagnotto, M.E., Kirsch, H.E., Barbaro, N.M., and Baraban, S.C. (2007) Neocortical hyperexcitability in a human case of tuberous sclerosis complex and mice lacking neuronal expression of TSC1. *Ann. Neurol.*, **61**, 139–152.

63 Ehninger, D., Han, S., Shilyansky, C., Zhou, Y., Li, W., Kwiatkowski, D.J., Ramesh, V., and Silva, A.J. (2008) Reversal of learning deficits in a $Tsc2^{+/-}$ mouse model of tuberous sclerosis. *Nat. Med.*, **14**, 843–848.

64 Lee, L., Sudentas, P., Donohue, B., Asrican, K., Worku, A., Walker, V., Sun, Y., Schmidt, K., Albert, M.S., El-Hashemite, N., Lader, A.S., Onda, H., Zhang, H., Kwiatkowski, D.J., and Dabora, S.L. (2005) Efficacy of a rapamycin analog (CCI-779) and IFN-gamma in tuberous sclerosis mouse models. *Genes Chromosomes Cancer*, **42**, 213–227.

65 Dabora, S.L., Roberts, P., Nieto, A., Perez, R., Jozwiak, S., Franz, D., Bissler, J., Thiele, E.A., Sims, K., and Kwiatkowski, D.J. (2002) Association between a high-expressing interferon-gamma allele and a lower frequency of kidney angiomyolipomas in TSC2 patients. *Am. J. Hum. Genet.*, **71**, 750–758.

66 Lee, L., Sudentas, P., and Dabora, S.L. (2006) Combination of a rapamycin analog (CCI-779) and interferon-gamma is more effective than single agents in treating a mouse model of tuberous sclerosis complex. *Genes Chromosomes Cancer*, **45**, 933–944.

67 Messina, M.P., Rauktys, A., Lee, L., and Dabora, S.L. (2007) Tuberous sclerosis preclinical studies: timing of treatment, combination of a rapamycin analog (CCI-779) and interferon-gamma, and comparison of rapamycin to CCI-779. *BMC Pharmacol.*, **7**, 14.

68 Meikle, L., Pollizzi, K., Egnor, A., Kramvis, I., Lane, H., Sahin, M., and Kwiatkowski, D.J. (2008) Response of a neuronal model of tuberous sclerosis to mammalian target of rapamycin (mTOR) inhibitors: effects on mTORC1 and Akt signaling lead to improved survival and function. *J. Neurosci.*, **28**, 5422–5432.

69 Zeng, L.H., Xu, L., Gutmann, D.H., and Wong, M. (2008) Rapamycin prevents epilepsy in a mouse model of tuberous sclerosis complex. *Ann. Neurol.*, **63**, 444–453.

70 Uhlmann, E.J., Apicelli, A.J., Baldwin, R.L., Burke, S.P., Bajenaru, M.L., Onda, H., Kwiatkowski, D., and Gutmann, D.H. (2002) Heterozygosity for the tuberous sclerosis complex (TSC) gene products results in increased astrocyte numbers and decreased p27-Kip1 expression in TSC2 + /− cells. *Oncogene*, **21**, 4050–4059.

71 Pollizzi, K., Malinowska-Kolodziej, I., Doughty, C., Betz, C., Ma, J., Goto, J., and Kwiatkowski, D.J. (2009) A hypomorphic allele of Tsc2 highlights the role of TSC1/TSC2 in signaling to AKTand models mild human TSC2 alleles. *Hum. Mol.* Genet., **18**, 2378–2387.

8
Animal Models of TSC: Insights from *Drosophila*

Duojia Pan

8.1
Introduction

The *TSC1* and *TSC2* genes were initially mapped to human chromosomes 9q34 and 16p13, subsequently identified by positional cloning, and shown to encode a coiled-coil domain-containing protein and a putative GTPase activating protein (GAP), respectively (Chapter 4) [1, 2]. The molecular identification of TSC1 (also called hamartin) and TSC2 (also called tuberin) unleashed an era of mechanistic studies into their biochemical and cellular functions. Early studies in mammalian cells implicated these tumor suppressor proteins in numerous cellular functions, including cell cycle [3, 4], endocytosis [5], cell adhesion [6], and transcription [7]. However, it was not clear how these diverse effects of TSC1 and TSC2 could account for their tumor suppressor function, nor what was the precise mechanism of that function. Despite the lack of a unifying molecular mechanism for TSC1 and TSC2, one consistent finding from these early studies was that TSC1 and TSC2 formed a TSC1–TSC2 protein complex in mammalian cells [8, 9], providing a tentative explanation for the similar disease phenotype when either gene is mutated in TSC patients.

A powerful approach to understanding gene function in humans is to take advantage of the evolutionary conservation of many gene products. For example, despite the ~700 million year evolutionary divergence, ~70% of human disease genes have direct counterparts in the *Drosophila* genome [10], thus making *Drosophila* an ideal model system to understand the detailed molecular mechanisms of the conserved human disease genes. Indeed, simple model organisms such as the fruit fly *Drosophila* and the roundworm *Caenorhabditis elegans* have been instrumental in unraveling many important signaling pathways in animal development and physiology due to their small genome size, minimal genetic redundancy, ease of handling, rapid development, and short breeding cycle.

In the last decade, studies in *Drosophila* have provided key insights into the molecular mechanisms of the TSC1–TSC2 tumor suppressor complex [11]. Most notably, these studies led to the discovery a signaling pathway that connects TSC–TSC2 to the small GTPase Rheb (*R*as *h*omologue *e*nriched in *b*rain) and the

Tuberous Sclerosis Complex: Genes, Clinical Features, and Therapeutics
Edited by David J. Kwiatkowski, Vicky Holets Whittemore, and Elizabeth A. Thiele

ISBN: 978-3-527-32201-5

Ser/Thr kinase TOR (target of rapamycin). Here, I provide a historical review on how *Drosophila* genetics has facilitated our understanding of the TSC1–TSC2 tumor suppressor complex in humans.

8.2 Connecting TSC1–TSC2 to the Insulin/PI3K Signaling Pathway

The first *Drosophila* model of TSC was reported in 1999. In a forward genetic screen for mutations that affect eye development, Ito and Rubin [12] isolated a group of mutations that resulted in a dramatic increase in cell size. Molecular characterization demonstrated that these mutations occurred in a single gene representing the *Drosophila TSC2* homologue. They further reported the existence of a *TSC1* homologue in the *Drosophila* genome [12]. For simplicity, we shall refer to the *Drosophila TSC1* and *TSC2* homologues as *Tsc1* and *Tsc2* to be distinguished from their mammalian counterparts. Due to an error in DNA quantitation, the authors mistakenly concluded that the increased size of the *Tsc2* mutant cells is caused by endoreplication, in which the mutant cells undergo repeated S-phase without entering M-phase. Despite this imperfection, this study is significant since it demonstrated that increased cell size, a phenotype that is also manifest in subependymal giant cell astrocytoma (SEGA) and other tumors in TSC patients, might represent a fairly direct consequence of the loss of the TSC1–TSC2 complex. Following this study, three groups independently reported that mutations in the *Drosophila Tsc1* gene cause a similar increase in cell size as the *Tsc2* mutations (Figure 8.1) [13–15], in agreement with the fact that mutations of either *TSC1* or *TSC2* lead to identical disease phenotypes in humans.

Concurrent with the report by Ito and Rubin and in an independent line of research, a number of studies in *Drosophila* have implicated the insulin pathway as a key mediator of cell growth control. The insulin signaling pathway, which was elucidated largely by biochemical studies in mammalian cells, is initiated by insulin receptor or insulin-like growth factor receptors (Chapter 6) [16]. These receptors activate PI3K either directly or through IRS proteins. Phosphorylation of the membrane lipid PIP2 by PI3K produces the second messenger PIP3, which activates PDK1 and Akt. The PTEN tumor suppressor protein is a negative regulator of the insulin pathway by acting as a phosphatase that converts PIP3 to PIP2 [17]. The insulin pathway was thought to increase protein translation by phosphorylating the ribosomal subunit S6 kinase (S6K) and initiation factor 4E binding protein (4E-BP; also known as PHAS). However, despite the demonstration that the insulin pathway is required for mammalian tissue and organismal growth [18, 19], these earlier studies largely did not distinguish whether the major effect of the insulin pathway was on cell growth or cell cycle control. In contrast, genetic studies in *Drosophila* clearly implicated insulin signaling in the control of cell size. Thus, loss-of-function mutations in positive components of the pathway, such as *PI3K* [20], *IRS/chico* [21], and *Akt* [22] lead to decreased cell size, while mutations of the negative regulator *PTEN* lead to an increase in cell size [23–25]. A unique feature of the *Drosophila* system is the ability to monitor cell growth in

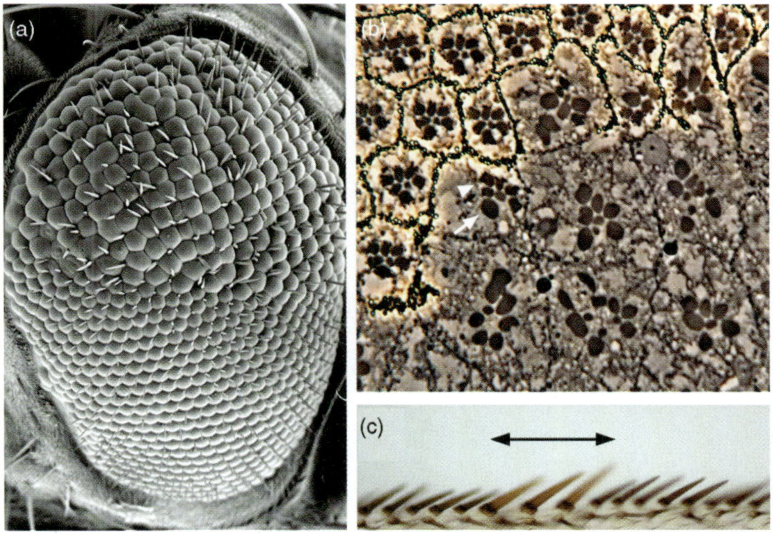

Figure 8.1 Loss of *Tsc1* causes an increase in cell size in *Drosophila*. *Tsc2* mutant cells show a similar cell size phenotype (not shown). Modified from Gao and Pan [15]. (a) An adult eye carrying a *Tsc1*$^{-/-}$ clone. The mutant clone is located at the upper portion of the eye. Note that *Tsc1*$^{-/-}$ ommatidia are larger and bulge out of the eye surface. (b) Section through a *Tsc1*$^{-/-}$ clone. The mutant clone is marked by the absence of pigment. At the clone border, mosaic ommatidia containing normal-sized heterozygous cells (arrowhead) and enlarged homozygous *Tsc1* mutant cells (arrow) can be seen, indicating that *Tsc1* controls cell size autonomously. (c) *Tsc1*$^{-/-}$ bristles on the wing margin are larger. Mutant bristles, genetically marked by the absence of the *y* gene product and thus displaying lighter pigmentation, are indicated by a line above the wing margin.

mosaic animals containing genetically mutant cells in an otherwise wild-type genetic background, which greatly facilitates a side-by-side comparison of cell size between different genotypes.

The observation that mutations in *Drosophila Tsc1* or *Tsc2* lead to a similar increase in cell size as mutations in *PTEN* immediately suggested a possible functional connection between Tsc1–Tsc2 and the insulin pathway. This hypothesis was tested by genetic epistasis analysis, which showed that double mutants of *Tsc1* and *Akt* displayed a *Tsc1* mutant phenotype, thus placing *Tsc1*genetically downstream of *Akt* (Figure 8.2) [14, 15]. It is worth noting that such genetic epistasis cannot distinguish whether the Tsc1–Tsc2 complex acts downstream of Akt in a linear biochemical pathway or functions in a parallel pathway that converges onto the insulin pathway at a point downstream of Akt. The latter was favored by Gao and Pan, who found that cells that are doubly mutant for *Tsc1* and *PTEN* show an additive increase in cell size compared to cells that are mutant for either gene alone (Figure 8.2) [15].

Thus, starting as two "orphan" tumor suppressor proteins, genetic studies in *Drosophila* provided the first clues linking Tsc1–Tsc2 to a known growth-regulatory pathway and set the stage for further studies into the biochemical functions of this tumor suppressor complex.

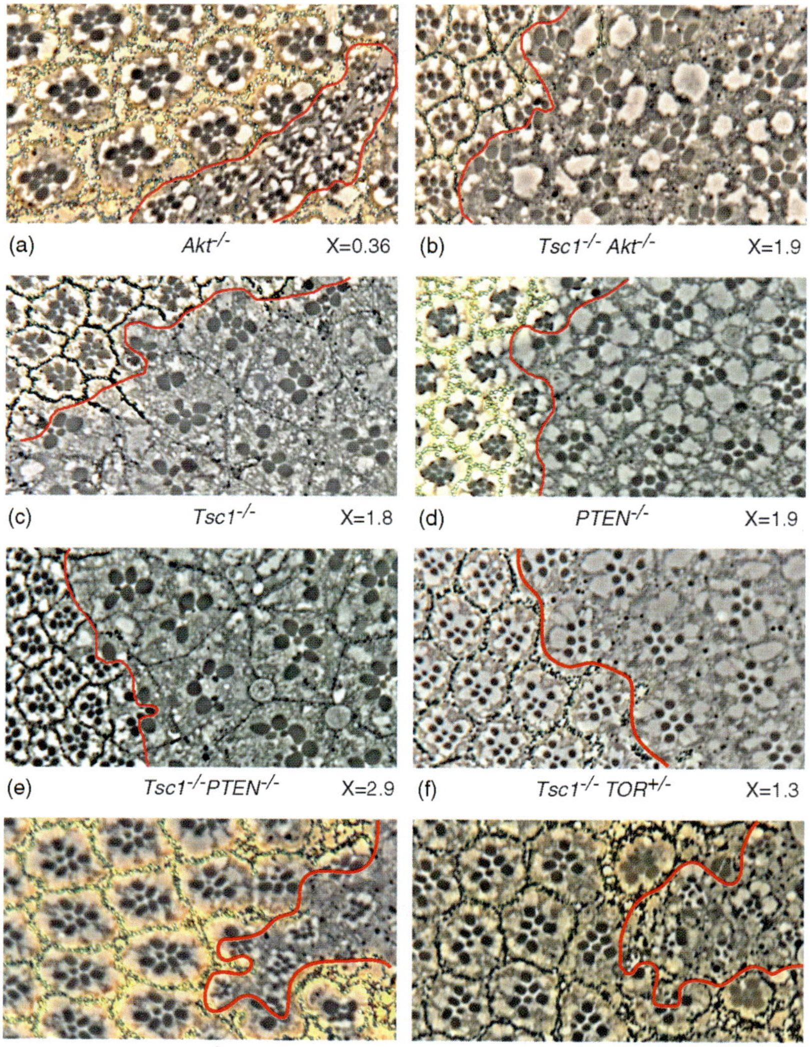

Figure 8.2 Genetic epistasis studies link *Tsc1–Tsc2* to the insulin and TOR signaling pathways. Modified from Gao and Pan [15] and Gao *et al.* [31]. Sections of *Drosophila* compound eyes containing clones of various genotypes. The genotype of the mutant clone is labeled below each section, as well as the relative size of the mutant photoreceptor rhabdomeres as compared to their wild-type counterparts (X value). Mutant clones were marked by the absence of pigment, and the borders of mutant clones were outlined with red lines. In all panels, the mutant clone is located at the right side of the section: (a) $Akt^{-/-}$, (b) $Tsc1^{-/-}Akt^{-/-}$, (c) $Tsc1^{-/-}$, (d) $PTEN^{-/-}$, (e) $Tsc1^{-/-}PTEN^{-/-}$, (f) $Tsc1^{-/-}TOR^{+/-}$, (g) $TOR^{-/-}$, and (h) $Tsc1^{-/-}TOR^{-/-}$.

8.3
The Tsc1–Tsc2 Complex as a Negative Regulator of TORC1

Besides the insulin pathway, translational regulators such as S6K and 4E-BP were also known to be regulated by a nutrient-sensitive pathway in mammalian cells involving the Ser/Thr protein kinase mammalian target of rapamycin (mTOR). mTOR exists in two distinct complexes, TORC1 and TORC2, which share mTOR and mLST8, but each has its unique subunits [26, 27]. TORC1, but not TORC2, is specifically inhibited by the immunosuppressant rapamycin. TORC1 couples a wide range of signals, such as amino acids and cellular energy levels, to the phosphorylation of S6K and 4E-BP. For example, amino acid starvation results in a rapid dephosphorylation of S6K and 4E-BP, while readdition of amino acids restores S6K and 4E-BP phosphorylation in a TORC1-dependent manner [28].

That both TORC1 and Tsc1–Tsc2 represent parallel inputs into the insulin pathway prompted investigation into the relationship between Tsc1–Tsc2 and TORC1 in *Drosophila*. This effort was facilitated by previous genetic analysis of *Drosophila TOR* mutants, which display decreased cell size and phenotypes similar to those caused by defective insulin signaling [29, 30]. Strikingly, heterozygosity of *TOR* suppressed the early lethality and cell size phenotype of loss-of-function *Tsc1* or *Tsc2* mutant flies (Figure 8.2) [31]. Such dominant dosage-sensitive genetic interactions suggested a functional link between *Tsc1–Tsc2* and *TOR*. Furthermore, double mutants of *Tsc1–Tsc2* and *TOR* showed a *TOR* mutant-like cell size phenotype, thus genetically placing *Tsc1–Tsc2* upstream of *TOR* (Figure 8.2) [31]. Taken together, these genetic studies suggested that Tsc1–Tsc2 acts upstream of and negatively regulates TORC1 in cell growth control in *Drosophila*.

The genetic relationship between *Tsc1–Tsc2* and *TOR* was further substantiated by biochemical analyses in *Drosophila* cell culture [31, 32]. Consistent with the genetic model, RNAi knockdown of Tsc1 or Tsc2 leads to increased S6K activity in a TOR-dependent manner [31, 32]. Interestingly, while S6K activity in normal *Drosophila*Schneider 2 (S2) cells is dependent on the presence of amino acids such that amino acid starvation leads to rapid inactivation of S6K, cells treated with Tsc1 or Tsc2 RNAi are strongly resistant to amino acid starvation such that S6K activity remains high after a long period of amino acid starvation [31], further implicating the Tsc1–Tsc2 complex as a negative regulator of the TORC1 signaling pathway.

8.4
Identification of the Small GTPase Rheb as a Direct Target of the Tsc1–Tsc2 Complex

Based on sequence homology, the TSC2 protein was predicted to function as a GTPase activating protein toward Ras family GTPases. GAP proteins normally stimulate the intrinsic GTPase activity of Ras proteins, facilitating the transition from the GTP-bound active form to the GDP-bound inactive form [33]. The GAP homology domain of TSC2 is clearly important for its biological function, since *TSC2* mutations that cause truncation of this domain are common in TSC patients

(Chapter 4), and *TSC2* missense mutations affecting this domain are also relatively common in TSC patients (Chapter 4) [36, 37]. However, the physiological substrate of TSC2's GAP domain had been elusive.

Genetic studies in *Drosophila* convincingly identified the Ras family GTPase Rheb as the elusive GTPase target of the Tsc2 GAP domain [38–40]. Loss of Rheb decreased S6K activity and inhibited cell growth, while overexpression of Rheb increased S6K activity and promoted cell growth. Furthermore, overexpression of Rheb conferred resistance to amino acid starvation similar to that caused by loss of Tsc1 or Tsc2 in *Drosophila* S2 cells [38]. Genetic analyses placed Rheb downstream of the Tsc1–Tsc2 complex. These observations strongly suggested that Rheb is a target of the Tsc2 GAP activity [38, 39]. Consistent with this model, cooverexpression of Tsc1 and Tsc2 dramatically decreased the relative ratio of GTP- to GDP-bound Rheb, while RNAi inhibition of Tsc2 resulted in an increase in Rheb GTP loading [41]. *In vitro*, the purified Tsc2 GAP domain displayed specific GAP activity toward Rheb, but not the closely related small GTPase Ras, demonstrating the specificity of the Tsc2 GAP activity [41]. These conclusions were further extended by studies in mammalian cells showing that overexpression of Rheb activated S6K and that the TSC1–TSC2 complex displayed specific GAP activity toward Rheb [42–45]. Taken together, these studies suggested an evolutionarily conserved role for Rheb as a direct target of TSC1–TSC2 in regulating TORC1 signaling.

Small GTPases cycle between an active GTP-bound form and an inactive GDP-bound form. While GAP proteins promote the transition of GTPases from GTP-bound to the GDP-bound forms, guanine–nucleotide exchange factors (GEFs) act in the opposite direction, facilitating the exchange of bound GDP by GTP. Recently, Choi and coworkers identified the *Drosophila* homologue of translationally controlled tumor protein (TCTP), a highly conserved protein upregulated in various tumors, as a putative GEF for Rheb [46]. RNAi knockdown of TCTP resulted in a similar decrease in cell size as that observed in *Rheb* mutant cells. Furthermore, TCTP directly associated with Rheb and displayed GEF activity against Rheb *in vivo* and *in vitro*. It will be important to investigate whether TCTP also functions as a physiological Rheb GEF in mammalian cells. If so, TCTP might offer a potential therapeutic target against TSC.

How does Rheb signal to TORC1? While this has yet to be reported in *Drosophila*, studies in mammalian cells have shown that Rheb can directly bind to and stimulate TORC1 kinase activity [47]. This result is consistent with the observation that Tsc2 and TOR can be coimmunoprecipitated when overexpressed in *Drosophila* S2 cells [31], suggesting that Tsc1–Tsc2, Rheb, and TOR might function in closely linked protein complexes.

8.5
Control of Autophagy by the Tsc–Rheb–TORC1 Pathway

In addition to stimulating protein translation, TORC1 signaling is also known to inhibit autophagy, a catabolic process involving the degradation of a cell's own components through the formation of a specialized double-membrane structure called the autophagosome [48]. While the molecular machinery controlling autophagy has been well defined in the budding yeast *Saccharomyces cerevisiae*, the relationships between TORC1 signaling and autophagy and its relevance to cell growth control have only recently been studied in *Drosophila* [49].

Neufeld and coworkers found that overexpression of the *Drosophila* homologue of Atg1, a Ser/Thr kinase and a critical regulator of autophagy in yeast, leads to a rapid induction of autophagy in multiple cell types including the larval fat body [49]. This autophagic response is similar to that induced by starvation or rapamycin treatment, and was strongly inhibited by loss of *Tsc1* or overexpression of *Rheb*, in agreement with the negative regulation of Atg1 by TORC1 signaling as described in yeast. Unexpectedly, Atg1 overexpression also leads a decrease in TORC1 activity, suggesting that TORC1 and Atg1 may mutually antagonize each other.

To examine the role of autophagy in TORC1-dependent growth control, Neufeld and coworkers examined the growth property of *Atg1* null mutant cells under different genetic background and nutritional conditions [49]. In normally fed animals, loss of *Atg1* has no effect on cell growth, consistent with the low rates of autophagy under such conditions. In contrast, *Atg1* null mutant cells display a significant growth advantage over surrounding wild-type cells under starvation or rapamycin treatment, suggesting that autophagy contributes to growth inhibition under these conditions. Unexpectedly, loss of *Atg1* does not ameliorate the growth defect exhibited by *TOR* mutant cells. Thus, autophagy is partly responsible for the growth inhibition of cells with reduced but not complete block of TORC1 signaling.

8.6
Cross Talk Between the Tsc–Rheb–TORC1 Pathway and the Insulin Pathway

Several studies have been conducted in *Drosophila* to investigate how the Tsc–Rheb–TORC1 pathway intersects with the insulin signaling pathway. In an earlier study, it was reported that Akt can phosphorylate and inactivate TSC2, thus providing an attractive and simple linear pathway linking the canonical insulin pathway and the Tsc–Rheb–TORC1 pathway [50]. However, this conclusion was challenged by the observation that a mutant form of Tsc2 carrying mutations of the presumed Akt phosphorylation sites could completely rescue the lethality and cell growth defects of *Tsc2* null mutant flies [51]. Thus, unless there are additional Akt phosphorylation sites on Tsc2, Tsc2 is not likely a critical substrate of Akt during normal *Drosophila* development, in agreement with previous models that Tsc1–Tsc2 functions in parallel with the canonical insulin pathway [15].

Genetic studies in *Drosophila* further revealed an important cross talk between the Tsc–Rheb–TORC1 and the insulin pathways. Radimerski *et al.* found that in both *Drosophila* larvae and cultured cells, loss of Tsc1–Tsc2 not only leads to increased S6K activity but also to inhibition of Akt activity [32]. Interestingly, the latter effect can be relieved by loss of S6K. These observations provided *in vivo* evidence for an important negative feedback between TORC1 and the insulin pathway whereby hyperactivation of the TORC1 leads to suppression of insulin signaling. An analogous feedback inhibition of insulin signaling induced by loss of Tsc1–Tsc2 was also observed in mammals, which provides a plausible explanation for the benign nature and limited proliferative capacity of the tumors arising in TSC patients as compared to those caused by loss of PTEN [52]. The molecular mechanism underlying this negative feedback has also been probed. In mammalian cells, this feedback loop appears to involve S6K-mediated phosphorylation and downregulation of IRS proteins and reduced expression of PDGFR [53,54a,54b]; both of these effects are reversed by rapamycin treatment. In *Drosophila*, the limited growth of *Tsc1* mutant tissues can be at least partially accounted for by increased protein levels of the FOXO transcription factor, which activates the transcription of negative growth regulators such as 4E-BP [55].

8.7 Relationship Between Tsc1–Tsc2 and Amino Acids-Mediated TORC1 Activation

A long-standing question in TOR signaling concerns how amino acids signal to stimulate TORC1. While RNAi knockdown of Tsc1 or Tsc2 leads to resistance to amino acid starvation in S2 cells [31], the GTP loading of Rheb is insensitive to amino acid levels [41]. These results favor a model in which Tsc1–Tsc2–Rheb and amino acids provide parallel inputs into TORC1.

Very recently, two exciting reports identified the Rag GTPases as obligatory intermediates in relaying the amino acid signal to TORC1 [56,57a]. In one study, Rag GTPases were purified as novel TORC1 components in mammalian cells [57a], while in the other study, Rag GTPases were identified in an RNAi screen for GTPases that affected S6K activation in *Drosophila* S2 cells [56]. RNAi knockdown of Rag expression suppressed the stimulatory effect of amino acids on TORC1. Conversely, expression of GTP-bound and constitutively active Rag mutants activated TORC1 in the absence of amino acids. Inactivation of Rag GTPases resulted in a reduction in cell size in fed but not starved *Drosophila* larvae, further supporting a requirement for Rag proteins in nutrient response *in vivo* (Figure 8.3) [56]. In mammalian cells, it was further shown that amino acids increase the GTP loading of Rag GTPases, which in turn promotes the translocation of TOR to Rab-7 positive structures [57a]. How amino acids activate Rag GTPases remains to be determined.

8.8 Upstream of the Tsc1–Tsc2 Complex

In contrast to our knowledge of its downstream effectors, relatively less is known about upstream regulators of the Tsc1–Tsc2 complex. In mammalian cells, multiple

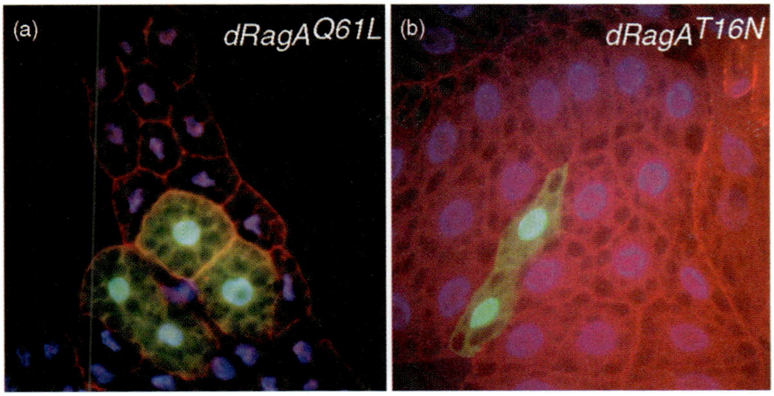

Figure 8.3 Nutrient-sensitive effect of Rag GTPases on cell size in *Drosophila*. Modified from Kim *et al.* [56]. Fat body cell clones overexpressing *dRagA* transgenes are positively marked by the expression of GFP (green). *Drosophila* larvae were subjected to standard fly food (fed) or food containing agar only (starved) for 48 h before analysis. Cell membrane and nuclei were highlighted by red and blue staining, respectively. (a) Clones overexpressing *dRagA*^{Q61L}, a constitutively active form of *dRagA*, under starvation condition. *dRagA*^{Q61L}-expressing cells were larger than their neighboring wild-type cells under starved (shown here) but not fed (not shown) condition. (b) Clones overexpressing *dRagA*^{T16N}, a dominant negative form of *dRagA*, under fed condition. *dRagA*^{T16N}-expressing cells were smaller than their neighboring wild-type cells under fed (shown here) but not starved (not shown) condition.

kinases can phosphorylate TSC1 or TSC2 at multiple sites, but the precise integration of these potential signals to regulate the GAP activity of TSC2 is difficult to dissect (Chapter 6) [57b]. One notable upstream regulator of TSC2 is AMP-activated protein kinase (AMPK), which directly senses cellular AMP/ATP ratio. AMPK can phosphorylate TSC2 and enhance its activity in mammalian cells, thus coupling cellular energy levels to TORC1 activity [58]. It is unclear whether this mechanism is conserved in *Drosophila*, since the AMPK phosphorylation site on TSC2 is not conserved in its *Drosophila* counterpart.

The best characterized upstream regulators of Tsc1–Tsc2 in *Drosophila* are Scylla and Charybdis, two related proteins whose expression is induced by hypoxia and starvation [59]. Overexpression of Scylla/Charybdis reduces cell growth, while simultaneous loss of both genes causes mild overgrowth and elevated susceptibility to hypoxia and starvation. More important, overexpression of Scylla/Charybdis does not suppress the cell growth phenotype caused by loss of Tsc2 or activation of Rheb, and triple mutants of *Tsc2, scylla,* and *charybdis* show a similar increase in cell size as *Tsc2* alone, suggesting that Scylla/Charybdis function upstream of Tsc1–Tsc2. The requirement for Scylla/Charybdis in the hypoxia response appears to be conserved in mammals, since mTOR inhibition by hypoxia also requires the TSC1/TSC2 tumor suppressor complex and the hypoxia-inducible gene REDD1/RTP801, a mammalian homologue of Scylla/Charybdis [60].

8.9 Summary

The fruit fly *Drosophila* has played a critical role in unraveling the molecular function of the Tsc1–Tsc2 tumor suppressor proteins. The elucidation of the Tsc–Rheb–TORC1 pathway provided a theoretical basis for clinical trials using rapamycin or its derivatives for the treatment of TSC patients [61]. Dissection of the Tsc1–Tsc2 complex in *Drosophila* has benefited not only from the power of genetic epistasis analyses and the evolutionary conservation of gene function but also from biochemical studies in *Drosophila* S2 cells, where the entire insulin/TOR signaling network is operational. Thus, insights learned from genetic studies at the organismal level can be quickly tested at the molecular and cellular levels in S2 cells, and conversely, proteins identified in cell-based assays can be readily tested for their physiological function by genetic manipulations. It is anticipated that *Drosophila* will continue to serve as an important model system to understand the biology of TSC.

Acknowledgments

TSC research in the Pan laboratory was conducted when the author was a Virginia Murchison Linthicum Endowed Scholar in Medical Science at UT Southwestern Medical Center, and was supported by National Institutes of Health (GM62323), American Heart Association (0130222N), American Cancer Society (RSG0303601DDC), and Tuberous Sclerosis Alliance.

References

1 van Slegtenhorst, M., de Hoogt, R., Hermans, C., Nellist, M., Janssen, B., Verhoef, S., Lindhout, D., van den Ouweland, A., Halley, D., Young, J., Burley, M., Jeremiah, S., Woodward, K., Nahmias, J., Fox, M., Ekong, R., Osborne, J., Wolfe, J., Povey, S., Snell, R.G., Cheadle, J.P., Jones, A.C., Tachataki, M., Ravine, D., Sampson, J.R., Reeve, M.P., Richardson, P., Wilmer, F., Munro, C., Hawkins, T.L., Sepp, T., Ali, J.B.M., Ward, S., Green, A.J., Yates, J.R.W., Kwiatkowska, J., Henske, E.P., Short, M.P., Haines, J.H., Jozwiak, S., and Kwiatkowski, D.J. (1997) Identification of the tuberous sclerosis gene *TSC1* on chromosome 9q34. *Science*, **277**, 805–808.

2 The European Chromosome 16 Tuberous Sclerosis Consortium (1993) Identification and characterization of the tuberous sclerosis gene on chromosome 16. *Cell*, **75**, 1305–1315.

3 Soucek, T., Yeung, R.S., and Hengstschläger, M. (1998) Inactivation of the cyclin-dependent kinase inhibitor p27 upon loss of the tuberous sclerosis complex gene-2. *Proc. Natl. Acad. Sci. USA*, **95**, 15653–15658.

4 Soucek, T., Pusch, O., Wienecke, R., DeClue, J.E., and Hengstschläger, M. (1997) Role of the tuberous sclerosis gene-2 product in cell cycle control. Loss of the tuberous sclerosis gene-2 induces quiescent cells to enter S phase. *J. Biol. Chem.*, **272**, 29301–29308.

5 Xiao, G.-H., Shoarinejad, F., Jin, F., Golemis, E.A., and Yeung, R.S. (1997) The tuberous sclerosis 2 gene product, tuberin, functions as a Rab5 GTPase activating protein (GAP) in modulating

endocytosis. *J. Biol. Chem.*, **272**, 6097–6100.

6 Lamb, R.S., Roy, C., Diefenbach, T.J., Vinters, H.V., Johnson, M.W., Jay, D.G., and Hall, A. (2000) The TSC1 tumor suppressor hamartin regulates cell adhesion through ERM proteins and the GTPase Rho. *Nat. Cell Biol.*, **2**, 281–287.

7 Henry, K.W., Yuan, X., Koszewski, N.J., Onda, H., Kwiatkowski, D.J., and Nordstrom, W. (1998) Tuberous sclerosis gene 2 product modulates transcription mediated by steroid hormone receptor family members. *J. Biol. Chem.*, **273**, 20535–20539.

8 Nellist, M., van Slegtenhorst, M., Goedbloed, M., van den Ouweland, A., Halley, D., and van der Sluijs, P. (1999) Characterization of the cytosolic tuberin–hamartin complex. Tuberin is a cytosolic chaperone for hamartin. *J. Biol. Chem.*, **274**, 35647–35652.

9 van Slegtenhorst, M., Nellist, M., Nagelkerken, B., Cheadle, J.P., Snell, R.G., van den Ouweland, A., Reuser, A., Sampson, J.R., Halley, D., and van der Sluijs, P. (1998) Interaction between hamartin and tuberin, the *TSC1* and *TSC2* gene products. *Hum. Mol. Genet.*, **7**, 1053–1057.

10 Rubin, G.M., Yandell, M.D., Wortman, J.R., Gabor Miklos, G.L., Nelson, C.R., Hariharan, I.K., Fortini, M.E., Li, P.W., Apweiler, R., Fleischmann, W., Cherry, J.M., Henikoff, S., Skupski, M.P., Misra, S., Ashburner, M., Birney, E., Boguski, M.S., Brody, T., Brokstein, P., Cellniker, S.E., Chervitz, S.A., Coates, D., Cravchik, A., Gabrielian, A., Galle, R.F., Gelbart, W.M., George, R.A., Goldstein, L.S., Gong, F., Guan, P., Harris, N.L., Hay, B.A., Hoskins, R.A., Li, J., Li, Z., Hynes, R.O., Jones, S.J., Kiehl, P.M., Lemaitre, B., Littleton, J.T., Morrison, D.K., Mungall, C., O'Farrell, P.H., Pickeral, O.K., Shue, C., Vosshall, L.B., Zhang, J., Zhao, Q., Zheng, X.H., Zhong, F., Zhong, W., Gibbs, R., Venter, J.C., Adams, M.D., and Lewis, S. (2000) Comparative genomics of the eukaryotes. *Science*, **287**, 2204–2215.

11 Pan, D., Dong, J., Zhang, Y., and Gao, X. (2004) Tuberous sclerosis complex: from Drosophila to human disease. *Trends Cell Biol.*, **14**, 78–85.

12 Ito, N. and Rubin, G.M. (1999) *gigas*, A *Drosophila* homolog of tuberous sclerosis gene product-2, regulates the cell cycle. *Cell*, **96**, 529–539.

13 Tapon, N., Ito, N., Dickson, B.J., Treisman, J.E., and Hariharan, I.K. (2001) The Drosophila tuberous sclerosis complex gene homologs restrict cell growth and cell proliferation. *Cell*, **105**, 345–355.

14 Potter, C.J., Huang, H., and Xu, T. (2001) Drosophila Tsc1 functions with Tsc2 to antagonize insulin signaling in regulating cell growth, cell proliferation, and organ size. *Cell*, **105**, 357–368.

15 Gao, X. and Pan, D. (2001) TSC1 and TSC2 tumor suppressors antagonize insulin signaling in cell growth. *Genes Dev.*, **15**, 1383–1392.

16 Proud, C.G. and Denton, R.M. (1997) Molecular mechanisms for the control of translation by insulin. *Biochem. J.*, **328**, 329–341.

17 Cantley, L.C. and Neel, B.G. (1999) New insights into tumor suppression: PTEN suppresses tumor formation by restraining the phosphoinositide 3-kinase/AKT pathway. *Proc. Natl. Acad. Sci. USA*, **96**, 4240–4245.

18 Shima, H., Pende, M., Chen, Y., Fumagalli, S., Thomas, G., and Kozma, S.C. (1998) Disruption of the p70(s6k)/p85(s6k) gene reveals a small mouse phenotype and a new functional S6 kinase. *EMBO J.*, **17**, 6649–6659.

19 Sun, H., Lesche, R., Li, D.M., Liliental, J., Zhang, H., Gao, J., Gavrilova, N., Mueller, B., Liu, X., and Wu, H. (1999) PTEN modulates cell cycle progression and cell survival by regulating phosphatidylinositol 3,4,5,-trisphosphate and Akt/protein kinase B signaling pathway. *Proc. Natl. Acad. Sci. USA*, **96**, 6199–6204.

20 Leevers, S.J., Weinkove, D., MacDougall, L.K., Hafen, E., and Waterfield, M.D. (1996) The *Drosophila* phosphoinositide 3-kinase Dp110 promotes cell growth. *EMBO J.*, **15**, 6584–6594.

21 Böhni, R., Riesgo-Escovar, J., Oldham, S., Brogiolo, W., Stocker, H., Andruss, B.F., Beckingham, K., and Hafen, E. (1999) Autonomous control of cell and organ size

by CHICO, a *Drosophila* homolog of vertebrate IRS1-4. *Cell*, **97**, 865–875.

22 Verdu, J., Buratovich, M.A., Wilder, E.L., and Birnbaum, M.J. (1999) Cell-autonomous regulation of cell and organ growth in Drosophila by Akt/PKB. *Nat. Cell Biol.*, **1**, 500–506.

23 Huang, H., Potter, C.J., Tao, W., Li, D.-M., Brogiolo, W., Hafen, E., Sun, H., and Xu, T. (1999) PTEN affects cell size, cell proliferation and apoptosis during *Drosophila* eye development. *Development*, **126**, 5365–5372.

24 Goberdhan, D.C.I., Paricio, N., Goodman, E.C., Mlodzik, M., and Wilson, C. (1999) *Drosophila* tumor suppressor *PTEN* controls cell size and number by antagonizing the Chico/PI3-kinase signaling pathway. *Genes Dev.*, **13**, 3244–3258.

25 Gao, X., Neufeld, T.P., and Pan, D. (2000) Drosophila PTEN regulates cell growth and proliferation through PI3K-dependent and -independent pathways. *Dev. Biol.*, **221**, 404–418.

26 Wullschleger, S., Loewith, R., and Hall, M.N. (2006) TOR signaling in growth and metabolism. *Cell*, **124**, 471–484.

27 Guertin, D.A. and Sabatini, D.M. (2007) Defining the role of mTOR in cancer. *Cancer Cell*, **12**, 9–22.

28 Hara, K., Yonezawa, K., Weng, Q.P., Kozlowski, M.T., Belham, C., and Avruch, J. (1998) Amino acid sufficiency and mTOR regulate p70 S6 kinase and eIF-4E BP1 through a common effector mechanism. *J. Biol. Chem.*, **273**, 14484–14494.

29 Zhang, H., Stallock, J.P., Ng, J.C., Reinhard, C., and Neufeld, T.P. (2000) Regulation of cellular growth by the *Drosophila* target of rapamycin *dTOR*. *Genes Dev.*, **14**, 2712–2724.

30 Oldham, S., Montagne, J., Radimerski, T., Thomas, G., and Hafen, E. (2000) Genetic and biochemical characterization of dTOR, the *Drosophila* homolog of the target of rapamycin. *Genes Dev.*, **14**, 2689–2694.

31 Gao, X., Zhang, Y., Arrazola, P., Hino, O., Kobayashi, T., Yeung, R.S., Ru, B., and Pan, D. (2002) Tsc tumour suppressor proteins antagonize amino-acid-TOR signalling. *Nat. Cell Biol.*, **4**, 699–704.

32 Radimerski, T., Montagne, J., Hemmings-Mieszczak, M., and Thomas, G. (2002) Lethality of Drosophila lacking TSC tumor suppressor function rescued by reducing dS6K signaling. *Genes Dev.*, **16**, 2627–2632.

33 Boguski, M.S. and McCormick, F. (1993) Proteins regulating Ras and its relatives. *Nature*, **366**, 643–654.

34 Kobayashi, T., Hirayama, Y., Kobayashi, E., Kubo, Y., and Hino, O. (1995) A germline insertion in the tuberous sclerosis (Tsc2) gene gives rise to the Eker rat model of dominantly inherited cancer. *Nat. Genet.*, **9**, 70–74.

35 Yeung, R.S., Xiao, G.H., Jin, F., Lee, W.C., Testa, J.R., and Knudson, A.G. (1994) Predisposition to renal carcinoma in the Eker rat is determined by germ-line mutation of the tuberous sclerosis 2 (TSC2) gene. *Proc. Natl. Acad. Sci. USA*, **91**, 11413–11416.

36 Dabora, S.L., Jozwiak, S., Franz, D.N., Roberts, P.S., Nieto, A., Chung, J., Choy, Y.S., Reeve, M.P., Thiele, E., Egelhoff, J.C., Kasprzyk-Obara, J., Domanska-Pakiela, D., and Kwiatkowski, D.J. (2001) Mutational analysis in a cohort of 224 tuberous sclerosis patients indicates increased severity of TSC2, compared with TSC1, disease in multiple organs. *Am. J. Hum. Genet.*, **68**, 64–80.

37 Maheshwar, M.M., Cheadle, J.P., Jones, A.C., Myring, J., Fryer, A.E., Harris, P.C., and Sampson, J.R. (1997) The GAP-related domain of tuberin, the product of the TSC2 gene, is a target for missense mutations in tuberous sclerosis. *Hum. Mol. Genet.*, **6**, 1991–1996.

38 Saucedo, L.J., Gao, X., Chiarelli, D.A., Li, L., Pan, D., and Edgar, B.A. (2003) Rheb promotes cell growth as a component of the insulin/TOR signalling network. *Nat. Cell Biol.*, **5**, 566–571.

39 Stocker, H., Radimerski, T., Schindelholz, B., Wittwer, F., Belawat, P., Daram, P., Breuer, S., Thomas, G., and Hafen, E. (2003) Rheb is an essential regulator of S6K in controlling cell growth in Drosophila. *Nat. Cell Biol.*, **5**, 559–566.

40 Patel, P.H., Thapar, N., Guo, L., Martinez, M., Maris, J., Gau, C.L., Lengyel, J.A., and

Tamanoi, F. (2003) Drosophila Rheb GTPase is required for cell cycle progression and cell growth. *J. Cell Sci.*, **116**, 3601–3610.
41 Zhang, Y., Gao, X., Saucedo, L.J., Ru, B., Edgar, B.A., and Pan, D. (2003) Rheb is a direct target of the tuberous sclerosis tumour suppressor proteins. *Nat. Cell Biol.*, **5**, 578–581.
42 Garami, A., Zwartkruis, F.J., Nobukuni, T., Joaquin, M., Roccio, M., Stocker, H., Kozma, S.C., Hafen, E., Bos, J.L., and Thomas, G. (2003) Insulin activation of Rheb, a mediator of mTOR/S6K/4E-BP signaling, is inhibited by TSC1 and 2. *Mol. Cell*, **11**, 1457–1466.
43 Tee, A.R., Manning, B.D., Roux, P.P., Cantley, L.C., and Blenis, J. (2003) Tuberous sclerosis complex gene products, tuberin and hamartin, control mTOR signaling by acting as a GTPase-activating protein complex toward Rheb. *Curr. Biol.*, **13**, 1259–1268.
44 Inoki, K., Li, Y., Xu, T., and Guan, K.L. (2003) Rheb GTPase is a direct target of TSC2 GAP activity and regulates mTOR signaling. *Genes Dev.*, **17**, 1829–1834.
45 Castro, A.F., Rebhun, J.F., Clark, G.J., and Quilliam, L.A. (2003) Rheb binds tuberous sclerosis complex 2 (TSC2) and promotes S6 kinase activation in a rapam. *J. Biol. Chem.*, **278**, 32493–32496.
46 Hsu, Y.C., Chern, J.J., Cai, Y., Liu, M., and Choi, K.W. (2007) Drosophila TCTP is essential for growth and proliferation through regulation of dRheb GTPase. *Nature*, **445**, 785–788.
47 Long, X., Lin, Y., Ortiz-Vega, S., Yonezawa, K., and Avruch, J. (2005) Rheb binds and regulates the mTOR kinase. *Curr. Biol.*, **15**, 702–713.
48 Klionsky, D.J. (2007) Autophagy: from phenomenology to molecular understanding in less than a decade. *Nat. Rev. Mol. Cell Biol.*, **8**, 931–937.
49 Scott, R.C., Juhasz, G., and Neufeld, T.P. (2007) Direct induction of autophagy by Atg1 inhibits cell growth and induces apoptotic cell death. *Curr. Biol.*, **17**, 1–11.
50 Potter, C.J., Pedraza, L.G., and Xu, T. (2002) Akt regulates growth by directly phosphorylating Tsc2. *Nat. Cell Biol.*, **4**, 658–665.
51 Dong, J. and Pan, D. (2004) Tsc2 is not a critical target of Akt during normal Drosophila development. *Genes Dev.*, **18**, 2479–2484.
52 Manning, B.D., Logsdon, M.N., Lipovsky, A.I., Abbott, D., Kwiatkowski, D.J., and Cantley, L.C. (2005) Feedback inhibition of Akt signaling limits the growth of tumors lacking Tsc2. *Genes Dev.*, **19**, 1773–1778.
53 Harrington, L.S., Findlay, G.M., Gray, A., Tolkacheva, T., Wigfield, S., Rebholz, H., Barnett, J., Leslie, N.R., Cheng, S., Shepherd, P.R., Gout, I., Downes, C.P., and Lamb, R.F. (2004) The TSC1–2 tumor suppressor controls insulin-PI3K signaling via regulation of IRS proteins. *J. Cell Biol.*, **166**, 213–223.
54 (a) Shah, O.J., Wang, Z., and Hunter, T. (2004) Inappropriate activation of the TSC/Rheb/mTOR/S6K cassette induces IRS1/2 depletion, insulin resistance, and cell survival deficiencies. *Curr. Biol.*, **14**, 1650–1656;(b) Zhang, H., Cicchetti, G., Onda, H., Koon, H.B., Asrican, K., Bajraszewski, N., Vazquez, F., Carpenter, C.L., and Kwiatkowski, D.J. (2003) Loss of Tsc1/Tsc2 activates mTOR and disrupts PI3K–Akt signaling through downregulation of PDGFR. *J. Clin. Invest.*, **112**, 1223–1233
55 Harvey, K.F., Mattila, J., Sofer, A., Bennett, F.C., Ramsey, M.R., Ellisen, L.W., Puig, O., and Hariharan, I.K. (2008) FOXO-regulated transcription restricts overgrowth of Tsc mutant organs. *J. Cell Biol.*, **180**, 691–696.
56 Kim, E., Goraksha-Hicks, P., Li, L., Neufeld, T.P., and Guan, K.L. (2008) Regulation of TORC1 by Rag GTPases in nutrient response. *Nat. Cell Biol.*, **10**, 935–945.
57 (a) Sancak, Y., Peterson, T.R., Shaul, Y.D., Lindquist, R.A., Thoreen, C.C., Bar-Peled, L., and Sabatini, D.M. (2008) The Rag GTPases bind raptor and mediate amino acid signaling to mTORC1. *Science*, **320**, 1496–1501; (b) Huang, J. and Manning, B.D. (2008) The TSC1–TSC2 complex: a molecular switchboard controlling cell growth. *Biochem. J.*, **412**, 179–190.
58 Inoki, K., Zhu, T., and Guan, K.L. (2003) TSC2 mediates cellular energy response to

control cell growth and survival. *Cell*, **115**, 577–590.

59 Reiling, J.H. and Hafen, E. (2004) The hypoxia-induced paralogs Scylla and Charybdis inhibit growth by down-regulating S6K activity upstream of TSC in Drosophila. *Genes Dev.*, **18**, 2879–2892.

60 Brugarolas, J., Lei, K., Hurley, R.L., Manning, B.D., Reiling, J.H., Hafen, E., Witters, L.A., Ellisen, L.W., and Kaelin, W.G., Jr. (2004) Regulation of mTOR function in response to hypoxia by REDD1 and the TSC1/TSC2 tumor suppressor complex. *Genes Dev.*, **18**, 2893–2904.

61 Bissler, J.J., McCormack, F.X., Young, L.R., Elwing, J.M., Chuck, G., Leonard, J.M., Schmithorst, V.J., Laor, T., Brody, A.S., Bean, J., Salisbury, S., and Franz, D.N. (2008) Sirolimus for angiomyolipoma in tuberous sclerosis complex or lymphangioleiomyomatosis. *N. Engl. J. Med.*, **358**, 140–151.

Part IV
Brain Involvement

Tuberous Sclerosis Complex: Genes, Clinical Features, and Therapeutics
Edited by David J. Kwiatkowski, Vicky Holets Whittemore, and Elizabeth A. Thiele

ISBN: 978-3-527-32201-5

9
Pathogenesis of TSC in the Brain

Peter B. Crino, Rupal Mehta, and Harry V. Vinters

9.1
Introduction

The neurologic manifestations of tuberous sclerosis complex (TSC) are believed to result from structural abnormalities in the brain that occur as a consequence of mutations in the *TSC1* or *TSC2* genes [1]. The neuropathologic features of TSC include tubers, subependymal nodules (SENs), and subependymal giant cell astrocytomas (SEGAs), also known as subependymal giant cell tumors (SGCTs). These three lesions exhibit some common histopathologic features including altered regional architecture, abnormal cellular morphology, and excessive numbers of astrocytes. From a clinical perspective, tubers are highly associated with infantile spasms, epilepsy, cognitive disability, and autism in TSC, while growth of SEGAs can cause hydrocephalus, increased intracranial pressure, and focal neurological signs. For many TSC patients who do not respond to antiepileptic drugs (AEDs), surgical resection of a tuber is often necessary to achieve seizure control. Surgical resection remains a mainstay of therapy for symptomatic SEGAs. Indeed, resected tubers and SEGAs yield numerous tissue samples for study at many tertiary care medical centers. Neuropathological examination of tubers and SEGAs is an important diagnostic tool for TSC patients. In addition, studying the cellular pathology of these lesions has led to pivotal insights into the developmental pathogenesis of TSC.

9.2
Tubers

Tubers are focal malformations of cortical development associated with TSC. They can be detected by fetal magnetic resonance imaging (MRI) as early as 20 weeks' gestation [2–4]. On gross pathological examination, tubers are well-circumscribed regions of cerebral cortex that can often be readily identified on visual inspection. They tend to be confined to a single gyrus but may span two or more gyri. In rare cases, tubers may have a lobar or even hemispheric distribution, sometimes even resulting in hemimegalencephaly (HME). The majority of tubers examined by

Tuberous Sclerosis Complex: Genes, Clinical Features, and Therapeutics
Edited by David J. Kwiatkowski, Vicky Holets Whittemore, and Elizabeth A. Thiele

ISBN: 978-3-527-32201-5

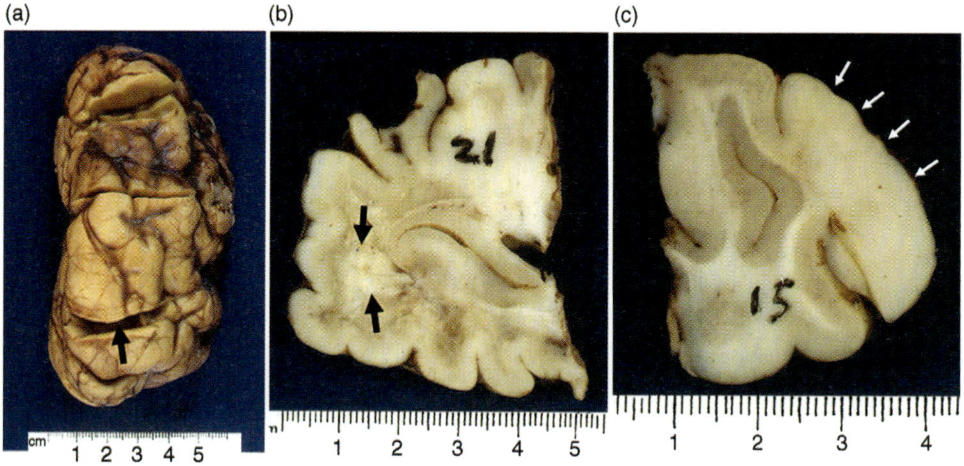

Figure 9.1 Resected cortical tubers in TSC. Panel (a) shows external appearance of a specimen. Arrow indicates large flat superficial tuber. (b and c) Two different sections of a fixed tuber. Note blurring of the cortex–white matter junction in many gyri. Arrows in (b) indicate an area that has a chalky white appearance, probably representing calcification. The number "21" on the specimen represents a unique identifier for this slice (as does "15" for the specimen in panel (c)). White arrows in panel (c) indicate a single large tuber in a slice that otherwise shows many regions of normal cortex–white matter.

neuropathologists are removed during resective epilepsy surgery. When seen as cortical resection specimens, tubers have a variable appearance, usually with mushroom-shaped gyri and loss of the cortex–white matter junction (Figure 9.1). However, it may be more correct to state that focal corticotomies are now frequently performed on TSC patients with intractable seizures, and the resulting tissue specimen may not have the dramatic appearance of a tuber as described above. Upon direct palpation, the root-like or tuberous consistency of a tuber is firmer than that of the surrounding cortex. On thick sectioning, tubers often are more pale than the surrounding cortex and may exhibit irregular thickening and blurring of the cortical mantle and gray and white matter junction. Tubers can extend deep into the subcortical white matter and measure several centimeters in cross-sectional area or they can be confined to the cortical mantle and measure less than a centimeter in size. In older children and adult TSC patients, tubers may calcify or even undergo cystic degeneration [5, 6]. Using Luxol fast blue stains, there is clear evidence for abnormal myelination in the subcortical white matter beneath tubers as evidenced by patchy pallor and vacuolization of the white matter [7].

On microscopic pathological examination using either Nissl or hematoxylin/eosin staining, there is marked disorganization of cortical lamination (Figure 9.2). Typically, the normal hexalaminar architecture of the cortex is lost or persists in only a rudimentary fashion. The boundaries between white matter and cortical mantle are blurred (this change is especially dramatic on Nissl or Kluver-Barrera stained sections). Interestingly, the junction between layer I and deeper portions of the cortex is often preserved. Other histologic features frequently encountered include

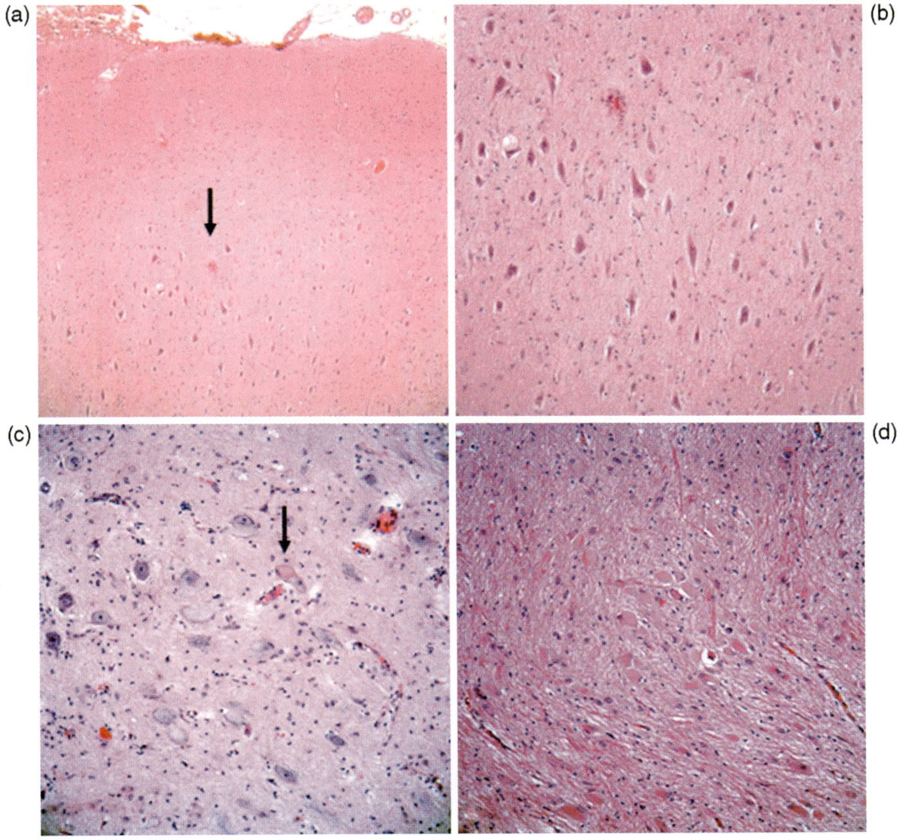

Figure 9.2 Microscopic pathology of cortical tubers in TSC. All sections are from cortical resections for intractable seizures in TSC. At low magnification (a), there is subtle yet easily recognizable neuronal disorganization, with some apparent neuronal "cytomegaly" (abnormal region indicated by arrow). At higher magnification (b), cytomegaly and disorganization are seen more clearly. Neuronal disorganization can be more profound (c), which also shows an admixture of enlarged ("cytomegalic") and slightly dysmorphic neurons, and at least one balloon cell (arrow). In another specimen (d), several balloon cells are seen in a coarsely fibrillar background. All micrographs are from sections stained with H&E, original magnifications: (a) 10×; (b and c) 40×; (d) 20×.

punctuate or vascular calcinosis, often with large aggregates of calcium distributed throughout the specimen (Figure 9.3). Angiomatoid vascular structures (Figure 9.4) are less commonly observed. Recent studies have demonstrated that there are enhanced numbers of CD68 immunoreactive microglial within tubers [8, 9].

There are several unique histological features that can be used to define a tuber [10]. First, there are excessive numbers of astrocytes that can be detected immunohistochemically by expression of glial fibrillary acidic protein (GFAP) across the thickness of tubers [11]. Increased expression of the glial adhesion molecule CD44 is also observed in tubers [12]. Second, there is a highly heterogeneous population of

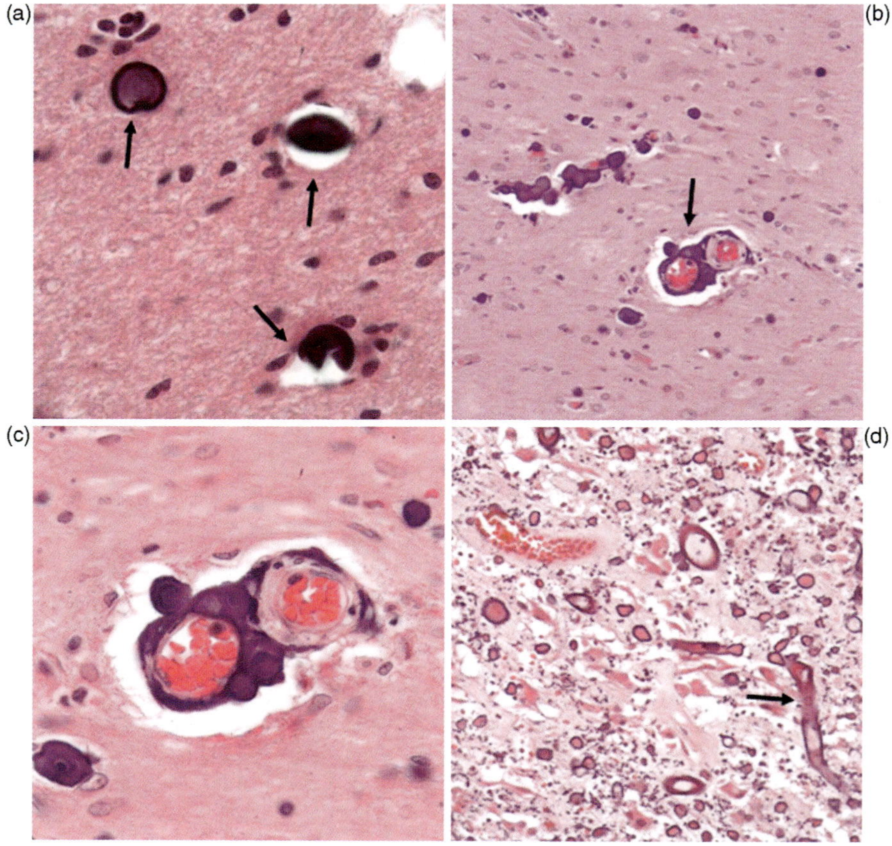

Figure 9.3 Calcification in tubers of TSC. This may be seen in or immediately adjacent to a tuber. (a) Punctate calcification (arrows) within brain parenchyma. (b) low power and (c) high power. Calcification centered on blood vessels and involving vessel walls. (d) Large microscopic field dominated by calcifications, including some in vessel walls (arrow). All panels from sections stained with H&E, original magnifications: (a and c) 100×; (b and d) 40×.

neurons within tubers, some exhibiting bizarre and aberrant morphologies (dysmorphic neurons; Figure 9.5) while others appear more like normal cortical neurons. These cells often exhibit a roughly pyramidal shape and a more definitive neuronal morphology with dendritic and axonal projections. Dysmorphic neurons within a tuber often show loss of radial orientation with respect to the pial surface. Dysmorphic neurons often exhibit an enlarged cell body and contain cytoplasmic Nissl substance. Unlike giant cells (see below), multinucleation is rare in dysmorphic neurons. Analysis of tubers with silver impregnation techniques such as Bodian or modified Bielschowsky stains reveals cytoskeletal abnormalities (coarse cytoplasmic fibrillar inclusions) that are almost identical to those seen in the neurofibrillary tangle-containing neurons of Alzheimer disease cortex. Immunohistochemical analysis reveals that many dysmorphic neurons are akin to pyramidal neurons in

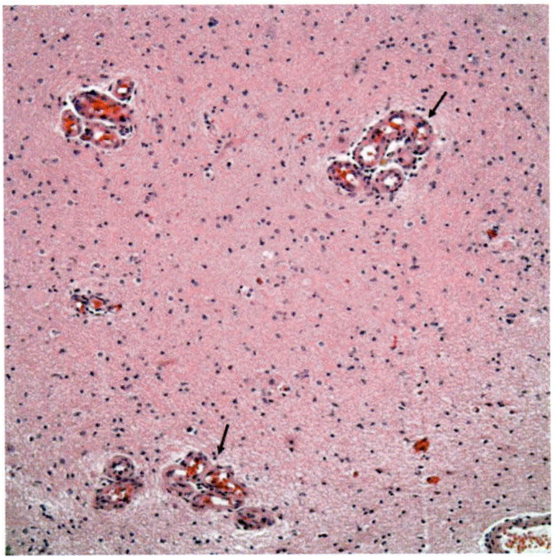

Figure 9.4 Angiomatosis in TSC. Clusters of thin-walled blood vessels (arrows) are rarely seen in tubers or cortical resections from TSC patients. Micrograph is from an H&E stained section, original magnification: 100×.

that they express glutamate transporters such as EAAC1 [13], or cytoskeletal proteins such as MAP2 and SMI32 within the appropriate subcellular compartment. The integrity and extent of efferent projections from dysmorphic neurons to other cortical or subcortical targets is unknown. Third, there are cells of indeterminate neuronal versus glial phenotype (Figure 9.6) that share some features of both neurons and astrocytes; for example, they may have a characteristic nucleolated nucleus (as would be expected in a neuron) but glassy amphophilic or even eosinophilic cytoplasm (as would be anticipated in an astrocyte) [10].

However, the *sine qua non* feature of tubers in TSC is the presence of dramatically enlarged cells with unique and abnormal morphologies known in the literature as either "balloon" (BCs) or "giant" (GCs) cells (Figures 9.7 and 9.8). These cells are morphologically similar to BCs identified in sporadic focal cortical dysplasia (FCD) type 2B in that they resemble enormous gemistocytic astrocytes with abundant opalescent, eosinophilic cytoplasm and a laterally displaced nucleus containing coarse chromatin. There is some debate as to the appropriate terminology for these cell types in FCD versus TSC. For the purposes of this chapter, GCs will refer to cytomegalic cells in tubers [14, 15] whereas BCs refer to FCD. GCs can achieve 120 μm in maximal dimension when measured with three-dimensional computer morphometry but usually range from 80 to 100 μm. GCs can be found across the thickness of the tuber as well as within the subcortical white matter and in layer I as well. In most tuber specimens, dysmorphic neurons are interspersed with GCs throughout the thickness of the tuber. GCs can be observed as single cells or clumped together in clusters or chains (Figure 9.9), often near small blood vessels. In virtually every tuber sample, some of the GCs are multinucleated. One quantitative analysis [16] counted GCs within 10 tuber specimens and found that between 20 and

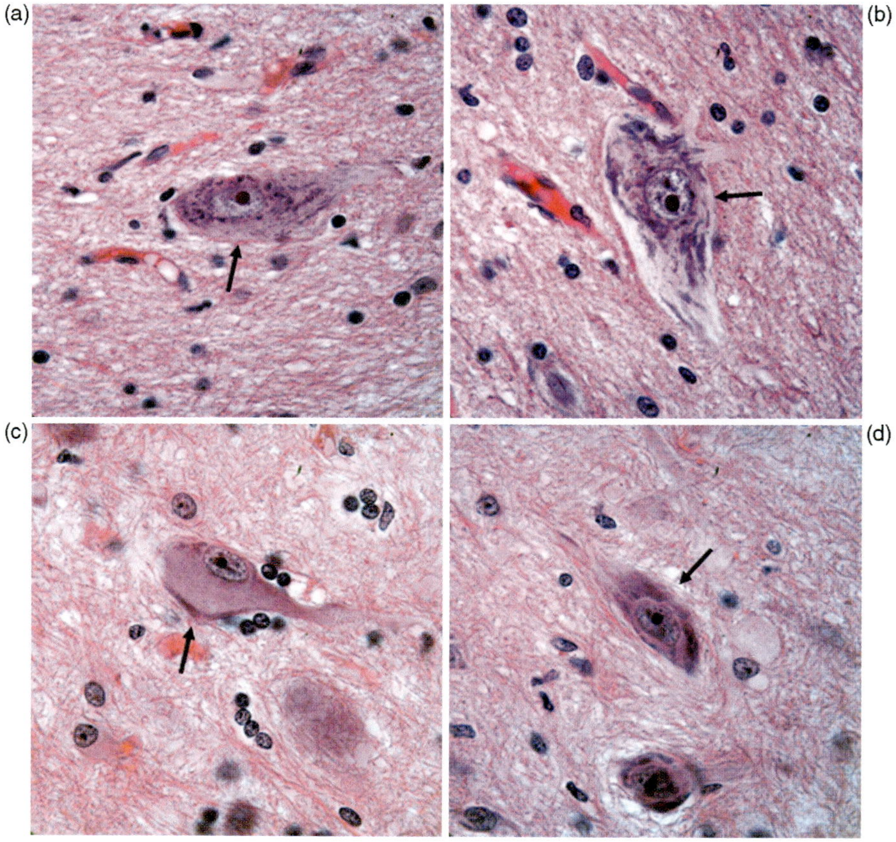

Figure 9.5 Dysmorphic neurons in TSC. Dysmorphic neurons (arrows) often show coarse or clumped Nissl substance (panels a, b, and d) or Nissl that is displaced to the periphery (panel c). All panels from H&E stained sections, original magnifications: 100× for all panels.

30% of the cells in this sample met morphological criteria for a GC. In contrast, experience dictates that in some tuber samples, there are sheets of GCs comprising most of its cellular constituents. Unlike dysmorphic neurons, GCs do not exhibit prominent Nissl substance and many are periodic acid Schiff (PAS) positive.

In the absence of specialized staining techniques or marker protein expression, the distinction between axons and dendrites or other process extensions from GCs is difficult to make. GCs extend short locally ramifying processes. Golgi–Cox staining analysis of tuber samples demonstrates that GCs extend small multipolar processes that are not easily defined as axons or dendrites. Immunohistochemical analysis using axon (neurofilament protein) or dendrite (microtubule associated protein-2)-specific markers has demonstrated that GCs are a highly heterogeneous population with some extending MAP2 positive dendrites while others extend several small neurofilament positive processes, suggestive of axons (Figures 9.10 and 9.11).

Cortical tubers may rarely be found in the cerebellum, although the clinical significance of these lesions is unknown. In one reported pathologic analysis from

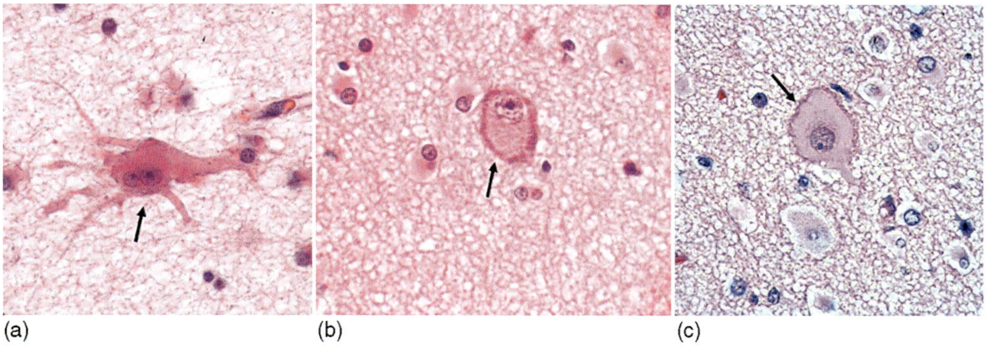

Figure 9.6 Cells of "indeterminate phenotype," with features of both astrocytes and neurons, in TSC. (a) Binucleate cell with a neuronal shape and short stubby processes emanating from its cytoplasm; note that the cytoplasm lacks Nissl substance and is eosinophilic and glassy, as would be expected in a balloon cell. (b) Cell with neuronal phenotype but peripherally displaced Nissl substance and fairly glassy eosinophlic cytoplasm. (c) A larger cell (arrow) with neuronal shape and characteristic nucleolated nucleus, but glassy amphophilic cytoplasm. All micrographs from sections stained with H&E, original magnifications: 100×.

a 5-year-old boy [16], a cerebellar tuber exhibited marked disorganization of neuronal architecture with ectopic neurons in the molecular and granule cell layers and white matter, along with calcification, gliosis, and Rosenthal fiber deposition. Giant cells with glassy, pale, eosinophilic cytoplasm were also detected. There was a marked loss of myelin in the white matter. Electron microscopic analysis of cerebellar GCs documented abundant lysosomal inclusions, prominent rough endoplasmic reticulum and Golgi complexes, microtubules, intermediate filaments, and synaptic contacts [17].

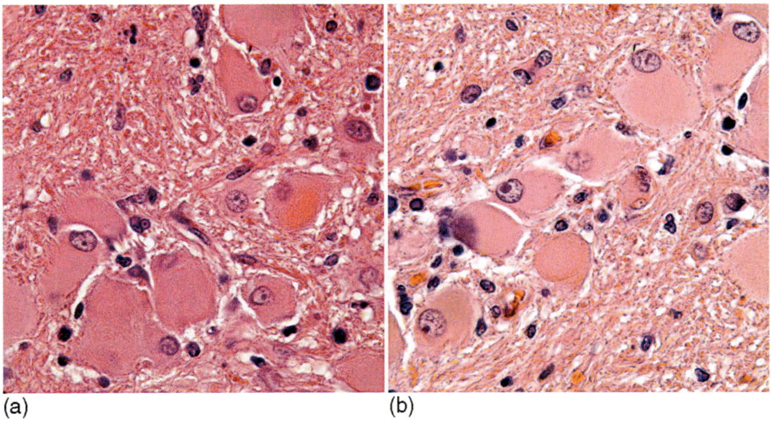

Figure 9.7 Balloon and giant cells in TSC. Balloon cells (BCs) or giant cells (GCs) may be found singly, in small clusters (a), or in an almost linear arrangement (b). Note that some of the BCs/GCs in (b) appear to have large nucleoli, suggesting neuronal lineage. H&E stain, original magnification of both panels: 100×.

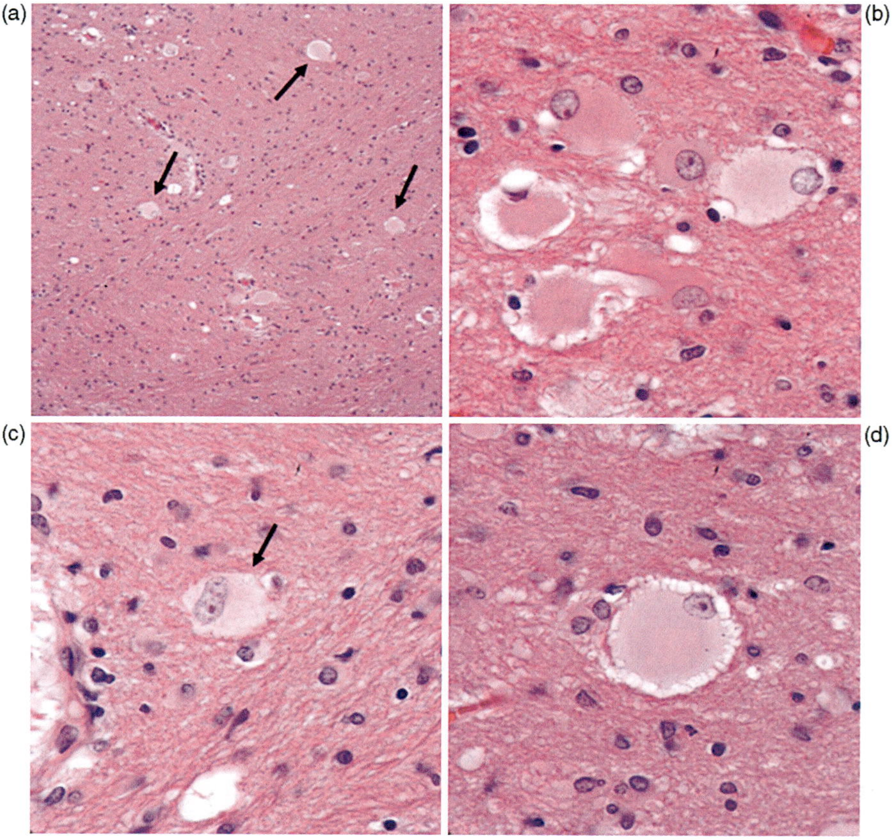

Figure 9.8 Balloon cells in TSC. (a) Balloon/giant cells are seen in the white matter, as small opalescent eosinophilic structures (arrows) in a "sea of oligodendroglia." (b–d) At higher magnification, they resemble gemistocytic astrocytes, with glassy eosinophilic cytoplasm and coarse nuclear chromatin. Balloon cells may be bi- or multinucleated (arrow in (c)). All micrographs are from sections stained with H&E, original magnifications: (a) 10×; (b–d) 100×.

9.3 SENs and SEGAs

Subependymal nodules are present in about 80% of TSC patients and are believed to be asymptomatic, that is, not related to cognitive deficits or epilepsy in TSC. SENs are nodular lesions (typically less then 1 cm in size) located on the surfaces of the lateral and third ventricles (Figure 9.12). SENs are typically covered by a thin layer of ependyma and can exhibit extensive vascularization. SENs can extend into the periventricular white matter and the basal ganglia. These lesions develop in fetal life and often degenerate or calcify during later life.

Based on a few studies using serial neuroimaging, it is widely believed that SENs grow to form SEGAs although the molecular mechanisms governing transformation

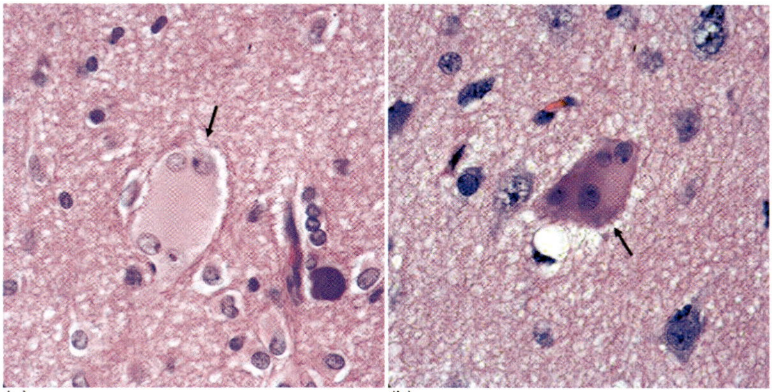

(a) (b)

Figure 9.9 Binucleated and multinucleated cells are commonly encountered in TSC tubers. (a and b) Multinucleated cells have the phenotype of a BC (a) or a more neuronal appearance (b). Both panels from sections stained with H&E, original magnifications: 100×.

from SEN to SEGA are unknown [18]. The molecular mechanisms governing their growth remain to be fully defined. SEGAs generally appear within the first 20 years of life. SEGAs generally exceed 1 cm in diameter but can grow to greater than 10 cm in size. SEGAs extend into the lateral ventricle and often obstruct the flow of CSF through the lateral ventricle and the foramen of Monro, causing hydrocephalus, focal neurological deficits, and even death [19]. Thus, in a select group of TSC patients, SEGAs require surgical removal. Overall, SEGAs are relatively rare and represent only about 1–2% of pediatric brain tumors. SEGAs can occur as sporadic tumors [20]; however, most of these likely represent somatic mosaic TSC cases, that is, TSC gene mutation occurring within a restricted population of cells within a limited number of organ systems.

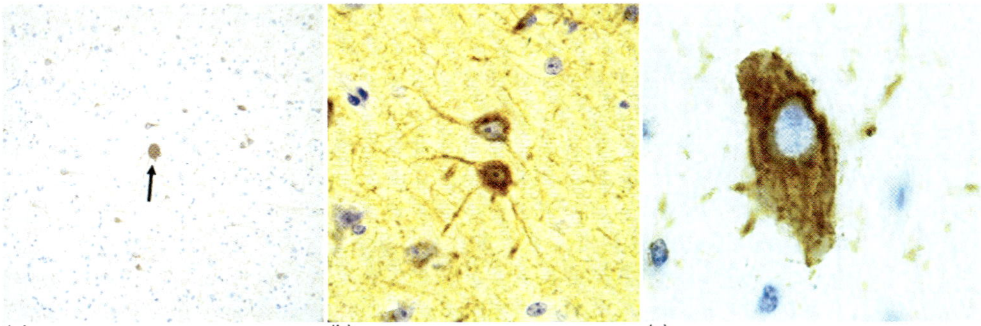

(a) (b) (c)

Figure 9.10 Neurofilament staining (recognizing 200 kDa isoform of neurofilament) of dysmorphic neurons in tubers. (a) Individual dysmorphic neurons are highlighted at low magnification (b and c). Neuronal dysmorphism including abnormal processes are highlighted. Original magnifications: (a) 40×; (b and c) 100×.

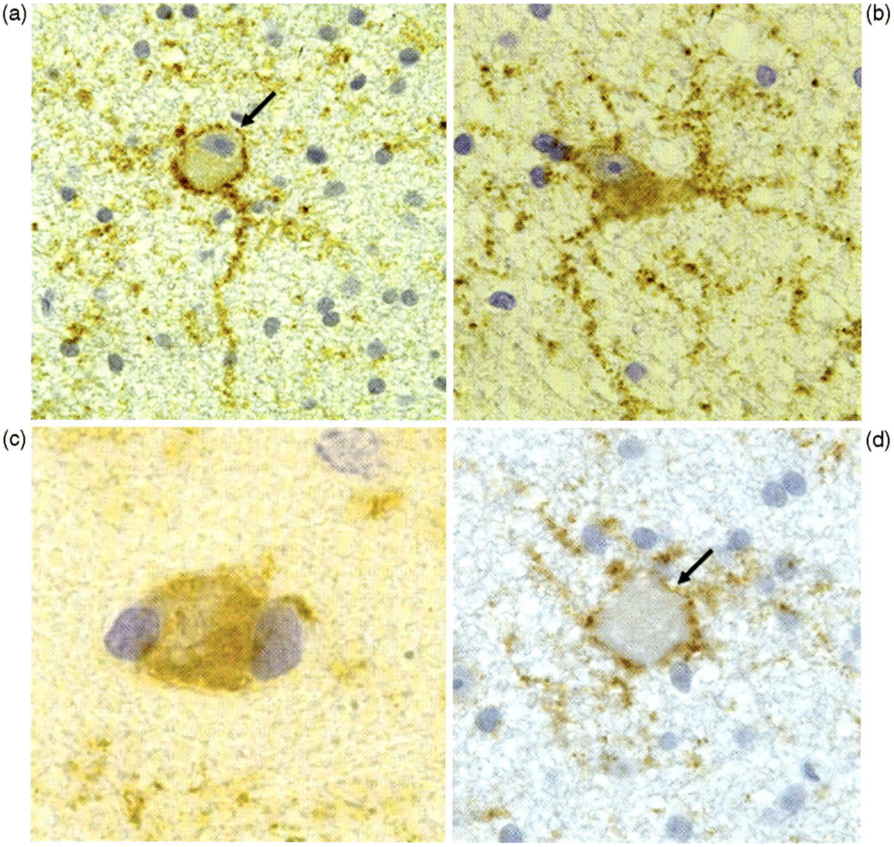

Figure 9.11 Antisynaptophysin staining identifies dysmorphic neurons in tubers. Synaptophysin is present in neuronal perikaryon and cell processes (a and d), or neuronal cytoplasm (panel b). Rarely a cell with balloon cell morphology shows cytoplasmic labeling (panel c). Original magnifications: all panels 100×.

Both SENs and SEGAs (Figure 9.13) consist of dysmorphic glial cells as well as GCs [10, 14, 19]. SEGAs frequently calcify. On histologic examination, the cells appear most consistently to be swollen astrocytes arranged in sheets, clusters, or pseudorosettes. Individual cells range in shape from polygonal to epithelioid to spindle shaped with marked pleomorphism. SEGAs are highly vascular lesions. The rich vascular stroma within SEGAs likely contributes to a small incidence of spontaneous, intratumoral hemorrhage. Cellular immunoreactivity for GFAP, neurofilament, S-100, neuron-specific enolase, and synaptophysin proteins suggests that SCGTs contain both glial and neuronal cell types (Figure 9.14). In another study, there was minimal expression of p53 while there was high expression of bax protein [18]. Interestingly, unlike other TSC-associated lesions, SEGAs do not stain with HMB45 [21]. Variable reductions in TSC2 protein expression have been observed in SEGAs [22]. SEGAs exhibit low numbers of mitotic figures, and relatively

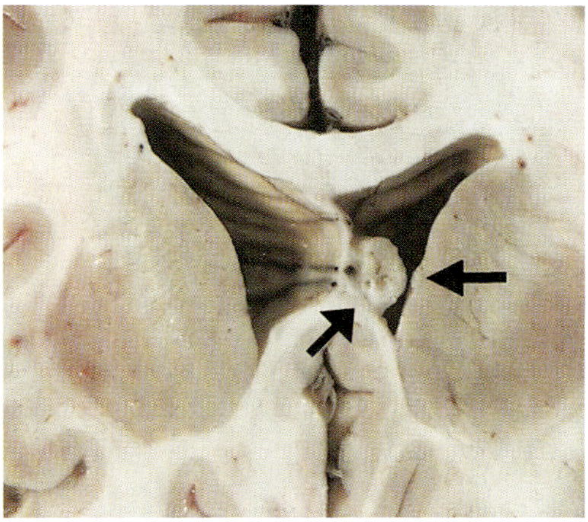

Figure 9.12 Subependymal nodule in a patient with TSC (autopsy specimen). Arrows indicate a SEN apparently attached to the septum pellucidum and adjacent brain parenchyma, but protruding into the lateral ventricle. Smaller nodules appear to be present in the subependymal region of the right caudate nucleus, which has a scalloped appearance. Photograph courtesy of Dr. Andres Kulla, Tartu University Medical Center, Estonia.

low labeling index when sections are stained with antibody to a proliferating cell antigen (PCNA) or Ki-67 (Figure 9.15). However, SEGAs are not malignant tumors although vascular proliferation and even necrosis may be observed within them. SEGAs exhibit a diploid DNA pattern and only occasional mitotic figures are noted. In a recent study of 23 SEGAs, investigators found that on microscopic examination there was heterogeneous histology consisting of bundles of spindle cells, gemistocytic and ganglion-like cells, as well as an interspersed inflammatory cell component, including mast cells and T lymphocytes [23]. In a few specimens, there were areas of necrosis and/or enhanced cellular mitoses. Using markers of mitotic activity, the mean MIB-1 labeling index (LI) was 3.0%, the mean topoisomerase II alpha (topo II alpha) LI was 2.9%, and mean the PCNA L1 was 32.5% [23]. These data were similar to a previous study demonstrating a MIB-1 LI of 0.1–3.8% [20]. The disparity between Ki-67 and PCNA expression reflects the fact that Ki-67 is a marker for active mitosis whereas PCNA demarcates cells in the S-phase of the cell cycle.

Associations of TSC with other brain neoplasms, for example, glioma, ependymoma, schwannoma, medulloblastoma, and ganglioma are largely viewed to be coincidental rather than syndromic associations although in two cases sacral chordomas were found to contain TSC gene mutations [24].

9.4
Cell Lineage

The cell lineage and phenotype of GCs and dysmorphic neurons in tubers have not been fully defined. Indeed, a pivotal issue regarding the pathogenesis of

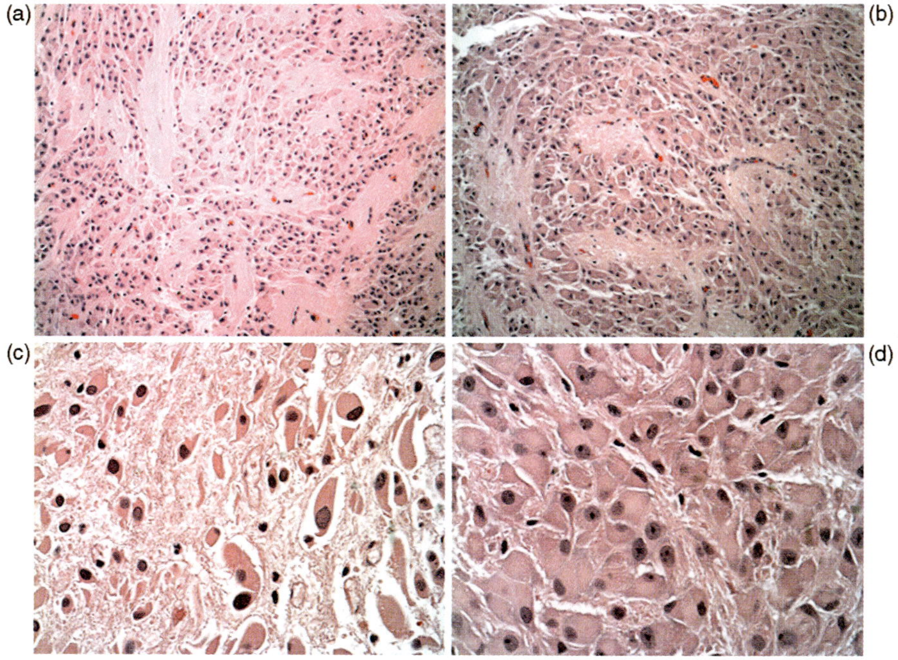

Figure 9.13 Microscopic pathology of subendymal giant cell astrocytoma/ subendymal giant cell tumor (SEGA/SGCT). All panels are from sections stained with hematoxylin and eosin. This tumor is composed of cells with variable morphology, ranging from spindle cell to polygonal, in a fibrillar stroma. Many cells have prominent nucleoli (d). Images courtesy of Dr. Negar Khanlou, Section of Neuropathology, UCLA Medical Center. Micrographs are from H&E stained sections, original magnifications: (a) 10×; (b) 20×; (c and d) 40×.

tubers is determining whether GC is derived from glial or neuronal cell precursors during early brain development since GCs express both glial and neuronal markers (Figure 9.16) and whether all tubers are derived from the same types of progenitor cells. Immunohistochemical analyses have demonstrated that GCs express GFAP and S100 proteins, markers of astrocytes [9]. In contrast, the identification of rough endoplasmic reticulum, intermediate filaments extending into processes of ambiguous morphology, prominent paranuclear Golgi zones, and dense core granules (secretory vesicles) in GC suggest neuronal features. GCs in tubers also express the proteins nestin, neurofilament, internexin, neuron-specific enolase, tubulin and MAP2C, which are typically enriched in neurons [25, 26]. In addition, a small proportion of GCs express met-enkephalin, β-endorphin, serotonin, and neuropeptide Y, also suggesting a neuronal phenotype [26]. Synaptophysin immunoreactivity has been reported along the cell membrane of GCs [27]. GCs were shown to express several $GABA_A$ receptor subunits, intermediate filament, and calcium channel subunit mRNAs [25] consistent with neuronal phenotype. Subsequent work has demonstrated that GCs express NMDA, GluR, and $GABA_A$

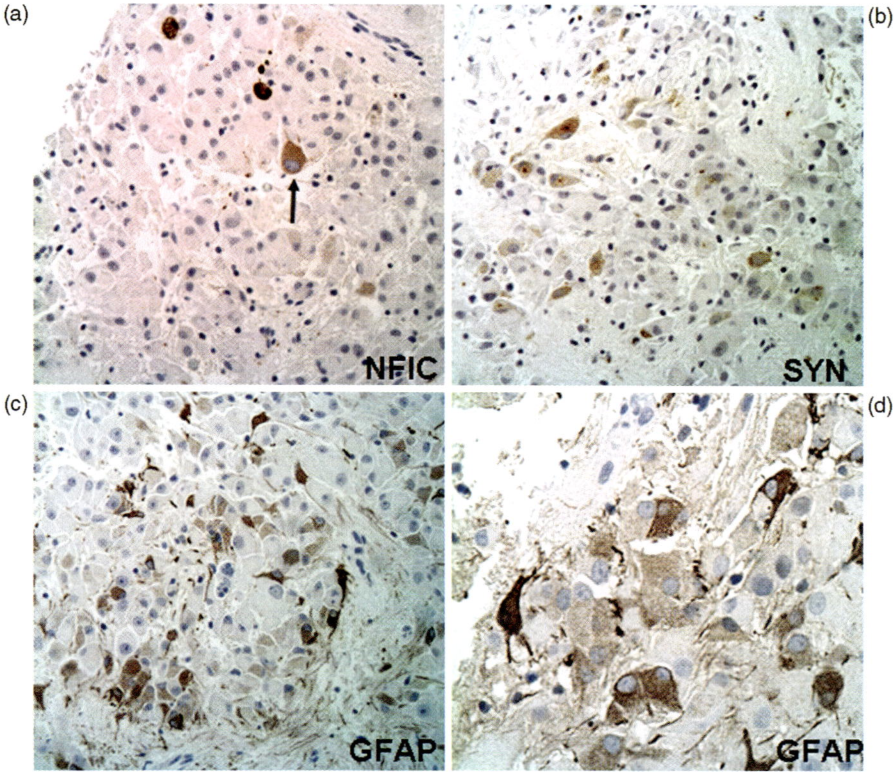

Figure 9.14 SEGA, immunohistochemical features. Tumor cells are variably immunoreactive for neurofilament (NFIC, arrow (a)) synaptophysin (b), or GFAP (c and d). Images courtesy of Dr. Negar Khanlou, Section of Neuropathology, UCLA Medical Center. Original magnifications: (a–c) 20×; (d) 40×.

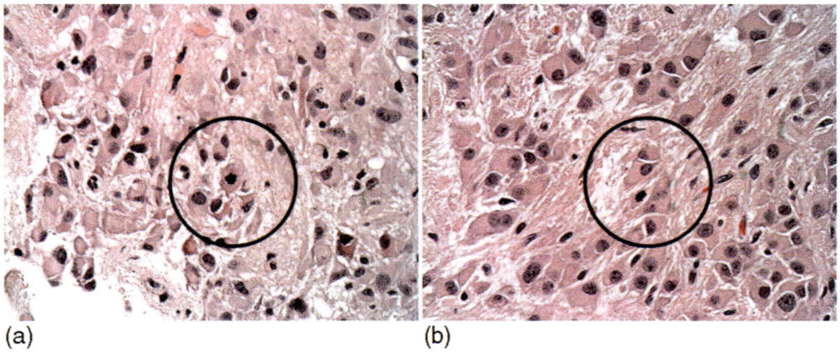

Figure 9.15 Cell proliferation in SEGA. Though SEGA are largely nonproliferative, mitotic figures may be found within them (circles in both panels) and the Ki-67 labeling index may be quite high, on the order of 5–10% (not shown here). Both panels are from H&E stained sections, original magnifications: 40×.

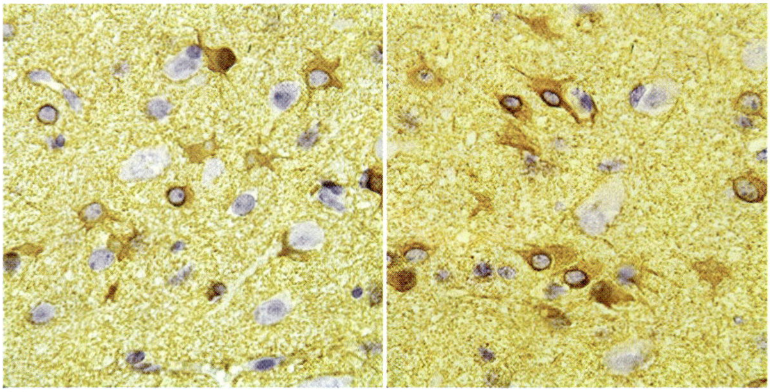

(a) (b)

Figure 9.16 GFAP expression by astrocytes and balloon/giant cells. GFAP immunohistochemistry is prominent in reactive astrocytes but may also be expressed in stains both balloon/giant cells as shown in these two panels. Original magnifications: (a) 60×; (b) 40×.

receptor subunit mRNAs [13, 28], both supporting a neural phenotype and suggesting the possibility of altered synaptic connectivity and aberrant excitability between cell types in tubers. Whether GCs make synaptic connections is unclear. However, based on analyses of human focal cortical dysplasias in slice preparations, it is likely that connections do exist between dysmorphic neurons but that GCs may be electrically silent. Perhaps most problematic is that within the same tuber, GC and cells with indeterminate phenotype and morphology can be found that coexpress or independently express both neuronal and astrocytic markers. Indeed, in the same patient, one tuber may contain GC expressing largely neuronal markers while another tuber may contain GC expressing largely astrocytic markers. An explanation for this observation is unknown, but it has been speculated that variable expression or coexpression of glial and neuronal markers may reflect aberrant differentiation of neuroglial progenitor cells resulting in the appearance of a mixed cellular phenotype [10, 15].

During normal cortical development, neurons destined for the cerebral cortex derive from progenitor cells in either the ventricular zone (VZ) or the ganglionic eminences (GE). Projection neurons in layers III and V derive from the VZ from radial glial progenitor cells whereas inhibitory interneurons are born within the GE and follow nonradial (tangential) pathways to the evolving cortical plate. Several marker proteins have been used to identify subsets of neurons from either VZ or GE, including OTX1 and Pax6 (VZ) and Dlx1 or Nkx (GE). These markers also identify subsets of glutamatergic and GABAergic neurons that derive from VZ and GE, respectively. Prior studies have demonstrated that dysmorphic neurons in tubers express the neuronal glutamate transporter (EAAC1) suggestive of a neuronal and potentially excitatory phenotype [13]. Conversely, these cells do not express GAD65 or the vesicular GABA transporter, suggesting that these cells are not inhibitory. Indeed, GCs were also found to be parvalbumin or calbindin-D(28k) negative [29]. Recent data have demonstrated that GCs and many larger dysmorphic neurons express brain lipid binding protein and vimentin, which are typically expressed by radial glial cells in the VZ [30]. In addition, GCs express ezrin, a protein that is normally abundantly expressed within radial glia and migrating cells in the intermediate zone in the developing cerebral cortex [31]. These data provide evidence that GCs and dysmorphic neurons derive from the VZ rather than the GEs.

SEGAs express cellular markers found in progenitor cells derived from the subventricular zone (SVZ) adjacent to the lateral ventricles [30, 32] and many of these marker proteins are also expressed in cortical tubers. For example, the expression of the marker proteins collapsing response mediator protein-4 (CRMP4) and doublecortin (DCX) in human tubers [33] and SGCT suggests that the cell lineage of tubers, SENs, and SEGAs may derive from progenitors in the embryonic ventricular zone and SVZ and that these lesions may be linked pathogenically [32]. These studies suggest that tubers and SENs/SEGAs may derive from a common progenitor cell during embryonic development. A large number of TSC patients exhibit both subependymal nodules and tubers radiographically. In fact, a compelling study demonstrated that the number of subependymal nodules in TSC patients can increase over time and that tuber number is positively correlated with the number of subependymal nodules [34]. One logical proposal is that a pool of progenitor cells within the SVZ populates SENs and may also yield new cells destined for tubers. Alternatively, newly generated cells in SENs or SGCT that express CRMP4 may migrate into tubers. Indeed, many of the DCX or CRMP4 immunoreactive cells were in the subcortical white matter, possibly *en route* to tubers. Thus, these cell types may also fail to shut off CRMP4 or DCX expression as a consequence of altered gene transcription or as a direct result of *TSC1* or *TSC2* mutations. Another possibility is that a population of dormant progenitor cells may exist within tubers that yield new cells expressing CRMP4 and DCX. While there is no current evidence to suggest that tubers can enlarge, the MRI characteristics of tubers can change over time, suggesting that these lesions may be more dynamic than static. Radial migration lines and white matter abnormalities [35] have been reported adjacent to tubers and raise the interesting possibility that these features of TSC reflect the pathway of migrating neurons from the SVZ or directly from SENs/SEGAs.

Another compelling and unexplained feature of GCs is the expression of protein markers typically found in immature neurons or progenitor cells. For example, GCs express nestin, vimentin, and brain lipid binding proteins, all of which are typically expressed by neuroglial progenitor cells in the developing cerebral cortex [25, 26, 30, 32]. GCs and dysmorphic neurons express FGF-2 [36] and the NMDA2D receptor subunit [13] also supportive of an embryonic phenotype. As discussed above, one hypothesis is that GCs represent a unique cell type in TSC that has failed to fully differentiate and thus exhibits persistent expression of certain developmentally inappropriate proteins.

9.5 mTOR Activation and Biallelic TSC Gene Inactivation

Mutations in *TSC1* or *TSC2* likely have a significant impact on neuroglial development [37]. TSC1 and TSC2 form a functional protein–protein heterodimeric complex that constitutively inhibits the activation (phosphorylation) of mTOR (mammalian target of rapamycin), p70-S6-kinase, and ribosomal S6 proteins that contribute to ribosomal assembly and protein translation (Chapter 6) [38]. The mTOR pathway proceeds downstream of the insulin-like growth factor-1 (IGF-1) receptors, PI3K, and Akt and serves as a key regulator of cell size via effects on ribosome biosynthesis and 5′-cap-dependent mRNA translation. Constitutive negative modulation of this cascade by TSC1–TSC2 results in growth suppression, diminished protein synthesis, and restricted cell size. However, in response to growth factor stimulation (e.g., IGF-1) and other stimuli, TSC2 is inactivated by phosphorylation (e.g., by AKT), leading to increased levels of Rheb-GTP (Ras homologue expressed in brain), which leads to activation of mTOR, and downstream phosphorylation and activation of p70S6 kinase, ribosomal S6, and 4E-BP1. There is cell-specific activation of the mTOR cascade in GCs in human tubers as evidenced by aberrant hyperphosphorylation of S6 kinase and S6 proteins [39, 40]. Abnormal enhanced phosphorylation of ribosomal S6 protein is also observed in SEGAs [41]. Interestingly, treatment with the mTOR-selective antagonist rapamycin leads to regression of SEGAs, likely by interrupting the constitutive activation of mTOR [42].

Loss of TSC1 or TSC2 function leading to mTOR cascade activation is believed to result from biallelic inactivation of either *TSC1* or *TSC2* [42]. In keeping with the Knudson "two-hit" mutational model, inactivation of both *TSC1* or *TSC2* alleles is necessary for lesion formation [43, 44]. By this mechanism, a somatic "second-hit" mutation superimposed on an existing germline mutation leads to loss of TSC1 or TSC2 function. Heterozygosity at either locus is insufficient to cause lesion formation. Recent studies demonstrate that SEGAs form as a consequence of LOH at either TSC gene locus [41]. Both germline and somatic mutations can be detected in SEGAs consistent with the robust expression of P-S6 protein in these tumors. Variable reductions in TSC1 or TSC2 expression have been reported in tubers and SEGAs, irrespective of mutational status [45].

A debate surrounds the mechanisms responsible for tuber formation [37, 46]. It has been proposed that biallelic *TSC* gene inactivation leads to mTOR cascade activation in GCs of tubers although LOH has been reported in only one tuber specimen [44] with several studies failing to identify two mutational hits [43, 47]. Reduced TSC1 or TSC2 expression [45] and hyperphosphorylation of p70S6 kinase and ribosomal S6 proteins [39, 40] support a two-hit mutational mechanism (Figure 9.17). If tubers form by mechanisms similar to SEGAs, then a somatic *TSC* gene mutation occurring in a single neuroepithelial progenitor cell in the embryonic brain, in combination with the effects of the germline mutation, would lead to TSC1 or TSC2 inactivation. Subsequent activation of mTOR signaling would then lead to cytomegaly (giant cells) in the progeny derived from this progenitor cell (Figure 9.18). The mTOR pathway is not activated in adjacent neuroepithelial cells that contain only

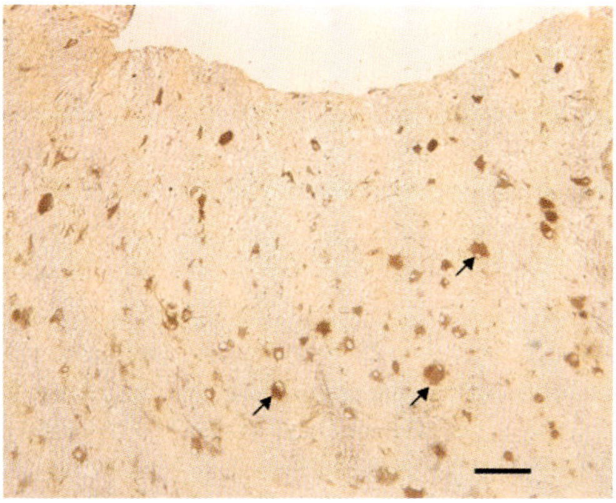

Figure 9.17 Phospho-S6 expression in TSC tubers. Tuber specimen exhibiting numerous phospho-S6 immunolabeled cells (arrows). Phospho-S6 is a downstream marker for mTOR cascade activation. Scale bar, 300 μm.

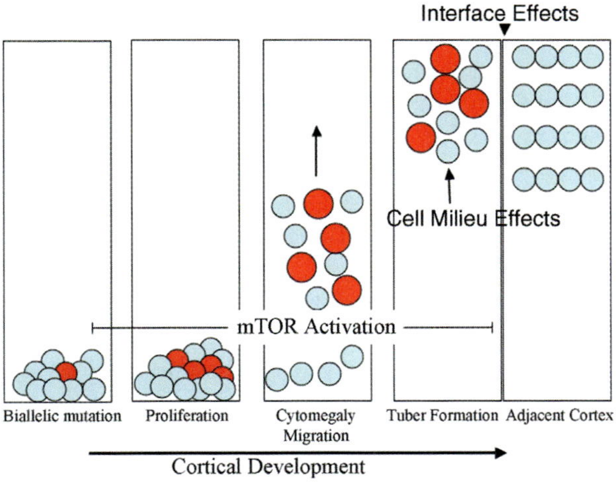

Figure 9.18 Two-hit model of tuber formation during brain development. From left to right, each panel represents a stage of cortical development. On the far left, progenitor cells in the ventricular zone (pale blue) undergo successive rounds of mitosis. A somatic second-hit mutation in either TSC1 or TSC2 occurs (red cell) leading to inactivation of the encoded TSC protein. Red cells undergo division leading to a pool of cells containing both a germline and a somatic TSC gene mutation. These cells exhibit mTOR activation and migrate aberrantly into the developing cortical plate. The focal collection of abnormal cells constitutes a tuber and is an admixture of cells with a single germline mutation (blue) and cells with two hits (red). The effects of red cells on adjacent blue cells may completely disrupt normal lamination cues. The adjacent cortex consisting of all blue cells is able to correctly form normal cortical lamina.

germline *TSC* gene mutations, for example, haploinsufficient cells, and these progeny migrate into the cortical plate and become DNs. These results also allowed us to propose a model for tuber formation during brain development in which a progenitor cell sustains a somatic "second-hit" mutation early in corticogenesis and continues to divide generating progeny lacking functional *TSC1* or *TSC2*. As a consequence, the mTOR cascade is activated, leading to cytomegaly and perhaps impaired migration or lamination. Tubers form as a mosaic lesion of null cells containing germline and somatic TSC gene mutations (in red) and haploinsufficient cells (e.g., dysplastic neurons, depicted in blue), containing only germline mutations. Interestingly, in this model the genotype of cells in adjacent nontuber cortex (depicted in blue to the right of the tuber) is the same as dysplastic neurons (also in blue) within the tuber. Alternatively, the effects of haploinsufficiency alone may lead to altered neuronal migration, glial proliferation, and cellular morphology characteristic of tubers.

9.6 Alternative Signaling Cascades in TSC Brain Lesions

Although activation of the mTOR cascade is a robust finding in TSC brain lesions, there is solid evidence that other cell signaling cascades may be activated as well. Future studies will be required to determine how lesion formation can be linked functionally to each aberrantly activated pathway. For example, several studies have demonstrated that MAPK is phosphorylated in tubers [8, 48]. MAPK activation appears to be a parallel event in mTOR pathway signaling and reflects activity of several upstream kinases. Abnormally enhanced expression of cyclin D1 protein has been shown to colocalize with P-S6 in GC within SEGAs suggesting that changes in signaling mediated via glycogen synthase kinase 3 (GSK3) or β-catenin may be altered in TSC as well [49].

Altered expression of neurotrophins 3 and 4 and trks B and C [16] suggests that guidance cues for appropriate neuronal migration may be disrupted in tubers. Specifically, neurotrophin-3 (NT3) and trkB mRNA expression was reduced while neurotrophin-4 (NT4) and trkC mRNA expression was increased in whole tuber sections. In single microdissected DNs and GCs, NT3 mRNA expression was reduced in GCs and trkB mRNA expression was reduced in DNs. NT4 mRNA expression was increased in DNs and trkC mRNA expression was increased in both DNs and GCs. Consistent with these observations, NT3 mRNA expression was reduced but trkC mRNA expression was increased *in vitro* in human NTera2 neurons transfected with a TSC2 antisense construct that reduced TSC2 expression and in neuroepithelial cells derived from the *Tsc2* knockout mouse brain [50]. Alterations in NT4/trkB and NT3/trkC expression likely contribute to tuber formation during brain development as downstream effects of the TSC1 and TSC2 pathways in TSC.

Expression of proinflammatory cytokines has been demonstrated in tubers, suggesting that there may be a subtle inflammatory component in tubers [8, 9] and may in part explain why some tubers calcify or become cystic over time. Using gene array analysis, an increase in ICAM-1 mRNA expression (and subsequently ICAM-1 protein) was detected in human tubers and the Tsc1cKO mouse cortex [8]. Further investigation demonstrated expression of molecules involved in ICAM-1 activation and signaling in tubers, including tumor necrosis factor alpha (TNFα), mitogen activated protein kinase (MAPK), and nuclear factor kappa B (NF-κB). Tubers exhibited increased numbers of CD68 immunoreactive macrophages clustered around GCs, suggesting that tuber formation may involve activation of proinflammatory cytokine signaling pathways. In a recent study [9], tubers were shown to contain microglial cells expressing class II antigens (HLA-DR) and perivascular and parenchymal T lymphocytes (CD3 + with a predominance of CD8 + T-cytotoxic/suppressor lymphoid cells). Activated microglia and reactive astrocytes expressed IL-1β and its signaling receptor IL-1RI, as well as components of the complement cascade, such as C1q, C3c, and C3d. Albumin extravasation, with uptake in astrocytes, was observed in both tubers and SGCT, suggesting that alterations in blood–brain barrier permeability are associated with inflammation in TSC-associated lesions.

9.7 Structural Alterations in Nontuber Brain Areas

A compelling observation is that some TSC patients exhibit profound neurological disorders, that is, infantile spasms or autism, but have near normal neuroimaging studies. Interestingly, whole-cell patch-clamp recordings from human nontuber brain tissue reveal enhanced neuronal excitability, for example, favoring seizures, supporting the idea that nontuber brain tissue in TSC may nonetheless be capable of generating seizures [51]. A growing body of evidence suggests that there are microscopic structural alterations not detectable by MRI that can disrupt neurological function. In the few reported neuropathological analyses of the postmortem TSC brain, disruption of normal brain architecture distinct from tubers includes small structural abnormalities such as heterotopias, subcortical nodules, radial migration lines, areas of hypomyelination, and small cortical dysplasias [10, 52]. Very rarely heterotopic collections of cells may be seen in the brain stem and spinal cord in TSC. These lesions differ from tubers in that they are smaller, GCs are an infrequent finding, cortical lamination is mildly altered, and they do not exhibit calcification. Recent MRI analyses in TSC patients have confirmed subtle structural abnormalities outside of tubers in the cortex and within subcortical structures such as the thalamus and basal ganglia [36] and suggest that these nontuber brain lesions, in addition to tubers, may contribute to autism and cognitive disability in TSC. The histopathology of these lesions has not been comprehensively investigated and the mechanistic relationship of these abnormalities to TSC gene mutations is unknown; that is, do these lesions form by similar processes as tubers, are they secondary events, or are they a unique phenotype of TSC? In addition, while activation of the mTOR cascade is a robust finding in tubers, it is unclear whether mTOR activation occurs in nontuber lesions.

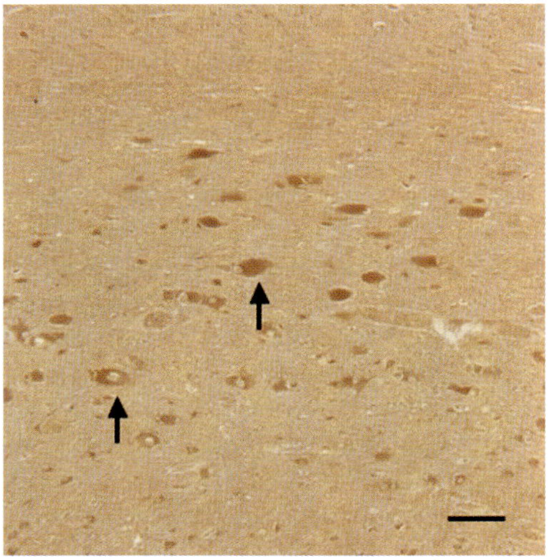

Figure 9.19 TSC "microtuber." An isolated nontuber associated cluster of phospho-S6 labeled GCs (arrows) is seen in the subcortical white matter. Scale bar, 300 μm.

Analysis of P-S6 expression by immunohistochemistry in nontuber cortical regions from postmortem TSC brain specimens reveals P-S6 labeled cells with abnormal morphology [52]. In fact, P-S6 immunoreactivity identifies numerous regions of aberrant cortical lamination in areas that are histopathologically distinct from tubers. In these areas, we find (1) small islands of GCs (we term these "microtubers"; Figure 9.19) that express P-S6, (2) only one or two GCs (expressing P-S6) surrounded by multiple P-S6 labeled dysplastic neurons ("dysplasias"; Figure 9.20), or (3) heterotopia (abnormal collections of cells in subcortical white matter). These findings are consistent with older studies using standard histological staining techniques that detected subtle alterations of brain architecture throughout the cortex of postmortem TSC specimens [52]. These pilot data suggest a potentially highly relevant mechanism by which nontuber lesions may result from enhanced mTOR cascade activation and loss of *TSC1* or *TSC2* in the absence of tuber formation. It is thus possible that other brain areas may contain cells that lack functional *TSC1* or *TSC2* and yet do not form tubers, perhaps due to their embryological origin or progenitor cell subtype.

Increased P-S6 protein labeling serves as a valuable marker for aberrant mTOR activation in cells lacking *TSC1* or *TSC2*. These data raise several pivotal questions: (1) Does P-S6 expression in these cells result from biallelic gene inactivation? (2) If so, then why are these lesions distinct from tubers? (3) What are the distinct mechanisms that determine formation of tubers versus more subtle structural abnormalities, for example, loss of TSC1 or TSC2 in specific embryonic brain regions or at specific developmental epochs or in a specific subset of progenitor cell types? Perhaps there are subsets of progenitor cells that are incapable of tuber formation or alternatively, perhaps tubers can form only at precise developmental epochs. Thus, an important new future perspective on neurological manifestations of TSC is to fully consider the

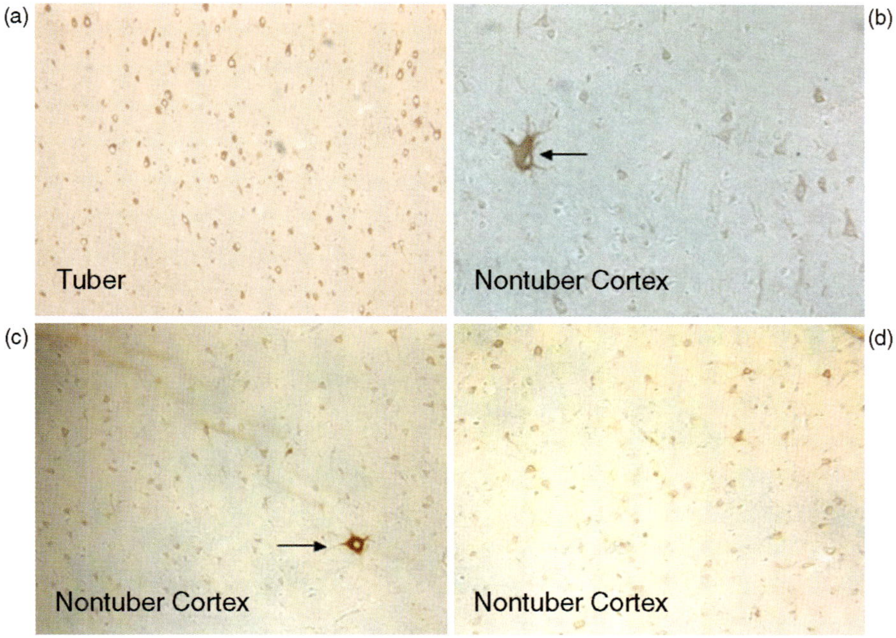

Figure 9.20 Phospho-S6 labeling identifies abnormal cells in a variety of locations in the TSC brain. (a) Typical tuber with many phospho-S6 labeled cells. (b and c) Single phospho-S6 labeled GC in nontuber cortex with only mild disruption of normal lamination. (d) Phospho-S6 labeled dysmorphic neurons in nontuber cortex. Scale bar 200 μm.

effects of radiographically visible lesions (tubers) as well as radiographically minimal or occult lesions on brain function.

9.8 Conclusions and Future Directions

Studying the neuropathological abnormalities in TSC has yielded important insights into the structural consequences of TSC mutations during brain development. The changes that occur at both the cellular and the regional level shed light on neurological manifestations of TSC including epilepsy, cognitive disability, and autism. Finally, the morphological alterations in the brain have allowed investigators to understand the normal functions of TSC1 and TSC2 in regulating cortical lamination, axon outgrowth, and cell size.

Clearly, ongoing study of TSC brain specimens will continue to open new avenues for inquiry. Several important directions can be followed in the future. First, understanding how loss of TSC1 or TSC2 alters cell migration may help define the loss of lamination in tubers. Second, defining the mechanisms that govern the transition from SEN to SEGA and then the unrestricted growth of SEGAs could yield new treatment strategies for TSC patients. Third, defining how loss of TSC1 or TSC2 changes the structural network properties of the brain to culminate in infantile spasms or intractable epilepsy would provide novel approaches to treat seizures. Finally, a comprehensive analysis of structural alterations in subcortical regions such as the thalamus or basal ganglia, or limbic system structures such as the hippocampus and amygdala in addition to the neocortex, will likely identify potential cytoarchitectural substrates that contribute to cognitive disabilities in TSC.

Acknowledgments

P.B.C. was supported by NINDS R01NS0450, Department of Defense, and the Tuberous Sclerosis Alliance. H.V.V. supported in part by the Daljit S. and Elaine Sarkaria Chair in Diagnostic Medicine.

References

1 Crino, P.B., Nathanson, K.L., and Henske, E.P. (2006) The tuberous sclerosis complex. *N. Engl. J. Med.*, **355**, 1345–1356.
2 Park, S.H., Pepkowitz, S.H., Kerfoot, C., De Rosa, M.J., Poukens, V., Wienecke, R., DeClue, J.E., and Vinters, H.V. (1997) Tuberous sclerosis in a 20-week gestation fetus: immunohistochemical study. *Acta Neuropathol.*, **94**, 180–186.
3 Levine, D., Barnes, P., Korf, B., and Edelman, R. (2000) Tuberous sclerosis in the fetus: second-trimester diagnosis of subependymal tubers with ultrafast MR imaging. *AJR Am. J. Roentgenol.*, **175**, 1067–1069.
4 Wortmann, S.B., Reimer, A., Creemers, J.W., and Mullaart, R.A. (2008) Prenatal diagnosis of cerebral lesions in tuberous sclerosis complex (TSC). Case report and review of the literature. *Eur. J. Paediatr. Neurol.*, **12**, 123–126.
5 Rott, H.D., Lemcke, B., Zenker, M., Huk, W., Horst, J., and Mayer, K. (2002) Cyst-like cerebral lesions in tuberous sclerosis. *Am. J. Med. Genet.*, **111**, 435–439.
6 Jurkiewicz, E., Jozwiak, S., Bekiesinska-Figatowska, M., Pakula-Kosciesza, I., and Walecki, J. (2006) Cyst-like cortical tubers in patients with tuberous sclerosis complex: MR imaging with the FLAIR sequence. *Pediatr. Radiol.*, **36**, 498–501.
7 Scheithauer, B.W. and Regan, T.J. (1999) Neuropathology, in *Tuberous Sclerosis Complex* (eds M.R. Gomez, J.R. Sampson, and V.H. Whittemore), Oxford University press, New York.
8 Maldonado, M., Baybis, M., Newman, D., Kolson, D.L., Chen, W., McKhann, G., 2nd, Gutmann, D.H., and Crino, P.B. (2003) Expression of ICAM-1, TNF-alpha, NF kappa B, and MAP kinase in tubers of the tuberous sclerosis complex. *Neurobiol. Dis.*, **14**, 279–290.
9 Boer, K., Jansen, F., Nellist, M., Redeker, S., van den Ouweland, A.M., Spliet, W.G., van Nieuwenhuizen, O., Troost, D. *et al.* (2008) Inflammatory processes in cortical tubers and subependymal giant cell tumors of tuberous sclerosis complex. *Epilepsy Res.*, **78**, 7–21.
10 Vinters, H.V. and Miyata, H. (2006) Neuropathologic features of tuberous sclerosis, in *Russell and Rubinstein's Pathology of Tumors of the Nervous System*, 7th edn (eds R.E. McLendon, M.K. Rosenblum, and D.D. Bigner), Hodder Arnold, London, UK, pp. 955–969.
11 Sosunov, A.A., Wu, X., Weiner, H.L., Mikell, C.B., Goodman, R.R., Crino, P.B., and McKhann, G.M., 2nd (2008) Tuberous sclerosis: a primary

pathology of astrocytes? *Epilepsia*, **49** (Suppl. 2), 53–62.

12 Arai, Y., Takashima, S., and Becker, L.E. (2000) CD44 expression in tuberous sclerosis. *Pathobiology*, **68**, 87–92.

13 White, R., Hua, Y., Lynch, D.R., Scheitauer, B., and Crino, P.B. (2002) Differential transcription of neurotransmitter receptor subunits and uptake sites in giant cells and dysplastic neurons in cortical tubers. *Ann. Neurol.*, **49**, 67–78.

14 Mizuguchi, M. and Takashima, S. (2001) Neuropathology of tuberous sclerosis. *Brain Dev.*, **23**, 508–515.

15 Mizuguchi, M. (2007) Abnormal giant cells in the cerebral lesions of tuberous sclerosis complex. *Congenit. Anom. (Kyoto)*, **47**, 2–8.

16 Kyin, R., Hua, Y., Baybis, M., Scheithauer, B., Kolson, D., Uhlmann, E., Gutmann, D., and Crino, P.B. (2001) Differential cellular expression of neurotrophins in cortical tubers of the tuberous sclerosis complex. *Am. J. Pathol.*, **159**, 1541–1554.

17 Jay, V., Edwards, V., Musharbash, A., and Rutka, J.T. (1998) Cerebellar pathology in tuberous sclerosis. *Ultrastruct. Pathol.*, **22**, 331–339.

18 Kim, S.K., Wang, K.C., Cho, B.K., Jung, H.W., Lee, Y.J., Chung, Y.S., Lee, J.Y., Park, S.H. *et al.* (2001) Biological behavior and tumorigenesis of subependymal giant cell astrocytomas. *J. Neurooncol.*, **52**, 217–225.

19 Goh, S., Butler, W., and Thiele, E.A. (2004) Subependymal giant cell tumors in tuberous sclerosis complex. *Neurology*, **63**, 1457–1461.

20 Takei, H., Adesina, A.M., and Powell, S.Z. (2009) Solitary subependymal giant cell astrocytoma incidentally found at autopsy in an elderly woman without tuberous sclerosis complex. *Neuropathology*, **29**, 181–186. Immunohistochemical characterization of subependymal giant cell astrocytomas. *Acta Neuropathol.*, **91**, 368–375.

21 Gyure, K.A. and Prayson, R.A. (1997) Subependymal giant cell astrocytoma: a clinicopathologic study with HMB45 and MIB-1 immunohistochemical analysis. *Mod. Pathol.*, **10**, 313–317.

22 Henske, E.P., Wessner, L.L., Golden, J., Scheithauer, B.W., Vortmeyer, A.O., Zhuang, Z., Klein-Szanto, A.J., and Kwiatkowski, D.J. (1997) Loss of tuberin in both subependymal giant cell astrocytomas and angiomyolipomas supports a two-hit model for the pathogenesis of tuberous sclerosis tumors. *Am. J. Pathol.*, **151**, 1639–1647.

23 Sharma, M., Ralte, A., Arora, R., Santosh, V., Shankar, S.K., and Sarkar, C. (2004) Subependymal giant cell astrocytoma: a clinicopathological study of 23 cases with special emphasis on proliferative markers and expression of p53 and retinoblastoma gene proteins. *Pathology*, **36**, 139–144.

24 Lee-Jones, L., Aligianis, I., Davies, P.A., Puga, A., Farndon, P.A., Stemmer-Rachamimov, A., Ramesh, V., and Sampson, J.R. (2004) Sacrococcygeal chordomas in patients with tuberous sclerosis complex show somatic loss of TSC1 or TSC2. *Genes Chromosomes Cancer*, **41**, 80–85.

25 Hirose, T., Scheithauer, B.W., Lopes, M.B., Gerber, H.A., Altermatt, H.J., Hukee, M.J., VandenBerg, S.R., and Charlesworth, J.C. (1995) Tuber and subependymal giant cell astrocytoma associated with tuberous sclerosis: an immunohistochemical, ultrastructural, and immunoelectron and microscopic study. *Acta Neuropath.*, **90**, 387–399.

26 Crino, P.B., Trojanowski, J.Q., Dichter, M.A., and Eberwine, J. (1996) Embryonic neuronal markers in tuberous sclerosis: single-cell molecular pathology. *Proc. Natl. Acad. Sci. USA*, **26**, 14152–14157.

27 Yamanouchi, H., Ho, M., Jay, V., and Becker, L.E. (1997) Giant cells in cortical tubers in tuberous sclerosis showing synaptophysin-immunoreactive halos. *Brain Dev.*, **19**, 21–24.

28 Talos, D.M., Kwiatkowski, D.J., Cordero, K., Black, P.M., and Jensen, F.E. (2008) Cell-specific alterations of glutamate receptor expression in tuberous sclerosis complex cortical tubers. *Ann. Neurol.*, **63**, 454–465.

29 Valencia, I., Legido, A., Yelin, K., Khurana, D., Kothare, S.V., and

Katsetos, C.D. (2006) Anomalous inhibitory circuits in cortical tubers of human tuberous sclerosis complex associated with refractory epilepsy: aberrant expression of parvalbumin and calbindin-D28k in dysplastic cortex. *J. Child Neurol.*, **21**, 1058–1063.

30 Ess, K.C., Kamp, C.A., Tu, B.P., and Gutmann, D.H. (2005) Developmental origin of subependymal giant cell astrocytoma in tuberous sclerosis complex. *Neurology*, **26**, 1446–1449.

31 Johnson, M.W., Miyata, H., and Vinters, H.V. (2002) Ezrin and moesin expression within the developing human cerebrum and tuberous sclerosis-associated cortical tubers. *Acta Neuropathol.*, **104**, 188–196.

32 Lee, A., Maldonado, M., Baybis, M., Walsh, C.A., Scheithauer, B., Yeung, R., Parent, J., Weiner, H.L., and Crino, P.B. (2003) Markers of cellular proliferation are expressed in cortical tubers. *Ann. Neurol.*, **53**, 668–673.

33 Mizuguchi, M., Yamanouchi, H., Becker, L.E., Itoh, M., and Takashima, S. (2002) Doublecortin immunoreactivity in giant cells of tuberous sclerosis and focal cortical dysplasia. *Acta Neuropathol.*, **104**, 418–424.

34 Hosoya, M., Naito, H., and Nihei, K. (1999) Neurological prognosis correlated with variations over time in the number of subependymal nodules in tuberous sclerosis. *Brain Dev.*, **21**, 544–547.

35 Ridler, K., Bullmore, E.T., De Vries, P.J., Suckling, J., Barker, G.J., Meara, S.J., Williams, S.C., and Bolton, P.F. (2001) Widespread anatomical abnormalities of grey and white matter structure in tuberous sclerosis. *Psychol. Med.*, **31**, 1437–1446.

36 Sugiura, C., Miyata, H., Ueda, M., Ohama, E., Vinters, H.V., and Ohno, K. (2008) Immunohistochemical expression of fibroblast growth factor (FGF)-2 in epilepsy-associated malformations of cortical development (MCDs). *Neuropathology*, **28**, 372–381.

37 Marcotte, L. and Crino, P.B. (2006) The neurobiology of the tuberous sclerosis complex. *Neuromolecular Med.*, **8**, 531–546.

38 Huang, J. and Manning, B.D. (2008) The TSC1-TSC2 complex: a molecular switchboard controlling cell growth. *Biochem. J.*, **412**, 179–190.

39 Baybis, M., Yu, J., Lee, A., Golden, J.A., Weiner, H., McKhann, I.I.G., Aronica, E., and Crino, P.B. (2004) Activation of the mTOR cascade distinguishes cortical tubers from focal cortical dysplasia. *Ann. Neurol.*, **56**, 478–487.

40 Miyata, H., Chiang, A.C., and Vinters, H.V. (2004) Insulin signaling pathways in cortical dysplasia and TSC-tubers: tissue microarray analysis. *Ann. Neurol.*, **56**, 510–519.

41 Chan, J.A., Zhang, H., Roberts, P.S., Jozwiak, S., Wieslawa, G., Lewin-Kowalik, J., Kotulska, K., and Kwiatkowski, D.J. (2004) Pathogenesis of tuberous sclerosis subependymal giant cell astrocytomas: biallelic inactivation of TSC1 or TSC2 leads to mTOR activation. *J. Neuropathol. Exp. Neurol.*, **63**, 1236–1242.

42 Franz, D.N., Leonard, J., Tudor, C., Chuck, G., Care, M., Sethuraman, G., Dinopoulos, A., and Thomas, G. (2006) Rapamycin causes regression of astrocytomas in tuberous sclerosis complex. *Ann. Neurol.*, **59**, 490–498.

43 Henske, E.P., Wessner, L.L., Golden, J., Scheithauer, B.W., Vortmeyer, A.O., Zhuang, Z., Klein-Szanto, A.J., Kwiatkowski, D.J., and Yeung, R.S. (1997) Loss of tuberin in both subependymal giant cell astrocytomas and angiomyolipomas supports a two-hit model for the pathogenesis of tuberous sclerosis tumors. *Am. J. Pathol.*, **151**, 1639–1647.

44 Sepp, T., Yates, J.R., and Green, A.J. (1996) Loss of heterozygosity in tuberous sclerosis hamartomas. *J. Med. Genet.*, **33**, 962–964.

45 Johnson, M.W., Emelin, J.K., Park, S.H., and Vinters, H.V. (1999) Co-localization of TSC1 and TSC2 gene products in tubers of patients with tuberous sclerosis. *Brain Pathol.*, **9**, 45–54.

46 Jozwiak, J. and Jozwiak, S. (2007) Giant cells: contradiction to two-hit model of tuber formation? *Cell. Mol. Neurobiol.*, **27**, 251–261.

47 Niida, Y., Stemmer-Rachamimov, A.O., Logrip, M., Tapon, D., Perez, R., Kwiatkowski, D.J., Sims, K., MacCollin, M. *et al.* (2001) Survey of somatic mutations in tuberous sclerosis complex (TSC) hamartomas suggests different genetic mechanisms for pathogenesis of TSC lesions. *Am. J. Hum. Genet.*, **69**, 493–503.

48 Han, S., Santos, T.M., Puga, A., Roy, J., Thiele, E.A., McCollin, M., Stemmer-Rachamimov, A., and Ramesh, V. (2004) Phosphorylation of tuberin as a novel mechanism for somatic inactivation of the tuberous sclerosis complex proteins in brain lesions. *Cancer Res.*, **64**, 812–816.

49 Jozwiak, J., Kotulska, K., Grajkowskam, W., Jozwiak, S., Zalewski, W., Oldak, M., Lojek, M., Rainko, K. *et al.* (2007) Upregulation of the WNT pathway in tuberous sclerosis-associated subependymal giant cell astrocytomas. *Brain Dev.*, **29**, 273–280.

50 Onda, H., Crino, P.B., Zhang, H., Murphey, R.D., Rastelli, L., Gould-Rothberg, B.E., and Kwiatkowski, D.J. (2002) Tsc2 null murine neuroepithelial cells are a model for human tuber giant cells, and show activation of an mTOR pathway. *Mol. Cell. Neurosci.*, **21**, 561–574.

51 Wang, Y., Greenwood, J.S., Calcagnotto, M.E., Kirsch, H.E., Barbaro, N.M., and Baraban, S.C. (2007) Neocortical hyperexcitability in a human case of tuberous sclerosis complex and mice lacking neuronal expression of TSC1. *Ann. Neurol.*, **61**, 139–152.

52 Boer, K., Troost, D., Jansen, F., Nellist, M., van den Ouweland, A.M., Geurts, J.J., Spliet, W.G., Crino, P. *et al.* (2008) Clinicopathological and immunohistochemical findings in an autopsy case of tuberous sclerosis complex. *Neuropathology*, **28**, 577–590.

10
Epilepsy in TSC

Elizabeth A. Thiele and Howard L. Weiner

10.1
Overview of Epilepsy in TSC

Neurologic involvement occurs in over 90% of individuals with tuberous sclerosis complex (TSC), making the brain one of the two most affected organ systems in the disorder. The neuropathologic hallmarks of TSC include cortical tubers, subependymal nodules, and subependymal giant cell tumors (also known as subependymal giant cell astrocytomas). Neurologic clinical manifestations include epilepsy, cognitive impairment, autism spectrum disorders, behavioral and psychiatric difficulties, and sleep disorders.

Epilepsy is the most common neurologic disorder in TSC, affecting approximately 75–90% of individuals [1, 2]. Although the onset of epilepsy in TSC can occur throughout an individual's lifetime, for the majority, seizure onset occurs during the first year of life. Approximately 70% of individuals with TSC will experience seizure activity before the age of 1 year, and there are many anecdotal reports of seizure onset *in utero*. Seizure onset can also occur in later childhood, adolescence, and rarely in adulthood.

Although the mechanisms of epilepsy in TSC are not well understood, seizures are thought typically to have a focal or partial onset. Therefore, partial and partial complex seizures are common, as well as secondarily generalized seizures. But individuals with TSC, particularly children, can also experience other seizure types, including myoclonic, atonic, and atypical absence. Approximately one-third of infants with TSC will develop infantile spasms, a catastrophic epilepsy syndrome of early childhood. A portion of children with TSC will develop Lennox–Gastaut syndrome (LGS), another catastrophic epilepsy syndrome. Both of these syndromes are discussed in detail below.

10.2
Role of Electroencephalography

As for epilepsy from other causes, the electroencephalogram (EEG) plays an important role in the diagnosis and characterization of epilepsy in TSC. Common

Tuberous Sclerosis Complex: Genes, Clinical Features, and Therapeutics
Edited by David J. Kwiatkowski, Vicky Holets Whittemore, and Elizabeth A. Thiele

ISBN: 978-3-527-32201-5

EEG features seen in individuals with TSC are diffuse background slowing, focal and multifocal epileptiform discharges, slow spike and wave discharges, and amplitude asymmetries (Figure 10.1) [3, 4].

The localization of interictal epileptiform activity appears to be consistent over time in most individuals with TSC, although in some individuals additional epileptiform regions may appear with time [5]. The degree of multifocality of epileptiform discharges may correlate with difficulty of seizure control in individuals with TSC, though this is not completely understood. EEG can be useful in localizing seizure onset in TSC, but it is not consistently reliable. Individuals with tonic seizures typically have a generalized electrodecrement at seizure onset, making lateralization and localization difficult unless there is a clear sign prior to the seizure onset. Additionally, some individuals with TSC and partial or partial complex seizures may not have a good electrographic correlate to their seizure activity, perhaps due to seizure onset in structures deeper than the cortical surface that is not easily "seen" on surface EEG.

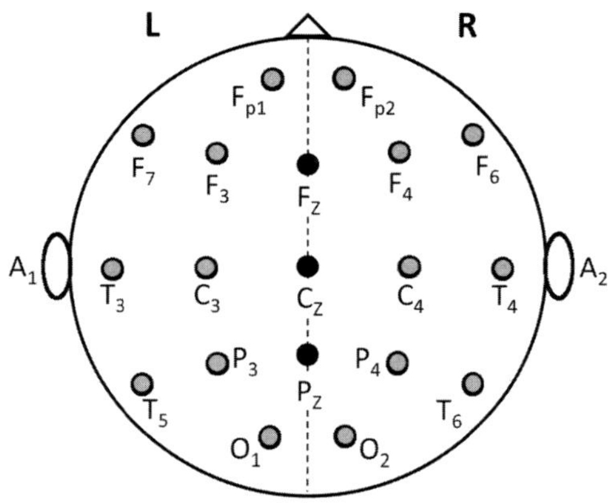

(a)

Figure 10.1 EEG findings in TSC. (a) International 10–20 electrode placement for EEG. (b) Normal background EEG with eyeblink artifact. Tracing is in bipolar longitudinal montage (as are all EEG figures in chapter), showing from top to bottom left temporal (Fp1 frontally to O1 occipitally), right temporal (Fp2 frontally to O2 occipitally), left parasagittal (Fp1 to O1), right parasagittal (Fp2 to O1), sagittal (Fz–Pz), and transverse (T1–T2). Tracing shows a normal anterior–posterior gradient, and the presence of a reactive posterior dominant rhythm. Tracing also shows several artifacts caused by eyeblinks (marked by asterisks). (c) EEG recorded during stage II sleep using bipolar longitudinal montage. Normal sleep spindles are seen bilaterally (marked by gray bars). Right central sharp waves (abnormal, epileptiform feature) are shown by asterisks. (d) Awake EEG recording showing frequent synchronous bioccipital sharp and spike waves, most prominent over left hemisphere (first four marked by asterisks). (e) Awake EEG recording showing run of semirhythmic left posterior quadrant spike and wave discharges (first four marked by asterisks). (f) EEG recording showing multifocal bilateral independent epileptiform discharges (examples marked by asterisks).

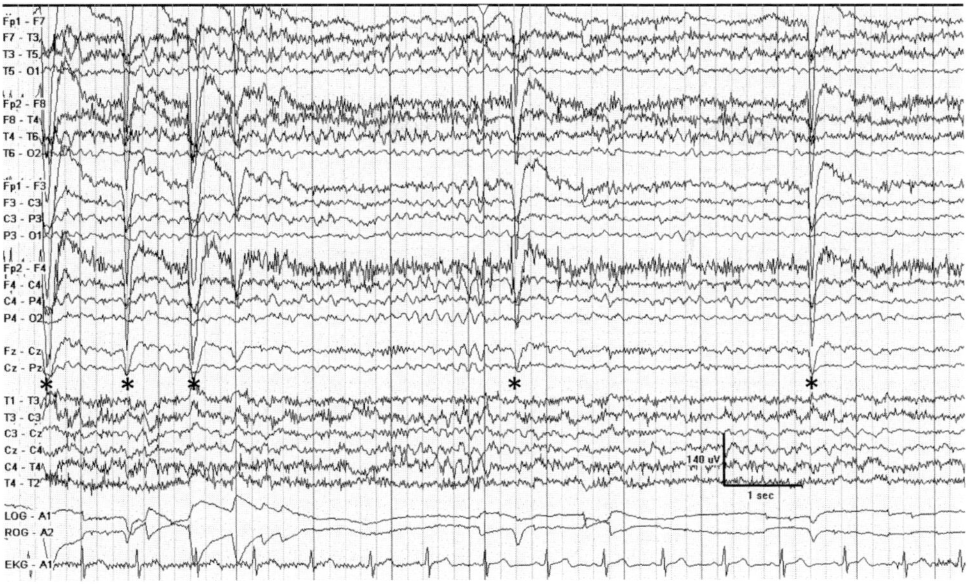

(b)

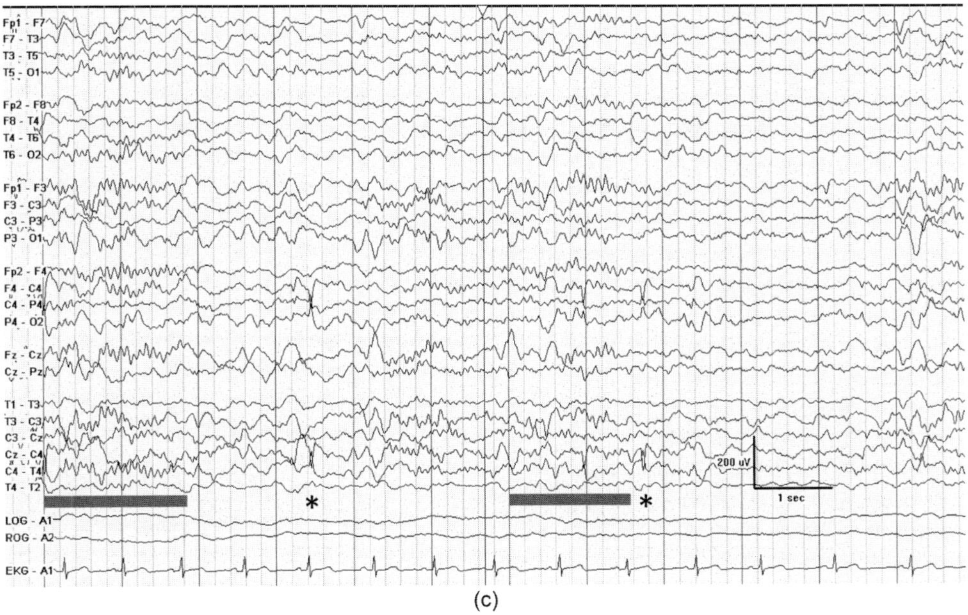

(c)

Figure 10.1 (*Continued*).

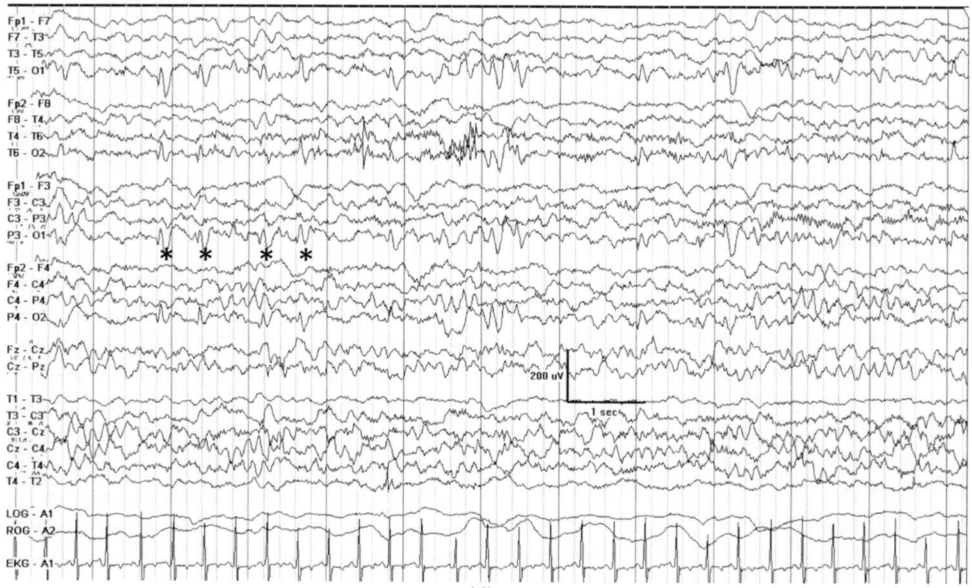

(d)

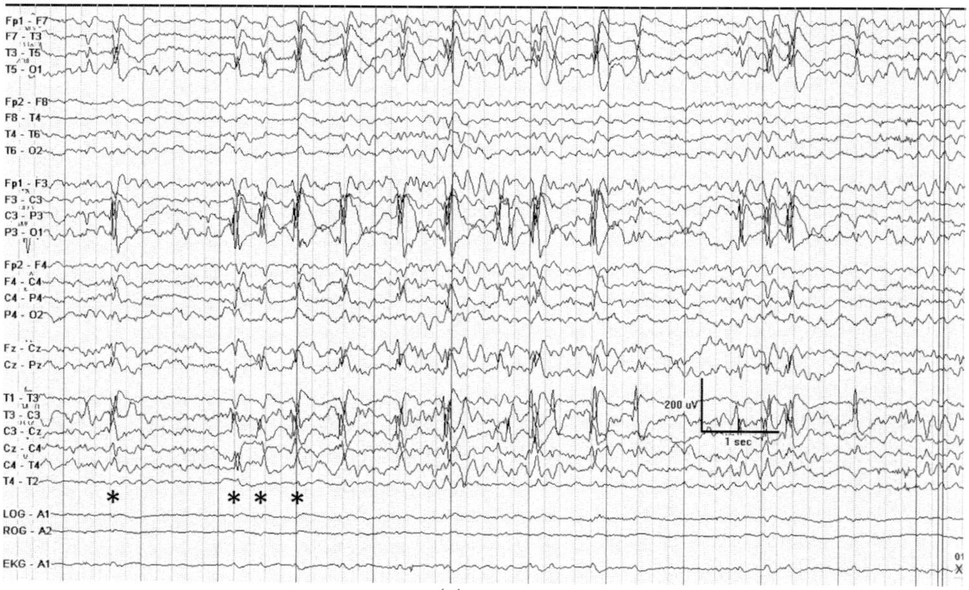

(e)

Figure 10.1 *(Continued)*.

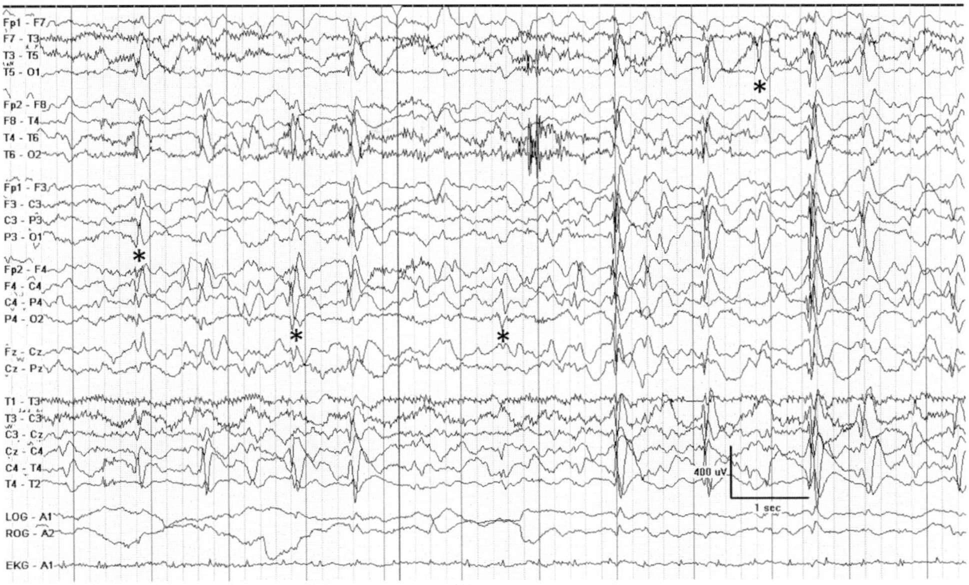

(f)

Figure 10.1 *(Continued)*.

10.3 Treatment of Epilepsy in TSC

10.3.1 Pharmacologic Treatment

Similar to epilepsy from other etiologies, anticonvulsant medications are considered the first-line treatment for seizures in the setting of TSC. As seizures in TSC are partial in onset, meaning that they start in a focal area of cortex, most available anticonvulsant medications can potentially be effective with the exception of treatment of IS and LGS, which are discussed elsewhere in this chapter. Data evaluating the efficacy and tolerability of particular anticonvulsant medications in large populations of individuals with TSC are not available although several small case series have been reported [6, 7]. The newer generation anticonvulsant drugs (ACDs), many of which have broad-spectrum efficacy and are therefore useful in individuals with mixed seizure disorders and generalized seizures, are considered overall safer and better tolerated than older generation ACDs. Therefore, use of newer generation ACDs in the treatment of epilepsy in the setting of TSC should be strongly considered. There is no evidence that any of the currently available ACDs is contraindicated in TSC. Since they can be effective at helping reduce abnormal brain activity, all of the ACDs have potential central nervous system side effects, and many can be associated with lethargy, somnolence, or behavioral changes. Therefore, individuals with TSC should be educated about potential side effects of a particular medication if started on the drug – especially since optimal epilepsy treatment means both minimal seizure activity and minimal drug side effects.

Unfortunately, many individuals with TSC develop medically refractory or intractable epilepsy, in other words, epilepsy that is not effectively controlled by medical therapy. Up to three-fourths of individuals with TSC who have seizures will develop refractory epilepsy, compared to about one-third of individuals with epilepsy due to all etiologies [8, 9].

10.3.2
Nonpharmacologic Treatment

If individuals with TSC have epilepsy that is not controlled by two or more appropriate medications, nonpharmacologic treatments should be considered. Current options include dietary therapy, vagus nerve stimulation, and epilepsy surgery.

Dietary therapies of epilepsy in current clinical practice include the classic ketogenic diet (KGD) as well as modern variations of the KGD, including the modified Atkins' diet (MAD) and low glycemic index treatment (LGIT) [10–12]. The ketogenic diet has been shown to be effective in treating intractable epilepsy in TSC [13], and it is likely that the two modified diets will be as well. The ketogenic diet is a high-fat, low-carbohydrate diet that was developed in the 1920s following the observation by many that while fasting, seizure frequency is often significantly reduced. The ketogenic diet was designed to mimic starvation: both in starvation and on a ketogenic diet, the body metabolizes fat into circulating ketone bodies, molecules that can substitute for glucose as a fuel for energy production in much of the body including the brain. The mechanism of action of the ketogenic diet is not known. However, over the past 80 years, the ketogenic diet has proved to be a very effective treatment for epilepsy, with many treated individuals becoming seizure free [14]. However, the ketogenic diet is very restrictive and is quite difficult for both the patient and family to implement and adhere to. In addition, access to the ketogenic diet as a treatment option may also be limited as its use requires an experienced and trained dietitian, and the number of individuals who can be started and followed on the diet by each dietitian further limits its availability due to the labor-intensive nature of calculating and maintaining each patient on the diet. With the goal of making dietary therapy more available and "doable," the two variations on the KGD mentioned above have been developed over the past few years, and both are currently being used nationally and internationally in the treatment of refractory pediatric epilepsy. The efficacy of these two diets, the LGIT and MAD, in TSC has not been established, but it is likely that they will both prove to play an important role in the treatment of refractory epilepsy in TSC. In our experience, the LGIT can be effective and well tolerated in treating epilepsy in children and adults with TSC.

The vagus nerve stimulator (VNS) is a pacemaker-like device providing continuous intermittent stimulation as well as magnet-activated stimulation of the left vagus nerve. The VNS has been used in the treatment of refractory partial onset epilepsy since FDA approval in 1997. Although the mechanisms of action of VNS therapy are not completely understood, the VNS has been shown to be effective in the treatment of partial epilepsies and LGS. Although limited data are available, VNS also appears to be effective in individuals with epilepsy due to TSC, as in two studies it resulted in a >50% seizure reduction in over half of TSC patients treated with VNS [15, 16].

10.3.3 Epilepsy Surgery in TSC

Surgery can be an effective therapeutic option for children with TSC who have seizures refractory to medical therapy. Several studies published by different investigators from major epilepsy centers around the world over the past 20 years have shown that, in the appropriately selected candidate, brain surgery can be quite successful in reducing seizures in patients with TSC (Figures 10.2 and 10.3) [17, 18].

Historically, there have been serious reservations regarding the role of epilepsy surgery in the management of refractory epilepsy in TSC, largely due to the concern that most individuals with TSC have multiple cortical tubers, and that each of these lesions might have epileptogenic potential. However, with a better understanding of epilepsy in TSC, it seems that in most individuals with TSC, only one or two tubers appear to be associated with epileptogenic foci. After resection of a site of seizure focus, individuals can experience long-term seizure control. Given the high rate of medically refractory epilepsy in TSC and its impact on neurocognitive development and function, epilepsy surgery has become a more utilized treatment in TSC. Improved neuroimaging, improved modern neurosurgical techniques, and improved pediatric and adult neuroanesthetic care have also helped make resective epilepsy surgery a more successful treatment option for controlling refractory epilepsy in TSC.

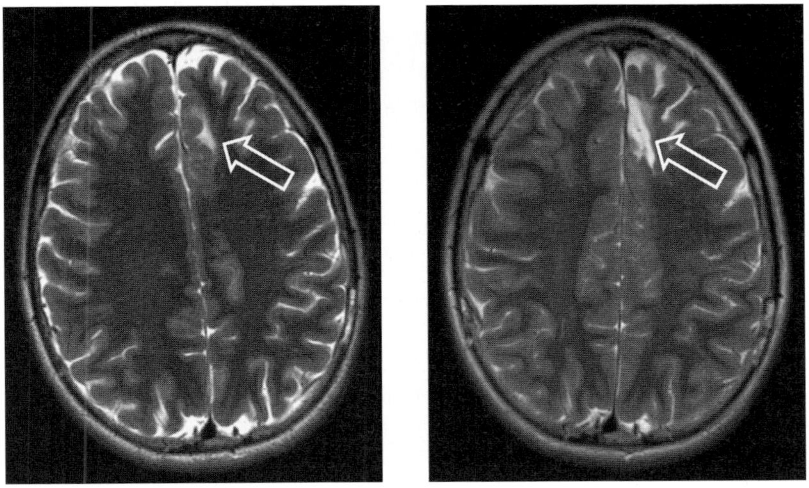

Figure 10.2 Epilepsy surgery in TSC: case 1. (a) Axial T2 FSE (fast spin echo) image from preoperative MRI with cortical tuber correlating with identified epileptogenic region identified by arrow. (b) Axial T2 FSE image from postoperative MRI from same child with arrow showing extent of surgical resection.

A 16.5-year-old right-handed girl presented with a history of tuberous sclerosis and intractable seizures. She was diagnosed with TSC after developing seizures at 18 months of age. Over the years, her seizures continued to be very stereotyped, and were characterized by episodes during which her face initially had the appearance "as if the shades are down" with a grin on her face. She then looked to the left with her head and eyes and they slowly slumped to the left. She would then make a vocalization like an "ah" sound or laughter, and may also have some chewing behaviors. She then often ran around the room or jumped. The seizures lasted in total for 30–60 s and were followed by a brief post ictal period. Prior to surgery, she had been on eight different anticonvulsant medications without effective control of seizures. Seizure frequency ranged from 5 to 40 per day. At times of higher seizure frequency, she had a relative regression in her cognitive abilities, and was in a special education program.

Her presurgical evaluation done when she was 14 years old included continuous video EEG monitoring and MRI and FDG-PET scanning. Her interictal EEG was abnormal, with frequent high-amplitude slow waves over the left frontal central leads that at times occurred in brief 1 Hz runs. Over 50 seizures were captured in 3 days of EEG monitoring, all arising from the left frontocentral region, and clinically were similar to episodes described above. Her MRI showed multiple bilateral cortical tubers, including a tuber in the left frontal region extending down to the frontal horn of the left lateral ventricle (Figure 10.2). FDG-PET scan showed a region of hypermetabolism in the left anterior cingulate gyrus.

The patient underwent surgical resection of the left frontal tuber. Interoperative electrocorticography was used to better define epileptogenic regions. Pathology of the resected tissue was consistent with cortical tuber.

Subsequent to surgery, the patient has remained seizure free for over 2 years, now off anticonvulsant medications. Her mother and teachers noted a significant improvement in her attention and cognitive functioning following surgery. A follow-up EEG 1 year following surgery showed mild slowing over the left frontal region, but no epileptiform activity.

In general, a TSC patient being evaluated for epilepsy surgery will have not responded to aggressive and appropriate anticonvulsant medical management and will have been considered or tried on nonpharmacologic therapies such as dietary therapy and VNS.

Evaluations for possible epilepsy surgery are typically performed at comprehensive epilepsy centers, with experience in managing highly refractory and complex patients. To determine if an individual is a good candidate for resective epilepsy surgery, the individual undergoes a presurgical evaluation that includes video EEG monitoring, characterizing both ictal and interictal behaviors and EEG features, as well as sophisticated neuroimaging. In addition, noninvasive functional testing is also usually performed, including brain positron emission tomography (PET), single photon emission computed tomography (SPECT), and magnetoencephalography (MEG). Results from these studies are reviewed in the context of the individual's

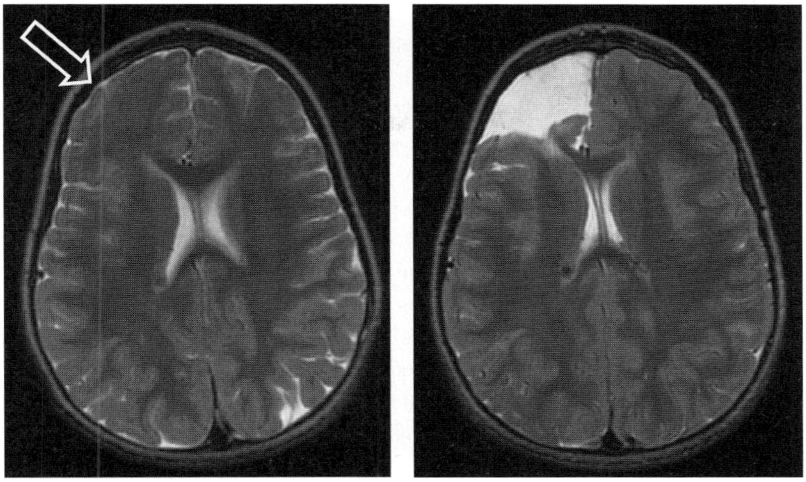

Figure 10.3 Epilepsy surgery in TSC: case 2. (a) Axial TS FSE image from preoperative MRI with arrow showing large cortical tuber in right frontal region correlating with identified epileptogenic region. (b) Axial TS FSE image from postoperative MRI showing extent of surgical resection.

A 6.5-year-old boy had been diagnosed with TSC after developing seizures at 6 months of age. At 4 months of age, he developed staring spells, which evolved into episodes of staring and movements of the left arm at 6 months of age. Seizures proved refractory to six medications, and longest period of time during which seizures were controlled was 2 months. In addition to seizures, he also had language delay and behavioral difficulties.

At the time of presurgical evaluation, he was on four anticonvulsant medications, and having daily seizures. Presurgical evaluation consisted of continuous video EEG monitoring, brain MRI, and FDG-PET scan.

His interictal EEG was abnormal, with frequent high-voltage sharp waves over the right frontal and less frequently right temporal areas. One electroclinical and seven electrographic seizures were captured, all arising from the right frontal temporal region. MRI showed multiple bilateral cortical tubers, with the most prominent tuber in the right inferior frontal lobe, which also showed evidence of calcification (Figure 10.3). FDG-PET scan showed hypometabolism in right frontal and temporal lobes.

He underwent surgical resection of the right frontal tuber, with intraoperative corticography to help guide extent of resection. Pathology on resected specimen was consistent with cortical tuber.

Postoperatively, he was seizure free, and had significant improvements in cognition, language, and behavior, allowing integration into regular classrooms. However, after remaining seizure free for over 2 years following surgery, he developed a new seizure type characterized by him developing a "wide-eyed frightened look," having garbled speech, and being disoriented for several seconds. Video EEG monitoring showed that these seizures arise from the right frontotemporal region. These seizures have been easily controlled with anticonvulsant medications.

epilepsy, particularly how frequent the seizures are, and how seizure activity and current medications are impacting the individual's quality of life and neurocognitive functioning. From this integrative approach, a recommendation is made regarding the potential value of resective surgery. Although the ideal goal in the treatment of epilepsy, including surgical treatment, is "no seizures, no side effects," the attainment of complete seizure elimination may not be possible in all cases. For some individuals with TSC and refractory epilepsy, a reduction in seizure frequency and severity, and simplification of the medication regimen may have a tremendous impact on the child's overall well-being and functional outcome.

Since the initial report of resective epilepsy surgery in TSC over 40 years ago [17], numerous studies have suggested that the best surgical outcome occurs when a single tuber and associated well-documented epileptogenic region can be identified as the source of seizures [19–22]. However, identification of the "epileptogenic tuber" or "epileptogenic region" remains a challenge in TSC, since most individuals have multiple bilateral tubers, many individuals have bilateral, multifocal epileptiform changes on EEG, and many individuals have ictal EEG patterns that make lateralization and localization of seizure onset difficult to determine. Given these difficulties, many new technologies have been developed with the goal of improving the ability to identify the seizure focus in TSC noninvasively. PET scanning using an alpha-[^{11}C]methyl-L-tryptophan ([^{11}C]AMT) has been used successfully by some investigators to identify "epileptogenic tubers" based on the hypothesis that serotonin synthesis might be increased in epileptic tubers [23]. Other investigators have examined the possible utility of fluorodeoxyglucose (FDG) PET/MRI coregistration and diffusion tensor imaging (DTI) in identifying which tubers are epileptogenic [24]. Several reports have also suggested that MEG may be more helpful in identifying the epileptogenic zone in TSC than surface EEG [25, 26].

If the epileptic focus or zone can be identified in an individual with TSC with a consensus of findings from the above studies, then the focus, usually including a cortical tuber and surrounding region, can often be resected. Depending on relationship of the identified region to eloquent cortex, additional preoperative testing may be necessary to identify potential postsurgical morbidities, such as intracarotid sodium amobarbital procedure (also known as Wada testing after the first physician who performed it, Dr. Juhn Wada) [27] to help lateralize language and memory. If the epileptogenic focus is localized to the region of a tuber (i.e., EEG ictal and interictal features as well as PET, SPECT, or MEG findings correlate with MRI localization of tuber), many centers further identify the zone by intraoperative corticography. However, if it is not possible to easily identify "the tuber" for resection, then invasive or phase 2 monitoring is usually recommended. This involves an initial craniotomy with placement of subdural electrodes and additional monitoring, often occurring over several days, to further characterize the ictal and interictal features. Subsequent to sufficient localizing data being collected, a second surgery is performed for resection of the identified region.

Traditionally, individuals who did not have a well-localized single tuber/epileptogenic region were excluded from resective surgery, based on the concept that they likely had multifocal epilepsy and therefore would not benefit from resective surgery.

Some individuals, particularly those with atonic seizures or drop attacks, are then considered for a corpus callosotomy. Corpus callosotomy is a surgical procedure that cuts the fibers connecting the two cerebral hemispheres with the thought that seizure propagation or generalization would be prevented. Even if seizures are not controlled by this approach, the generalization of those seizures, resulting in drops and potential significant physical injury, is prevented. In addition, it is possible that following a corpus callosotomy seizure onset might be lateralized allowing consideration of further resective surgery, if seizures continue to be difficult to control.

However, recently it has been demonstrated that a dominant seizure focus can sometimes be associated with seizure onset even when EEG features suggest a multifocal epilepsy, and that individuals with these features can benefit significantly from localized resection. This has been shown in bilateral temporal epilepsy, hypothalamic hamartoma, infantile spasms resulting from cryptogenic focal cortical dysplasia, and congenital focal brain lesions with generalized EEG findings [28–31]. This treatment philosophy has also been applied as a strategy in the management of TSC patients [32]. This approach is particularly attractive for young children with highly refractory, malignant epilepsy with goals of both improved seizure control and better neurocognitive development, since uncontrolled seizures are a significant risk factor for neurocognitive impairment in TSC.

To enable resection of dominant seizure foci, bilateral strip electrode studies are often performed to permit detection of one or more discrete areas of the brain that could be targeted for resective epilepsy surgery, and a multistaged surgical approach is used to identify and define the extent of seizure foci in individuals with TSC [33]. Through this approach, many children have shown improvements in seizure control as well as neurocognitive functioning [32].

Both this invasive EEG monitoring and intraoperative corticography have shown that for optimal surgical outcome, resection of the cortical tuber as well as surrounding additional epileptogenic tissue is required [34, 35]. Therefore, it is recommended that either intraoperative corticography or invasive EEG monitoring be done in individuals with TSC to better define the epileptogenic region.

It is clear that epilepsy surgery plays a very important role in the management of refractory epilepsy in TSC. Prospective studies are needed to help determine if such surgery improves long-term seizure control and neurocognitive outcomes, and to help determine the optimal timing of surgery in the course of an individual's epilepsy.

10.4 Infantile Spasms

At least one-third of infants with TSC will develop infantile spasms (IS), one of the catastrophic epilepsy syndromes of early childhood [36–41]. Although there are many different etiologies for IS, including neonatal encephalopathy, infection, and other genetic and dysgenetic causes, TSC is likely the single most common cause of IS and has accounted for up to 10% of cases in several retrospective clinical series [42].

10.4.1
Clinical Features of IS

Similar to IS from other causes, infants with TSC typically have onset of IS before 1 year of age, with peak onset between 4 and 6 months of age. IS have been recognized in many TSC infants of age <1 week.

Infantile spasms are characterized clinically by the infant having quick sudden flexion of the limbs, often described as a "salaam attack." In addition or instead of flexion of the limbs with IS, infants may have a quick extension of the limbs, or most commonly a mixed flexion and extension. Infants may have flexion of the neck, trunk, and arms with extension of the legs, or flexion of the legs and extension of the arms with varying flexion of the neck and trunk. Some infants have very subtle IS, characterized, for example, by slight head nods, upward eye deviation, or elevation of the shoulder. In TSC infants in particular, IS may also appear asymmetric, such as flexion or extension of the limbs on one side of the body or quick episodes of eye deviation to one side.

Classically, IS appear in clusters rather than in isolated fashion throughout the day. Although IS can occur at any time of the day, the most common time for the clusters to occur is at sleep–wake interfaces. During a cluster, an infant may have anywhere from a few to over 100 individual spasms occurring 5–30 s apart, and the number of clusters per day can range anywhere from 1 to over 50. Following an individual cluster of IS, the infant may appear lethargic, irritable, or hyperalert.

The clustering or repetitive character of IS is often key in considering the diagnosis. Particularly in TSC, as IS are very common and may be subtle or asymmetric, infants with very mild or atypical behaviors (such as head nodding, eye blinking, stereotyped movement of a single limb) that occur repetitively should be evaluated for possible IS. During a cluster of IS, other clinical features may also be seen, such as the infant having a change in breathing pattern or rate, crying at the end of the cluster, laughing, facial flushing, eye deviation, or smiling.

With the onset of IS, infants also often experience a relative plateauing or even regression in behaviors and developmental skills. Parents often report a disappearance of the infant's social smile as well as appearance of apathy with a loss of interest in surroundings and individuals. Infants often develop visual inattention and experience a relative autistic regression, which often precedes the onset of clinical spasms. Motor regression occurs less commonly, though the infant may stop reaching and grabbing, possibly related to a loss of interest in objects. However, in some infants with IS, personality and acquisition of developmental milestones are not affected and continue normally.

The factors influencing the development of IS in TSC are unclear, but likely include cortical tuber location, cortical tuber burden, and genotype. Similarly, factors influencing neurocognitive outcome of IS in TSC are unclear, but likely include genotype, duration of spasms before effective treatment control, and EEG features at time of IS presentation [36, 38, 39, 41, 43–47].

Unfortunately, the diagnosis of IS is often delayed, with reports showing an average of 4 months from onset to diagnosis [43]. The delay often results from a misinterpretation of the infant's behaviors – ranging from a failure of parents to recognize the sometimes subtle spasms as an unusual behavior (though this is rare) to misidentification of the seizures as a response to discomfort or colic, particularly in the absence of distinctive EEG features.

10.4.2
EEG Features of Infantile Spasms

The classic features of IS on EEG include the interictal pattern of hypsarrhythmia (Figure 10.4) [48]. Hypsarrhythmia, which is an awake interictal pattern, is characterized by three main features on EEG: high-voltage activity (often up to 500 μV compared to the normal 100–150 μV background activity), a disorganized background, and bilateral multifocal epileptiform discharges. Although considered the

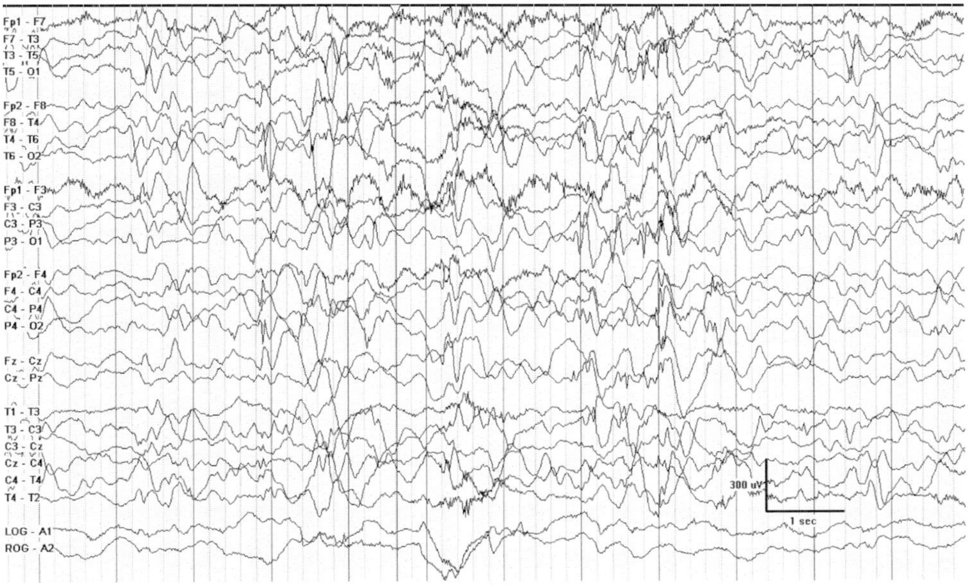

(a)

Figure 10.4 Infantile spasms: EEG features. (a) Hypsarrhythmia. Awake interictal EEG pattern often seen in IS, characterized by a disorganized background, high voltages, with multifocal epileptiform discharges. (b) EEG recording showing a cluster of markedly high-amplitude spike and slow waves followed by return to background rhythm, each correlating with clinical a spasm (inset a). High-frequency noise seen in several channels represents artifact. (c) EEG recording showing high-amplitude slow wave followed by several second relative decrement in background activity (shaded area). High-amplitude slow wave correlated with clinical spasm. (d) EEG recording showing more subtle EEG features of high-amplitude slow wave and very brief relative background decrement correlating with clinical spasm (shaded area). (e) EEG recording showing high-amplitude slow wave followed by background decrement correlating with infantile spasm (shaded area), followed by onset of partial seizure with onset in right temporal region (seizure onset marked by asterisk).

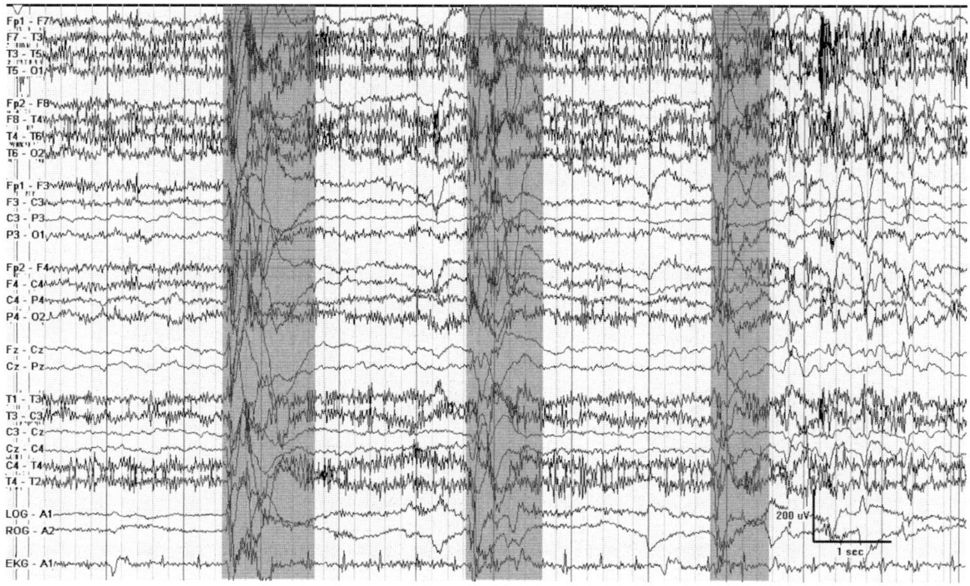

(b)

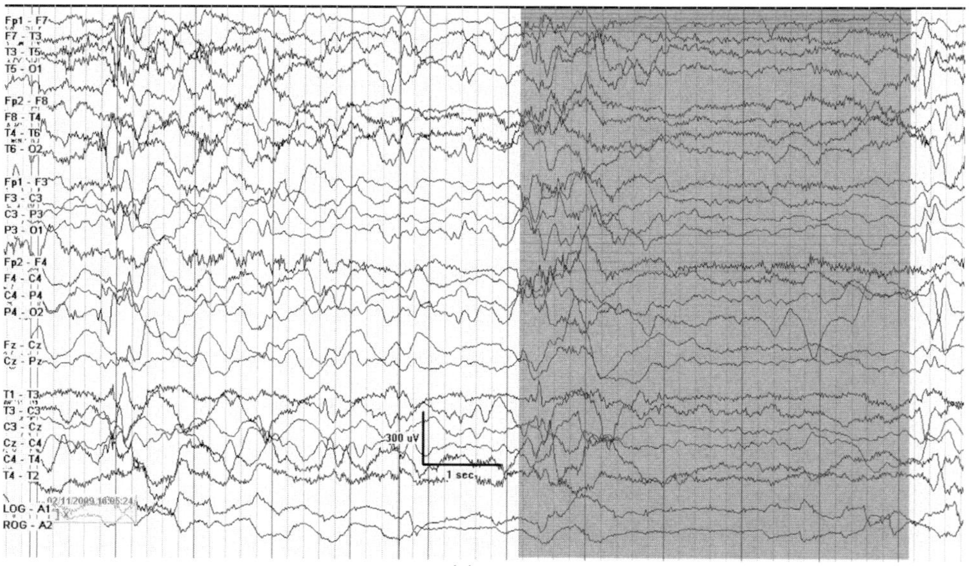

(c)

Figure 10.4 *(Continued)*.

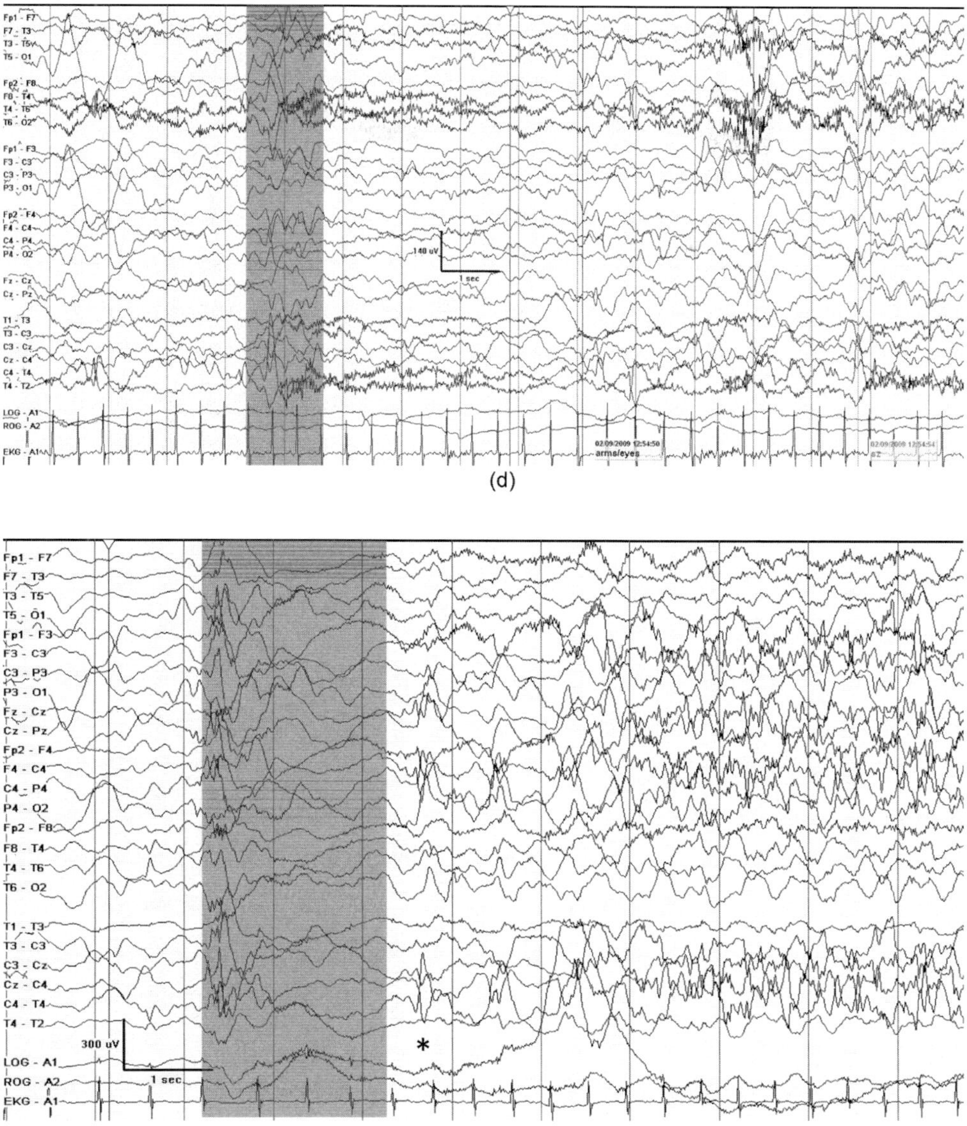

Figure 10.4 (*Continued*).

classic EEG pattern for IS, hypsarrhythmia is seen in only 40–70% of infants who develop IS associated with all etiologies, and is present more commonly early in the course of IS [49–51]. Its presence may be even less common with certain etiologies of IS, such as TSC. Therefore, video EEG is recommended for any infant with possible IS by clinical history who does not have features of hypsarrhythmia on EEG.

The most common ictal EEG pattern of a spasm is a high-voltage, frontal predominant generalized slow wave with a subsequent attenuation of the voltage, or electrodecrement (Figure 10.4). However, the ictal pattern can be the electrodecrement alone, generalized sharp wave and slow wave complexes, or a transient suppression of the hypsarrhythmic pattern (Figure 10.4).

10.4.3 Treatment of Infantile Spasms in TSC

Currently, there are unfortunately a limited number of effective treatments for IS. ACTH (adrenocorticotrophin hormone) and vigabatrin are presently the most effective anticonvulsant drugs available for IS. ACTH has been the preferred treatment since the 1950s, and for most cases of IS in the United States, it remains the preferred treatment as vigabatrin has not been FDA approved until recently. ACTH is delivered as a daily intramuscular injection given over a 1–2-month course, and the most common dosage is 20–40 IU daily although both higher and lower dose protocols are used. In infants with IS from all etiologies including TSC treated with ACTH, spasms are controlled in at least 50%, and the EEG shows marked improvement with normal features in approximately 30–40% [52–57]. Unfortunately, following taper of ACTH spasms recur in approximately 30%, most commonly within the first 2 months. If a second course of ACTH is then attempted, seizures are controlled in <70%. Side effects of ACTH treatment include infections (which can be a cause of mortality and include mucosal candidiasis and subcutaneous abscess at site of injection), arterial hypertension (which is seen in at least one-third of infants treated with ACTH), electrolyte imbalances, growth effects, irritability, and sleep disturbances.

Vigabatrin is an irreversible GABA transaminase inhibitor that has been used in the treatment of IS in Europe since 1990 [58–61], and was recently approved for use in the treatment of IS in the United States by the FDA. Like ACTH, it can lead to control of seizures within days to a month of treatment. Vigabatrin has been shown in several studies to be more effective in controlling IS in TSC patients than ACTH [58, 60, 62, 63]. Currently, vigabatrin is considered as the "treatment of choice" or first-line treatment of TSC IS.

The most significant side effect of vigabatrin is retinal toxicity that is thought to occur in up to 30–50% of adults who take the medication, though this may be much less common in children [64–67]. Vigabatrin-induced retinal toxicity can result in irreversible concentric visual field defects, which can be severe [64]. The mechanisms of this toxicity are not understood. Vigabatrin can also be effective in the treatment of refractory complex partial seizures, although there is no evidence of increased efficacy in treating complex partial seizures in TSC. If used to treat partial complex seizures, it is recommended that individuals be closely followed by an ophthalmologist with routine visual field testing by kinetic perimetry. Similar testing is not feasible in infants treated with vigabatrin for infantile spasms. Although electroretinography can be helpful in detecting vigabatrin-related retinal changes in infants, this technology is not readily available. However, due to the significant possible cognitive impairment resulting from ineffectively controlled IS, it is thought by many that the possibility of retinal toxicity and resulting constriction of visual peripheral fields is an acceptable side effect.

Another vigabatrin side effect of uncertain clinical significance is a signal change on MRI in the deep gray nuclei of the brain while infants are on high-dose vigabatrin treatment [68, 69]. It is thought that these signal changes may be seen in up to 20% of infants who take the medication, but they are thought to be reversible if vigabatrin is subsequently discontinued. Other possible side effects of vigabatrin treatment include somnolence, hyperactivity, and insomnia.

The ketogenic diet is also an effective nonpharmacologic treatment for IS [70, 71], and should be considered if anticonvulsant medications do not provide seizure control or are not available. Increasing experience with the ketogenic diet in IS may in the near future promote the ketogenic diet to a "first-line" therapy for IS. As discussed above, the ketogenic diet is also an effective treatment of other forms of refractory epilepsy in TSC [13].

Should an infant with TSC continue to have IS in spite of medical therapies, the possibility of epilepsy surgery should be considered, as discussed above.

10.4.4 Infantile Spasms in TSC: Outcome

Infants who develop IS, with TSC as well as all etiologies, are at significant risk of subsequent mental retardation, autism spectrum disorders, visual and auditory difficulties, attention-deficit hyperactivity disorder (ADHD), and epilepsy [2, 37, 45, 47, 50, 72–74]. Factors that may affect the infant's neurocognitive outcome include the etiology of IS, response to treatment, characteristics of EEG features, and presence of other seizure types. Studies of IS from all etiologies suggest that the likely most important variable in outcome is the etiology of IS, and in general, symptomatic causes such as TSC have a worse prognosis than cryptogenic [50, 73, 74]. However, up to one-third of children with TSC who develop IS will have a normal cognitive outcome, suggesting that other variables play a significant role in outcome [36, 43, 47]. The particular EEG features associated with IS may also be predictive of outcome, as many reports have suggested that the presence of hypsarrhythmia is a predictor of poor outcome in all IS patients, including TSC [46, 51, 73].

Other variables that may influence outcomes in IS and TSC include duration of IS, time from onset until treatment, duration of hypsarrhythmia on EEG, and TSC genotype, with a higher incidence of IS in those with a TSC2 disease-causing mutation [36, 43, 46, 47].

10.5 Lennox–Gastaut Syndrome

Onset of LGS occurs between 1 and 8 years of age with a mean age of onset of 3 years. The majority of TSC children who develop LGS have had prior infantile spasms or partial onset seizures. Lennox–Gastaut syndrome is a mixed seizure disorder characterized by the presence of myoclonic, atonic, and atypical absence seizures as well as EEG features of 1.5–2 Hz slow spike and wave activity [75–77]. The majority of children who have LGS develop a highly refractory seizure disorder and require polytherapy for optimal seizure control, including anticonvulsant medications, dietary therapy, vagus nerve stimulation, and occasionally epilepsy surgery including corpus callosotomy [78]. Broad-spectrum anticonvulsant medications are indicated for the treatment of LGS, as medications typically used for partial seizures can exacerbate seizure activity in LGS. Complete seizure control is often not achieved. Similar to infantile spasms, children with TSC who develop LGS are at a much greater risk of poor neurocognitive outcome.

10.6 Pathogenesis of Epilepsy in TSC

The causes of epilepsy in TSC are not well understood. It is thought that the tubers, which are areas of dysgenesis, are somehow associated with the pathogenesis of epilepsy. Evidence in favor of this model is that many individuals with TSC become seizure free following surgical resection of a tuber identified as being associated with seizure onset [19, 22, 79]. However, it is not understood if the tubers themselves are epileptogenic, or perhaps "irritating" to the neighboring neurons, or perhaps the areas surrounding discrete tubers are also abnormal and lead to epilepsy. There is some evidence that at least some tubers associated with the epileptogenic region are electrically silent [34, 80], suggesting that at least in some individuals the tubers themselves do not "cause" seizures.

Individuals with TSC often have multiple bilateral tubers, and yet may have a clear single epileptogenic focus as determined both from localization of EEG abnormalities and from focal onset to their seizures shown by video EEG monitoring during seizure activity. Other individuals with TSC have bilateral multifocal epileptiform discharges on EEG, yet all their clinical seizures appear to be associated with only one of these regions. For other individuals with TSC, it is not possible even to lateralize their seizure onset. This range of findings raises fundamental questions about whether all tubers "are created equal," and why some are associated with epileptogenic foci and others are not. It is clear that tubers do not have uniform pathologic characteristics, so it is perhaps not surprising that clinical effects of tubers might be highly variable. For example, some tubers undergo calcification or develop cyst-like changes. Recently, it has been shown that cyst-like tubers are associated with epileptogenic foci in TSC, although the mechanisms of that relationship are unclear [81]. In addition, some individuals with TSC and epilepsy (perhaps 5–10%) do not appear to have tubers on high-quality MRI imaging. This raises the possibility that epileptogenicity in some individuals with TSC may be related to diffuse abnormalities, or subtle cortical defects below the resolution of current imaging. TSC-causing mutations can clearly have effects on neuronal morphology and function that are distinct from tuber formation [82, 83].

The relationship between genotype and epilepsy phenotype in TSC is also not well understood. Similar to other TSC manifestations, a mutation in the TSC2 gene is associated with a significantly increased risk of refractory epilepsy, as well as infantile spasms, in comparison to individuals with TSC1 mutations [36, 84].

Many aspects of the basic function of the TSC1/TSC2 proteins in brain function are being elucidated, which may help to explain epileptogenesis in TSC [85, 86]. These range from rapamycin treatment affecting epilepsy in animal models [87] to alterations in glutamate receptor expression in cortical tubers [88], to effects of the TSC proteins on axon formation [83], to effects on neuronal morphology with subsequent altered glutamatergic synaptic activity [82]. Also see Chapter 9 for further discussion.

10.7 The Natural History of Epilepsy in TSC

The natural history of epilepsy in TSC has not been well characterized and is not well understood. Some individuals with TSC have seizure disorders that are readily controlled by medication, while others develop seizure disorders that are highly refractory to treatment, and yet others do not develop epilepsy during their lifetime. Some individuals with TSC are able to taper off medications after having seizure control for a period of time and remain seizure free, while others become seizure free following dietary therapy or epilepsy surgery and others continue to have frequent seizures without effective control in spite of multiple treatment strategies including resective epilepsy surgery.

The relationship between epilepsy and the other neurologic features of TSC including cognitive impairment, autism spectrum disorders, and mental health issues is also not well understood. However, several studies have implicated early age of onset of seizures, presence of infantile spasms, and refractory epilepsy as significant variables for both subsequent cognitive impairment and the development of autism spectrum disorders [89–92]. The relationship between the neuroanatomic features of TSC and these clinical symptoms is also not well understood. Previous work has shown a correlation between tuber number and refractory epilepsy as well as neurocognitive impairment. Current opinion suggests that rather than number of tubers it may be the overall tuber volume or tuber burden an individual has that may be associated with severity of neurologic phenotype. The relationship between EEG features, independent of clinical seizures, and their neurocognitive and neurobehavioral phenotype is also not well understood. It is likely that many individuals with TSC experience epileptic encephalopathy, related to the presence, location, and frequency of interictal epileptiform discharges, which affect the individual's neurocognitive and neurobehavioral functioning. Although these relationships are not well understood, it is clear that there is a strong association between epilepsy and the other neurologic symptoms of TSC, particularly cognitive impairment and autism spectrum disorders. Therefore, it is possible that early and effective seizure control may help improve overall neurologic outcome in TSC.

A better understanding of the process of epileptogenesis in TSC might also strongly impact neurologic outcome. If markers predicting seizures and IS in infants with TSC could be identified, then there may be a future role for prophylactic treatment of at-risk infants. Identification of such markers as well as determining appropriate prophylactic treatment should become a focus of clinical research in TSC in the next few years.

References

1 Webb, D.W., Fryer, A.E., and Osborne, J.P. (1991) On the incidence of fits and mental retardation in tuberous sclerosis. *J. Med. Genet.*, **28**, 395–397.

2 Thiele, E.A. (2004) Managing epilepsy in tuberous sclerosis complex. *J. Child Neurol.*, **19**, 680–686.

3 Ganji, S. and Hellman, C.D. (1985) Tuberous sclerosis: long-term follow-up and longitudinal electroencephalographic study. *Clin. Electroencephalogr.*, **16**, 219–224.

4 Pampiglione, G. and Moynahan, E.J. (1976) The tuberous sclerosis syndrome: clinical and EEG studies in 100 children. *J. Neurol. Neurosurg. Psychiatry*, **39**, 666–673.

5 Jansen, F.E., van Huffelen, A.C., Bourez-Swart, M., and van Nieuwenhuizen, O. (2005) Consistent localization of interictal epileptiform activity on EEGs of patients with tuberous sclerosis complex. *Epilepsia*, **46**, 415–419.

6 Franz, D.N., Glauser, T.A., Tudor, C., and Williams, S. (2000) Topiramate therapy of epilepsy associated with Angelman's syndrome. *Neurology*, **54**, 1185–1188.

7 Collins, J.J., Tudor, C., Leonard, J.M., Chuck, G., and Franz, D.N. (2006) Levetiracetam as adjunctive antiepileptic therapy for patients with tuberous sclerosis complex: a retrospective open-label trial. *J. Child Neurol.*, **21**, 53–57.

8 Chu-Shore, C.J., Major, P., Camposano, S., Muzykewicz, D., and Thiele, E.A. (2009) The natural history of epilepsy in tuberous sclerosis complex. *Epitepsia*, Dec 22 [Epub ahead of print], PMID: 20041940.

9 Kwan, P. and Brodie, M.J. (2000) Early identification of refractory epilepsy. *N. Engl. J. Med.*, **342**, 314–319.

10 Pfeifer, H.H. and Thiele, E.A. (2005) Low-glycemic-index treatment: a liberalized ketogenic diet for treatment of intractable epilepsy. *Neurology*, **65**, 1810–1812.

11 Kossoff, E.H., McGrogan, J.R., Bluml, R.M., Pillas, D.J., Rubenstein, J.E., and Vining, E.P. (2006) A modified Atkins diet is effective for the treatment of intractable pediatric epilepsy. *Epilepsia*, **47**, 421–424.

12 Muzykewicz, D.A., Lyczkowski, D.A., Memon, N., Conant, K.D., Pfeifer, H.H., and Thiele, E.A. (2009) Efficacy, safety, and tolerability of the low glycemic index treatment in pediatric epilepsy. *Epilepsia*, **50**, 1118–1126.

13 Kossoff, E.H., Thiele, E.A., Pfeifer, H.H., McGrogan, J.R., and Freeman F J.M. (2005) Tuberous sclerosis complex and the ketogenic diet. *Epilepsia*, **46**, 1684–1686.

14 Bailey, E.E., Pfeifer, H.H., and Thiele, E.A. (2005) The use of diet in the treatment of epilepsy. *Epilepsy Behav.*, **6**, 4–8.

15 Parain, D., Penniello, M.J., Berquen, P., Delangre, T., Billard, C., and Murphy, J.V. (2001) Vagal nerve stimulation in tuberous sclerosis complex patients. *Pediatr. Neurol.*, **25**, 213–216.

16 Major, P., Greenberg, E., Khan, A., and Thiele, E.A. (2008) Pyridoxine supplementation for the treatment of levetiracetam-induced behavior side effects in children: preliminary results. *Epilepsy Behav.*, **13**, 557–559.

17 Perot, P., Weir, B., and Rasmussen, T. (1966) Tuberous sclerosis. Surgical therapy for seizures. *Arch. Neurol.*, **15**, 498–506.

18 Jansen, F.E., Huffelen, A.C.V., Rijen, P.C.V., Leijten, F.S.S., Jennekens-Schinkel, A., Gosselaar, P., and Nieuwenhuizen, O.V. (2007) Epilepsy

surgery in tuberous sclerosis: the Dutch experience. *Seizure*, **16**, 445–453.
19 Bebin, E.M., Kelly, P.J., and Gomez, M.R. (1993) Surgical treatment for epilepsy in cerebral tuberous sclerosis. *Epilepsia*, **34**, 651–657.
20 Jarrar, R.G., Buchhalter, J.R., and Raffel, C. (2004) Long-term outcome of epilepsy surgery in patients with tuberous sclerosis. *Neurology*, **62**, 479–481.
21 Lachhwani, D.K., Pestana, E., Gupta, A., Kotagal, P., Bingaman, W., and Wyllie, E. (2005) Identification of candidates for epilepsy surgery in patients with tuberous sclerosis. *Neurology*, **64**, 1651–1654.
22 Jansen, F.E., van Huffelen, A.C., Algra, A., and van Nieuwenhuizen, O. (2007) Epilepsy surgery in tuberous sclerosis: a systematic review. *Epilepsia*, **48**, 1477–1484.
23 Chugani, D.C., Chugani, H.T., Muzik, O., Shah, J.R., Shah, A.K., Canady, A., Mangner, T.J., and Chakraborty, P.K. (1998) Imaging epileptogenic tubers in children with tuberous sclerosis complex using alpha-[^{11}C]methyl-L-tryptophan positron emission tomography. *Ann. Neurol.*, **44**, 858–866.
24 Chandra, P.S., Salamon, N., Huang, J., Wu, J.Y., Koh, S., Vinters, H.V., and Mathern, G.W. (2006) FDG-PET/MRI coregistration and diffusion-tensor imaging distinguish epileptogenic tubers and cortex in patients with tuberous sclerosis complex: a preliminary report. *Epilepsia*, **47**, 1543–1549.
25 Iida, K., Otsubo, H., Mohamed, I.S., Okuda, C., Ochi, A., Weiss, S.K., Chuang, S.H., and Snead, O.C. (2005) Characterizing magnetoencephalographic spike sources in children with tuberous sclerosis complex. *Epilepsia*, **46**, 1510–1517.
26 Jansen, F.E., Huiskamp, G., van Huffelen, A.C., Bourez-Swart, M., Boere, E., Gebbink, T., Vincken, K.L., and van Nieuwenhuizen, O. (2006) Identification of the epileptogenic tuber in patients with tuberous sclerosis: a comparison of high-resolution EEG and MEG. *Epilepsia*, **47**, 108–114.
27 Wada, J.A. (2008) A fateful encounter: sixty years later – reflections on the Wada test. *Epilepsia*, **49**, 726–727.
28 Chugani, H.T., Shields, W.D., Shewmon, D.A., Olson, D.M., Phelps, M.E., and Peacock, W.J. (1990) Infantile spasms. I. PET identifies focal cortical dysgenesis in cryptogenic cases for surgical treatment. *Ann. Neurol.*, **27**, 406–413.
29 Hirsch, L.J., Spencer, S.S., Spencer, D.D., Williamson, P.D., and Mattson, R.H. (1991) Temporal lobectomy in patients with bitemporal epilepsy defined by depth electroencephalography. *Ann. Neurol.*, **30**, 347–356.
30 Harvey, A.S. and Freeman, J.L. (2007) Epilepsy in hypothalamic hamartoma: clinical and EEG features. *Semin. Pediatr. Neurol.*, **14**, 60–64.
31 Wyllie, E., Lachhwani, D.K., Gupta, A., Chirla, A., Cosmo, G., Worley, S., Kotagal, P., Ruggieri, P., and Bingaman, W.E. (2007) Successful surgery for epilepsy due to early brain lesions despite generalized EEG findings. *Neurology*, **69**, 389–397.
32 Weiner, H.L., Carlson, C., Ridgway, E.B., Zaroff, C.M., Miles, D., LaJoie, J., and Devinsky, O. (2006) Epilepsy surgery in young children with tuberous sclerosis: results of a novel approach. *Pediatrics*, **117**, 1494–1502.
33 Bauman, J.A., Feoli, E., Romanelli, P., Doyle, W.K., Devinsky, O., and Weiner, H.L. (2005) Multistage epilepsy surgery: safety, efficacy, and utility of a novel approach in pediatric extratemporal epilepsy. *Neurosurgery*, **56**, 318–334.
34 Madhavan, D., Weiner, H.L., Carlson, C., Devinsky, O., and Kuzniecky, R. (2007) Local epileptogenic networks in tuberous sclerosis complex: a case review. *Epilepsy Behav.*, **11**, 140–146.
35 Bye, A.M., Matheson, J.M., Tobias, V.H., and Mackenzie, R.A. (1989) Selective epilepsy surgery in tuberous sclerosis. *Aust. Paediatr. J.*, **25**, 243–245.
36 Muzykewicz, D.A., Costello, D.J., Halpern, E.F., and Thiele, E.A. (2009) Infantile spasms in tuberous sclerosis complex: prognostic utility of EEG. *Epilepsia*, **50**, 290–296.
37 Riikonen, R. and Simell, O. (1990) Tuberous sclerosis and infantile spasms. *Dev. Med. Child Neurol.*, **32**, 203–209.
38 Curatolo, P., Cusmai, R., Cortesi, F., Chiron, C., Jambaque, I., and Dulac, O.

(1991) Neuropsychiatric aspects of tuberous sclerosis. *Ann. N.Y. Acad. Sci.*, **615**, 8–16.

39 Fukushima, K., Inoue, Y., Fujiwara, T., and Yagi, K. (2001) Long-term follow-up study of West syndrome associated with tuberous sclerosis. *Brain Dev.*, **23**, 698–704.

40 O'Callaghan, F.J., Noakes, M.J., Martyn, C.N., and Osborne, J.P. (2004) An epidemiological study of renal pathology in tuberous sclerosis complex. *BJU Int.*, **94**, 853–857.

41 Au, K.S., Williams, A.T., Roach, E.S., Batchelor, L., Sparagana, S.P., Delgado, M.R., Wheless, J.W., Baumgartner, J.E., Roa, B.B., Wilson, C.M., Smith-Knuppel, T.K., Cheung, M.-Y.C., Whittemore, V.H., King, T.M., and Northrup, H. (2007) Genotype/phenotype correlation in 325 individuals referred for a diagnosis of tuberous sclerosis complex in the United States. *Genet. Med.*, **9**, 88–100.

42 Curatolo, P., Seri, S., Verdecchia, M., and Bombardieri, R. (2001) Infantile spasms in tuberous sclerosis complex. *Brain Dev.*, **23**, 502–507.

43 Goh, S., Kwiatkowski, D.J., Dorer, D.J., and Thiele, E.A. (2005) Infantile spasms and intellectual outcomes in children with tuberous sclerosis complex. *Neurology*, **65**, 235–238.

44 Jambaqué, I., Cusmai, R., Curatolo, P., Cortesi, F., Perrot, C., and Dulac, O. (1991) Neuropsychological aspects of tuberous sclerosis in relation to epilepsy and MRI findings. *Dev. Med. Child Neurol.*, **33**, 698–705.

45 O'Callaghan, F.J.K., Harris, T., Joinson, C., Bolton, P., Noakes, M., Presdee, D., Renowden, S., Shiell, A., Martyn, C.N., and Osborne, J.P. (2004) The relation of infantile spasms, tubers, and intelligence in tuberous sclerosis complex. *Arch. Dis. Child.*, **89**, 530–533.

46 Primec, Z.R., Stare, J., and Neubauer, D. (2006) The risk of lower mental outcome in infantile spasms increases after three weeks of hypsarrhythmia duration. *Epilepsia*, **47**, 2202–2205.

47 Camposano, S.E., Major, P., Halpern, E., and Thiele, E.A. (2008) Vigabatrin in the treatment of childhood epilepsy: a retrospective chart review of efficacy and safety profile. *Epilepsia*, **49**, 1186–1191.

48 Gibbs, E.L., Fleming, M.M., and Gibbs, F.A. (1954) Diagnosis and prognosis of hypsarrhythmia and infantile spasms. *Pediatrics*, **13**, 66–73.

49 Hrachovy, R.A. and Frost, J.D. (2003) Infantile epileptic encephalopathy with hypsarrhythmia (infantile spasms/West syndrome). *J. Clin. Neurophysiol.*, **20**, 408–425.

50 Shields, W.D. (2006) Infantile spasms: little seizures, BIG consequences. *Epilepsy Curr.*, **6**, 63–69.

51 Kramer, U., Sue, W.C., and Mikati, M.A. (1997) Hypsarrhythmia: frequency of variant patterns and correlation with etiology and outcome. *Neurology*, **48**, 197–203.

52 Bobele, G.B. and Bodensteiner, J.B. (1990) Infantile spasms. *Neurol. Clin.*, **8**, 633–645.

53 Fois, A., Malandrini, F., Balestri, P., and Giorgi, D. (1984) Infantile spasms – long term results of ACTH treatment. *Eur. J. Pediatr.*, **142**, 51–55.

54 Hrachovy, R.A., Frost, J.D., Kellaway, P., and Zion, T. (1980) A controlled study of ACTH therapy in infantile spasms. *Epilepsia*, **21**, 631–636.

55 Hrachovy, R.A., Frost, J.D., Kellaway, P., and Zion, T.E. (1983) Double-blind study of ACTH vs prednisone therapy in infantile spasms. *J. Pediatr.*, **103**, 641–645.

56 Ito, M., Okuno, T., Fujii, T., Mutoh, K., Oguro, K., Shiraishi, H., Shirasaka, Y., and Mikawa, H. (1990) ACTH therapy in infantile spasms: relationship between dose of ACTH and initial effect or long-term prognosis. *Pediatr. Neurol.*, **6**, 240–244.

57 Sher, P.K. and Sheikh, M.R. (1993) Therapeutic efficacy of ACTH in symptomatic infantile spasms with hypsarrhythmia. *Pediatr. Neurol.*, **9**, 451–456.

58 Chiron, C., Dumas, C., Jambaqué, I., Mumford, J., and Dulac, O. (1997) Randomized trial comparing vigabatrin and hydrocortisone in infantile spasms due to tuberous sclerosis. *Epilepsy Res.*, **26**, 389–395.

59 Fejerman, N., Cersósimo, R., Caraballo, R., Grippo, J., Corral, S., Martino, R.H., Martino, G., Aldao, M., Caccia, P., Retamero, M., Macat, M.C., Blasi, M.A.D., and Adi, J. (2000) Vigabatrin as a first-choice drug in the treatment of West syndrome. *J. Child Neurol.*, **15**, 161–165.

60 Aicardi, J., Mumford, J.P., Dumas, C., and Wood, S. (1996) Vigabatrin as initial therapy for infantile spasms: a European retrospective survey. Sabril IS Investigator and Peer Review Groups. *Epilepsia*, **37**, 638–642.

61 Curatolo, P., Verdecchia, M., and Bombardieri, R. (2001) Vigabatrin for tuberous sclerosis complex. *Brain Dev.*, **23**, 649–653.

62 Mackay, M., Weiss, S., and Snead, O.C. (2002) Treatment of infantile spasms: an evidence-based approach. *Int. Rev. Neurobiol.*, **49**, 157–184.

63 Mackay, M.T., Weiss, S.K., Adams-Webber, T., Ashwal, S., Stephens, D., Ballaban-Gill, K., Baram, T.Z., Duchowny, M., Hirtz, D., Pellock, J.M., Shields, W.D., Shinnar, S., Wyllie, E., and Snead, O.C. (2004) Practice parameter: medical treatment of infantile spasms: report of the American Academy of Neurology and the Child Neurology Society. *Neurology*, **62**, 1668–1681.

64 Kinirons, P., Cavalleri, G.L., O'Rourke, D., Doherty, C.P., Reid, I., Logan, P., Liggan, B., and Delanty, N. (2006) Vigabatrin retinopathy in an Irish cohort: lack of correlation with dose. *Epilepsia*, **47**, 311–317.

65 Banin, E., Shalev, R.S., Obolensky, A., Neis, R., Chowers, I., and Gross-Tsur, V. (2003) Retinal function abnormalities in patients treated with vigabatrin. *Arch. Ophthalmol.*, **121**, 811–816.

66 Besch, D., Kurtenbach, A., Apfelstedt-Sylla, E., Sadowski, B., Dennig, D., Asenbauer, C., Zrenner, E., and Schiefer, U. (2002) Visual field constriction and electrophysiological changes associated with vigabatrin. *Doc. Ophthalmol.*, **104**, 151–170.

67 Harding, G.F.A., Robertson, K., Spencer, E.L., and Holliday, I. (2002) Vigabatrin: its effect on the electrophysiology of vision. *Doc. Ophthalmol.*, **104**, 213–229.

68 Pearl, P.L., Vezina, L.G., Saneto, R.P., McCarter, R., Molloy-Wells, E., Heffron, A., Trzcinski, S., McClintock, W.M., Conry, J.A., Elling, N.J., Goodkin, H.P., de Menezes, M.S., Ferri, R., Gilles, E., Kadom, N., and Gaillard, W.D. (2009) Cerebral MRI abnormalities associated with vigabatrin therapy. *Epilepsia*, **50**, 184–194.

69 Wheless, J.W., Carmant, L., Bebin, M., Conry, J.A., Chiron, C., Elterman, R.D., Frost, M., Paolicchi, J.M., Shields, W.D., Thiele, E.A., Zupanc, M.L., and Collins, S.D. (2009) Magnetic resonance imaging abnormalities associated with vigabatrin in patients with epilepsy. *Epilepsia*, **50**, 195–205.

70 Kossoff, E.H., Hedderick, E.F. Turner, Z., and Freeman F J.M. (2008) A case-control evaluation of the ketogenic diet versus ACTH for new-onset infantile spasms. *Epilepsia*, **49**, 1504–1509.

71 Nordli, D.R., Kuroda, M.M., Carroll, J., Koenigsberger, D.Y., Hirsch, L.J., Bruner, H.J., Seidel, W.T., and Vivo, D.C.D. (2001) Experience with the ketogenic diet in infants. *Pediatrics*, **108**, 129–133.

72 Curatolo, P., D'Argenzio, L., Cerminara, C., and Bombardieri, R. (2008) Management of epilepsy in tuberous sclerosis complex. *Expert Rev. Neurother.*, **8**, 457–467.

73 Koo, B., Hwang, P.A., and Logan, W.J. (1993) Infantile spasms: outcome and prognostic factors of cryptogenic and symptomatic groups. *Neurology*, **43**, 2322–2327.

74 Matsumoto, A., Watanabe, K., Negoro, T., Sugiura, M., Iwase, K., Hara, K., and Miyazaki, S. (1981) Long-term prognosis after infantile spasms: a statistical study of prognostic factors in 200 cases. *Dev. Med. Child Neurol.*, **23**, 51–65.

75 Crumrine, P.K. (2002) Lennox–Gastaut syndrome. *J. Child Neurol.*, **17** (Suppl. 1), S70–S75.

76 Aicardi, J. (1992) Epileptic encephalopathies of early childhood. *Curr. Opin. Neurol. Neurosurg.*, **5**, 344–348.

77 Aicardi, J. and Gomes, A.L. (1992) Clinical and electroencephalographic symptomatology of the 'genuine' Lennox–Gastaut syndrome and its

differentiation from other forms of epilepsy of early childhood. *Epilepsy Res. Suppl.*, **6**, 185–193.

78 Conry, J.A. (2004) Pharmacologic treatment of the catastrophic epilepsies. *Epilepsia*, **45** (Suppl. 5), 12–16.

79 Koh, S., Jayakar, P., Dunoyer, C., Whiting, S.E., Resnick, T.J., Alvarez, L.A., Morrison, G., Ragheb, J., Prats, A., Dean, P., Gilman, J., and Duchowny, M.S. (2000) Epilepsy surgery in children with tuberous sclerosis complex: presurgical evaluation and outcome. *Epilepsia*, **41**, 1206–1213.

80 Major, P., Rakowski, S., Simon, M.V., Cheng, M.L., Eskandar, E., Baron, J., Leeman, B.A., Frosch, M.P., and Thiele, E.A. (2008) Are cortical tubers epileptogenic? Evidence from electrocorticography. *Epilepsia*, **50**, 147–154.

81 Chu-Shore, C., Major, P., Montenegro, M., and Thiele, E. (2009) Cyst-like tubers are associated with TSC2 and epilepsy in tuberous sclerosis complex. *Neurology*, **72**, 1165–1169.

82 Tavazoie, S.F., Alvarez, V.A., Ridenour, D.A., Kwiatkowski, D.J., and Sabatini, B.L. (2005) Regulation of neuronal morphology and function by the tumor suppressors Tsc1 and Tsc2. *Nat. Neurosci.*, **8**, 1727–1734.

83 Choi, Y.-J., Nardo, A.D., Kramvis, I., Meikle, L., Kwiatkowski, D.J., Sahin, M., and He, X. (2008) Tuberous sclerosis complex proteins control axon formation. *Genes Dev.*, **22**, 2485–2495.

84 Dabora, S.L., Roberts, P., Nieto, A., Perez, R., Jozwiak, S., Franz, D., Bissler, J., Thiele, E.A., Sims, K., and Kwiatkowski, D.J. (2002) Association between a high-expressing interferon-gamma allele and a lower frequency of kidney angiomyolipomas in TSC2 patients. *Am. J. Hum. Genet.*, **71**, 750–758.

85 Holmes, G.L., Stafstrom, C.E., and Tuberous Sclerosis Study Group (2007) Tuberous sclerosis complex and epilepsy: recent developments and future challenges. *Epilepsia*, **48**, 617–630.

86 Wong, M. (2008) Mechanisms of epileptogenesis in tuberous sclerosis complex and related malformations of cortical development with abnormal glioneuronal proliferation. *Epilepsia*, **49**, 8–21.

87 Zeng, L.-H., Xu, L., Gutmann, D.H., and Wong, M. (2008) Rapamycin prevents epilepsy in a mouse model of tuberous sclerosis complex. *Ann. Neurol.*, **63**, 444–453.

88 Talos, D.M., Kwiatkowski, D.J., Cordero, K., Black, P.M., and Jensen, F.E. (2008) Cell-specific alterations of glutamate receptor expression in tuberous sclerosis complex cortical tubers. *Ann. Neurol.*, **63**, 454–465.

89 Winterkorn, E.B., Pulsifer, M.B., and Thiele, E.A. (2007) Cognitive prognosis of patients with tuberous sclerosis complex. *Neurology*, **68**, 62–64.

90 Bolton, P.F., Park, R.J., Higgins, J.N.P., Griffiths, P.D., and Pickles, A. (2002) Neuro-epileptic determinants of autism spectrum disorders in tuberous sclerosis complex. *Brain*, **125**, 1247–1255.

91 Bolton, P.F. (2004) Neuroepileptic correlates of autistic symptomatology in tuberous sclerosis. *Ment. Retard. Dev. Disabil. Res. Rev.*, **10**, 126–131.

92 Curatolo, P., Porfirio, M.C., Manzi, B., and Seri, S. (2004) Autism in tuberous sclerosis. *Eur. J. Paediatr. Neurol.*, **8**, 327–332.

11
Subependymal Giant Cell Astrocytomas

David Neal Franz, Darcy A. Krueger, and M. Gregory Balko

11.1
Introduction

Subependymal giant cell astrocytomas (SEGAs) occur in a minority (5–20%) of tuberous sclerosis complex (TSC) patients, but represent an important potential source of both morbidity and mortality. SEGAs, along with subependymal nodules (SENs) and cortical tubers, are one of the classic hallmark features of TSC. These lesions are histologically indistinguishable from SENs and are characterized by a progressive increase in size that is typically slow and insidious, but can be rapid and progressive, leading to serious neurologic symptoms from obstructive hydrocephalus [1]. Many attempts have been made to differentiate SEGAs from SENs on the basis of a variety of criteria, including contrast enhancement, location, lesion volume, and magnetic resonance spectroscopy (MRS) [2, 3]. However, with the possible exception of MRS, none of these, or indeed any criteria, besides serial growth or evidence of mass effect/CSF obstruction, can reliably distinguish a SEGA from an SEN [2]. As SEGAs progress, patients develop clinical symptoms of altered mental status and behavior, headache, ataxia, and visual loss. These symptoms lead to more obvious signs of postural headache, vomiting, coma, and, if untreated, death, resulting from progressive development of hydrocephalus [1, 4–6]. By the time clinical symptoms from a SEGA are evident, successful treatment and complete neurologic recovery is often not possible. This has led to recommendations for periodic neuroimaging screening exams for TSC patients [7–9]. The goal of screening is to identify SEGAs at a presymptomatic stage, to allow for successful treatment and avoidance of morbidity. This chapter will review current concepts on the pathology and pathogenesis of SEGAs, their natural history and clinical features, recommendations for radiographic and clinical diagnosis, neuroimaging surveillance, and finally discuss strategies for surgical and medical management.

Tuberous Sclerosis Complex: Genes, Clinical Features, and Therapeutics
Edited by David J. Kwiatkowski, Vicky Holets Whittemore, and Elizabeth A. Thiele

ISBN: 978-3-527-32201-5

11.2 Pathology and Pathogenesis of SEGA

Pathologically subependymal giant cell astrocytomas are typically low-grade (WHO grade I) minimally to noninvasive neoplasms of variable morphology. Since they often contain both astrocytic and neuronal features, these neoplasms are sometimes referred to as "subependymal giant cell tumors." Microscopically, SEGAs are composed of neoplastic cells with a vascular stroma in a fibrillary background. Because of the lack of invasiveness, brain parenchyma is absent. In spite of their name, SEGAs lack the prominent giant cells that are frequently encountered in other giant cell neoplasms such as pleomorphic xanthoastrocytoma (PXA) or giant cell glioblastoma. The neoplastic cells are characterized by medium to large epithelioid to spindle-shaped cells. Morphologically, they appear to be intermediate between neoplastic astrocytic gemistocytes and neuronal ganglion cells. While superficially resembling gemistocytes that possess glassy cytoplasm and an eccentrically disposed, dense, hyperchromatic nucleus, the neoplastic cells of SEGA are larger and contain more abundant "brick-red" cytoplasm and an eccentric vesicular nucleus, often with a prominent nucleolus (Figure 11.1). The cell processes radiate asymmetrically often from a location opposite to the nucleus, while the cell processes of neoplastic gemistocytes radiate evenly from the cell surface [10, 11].

A common feature of SEGA is ill-defined perivascular rosettes similar to those of ependymoma (Figure 11.2). The spindle-shaped cells of the tumor may be disposed in broad fascicles. Parenchymal and vascular calcifications are common. Mitoses, atypia, focal necrosis, and endothelial proliferation are rare. When encountered these features are not associated with aggressive behavior.

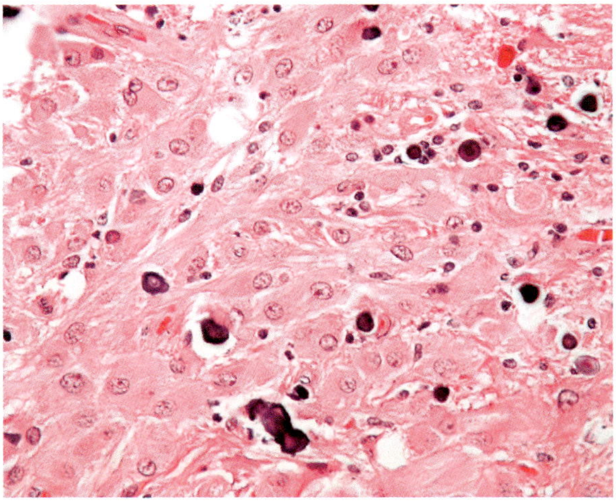

Figure 11.1 Subependymal giant cell astrocytoma characterized by large cells with glassy "brick-red" cytoplasm, eccentric vesicular nuclei and conspicuous nucleoli. Calcifications are common (H&E, 400×).

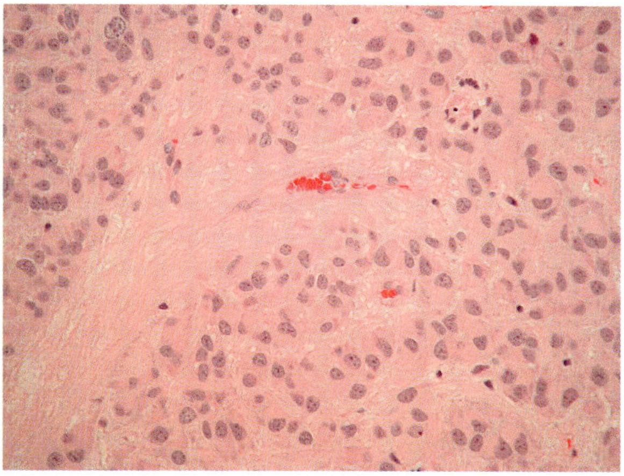

Figure 11.2 Perivascular fibrillarity similar to pseudorosettes of ependymoma may be present in SEGA (H&E, 400×).

Immunohistochemically, SEGAs are strongly positive for S-100 protein. Focal immunoreactivity is seen with glial fibrillary acidic protein (GFAP) (Figure 11.3). Neuronal markers, synaptophysin, neurofilament, and class III b-tubulin (TUJ-1) may also be present in some cells (Figure 11.4). The MIB-1 (Ki-67) proliferation index is low (Figure 11.5). While the balloon cells seen in TSC cortical tubers often contain tuberin, the TSC2 protein, SEGA cells typically do not express tuberin [12–16].

SEGAs should not be confused with subependymomas, as the latter are well-differentiated, low-grade glial neoplasms that occur in the walls of the ventricles. Cytologically, subependymomas are composed of clusters and nodules of bland nuclei separated by a hypocellular fibrillary matrix [17–19]. Most subependymomas

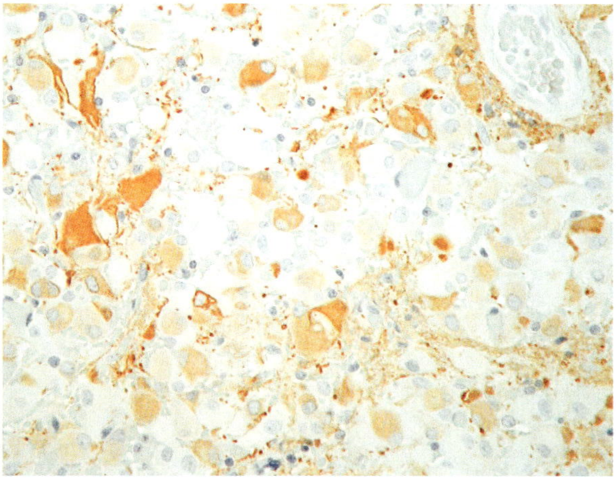

Figure 11.3 The tumor cells are variably reactive for glial fibrillary acidic protein (GFAP immunostain, 400×).

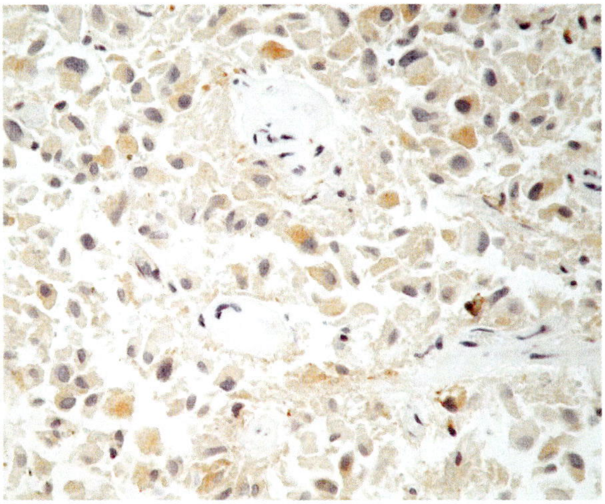

Figure 11.4 Synaptophysin immunoreactivity is common in scattered cells (Synaptophysin, 400×).

are asymptomatic, being found incidentally at autopsy, and are not seen at increased frequency in TSC patients.

SEGAs are likely to derive from neural progenitor cells, as their precursor lesions, SEN, are often seen early in life [23]. SEGAs have been shown to develop through a classic two-hit mechanism, following the Knudsen model for tumor suppressor genes [22]. Five of six SEGAs showed evidence for a second hit event that in four cases was loss of the wild-type allele (loss of heterozygosity, LOH) at the site of mutation, consistent with a genomic deletion event. In the fifth case, the second hit was a second point mutation in TSC2 [22]. However, SEGAs have been shown to express phosphorylated ERK as well [20, 21]. It is possible that this phenomenon occurs

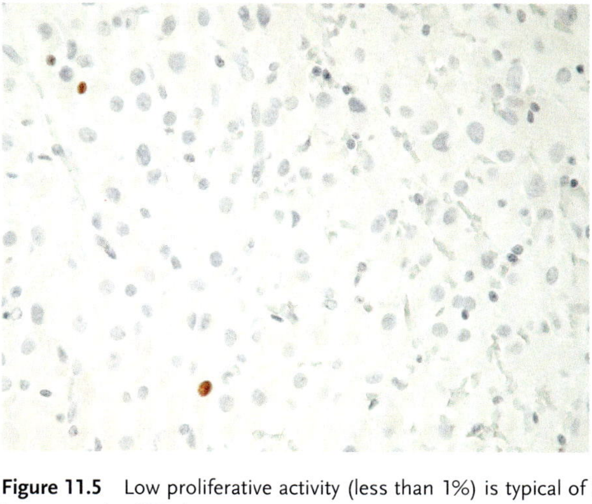

Figure 11.5 Low proliferative activity (less than 1%) is typical of SEGA (MIB-1 immunostain, 400×).

secondary to complete loss of TSC1 or TSC2 in the SEGA cells. Alternatively, phosphorylated ERK may reflect the occurrence of a third genetic hit that is also critical for SEGA development. This is unknown at this time. Both cortical tuber giant cells and SEGA cells show evidence of loss of normal regulation of the mammalian target of rapamycin complex 1 (mTORC1), consistent with genetic and/or functional loss of TSC1 or TSC2 [18, 22, 23]. For further discussion, the reader is referred to Chapters 6 and 9.

11.3 SENs Versus SEGAs

SENs are seen in the majority (70–80%) of tuberous sclerosis patients. These lesions typically occur in the area of the foramen of Monro in the lateral wall of the lateral ventricle, but they are also seen anywhere throughout the body of the lateral ventricle. Occasionally, they are also seen in other locations, including the hypothalamus, retina, and pineal region [24–26]. SENs often become confluent on the surface of the ventricle leading to their original description as "candle gutterings". By this is meant a similarity to the appearance of wax that has been dripped upon a surface and then solidified in an irregular confluent pattern. The majority of individuals with TSC have subependymal nodules in the region of the foramen of Monro that remain static throughout the individual's lifetime. However, up to one-fifth of TSC patients demonstrate progressive growth of a SEN, such that it becomes a SEGA, which may produce clinical symptoms and obstructive hydrocephalus.

In addition, individuals with tuberous sclerosis will occasionally experience gliomatous lesions in the region of the pineal gland or hypothalamus [31, 32]. Aggressive retinal astrocytomas have also been reported in TSC patients [25]. In one case, cranial and spinal metastases were reported from a SEGA [33, 34]. Regardless of where they arise, these lesions all have a similar histologic appearance to both SEGAs and SENs found in the lateral ventricles. When SEGAs are present in other locations, they have a tendency to exhibit more rapid growth than lesions in the area of the foramen of Monro. Rapid growth of retinal SEGAs has required ocular enucleation in a number of individuals [25, 26].

SEGAs develop in patients with each of TSC1 and TSC2 germline mutations. In different series, a greater frequency of SEGAs in patients with either TSC1 or TSC2 has been reported (Chapter 5). Thus, overall, it is uncertain whether these lesions occur at any greater frequency for one gene or the other, or for a certain class of mutations over another.

11.4 Diagnosis of SEGA Versus SEN

A common and important clinical issue in the care of TSC patients is to distinguish between a SEN and a SEGA, when a lesion is identified on a brain MRI or CT scan.

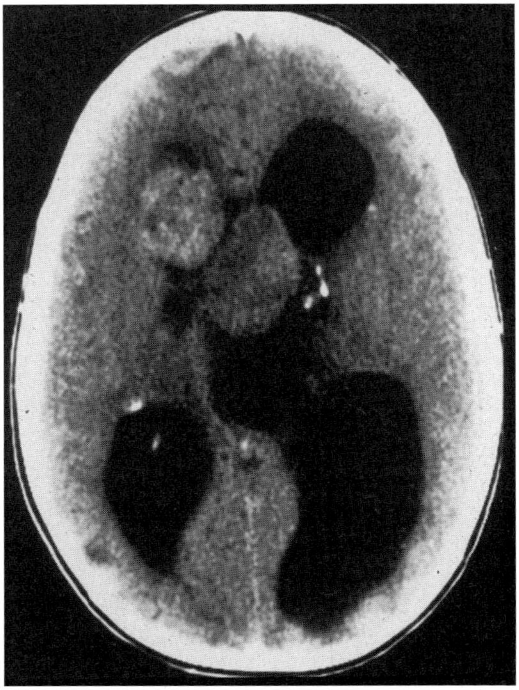

Figure 11.6 Bilateral SEGA's at presentation, Patient had been lost to follow-up for 3 years. He developed visual loss and ataxia less then 24 hours prior to this study (Non-contrasted cranial CT).

However, making this distinction can be extremely difficult [8, 9]. A variety of diagnostic criteria have been proposed to distinguish between SENs and SEGAs based on imaging characteristics, including diameter greater than 1 cm, location at the foramen of Monro, contrast enhancement, parenchymal involvement, presence of CSF obstruction, or serial growth. However, with the exception of the demonstration of growth on serial imaging, these criteria have been proven to be unreliable [35, 36]. One criterion that appears to be a useful adjunct to the presence of serial growth is the development of characteristic changes in the MRS characteristics of a subependymal nodule (Figure 11.6) [3]. A SEN likely to grow exhibits an increase in choline resonance and a decrease in myoinositol (mI) resonance, typical of the spectra of a low-grade glioma, and the opposite of that is seen in a cortical tuber. Such a change warrants closer surveillance of a putative SEGA, but by itself is insufficient to make the diagnosis. Note that therapy with an mTOR inhibitor such as rapamycin, discussed later in this chapter, causes a reversion of the MRS findings to that of a SEN. Unlike the gliomas of neurofibromatosis type I, SEGAs do not regress in size or stabilize once they have demonstrated serial growth [8, 9, 37]. SEGAs considerably larger than 1 cm may not produce CSF obstruction or demonstrate serial growth owing to individual variation in the configuration of the ventricle system and lesion location. Subependymal nodules frequently enhance regardless of their location, and

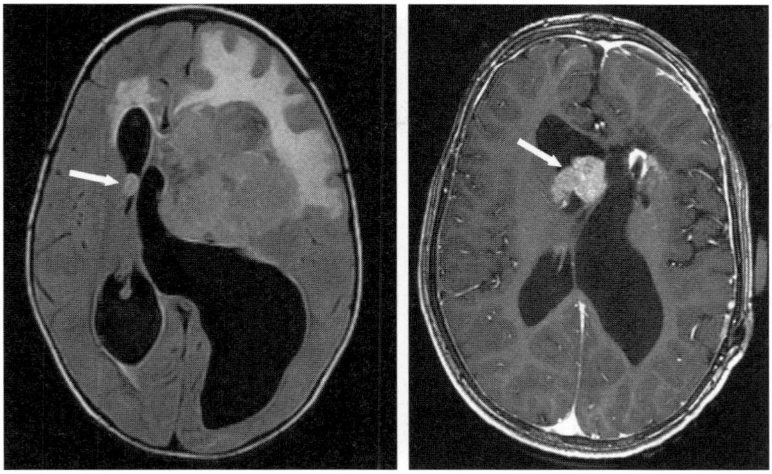

Figure 11.7 Left image – large SEGA with mass effect and hydrocephalus, patient had been symptomatic for <36 hours prior to this study. Right image – same patient $\sim$3 years status post successful resection of original lesion, now with progression of contralateral lesion (arrows) (T1 weighted MRI).

this factor alone does not indicate neoplastic potential. SEGAs can occur throughout the ventricular system and there is no particular location, which implies a higher probability of progressive growth. A possible exception to this is previously described examples of SEGAs in atypical locations, such as the retina, pineal region, or hypothalamus. The only reliable radiographic criterion for identifying a SEGA is the demonstration of obvious CSF obstruction (Figure 11.7) and mass effect, or the occurrence of growth on serial neuroimaging (Figure 11.8) [7, 36]. SEGA are usually unilateral, but can also be seen bilaterally, presenting simultaneously or sequentially (Figures 11.7 and 11.8). Approximately one-thirds of SEGAs have significant vascularity, demonstrated by increased contrast enhancement on CT or MRI. This vascularity gives these SEGA a propensity for both spontaneous hemorrhage and hemorrhage during attempted resection [38]. This management issue also supports a strategy of early operative or other intervention for growing SEGAs.

SEGAs are seen most frequently in childhood and adolescence (ages 5–18 years) with a lower incidence after age 21 years. However, SEGAs have been reported in individuals in their 30s and 40s, long past the time when routine screening for their appearance has been dropped [39–41]. SEGAs have occasionally been noted in neonates, or even prenatally. When present in the very young, SEGAs are more likely to exhibit rapid growth, as well as spontaneous hemorrhage [27–30].

When SEGAs are not recognized, they can produce focal neurologic symptoms owing to mass effect and/or hydrocephalus. A variety of symptoms are seen, including subtle cognitive impairment, nonspecific headache, increase in seizure frequency, or subtle behavior or personality changes that may be recognized only in retrospect. By the time overt signs of increased intracranial pressure, such as postural headache, vomiting, optic neuropathy, ataxia, and changes in mental status are

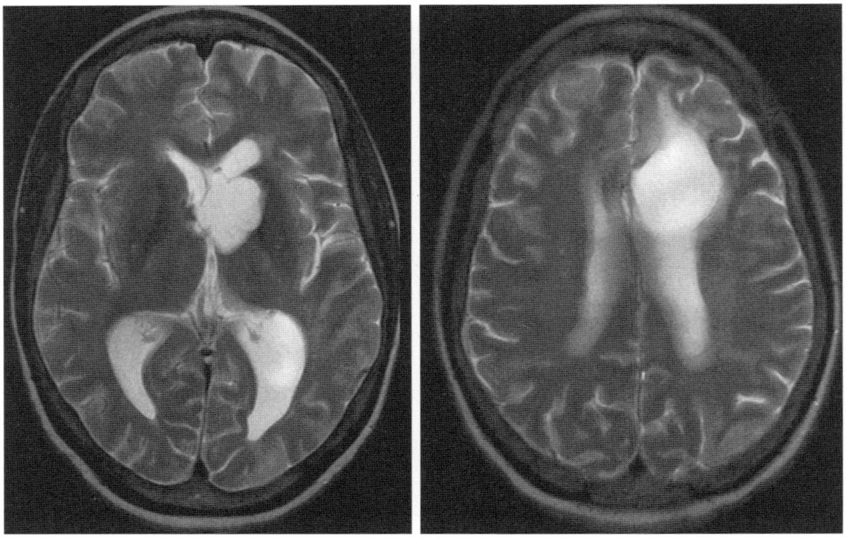

Figure 11.8 Large cavity status post transfrontal SEGA resection. Patient, previously of normal intelligence, developed symptoms ~12 hours prior to presentation at hospital. Postoperatively retarded with full scale IQ ~ 50 (T2 weighted MRI).

present, the likelihood that these symptoms can be fully reversed is markedly diminished [35]. Since the neurologic consequences of progressive SEGAs are potentially so serious, periodic surveillance with MRI or CT to identify lesions that are both >1 cm and exhibit growth over time is strongly recommended [8, 9].

11.5 Current Management of SEGASs

Current management recommendations for SEGAs in TSC are to either continue with short interval follow-up imaging, or proceed to surgical resection. Resection is recommended once serial growth is demonstrated based on the known propensity of these lesions to continue to grow leading to serious clinical symptoms, as described above [35, 42].

Operative resection is typically performed by a transfrontal or transcallosal approach (Figure 11.9). Potential postoperative complications include hemorrhage, CSF obstruction, seizures, and infection. It is important that gross total resection be achieved as SEGAs invariably recur if residual tissue is left behind. We have found the creation of a transfrontal operative corridor by the technique of balloon dilatation to be an effective approach enabling complete SEGA resection with reduced postoperative morbidity (Figure 11.10) [43]. In this method, an angioplasty balloon is stereotactically inserted through the frontal lobe into the proximity of the SEGA. The balloon is gradually inflated over a period of 5–7 days, creating an operative corridor. At the time of definitive surgery, the balloon catheter is deflated and removed and the

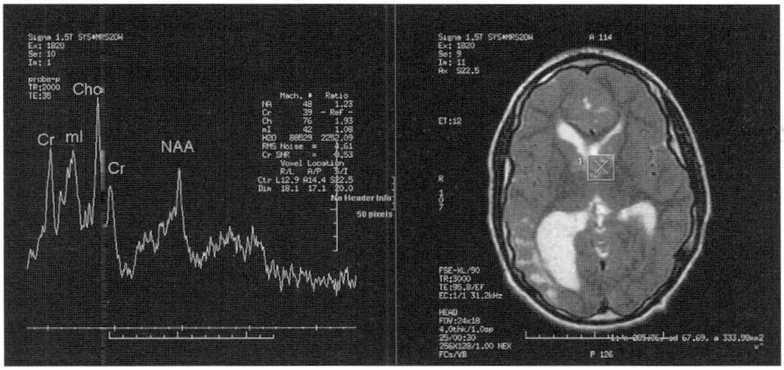

Figure 11.9 Single voxel MR Spectroscopy. Spectra demonstrates elevation of choline (Cho) and depression of myoinositol (mI) relative to creatine, indicative of increased membrane turnover and likely future growth.

tumor resected through the resultant operative channel. This increases the chance of achieving a gross total resection, as well as minimizing potential cognitive impairments or worsening seizures as a consequence of frontal lobe resection to achieve access to the tumor (Figure 11.11). In our experience, this approach is better tolerated in comparison to either a transfrontal or a transcallosal approach, which can be complicated by disconnection syndromes, cognitive impairments, and other issues [43].

Standard oncologic treatment modalities such as radiation therapy or conventional chemotherapy are contraindicated as treatment for SEGAs in TSC. Unfortunately,

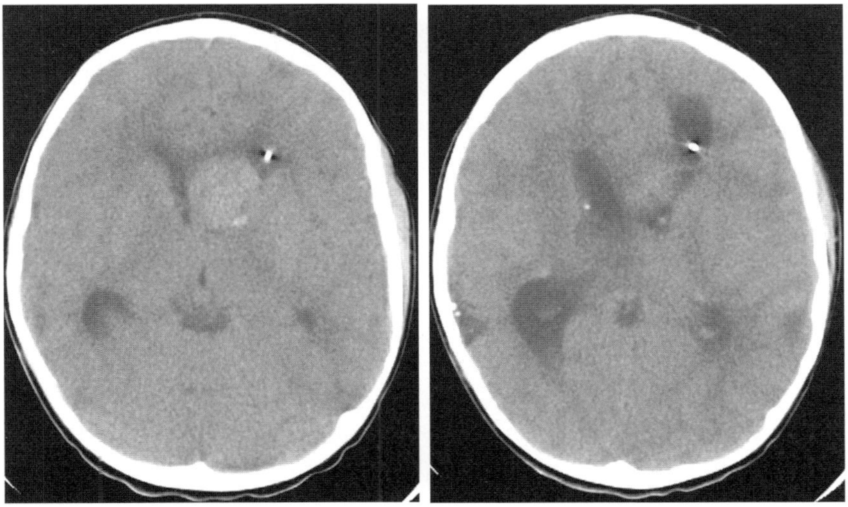

Figure 11.10 Balloon Tractotomy. Left – A balloon angioplasty catheter (arrow) is inserted stereotactically to the region of the SEGA. Right – the balloon is gradually inflated over 5–7 days creating an operative corridor for resection (Non-contrast cranial CT).

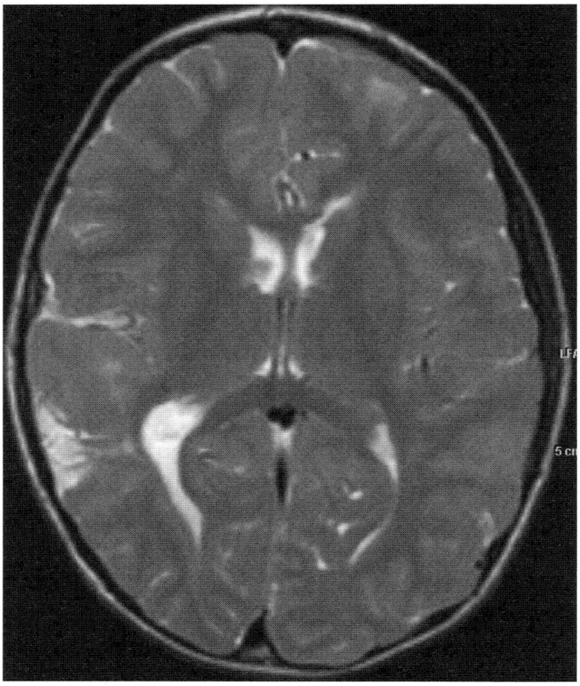

Figure 11.11 Balloon Tractotomy – Result. Same patient as figure 8 showing gross total resection of SEGA without disruption of overlying frontal lobe (T2 weighted MRI).

these standard treatment modalities interact badly with the germline TSC gene mutations present in TSC patients and can result in development of more aggressive neoplasia or lesion recurrence, even if an initial reduction in tumor volume is noted. Glioblastoma has been reported following radiation therapy for an individual with a SEGA [44, 45], a neoplasm that is otherwise extremely rare in TSC. In addition, radiation therapy and standard chemotherapy treatments have a risk of causing cognitive impairment or development of more aggressive neoplasms in other organ systems, including renal cell carcinoma, malignant epithelioid angiomyolipoma, or progressive lymphangioleiomyomatosis (LAM).

11.6 Medical Management of SEGAs

Until recently, no specific medical therapy was available for SEGAs. However, in 2006 we reported regression of SEGAs in five consecutive patients treated with the mTORC1 inhibitor rapamycin (sirolimus, Rapamune®) [46]. At the time we initiated this approach, it was unclear to what extent rapamycin crossed the blood–brain barrier. Subsequent studies with both rapamycin and its congener RAD001 (everolimus, Certican®) have shown that these agents enter the central nervous system and achieve clinically relevant levels [49].

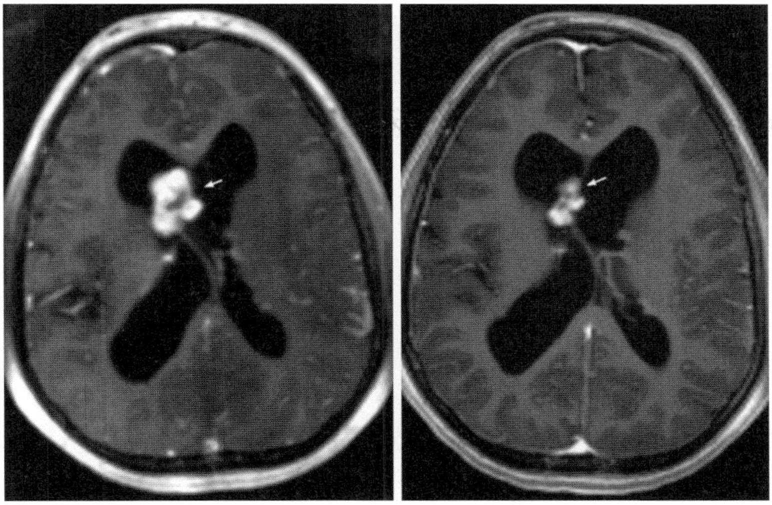

Figure 11.12 Response of a SEGA to rapamycin after 3 months therapy (Contrast enhanced T1 weighted MRI).

Subsequently, we have treated an additional 15 SEGA patients with rapamycin, either as part of a clinical trial or for compelling clinical indications. In every case, regression of the SEGA has been noted (Figure 11.12). The average reduction in lesion volume, as assessed by volumetric measurements on 1 mm coronal reformatted images from either postcontrast sagittal 3D SPGR or T1 MPRAGE MRI sequences, has been approximately 65%. Reduction in volume has been seen after as little as 2 weeks of therapy. Rapamycin treatment also caused involution of a large hypothalamic pilocytic astrocytoma in one patient with TSC (Figure 11.13), and in

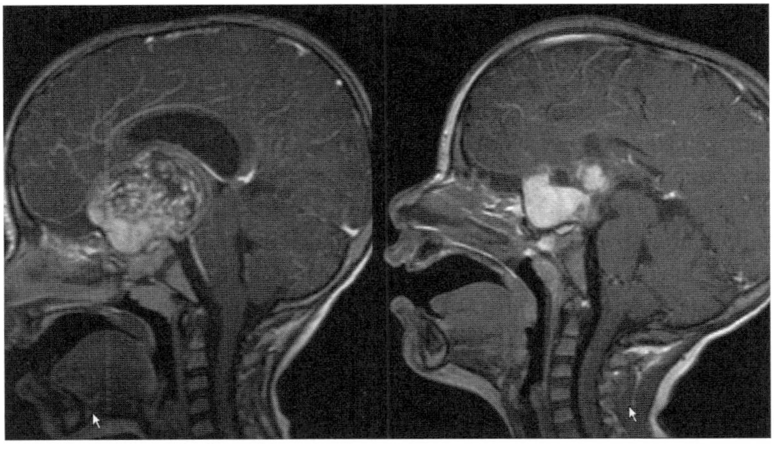

Figure 11.13 Involution of hypothalmic SEGA after 9 months rapamycin therapy (Contrast enhanced T1 weighted MRI).

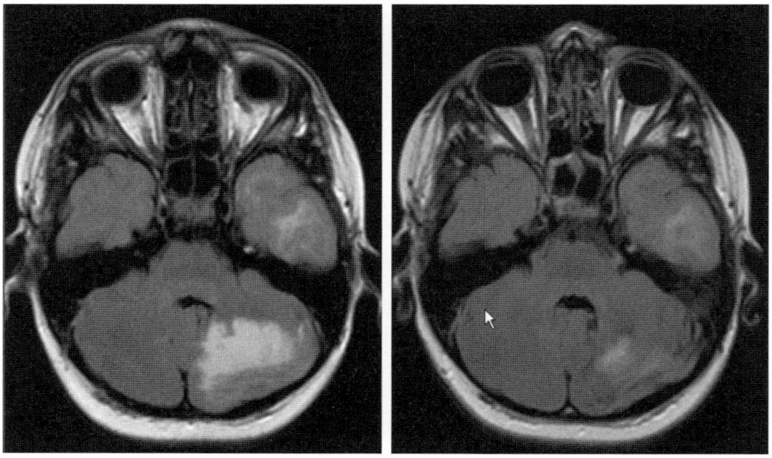

Figure 11.14 Reduction in signal intensity of a cerebellar tuber after 3 months rapamycin therapy for a SEGA (T1 weighted MRI).

another patient reduction of a large, prominent cerebellar tuber (Figure 11.14), indicating that rapamycin may have benefit use for other TSC lesions in addition to SEGA.

An open-label, prospective trial of RAD001 in SEGAs has shown similar findings to rapamycin, with an average reduction in lesion volume of 33% after 3 months and 55% after 18 months (Krueger, D. and Franz, D.N. (2008) Presented TSC International Research Conference, Brighton UK, unpublished data). (Table 11.1)

Table 11.1 Larger SEGA's exhibit more dramatic reductions in Volume.

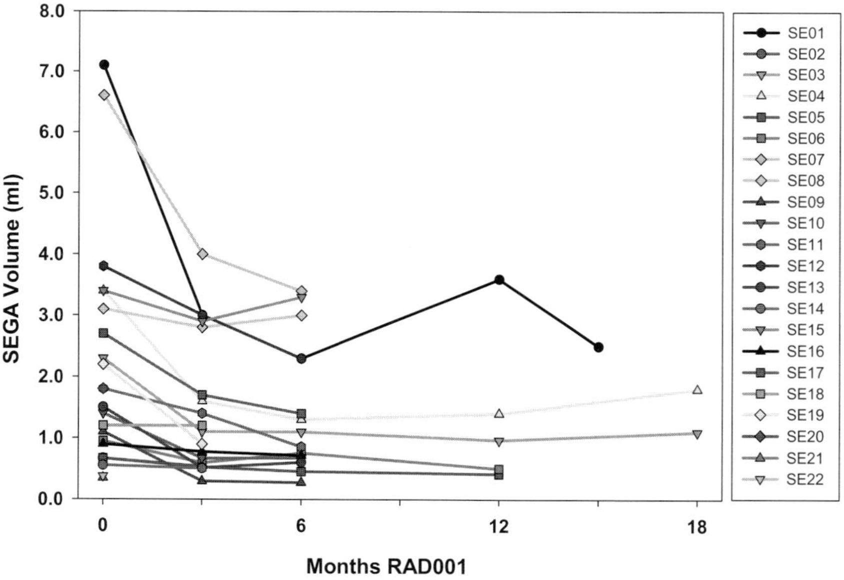

Table 11.2 SEGA Dose Response (mg/m^2/d) No correlation is seen serum level and amount of change from baseline volume.

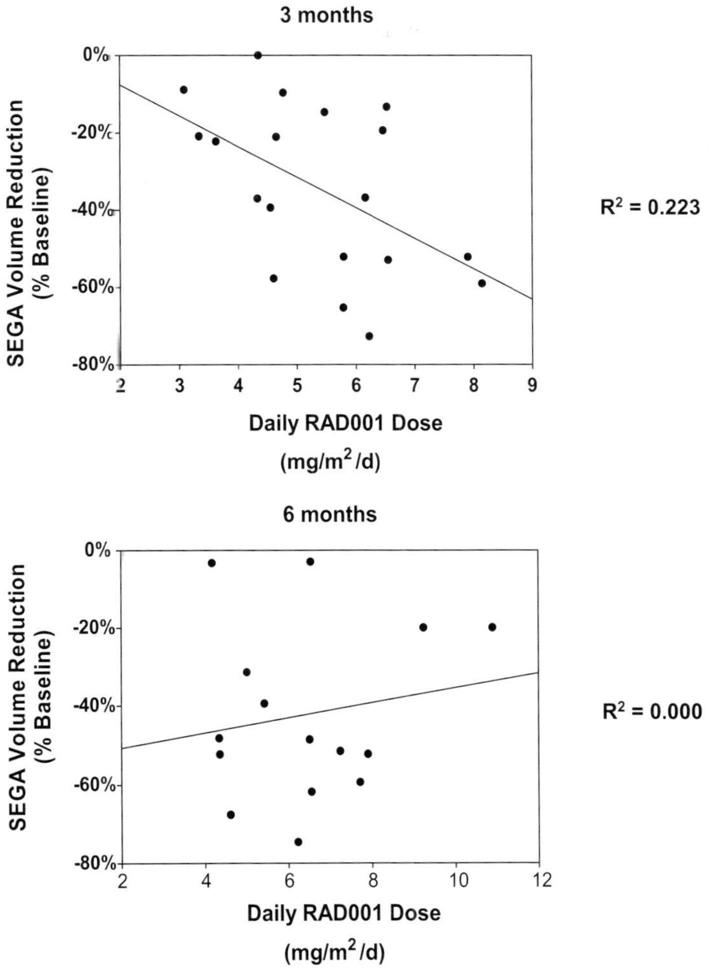

Interestingly, the response of SEGAs has not been clearly related to serum trough levels (Table 11.2), suggesting that some SEGAs may be more responsive than others. Furthermore, larger lesions appear to exhibit a greater response despite the achievement of relatively low serum levels. This raises the possibility that larger SEGA lesions are more "biologically active" and hence more responsive to the effects of mTORC1 inhibition or that they are associated with disruption of the blood–brain barrier leading to higher concentrations in the lesions, but the cause for this association between size and response is currently unknown.

Use of mTORC inhibitors such as rapamycin or RAD001 for the treatment of SEGAs continues to be the focus of much investigation. Issues such as optimal dosing and treatment duration have yet to be worked out. Furthermore, treatments with mTORC1 inhibitors such as rapamycin or RAD001 are associated with potentially significant side effects [47]. These agents have been used clinically as immunosuppressant drugs based on their ability to inhibit the proliferation of lymphocytes. This renders treated patients more susceptible to infections, though the clinical significance of this risk is uncertain. A rather uncommon, but potentially serious complication, is development of an allergic interstitial pneumonitis that usually, but not always, reverses upon cessation of therapy. TSC patients have been treated with rapamycin for over 4 years continuously at this point and with RAD001 for a duration of 20 months. In our initial series of SEGA patients treated with rapamycin, as well as subsequent patients who received off-label therapy ($n = 20$), side effects included lipid elevation in 7 (35%), aphthous ulcers in 6 (30%), acneiform rash in 3 (15%), and ankle edema in 1 (5%). Blood dyscrasias, specifically anemia, neutropenia, and/or thrombocytopenia, have also been reported with rapamycin in other studies, but were not noted in this treatment population. The most serious complication seen with rapamycin included a fatal pseudomonas brain abscess in a 40-year-old TSC patient who had undergone multiple, incomplete resections of bilateral SEGAs. He had undergone additional neurosurgical procedures for treatment of obstructive hydrocephalus and shunt malfunction. Rapamycin therapy was offered as he was not felt to be a candidate for further neurosurgical procedures. In addition, the patient had been treated with high-dose dexamethasone for over 1 year to control persistent cerebral edema. Nonetheless, a reduction in tumor volume was noted in response to rapamycin therapy. Another TSC patient with a SEGA developed a pulmonary embolism while on rapamycin. This may have been partially due to the rapamycin, but the patient was also found to have factor V Leiden deficiency. The patient was placed on warfarin and continues on rapamycin at the present time (Krueger, D. and Franz, D.N. (2008) Presented TSC International Research Conference, Brighton UK, unpublished data).

At this writing, we have 26 additional patients enrolled in a prospective trial of RAD001 for SEGA. 108 adverse events (AE) have been reported (average 0.64 AE/patient month of treatment) in these patients, with infections/upper respiratory illness and oral mucositis/aphthous ulcers comprising the majority of reported AE (35 and 25%, respectively). Only three AE were rated as severe (cough, seizure due medication noncompliance, and fever with neutropenia), and in each case treatment with RAD001 could be resumed. None of the patients have discontinued RAD001 because of AE, and in fact all patients remain on therapy except one who failed to continue with follow-up. Thus, despite the potential for serious side effects, both rapamycin and RAD001 (total $n = 41$) have been fairly well tolerated by TSC patients with SEGAs who have been treated to date.

Several recent studies have demonstrated the beneficial effects of rapamycin or RAD001 in the treatment of murine brain models of TSC [48–50]. In one of these studies, learning and memory deficits present in Tsc2 +/− mice responded to short-term treatment with rapamycin [50]. We have observed similar, albeit preliminary, promising results in neurocognitive function in TSC patients. In a clinical trial with RAD001, caregivers noted subjective improvements in behavior and social interactions in 19 of 21 subjects on whom follow-up data are available. Neuropsychologic testing pre- and posttreatment not only has shown any evidence of deterioration, but has also not shown any clear objective evidence of improvement. This may be due the high incidence of baseline cognitive impairments in the study group, rendering measurement of subtle changes difficult (Krueger, D. and Franz, D.N. (2008) Presented TSC International Research Conference, Brighton UK, unpublished data).

11.7
Conclusion and Summary

SEGAs are a clinically important source of both morbidity and potential mortality in TSC patients. Routine serial neuroimaging to screen for SEGAs in TSC patients, combined with early surgery, has been widely adopted and markedly reduces both the morbidity and the mortality due to these lesions. The precise role of newer therapies such as mTORC1 inhibitors is subject to further analysis and clinical investigation. However, it is already clear that these agents are effective in reducing SEGA volume in the majority of patients, providing the possibility of avoidance of surgery and operative complications. Nonetheless, resection by an experienced surgeon at an experienced institution remains the standard of care for SEGAs at this time. mTORC1 inhibitor therapy (with rapamycin of RAD001) requires further study and more detailed characterization of the risks and benefits of this approach before this can be broadly adopted as an alternative to surgical resection. The potential additional benefit of these drugs in seizure control and cognitive function makes them particularly attractive. However, these potential benefits must be considered in the context of the potential risks of treatment with these immunosuppressant drugs. In addition, many questions remain in regard to the long-term benefits and side effects of this approach, and the potential necessity for long-term treatment.

Acknowledgments

The authors wish to acknowledge the assistance of Melody Gulleman, Cynthia Tudor, PNP, Karen Agricola, FNP, and Prajakta Mangreshkar, MS in the preparation of this manuscript and clinical trials for SEGA at Cincinnati Children's Hospital.

References

1 Burger, P.C. and Scheithauer, B.W. (2007) *Tumors of the Central Nervous System*, AFIP Atlas of Tumor Pathology, Series 4, AFIP, pp. 114–118.

2 O'Callaghan, F.J., Martyn, C.N., Renowden S. *et al.* (2008) Subependymal nodules, giant cell astrocytomas and the tuberous sclerosis complex: a population-based study. *Arch. Dis. Child*, **93** (9), 751–754.

3 de Carvalho, N.A. Gasparetto, E.L. *et al.* (2006) Subependymal giant cell

astrocytoma with high choline/creatine ratio on proton MR spectroscopy. *Arq. Neuropsiquiatr.*, **64** (3B), 877–880.
4 Kim S.K., Wang, K.C., Cho, B.K. *et al.* (2001) Biological behavior and tumorigenesis of subependymal giant cell astrocytomas. *J. Neurooncol.*, **52** (3), 317–325.
5 Shepherd, C.W., Scheithauer, B., and Gomez, M.R. (1991) Brain tumors in tuberous sclerosis: a clinicopathologic study of the Mayo Clinic experience. *Ann. N.Y. Acad. Sci.*, **615**, 378–379.
6 Nabbout, R., Santos, M., Rolland, Y. *et al.* (2000) Early diagnosis of subependymal giant cell astrocytoma in children with tuberous sclerosis. *J. Neurol. Neurosurg. Psychiatry*, **66** (3), 370–375.
7 Goh, S., Butler, W., and Thiele, E.A. (2004) Subependymal giant cell tumors in tuberous sclerosis complex. *Neurology*, **63** (8), 1457–1461.
8 Roach, E.S., Gomez, M.R., and Northrup, H. (1998) Tuberous sclerosis complex consensus conference: revised clinical diagnostic criteria. *J. Child Neurol.*, **13** (12), 624–628.
9 Torres, O.A., Roach, E.S., Delgado, M.R. *et al.* (1998) Early diagnosis of subependymal giant cell astrocytoma in patients with tuberous sclerosis. *J. Child Neurol.*, **13** (4), 173–177.
10 Sharma, M., Ralte, A., Arora, R. *et al.* (2004) Subependymal giant cell astrocytoma: a clinicopathological study of 23 cases with special emphasis on proliferative markers and expression of p53 and retinoblastoma gene proteins. *Pathology*, **36** (2), 139–144.
11 Mizuguchi, M. and Takashima, S. (2001) Neuropathology of tuberous sclerosis. *Brain Dev.*, **23** (7), 508–515.
12 Jóźwiak, S., Kwiatkowski, D., Kotulska, K. *et al.* (2004) Tuberin and hamartin expression is reduced in the majority of subependymal giant cell astrocytomas in tuberous sclerosis complex consistent with a two-hit model of pathogenesis. *J. Child Neurol.*, **19** (2), 102–106.
13 Hirose, T., Scheithauer, B.W., Lopes, M.B. *et al.* (1995) Tuber and subependymal giant cell astrocytoma associated with tuberous sclerosis: an immunohistochemical, ultrastructural, and immunoelectron and microscopic study. *Acta Neuropathol.*, **90** (4), 387–399.
14 Scheithauer, B.W. (1992) The neuropathology of tuberous sclerosis. *J. Dermatol.*, **19** (11), 897–903.
15 Buccoliero, A.M., Franchi, A., Castigione, F. *et al.* (2009) Subependymal giant cell astroycytoma (SEGA): Is it an astrocytoma? Morphological, immunohistochemical and ultrastructural study. *Neuropathology* **29** (1), 25–30.
16 Sharma, M.C., Ralte, A.M., Arora, R. *et al.* (2004) Subependymal giant cell astrocytoma: a clinicopathological study of 23 cases with special emphasis on proliferative markers and expression of p53 and retinoblastoma gene proteins. *Pathology*, **35** (2), 139–144.
17 Kawahara, I., Tsutsumi, K., Hirose, M. *et al.* (2004) Solitary subependymal giant cell astrocytoma: a forme fruste of tuberous sclerosis complex? *No To Shinkei.*, **56** (7), 585–591, Japanese.
18 Jozwiak, J., Jozwiak, S., and Wlodarski, P. (2008) Possible mechanisms of disease development in tuberous sclerosis. *Lancet Oncol.*, **9** (1), 73–79.
19 Prahlow, J.A. and Teot, L.A., Lantz, P.E., and Stanton, C.A. (1995) Sudden death in epilepsy due to an isolated subependymal giant cell astrocytoma of the septum pellucidum. *Am. J. Forensic Med. Pathol.*, **16** (1), 30–37.
20 Jozwiak, J., Grajkowska, W., Kotulska, K. *et al.* (2007) Brain tumor formation in tuberous sclerosis depends on Erk activation. *Neuromolecular Med.*, **9** (2), 117–127.
21 Ma, L., Teruva-Feldsten, J., Bonner, P. *et al.* (2007) Identification of S664 TSC2 phosphorylation as a marker for extracellular signal regulated kinase mediated mTOR activation in tuberous sclerosis and human cancer. *Cancer Res.*, **67** (15), 7106–7112.
22 Chan, J.A., Zhang, H., Roberts, P.S. *et al.* (2004) Pathogenesis of tuberous sclerosis subependymal giant cell astrocytomas: biallelic inactivation of TSC1 or TSC2

leads to mTOR activation. *J. Neuropathol. Exp. Neurol.*, **63** (12), 1236–1252.

23 Ess, K.C., Kamp, C.A., Tu, B.P. *et al.* (2005) Developmental origin of subependymal giant cell astrocytoma in tuberous sclerosis complex. *Neurology*, **64** (8), 1446–1449.

24 Gunduz, K., Eagle, R.C., Jr., Shields, C.L. *et al.* (1999) Invasive giant cell astrocytoma of the retina in a patient with tuberous sclerosis. *Ophthalmology*, **106** (3), 639–642.

25 Shields, J.A., Eagle, R.C., Jr., Shields, C.L. *et al.* (2004) Aggressive retinal astrocytomas in four patients with tuberous sclerosis complex. *Trans. Am. Ophthalmol. Soc.*, **102**, 139–147, discussion 147–148.

26 Shields, J.A., Eagle, R.C., Jr., Shields, C.L. *et al.* (2005) Aggressive retinal astrocytomas in 4 patients with tuberous sclerosis complex. *Arch. Ophtalmol.*, **123** (6), 856–863.

27 Medhkour, A., Traul, D., and Husain, M. (2002) Neonatal subependymal giant cell astrocytoma. *Pediatr. Neurosurg.*, **36** (5), 271–274.

28 Hussain, N., Curran, A., Pilling, D. *et al.* (2006) Congenital subependymal giant cell astrocytoma diagnosed on fetal MRI. *Arch. Dis. Child*, **91** (6), 520.

29 Bhende, P., Babu, K., Kumari, P. *et al.* (2004) Solitary retinal astrocytoma in an infant. *J. Pediatr. Ophthalmol. Strabismus*, **41** (5), 305–307.

30 Hahn, J.S., Bejar, R., and Gladson, C.L. (1991) Neonatal subependymal giant cell astrocytoma associated with tuberous sclerosis: MRI, CT, and ultrasound correlation. *Neurology*, **41** (1), 124–128.

31 Raju, G.P., Urion, D.K., and Sahin, M. (2007) Neonatal subependymal giant cell astrocytoma: new case and review of literature. *Pediatr. Neurol.*, **36** (2), 128–131.

32 Dashti, S.R., Robinson, S., and Rodgers, M. (2005) Pineal region giant cell astrocytoma associated with tuberous sclerosis: case report. *J. Neurosurg.*, **102** (3 Suppl), 322–325.

33 Telfeian, A.E., Judkins, A., Younkin, D. *et al.* (2004) Subependymal giant cell astrocytoma with cranial and spinal metastases in a patient with tuberous sclerosis. Case report. *J. Neurosurg*, **100** (5 Suppl, Pediatrics), 498–500.

34 Au, K.S., Williams, A.T., Roach, E.S. *et al.* (2007) Genotype/phenotype correlation in 325 individuals referred for a diagnosis of tuberous sclerosis complex in the United States. *Genet. Med.*, **9** (2), 88–100.

35 de Ribaupierre, S., Dorfmuller, G., Bulteau, C. *et al.* (2007) Subependymal giant-cell astrocytomas in pediatric tuberous sclerosis disease: when should we operate? *Neurosurgery*, **60** (1), 83–89.

36 Clarke, M.J., Foy, A.B., Wetjen, N. *et al.* (2006) Imaging characteristics and growth of subependymal giant cell astrocytomas. *Neurosurg. Focus*, **20** (1), E5.

37 Kalina, P., Drehobl, K.E., Greenberg, R.W. *et al.* (1995) Hemorrhagic subependymal giant cell astrocytoma. *Pediatr. Radiol.*, **25** (1), 66–67.

38 Cuccia, V., Zuccaro, G., Sosa, F. *et al.* (2003) Subependymal giant cell astrocytoma in children with tuberous sclerosis. *Childs Nerv. Syst.*, **19** (4), 232–243.

39 Kawasaki, S., Yamamoto, Y., Sunami, N. *et al.* (1999) Subependymal giant cell astrocytoma associated with tuberous sclerosis exhibiting rapid regrowth after 17 years: a case report. *No Shinkei Geka*, **27** (6), 550–556, Japanese.

40 Kashiwagi, N., Yoshihara, W., Shimada, N. *et al.* (2000) Solitary subependymal giant cell astrocytoma: case report. *Eur. J. Radiol.*, **33** (1), 55–58.

41 Yamamoto, K., Yamada, K., Nakahara, T. *et al.* (2002) Rapid regrowth of solitary subependymal giant cell astrocytoma-case report. *Neurol. Med. Chir. (Tokyo)*, **42** (5), 224–227.

42 Roszkowski, M., Drabik, K., Barszcz, S. *et al.* (1995) Surgical treatment of intraventricular tumors associated with tuberous sclerosis. *Childs Nerv. Syst.*, **11** (6), 335–339.

43 Levine, N.B., Collins, J., Franz, D.N. *et al.* (2006) Gradual formation of an operative corridor by balloon dilation for resection of subependymal giant cell astrocytomas in children with tuberous sclerosis: specialized minimal access technique of balloon dilation. *Minim. Invasive Neurosurg.*, **49** (5), 317–320.

44 Matsumura, H., Takimoto, H., Shimada, N. *et al.* (1998) Glioblastoma following radiotherapy in a patient with tuberous sclerosis. *Neurol. Med. Chir. (Tokyo)*, **38** (5), 287–291.
45 Riemenschneider, M.J., Betensky, R.A., Pasedag, S.M. *et al.* (2006) AKT activation in human glioblastomas enhances proliferation via TSC2 and S6 kinase signaling. *Cancer Res.*, **66** (11), 5618–5623.
46 Franz, D.N., Leonard, J., Tudor, C. *et al.* (2006) Rapamycin causes regression of astrocytomas in tuberous sclerosis complex. *Ann. Neurol.*, **59** (3), 490–498.
47 Bissler, J.J., McCormack, F.X., Young, L.R. *et al.* (2008) Sirolimus for angiomyolipoma in tuberous sclerosis complex or lymphangioleiomyomatosis. *N. Engl. J. Med.*, **358** (2), 190–192; *N. Engl. J. Med.* **358** (19), 1963–4; author reply 1964.
48 Zeng, L.H., Xu, L., Gutmann, D.H. *et al.* (2008) Rapamycin prevents epilepsy in a mouse model of tuberous sclerosis complex. *Ann. Neuro.*, **63** (4), 444–453.
49 Meikle, L., Pollizzi, K., Egnor, A. *et al.* (2008) Response of a neuronal model of tuberous sclerosis to mammalian target of rapamycin (mTOR) inhibitors: effects on mTORC1 and Akt signaling lead to improved survival and function. *J. Neurosci.*, **28** (21), 5422–5432.
50 Ehninger, D., Hans, S., Shilyansky, C. *et al.* (2008) Reversal learning deficits in a TSC 2 + /− mouse model of tuberous sclerosis. *Nat. Med.*, **14** (8), 843–848.

12
Neurodevelopmental, Psychiatric and Cognitive Aspects of Tuberous Sclerosis Complex

Petrus J. de Vries

12.1
Introduction

Tuberous sclerosis complex is a multi-system disorder that can affect almost any organ system in the body. When parents and carers are asked which aspects of tuberous sclerosis complex (TSC) they are most worried about, the answer is almost invariably that the greatest concern is about the neurodevelopmental, psychiatric, or cognitive difficulties of their child, adolescent, or adult family member with TSC.

This chapter will focus on these aspects. The chapter will start with a description of the range of behavioral, psychiatric, intellectual, academic, neuropsychological, and psycho-social issues that are likely to be seen in individuals with TSC. After describing the potential difficulties, some principles for clinical assessment and management will be provided. Toward the end of the chapter, we will consider the possible causes of the learning and behavioral problems associated with TSC and will examine how research and clinical studies are developing to help understand and treat these difficulties.

12.2
Different Levels of Investigation

One of the first challenges of clinical and research work in the neurodevelopmental, psychiatric and cognitive aspects of TSC is to understand exactly which aspects of someone's learning and behavioral profile we are talking about. For instance, there is sometimes a misunderstanding that TSC is only about autism and mental retardation (global learning disability). Many families and professionals will not understand the difference between "neurodevelopment" and "cognition" or how "psychiatric" disorders may be different from "behavioral" problems. These terms are all related, but refer to different levels of investigation. In order to help individuals with TSC make best progress in their learning, development, and life, professionals and families have to understand the complex combination of strengths and weaknesses of every person across these different levels.

Tuberous Sclerosis Complex: Genes, Clinical Features, and Therapeutics
Edited by David J. Kwiatkowski, Vicky Holets Whittemore, and Elizabeth A. Thiele

ISBN: 978-3-527-32201-5

12.2.1 The Behavioral Level

This is the first and most "visible" level that parents and professionals are presented with. The "behavioral level" refers to the behavioral problems we observe at home, in clinic, or in school and may include positive or appropriate as well as challenging or unhelpful behaviors. In research settings, the behavioral level is usually studied through direct observation, parent and carer surveys, or rating scale measures.

From the earliest descriptions of the disorder [1], TSC has been recognized as being associated with a range of behavioral problems. In the 1980s, Ann Hunt, a British researcher and TSC parent, started comprehensive and systematic surveys of such difficulties [2, 3]. The most recent surveys in children and adolescents with TSC confirm rates between 44 and 69% of social-communication difficulties (including poor eye contact, repetitive and ritualistic behaviors, and speech and language delay), disruptive behaviors in 40–50% (including overactivity, restlessness, impulsivity, aggressive outbursts, temper tantrums, and self-injurious behaviors), as well as mood-related difficulties (including depressed mood, anxiety, and extreme shyness) and sleep problems in 20–50% [4].

These descriptive studies have also identified that some of these behavioral problems correlate with the level of global intellectual ability in TSC. Table 12.1 shows how the frequency of specific behavioral problems increases or decreases in relation to intellectual ability. The social-communication behaviors and disruptive behaviors show a strong link with intellectual ability. The more impaired individuals are, the more likely they will be to have some of these behavioral problems. Interestingly, the mood symptoms such as depressed mood, anxiety, or extreme shyness seem to be independent of an individual's overall ability.

However, it is also extremely important to note the high rates of behavioral problems in children and adolescents who have entirely normal global intellectual ability. In a recent survey, about one-third of normally intelligent children and adolescents with TSC were overactive, restless, and impulsive; more than 50% had aggressive behaviors and almost half had significant sleep problems [4].

In adults with TSC, surveys in the United Kingdom (de Vries, unpublished data) have also shown high rates of behavioral problems, in particular, high rates of anxiety and depressed mood. Using the hospital anxiety and depression scale (HADS), Lewis *et al.* showed that 56% (20/36) of able adults with TSC had abnormal scores suggestive of an anxiety disorder, and 19% (7/36) had abnormal scores suggestive of a depressive disorder [5]. A rating scale survey of 42 adults with TSC in the United States using the Symptom Checklist-90-Revised (SCL-90-R) found that 41% of the adults had abnormally high anxiety ratings and that 43% had abnormally high depression scores [6]. In the general population, we would expect abnormal rating scale scores on these measures in 5–10% of individuals. The findings above clearly suggest that all adults with TSC, even if they have entirely normal global intellectual ability, are at a significantly higher risk of a range of mood and other emotional disorders.

In summary, TSC is associated with a range of behavioral problems. In children and adolescents, these include social-communication and attention-related behaviors.

Table 12.1 The relationship between global intellectual ability and behavioral difficulties in children and adolescents with TSC.

Level of global intellectual ability	Severe to profound ID (IQ under 35)	Mild to moderate ID (IQ 35–69)	No intellectual disability (IQ 70 and over)
Social communication and language			
Autism spectrum disorder	78%	52%	17%
Poor eye contact	71%	38%	23%
Repetitive and ritualistic behaviors	83%	57%	20%
Speech and language delay	86%	86%	32%
Disruptive behaviors and sleep			
Overactivity	73%	58%	36%
Restlessness	69%	56%	36%
Impulsivity	58%	62%	36%
Aggression	58%	66%	51%
Temper tantrums	53%	70%	47%
Self-injury	69%	34%	17%
Sleep difficulties	74%	47%	41%
Mood			
Anxiety	32%	49%	35%
Depressed mood	20%	28%	20%
Extreme shyness	12%	14%	19%

The table shows the rates of behavioral problems (in percentage) in 265 children and adolescents with TSC across three intellectual ability groups. Data modified with permission from Ref. [4]. ID = Intellectual Disability.

In adults, these are particularly mood and anxiety related. Some of the behavioral problems are associated with global intellectual ability. However, even in children and adults who have normal intellectual ability, the rates of behavioral problems are much higher than seen in the general population. Families and clinicians should therefore have a high index of suspicion for such problems and should identify these early and seek further assessment and management.

12.2.2 The Psychiatric Level

At this level, people who present with worrying behaviors ("symptoms") are assessed to see if they meet criteria for any psychiatric disorder. Psychiatric disorders are defined as clusters of symptoms and signs that are present for a particular period of time, that cause impairment in the daily functioning of an individual, and that cannot be better explained by a different disorder or cause. The two main systems used for the classification of psychiatric disorders are the International Classification of Diseases, 10th edition (ICD-10) produced by the World Health Organization [7], and the Diagnostic and Statistical Manual of Mental Disorders, 4th edition (DSM-IV) produced by the American Psychiatric Association [8]. Both these systems are under constant review and newer versions of the ICD and DSM are underway.

Rating scale measures (as described above) may show high scores *suggestive* of a psychiatric disorder, but a rating scale cannot make a "diagnosis". Individuals have to be assessed by a mental health expert who will determine whether someone "meets criteria" for any particular disorder. Psychiatric disorders include autism and autism spectrum disorders (ASD), attention deficit hyperactivity disorder (ADHD), depressive disorder, and anxiety disorder.

In clinical practice, a psychiatrist or other mental health expert would typically review previous case notes, study rating scale measures that may have been used, meet with the family or individual to take a history, and perform a mental state and/or neurodevelopmental examination. The mental health professional would then use their clinical expertise to make a diagnosis based on the ICD or DSM classification systems.

In a research setting, there is an acknowledgment that in order to ensure systematic, accurate, and reliable diagnoses, additional tools and careful research diagnostic criteria are required. A number of research tools have therefore been developed over the years to aid in the diagnosing of psychiatric disorders such as autism and depressive disorders. Tools include the ADI-R (Autism Diagnostic Interview-Revised), ADOS (Autism Diagnostic Observational Schedule), SADS (Schedule for Affective Disorders and Schizophrenia), and many others [9–11]. Researchers are trained on such tools to reach "reliability"; that is, to ensure everyone using the tools will look for the same phenomena and code or record them in the same way. A research psychiatric assessment would typically include collecting a range of rating scale measures, standardized interviews, recording and re-checking of assessments to ensure accurate assessment. Often, research groups would have consensus meetings where all the research data are brought together to reach an overall consensus diagnosis. This detailed, systematic, and comprehensive process is very different from standard clinical practice.

12.2.2.1 Developmental Disorders

Developmental disorders are associated with abnormalities from early in the life of a person. They are typically identified and diagnosed in childhood, but there can be exceptions.

Autism and Autism Spectrum Disorders (ASD) The ASD are very strongly associated with TSC. The core diagnostic features of autism include three main domains: qualitative abnormalities in reciprocal social interaction (such as poorly modulated eye contact, limited facial expressions, difficulty in peer relationships, or lack of interest or subtle skills in reciprocal social interaction), qualitative impairment in communication (such as delay or lack of communicative language, difficulty with two-and-fro conversation, stereotyped and unusual use of language, limited or poorly integrated gestures), and stereotyped and repetitive patterns of behavior (such as a preoccupation with unusual objects, an intense interest in a topic or activity to the social exclusion of others, repetitive play, interest in parts of objects rather than the whole, stereotyped hand and finger mannerisms or sensory interests). Features have to be present by the age of 3 years in order to meet criteria for an ASD [7, 8]. There are no pathognomonic features of autism. That is, none of the features of autism on its own confirms or rules out autism. During an expert evaluation for an ASD, the diagnostic team will assess a child through history-taking and direct observation to determine whether the child have enough features to "meet criteria" for autism. Given the fact that there is a spectrum of autism that spans from very severe to mild, expert diagnostic teams often use interviewer-based schedules such as the ADOS and ADI-R to help in the process [9, 10]. The key to accurate diagnosis is to combine the information collected through interview and observation in different settings, not to depend on one or the other in isolation. Rating scale measures such as the CARS [12] and ABC [13] are used to suggest the likelihood of ASD, but are not sufficient to make a diagnosis of ASD.

In the general population, the rate of ASD is now around 1% [14]. In TSC, the most comprehensive and detailed studies using the ADOS and ADI-R in a population-based setting showed that 26% of children with TSC across all levels of intellectual ability met criteria for core infantile autism (14/53) and that a further 10% met criteria for an ASD (5/53) [15]. Other studies using rating scale and survey methods have reported rates of ASD in the order of 25–50% [3, 4, 16, 17]. There have been some suggestions that autism in TSC may be qualitatively different from autism in the general population. This view is based on a small-scale study [17] and no larger-scale studies have examined this question in further detail.

TSC is now accepted to be the medical or genetic condition most strongly associated with autism [18], more so than fragile X or neurofibromatosis, for instance. The likelihood of autism is strongly linked to the level of global intellectual ability of the individual. A child with TSC who has mild to moderate global intellectual disability (ID) may be 50 times more likely to have autism than a child without a diagnosis of TSC (52% in TSC versus 1% in the general population). However, even in children with normal intellectual ability, the rates of ASD may be 10–20 times higher in TSC than in the general population (17% in TSC versus 1% in the general population), as shown in Table 12.1.

Attention Deficit Hyperactivity Disorder (ADHD) ADHD is another developmental disorder very commonly associated with TSC. The core diagnostic features of ADHD include symptoms in three main domains: inattention (such as failure to attend to detail, getting easily distracted, difficulty sustaining attention, not following through on instructions, being forgetful and losing things), hyperactivity (such as being fidgety, squirmy, restless or overactive, always appearing on the go or driven like a motor), and impulsivity (such as difficulty in waiting his/her turn, butting into conversations, blurting out answers in school, talking excessively). To meet criteria for diagnosis, a range of these behavioral features has to be present before the age of 7 and should be observable in more than one setting (for instance, at home and in school). Importantly, the difficulties should be out of keeping with the overall developmental level of the child and should lead to impairment in their functional skills [7, 8]. There are different subtypes of ADHD, depending on which types of behavioral features are most prominent (e.g., predominantly inattentive type, predominantly hyperactive-impulsive type, combined type). The combined type of ADHD (DSM-IV code 314.01) is similar to hyperkinetic disorder as defined in ICD-10. There are various ADHD rating scales that can be used to identify children at risk of ADHD. However, similar to ASD, a diagnosis cannot be made on a rating scale alone. A comprehensive evaluation should include a developmental history, clinical observation of the child in more than one setting, and consideration of other developmental, educational, and social factors.

The rates of ADHD in the general population are between 3 and 5% [19]. In TSC not many systematic studies of ADHD have been performed. Gillberg *et al.* [20] performed neurodevelopmental histories and examination of 28 children in the west of Sweden, with the aim of diagnosing ADHD or ASD. Twelve of the 28 children (43%) met criteria for ADHD (subtypes not specified). Interestingly, 11 of the 12 also met criteria for an ASD. Only three children in this study did not meet criteria for ADHD or ASD or had no behavioral abnormalities on examination (3/28, 11%). In a study of children and adolescents in the East of England, de Vries identified that 11/20(55%) of the young people in the study met criteria for ADHD combined subtype/hyperkinetic disorder [21]. ADHD was not inevitably associated with ASD. The researchers observed that only one 1 of the 20 children in this study had been given a clinical diagnosis of ADHD and received treatment for the disorder. Muzykewicz *et al.* [22] used retrospective chart reviews and reported a diagnosis of ADHD in 30% of 241 patients. It was not clear from the paper whether all patients diagnosed were assessed using standardized criteria.

Adult ADHD is receiving increasing attention in the scientific literature but no systematic studies have been performed in TSC. The diagnostic criteria for adult ADHD are also under review. Future criteria are likely to include separate or additional symptoms for adults. Apart from inattention, and impulsivity symptoms, these may include poor organizational skills, internal feelings of restlessness or "edginess", unfocused mental activity or difficulty turning thoughts off, irritability, low frustration threshold, and stress intolerance [23].

ADHD in TSC is likely to be highly overrepresented in comparison to the general population. The disorder shows a strong association with the level of global intellectual ability in TSC so that individuals with global intellectual disability seem to be twice as likely to be diagnosed with ADHD than those with normal intelligence. Even in children and adolescents with normal intelligence, rates are likely to be 10 times higher in TSC than in the general population (30–55% in TSC versus 3–5% in the general population).

12.2.2.2 **Mood and Anxiety Disorders**

Mood disorders are diagnosed where a disturbance in mood is the predominant behavioral presentation. Diagnoses include major depressive disorder and bipolar disorder (with depressive and hypomanic or manic episodes). The anxiety disorders include panic attacks, agoraphobia, specific phobias, social phobia, and obsessive compulsive disorder (OCD). To meet criteria for these disorders, an individual must have had a number of symptoms for more than 2 weeks, usually associated with "biological" features (such as disturbed sleep, appetite, or energy), and the symptoms must have caused significant distress and impaired functional ability of the person. In addition, the symptoms should not be better explained by a general medical condition or medication [7, 8]. Given the complexity of making a clear psychiatric diagnosis of mood and anxiety disorders, a number of research tools have been developed to aid in the diagnostic process. These tools include the SADS for adults [11] and the K-SADS for children. Rating scale measures can be helpful to suggest individuals at risk of a mood or anxiety disorder, but these have to be followed by a comprehensive psychiatric interview and mental state examination in order to make a clinical diagnosis.

One of the first studies to suggest an increased risk of mood and, in particular, anxiety disorders in TSC was performed by Smalley *et al.* [24]. In the study of an extended kindred with *TSC2* mutations, the investigators used the SADS and K-SADS, as well as other tools, to study 37 individuals. Interviews were videotaped and re-rated by independent researchers. In addition, all cases with a positive psychiatric diagnosis were reviewed by a further diagnostic panel to ensure a very rigorous diagnostic process. Smalley's main finding was that 10/17 people affected by TSC met criteria for an anxiety disorder (59%) in contrast to only 13% of their unaffected family members. Six of the 17 (35%) met criteria for a mood disorder. Thirteen of the 17 (77%) of the individuals with TSC met criteria for one or more psychiatric disorders in contrast to 25% of their unaffected family members. Smalley suggested that the *TSC2* gene may have a specific association with anxiety disorders [24].

In a population-derived sample of 60 individuals with TSC in the United Kingdom, Raznahan *et al.* [25] used SADS as well as the SAPPA (Schedule for Assessment of Psychiatric Problems Associated with Autism and Other Developmental Disorders, Bolton and Rutter, unpublished) to determine the rates of mental illness. The study reported mood disorder in 30% and anxiety disorders in 5%, at rates lower than that of Smalley's study. Just focusing on individuals with TSC who had normal intellectual ability, Raznahan found mood disorders in 47% (15/32) and anxiety disorders in 9% (3/32). Retrospective case review from the Boston TSC clinic showed that a psychiatrist (without the use of research-based tools such as the SADS) diagnosed 28% of patients with an anxiety disorder and 26% with a mood disorder [22]. The Boston study did not present separate data for individuals with and without global intellectual impairment.

At present, the research evidence remains somewhat conflicting regarding the exact rates and nature of diagnosed mood and anxiety disorders in TSC. The majority of data do, however, suggest that the rates of mood and anxiety disorders are significantly increased in normally intelligent adults with TSC. Results from the Raznahan study suggest that there may be particular challenges in performing an accurate assessment of mood and anxiety disorders in individuals with learning disability who have TSC. It remains uncertain what the rates of bipolar mood disorders, OCD, and other anxiety disorders may be in TSC. Rating scale measures, however, suggest that rates may be far higher than have been identified in clinical or research settings to date.

12.2.2.3 Other Psychiatric Disorders

Outside the categories of developmental disorders and mood and anxiety disorders, psychiatric disorders also include schizophrenia and other psychotic disorders, eating disorders, dementias, sleep disorders and many other diagnostic categories. There is still extremely limited systematic information about these and other psychiatric disorders in individuals with TSC.

There have been case reports of anorexia nervosa [26], schizophreniform psychosis, and mania. Further examples of these and other psychiatric disorders reported in association with TSC are provided in Ref. [27]. In TSC, psychotic disorders with hallucinations or delusions are typically seen in association with seizure disorders. Temporal lobe seizures, in particular, are associated with "psychic" phenomena that can include simple or complex auditory hallucinations, simple or complex visual hallucinations, olfactory (smell) and gustatory (taste) hallucinations, and other perceptual disturbances such as micropsia, macropsia, and derealization/depersonalization phenomena [28]. All psychotic phenomena should therefore raise the suspicion of epileptiform discharges emanating from the temporal lobe or related mesial cortical brain structures [28]. The rates of psychotic disorders in TSC independent of seizure disorders seem to be around 1–2% [22, 25, 27].

12.2.2.4 **Are There Gender Differences in the Developmental and Psychiatric Disorders in TSC?**

It is well known that developmental disorders such as ASD and ADHD have a clear male predominance and that mood and anxiety disorders are typically overrepresented in women. In TSC the pattern is different. There is no difference in the rates of ASD or ADHD between male and female individuals [4, 5]. There are also to date no data to suggest that mood and anxiety disorders are more common in female individuals [4, 5]. It is possible that larger-scale studies may change this observation or that gender differences may exist in subgroups of individuals with TSC. The current evidence, however, suggests that the impact of the TSC mutations is sufficient to override any vulnerability or protective effects of gender.

12.2.2.5 **Psychiatric Level: Summary**

In many areas of psychiatric disorders robust, systematic studies using research diagnostic criteria are still lacking in TSC. Within the limitations of the available data on psychiatric disorders associated with TSC, there does seem to be a strong association of developmental disorders (ASD and ADHD) and of mood and anxiety disorders. It appears that schizophrenia and other psychotic disorders may not have an increased rate, given that the rates reported so far are very similar to the overall rate of schizophrenia in the general population (1%). The absence of an increased prevalence of psychosis in TSC might be an interesting negative finding, especially in comparison to other genetic disorders where psychosis is seen at high rates of up to 30% in velocardiofacial syndrome [29] and 20–50% in Prader–Willi syndrome [30], for instance. As we improve our understanding of developmental aspects of TSC it will be important to study the psychiatric disorders associated with older adults, and to improve our understanding of the lifespan trajectory of psychiatric illnesses in individuals with TSC.

12.2.3 The Intellectual Level

The "intellectual level" refers to the global intellectual ability or global intellectual functioning of an individual. In order to determine the global intellectual ability of an individual, we need to understand both their performance on "formal" or standardized tests of "IQ", as well as their ability to function in daily life [7, 8]. Many different tools or tests have been developed to evaluate the global intellectual abilities of individuals of all ages. The majority of individuals in the general population can be assessed directly through a range of verbal and nonverbal games or tasks. The examiner keeps a score of the individual's performance and then calculates "scaled" or "standardized" scores to show how well that individual did in comparison to others of that age and gender. Very young children and individuals with severe global intellectual disabilities cannot be given these sort of tasks and may need to be evaluated more indirectly by observing what they do and how they perform, or by interviewing parents or carers to find out exactly what the child or adult is able to do. Every person can be assessed or "tested". The important thing is to find the best tool to evaluate each individual.

For infants and children under the age of 6, tools include the Mullen Scales of Early Learning the Bayley Scales of Infant Development, and the WPPSI (Wechsler Preschool and Primary Scale of Intelligence). For school-aged children and adults, tools include the WISC (Wechsler Intelligence Scales for Children), the WAIS (Wechsler Adult Intelligence Scales), and the WASI (Wechsler Abbreviated Scales of Intelligence). Most of these tools evaluate the individual across a range of skills such as vocabulary, information, block design, and nonverbal reasoning. Even though tools may be similar, they are not identical, and comparison of performance on different tools should be made with caution.

When an individual cannot be assessed directly, a tool such as the Vineland Adaptive Behavior Scale (VABS) can be used. Using the Vineland schedule, a trained interviewer asks parents or carers a set of structured questions to determine the functional ability of that person across a number of domains such as communication, socialization, daily living skills, and motor skills. It is not uncommon that some people who may be able to perform relatively well on "formal" testing actually have significant difficulties functioning in daily life. The Vineland can be very helpful to identify such discrepancies and to confirm that such an individual requires significant support in real life, even if their "IQ scores" may seem relatively good.

In the general population, intellectual abilities as measured on IQ-type tests are broadly divided into the "normal" range (IQ $>$ 80), "borderline" range (IQ 70–80), and "low" range (IQ $<$ 70 or below 2 SD of the mean). Statistically, about 2% of the general population will fall in the "low" range [7, 8]. If an individual performs in the bottom 2% in comparison to others of his/her age and gender, and their ability to function in daily life is significantly less than expected for their age and gender, they are described as having "global intellectual impairment" or "intellectual disability". Until recently, international classification systems used the term "mental retardation" to describe this group of individuals. In both the United Kingdom and the United States, the term intellectual disability has now officially replaced the term "mental retardation". The ICD-10 divides intellectual disability into "mild" (IQ 50–69), "moderate" (IQ 35–49), "severe" (IQ 20–34), and "profound" (IQ < 20), depending on performance on direct and indirect assessments of an individual's abilities [7].

12.2.3.1 Two Intellectual Subgroups or Phenotypes in TSC

One of the striking features of TSC is the great variability (range) of intellectual abilities from very high to extremely low. The majority of individuals with TSC (about 70%) fall on a normal distribution, similar to that of the general population. The only difference is that the mean IQ score appears to be shifted slightly downward to a mean of 93 versus 100 in the general population [31]. Individuals in the normal distribution group can be assessed directly with IQ-type tests, and they show developmental gain over time, even though development may be at a rate that is slightly slower than seen in the general population. This group is referred to as the normal distribution or "ND" phenotype [32, 33]. The remaining 30% of individuals with TSC fall in the profoundly impaired range of intellectual disability. One cannot measure their abilities directly and has to depend on observations or parent interviews to get an indirect sense of their overall abilities. Individuals in this group typically show limited development of new skills over time. This group is referred to as the profound or "P" phenotype [32, 33].

Why may it be helpful to talk about two different subgroups of intellectual abilities in TSC? There are a number of clinical and research reasons to think separately of an "ND" and a "P" phenotype in TSC. Clinically, the two groups are associated with very different needs and demands in terms of care, and in terms of associated difficulties such as behavioral or psychiatric problems as previously discussed. Individuals in the ND phenotype are likely to make developmental gain, access educational programmes, and have a greater likelihood of independent or semi-independent living in adult life. They would typically have some functional communication skills that can be used to monitor their progress in terms of behavioral, psychiatric, or educational needs. In contrast, those in the profoundly impaired group tend to show little developmental gain over time and are likely to need very high levels of care into adult life. They are unlikely to have functional communication skills, and high levels of professional skill in observational and behavioral assessment may be required to determine whether they have additional mental health or physical care needs.

Many "new" families with a TSC child fear that their apparently able child may become profoundly impaired. This is sometimes the reason families avoid joining parent organizations such as the TSAlliance or TSA (UK). Some families who have children with profoundly impaired intellectual abilities believe that if they just tried harder to support and stimulate their child, their child would make great progress. They may therefore feel guilty and frustrated at the absence of progress and not realize that they are providing entirely appropriate care and nurturing to their child. It is extremely valuable for parents to know where their children are in terms of their intellectual abilities, in order to know how best to support them. It is equally important for parents to know that their child is not suddenly going to become profoundly impaired or show dramatic developmental gain.

From a research perspective, it is likely that the P and ND phenotypes will have different causal and contributory mechanisms. For parametric statistical modeling of causal or etiological contributions to IQ, a normal distribution is required. Separating these two groups may therefore help refine our understanding of the factors underlying intellectual abilities in TSC.

12.2.3.2 Is There a Predictable Pattern of Intellectual Strengths and Weaknesses in TSC?

A number of genetic disorders are associated with a specific pattern of intellectual strengths and weaknesses. In some, verbal skills (such as language skills) are typically better than performance skills (such as puzzle skills, block designs, map reading). In other disorders, the pattern may be the other way round.

In a consecutive series of children and adolescents in the Boston TSC clinic, Prather *et al.* used a range of standardized IQ-type tests to assess the intellectual abilities of 43 individuals [34]. Thirty-nine percent had intellectual disability, while the rest were in the normal range. In those who could be assessed directly, there was no specific pattern of verbal versus performance strengths. In fact, the most prominent finding was the great range of patterns among the children. Some children had verbal skills in the top 20% while their performance skills were in the bottom 20%. Others showed performance skills in the top 5–10%, but below average to low verbal skills. It is therefore important to realize that one cannot predict a particular pattern of intellectual strengths and weaknesses based on knowing that someone has TSC. Every individual's skills should be measured in order to know how best to support their learning needs and intellectual development.

12.2.3.3 The Association Between the Intellectual Level and the Behavioral/Psychiatric Levels

As shown in Table 12.1, the level of intellectual ability has a strong association with many but not all behavioral problems and is also strongly associated with an increased likelihood of many psychiatric disorders. Individuals with intellectual impairment are significantly more likely to develop a psychiatric disorder than people with normal intellectual abilities. However, in TSC even those with entirely normal intellectual abilities are at high risk of a number of behavioral problems and psychiatric disorders. Clinicians and families should therefore maintain a high index of suspicion and seek prompt expert assessment.

12.2.4 The Academic or Scholastic Level

In contrast to the "intellectual level" that evaluates global intellectual abilities, the focus here is on performance in specific academic skills. The DSM-IV uses the term "learning disorders" and ICD-10 uses the term "specific developmental disorders of scholastic skills" [7, 8]. For the purpose of this chapter, the terms "academic skills" or "academic disorder" will be used to refer to these ICD-10 and DSM-IV categories. Learning/scholastic disorders are diagnosed when a child presents with difficulties in academic skills such as reading, writing, spelling, or mathematics that are out of keeping with the expectation for their general intellectual level, their age, and their educational opportunities.

A reading disorder is characterized by difficulties in reading accuracy, speed, or comprehension that are out of keeping with the global intellectual ability and educational opportunities that the child has had. Children who have reading disorders (sometimes referred to as "dyslexia") often read with multiple errors that impact speed and comprehension of reading. Mathematics disorders can present with a range of maths-related difficulties including linguistic aspects (such as understanding of maths terms or concepts), perceptual maths skills (such as recognizing or reading maths symbols and the visuospatial aspects of maths), attentional skills (such as copying numbers, remembering what the task at hand is, and holding information in mind), or in learning maths facts (such as multiplication tables). Reading and maths disorders may be picked up in early school years, but in children with high global intellectual ability it is common for these disorders not to be identified till middle school years (age 8–12) when the real functional impact of the disorder is shown.

Disorders of written expression include difficulties with the production of written text, including grammar, punctuation, spelling, paragraph organization, and excessively poor handwriting. Problems with written expression are usually identified within the first few school years [8]. Developmental coordination disorder (DCD) is a disorder of motor skills and is sometimes referred to as "dyspraxia". It is characterized by unusual clumsiness and delays in achieving gross motor milestones (such as sitting, crawling, walking) and fine motor milestones (such as tying shoelaces, doing buttons, construction, etc.). DCD should only be diagnosed when the motor disorder is out of keeping with the expectation for age and global intellectual ability. In addition, it is a redundant diagnosis when a child has a known neurological disorder such as cerebral palsy or when a child meets criteria for an autism spectrum disorder (which is almost invariably associated with developmental coordination problems).

In TSC, there have to date not been any systematic studies of the rates and types of academic disorders across ages and ability levels. However, at a clinical level many children with TSC present with academic difficulties that have a significant impact on their ability to make academic progress. Using a standardized rating scale measure, de Vries showed that 36% of normally intelligent school-aged children with TSC were at high risk of academic disorders [21]. Maths disorders seem particularly common when children also have ADHD or other attention-related difficulties [7, 8]. Many children and adolescents with entirely normal intellectual abilities are found to have specific academic disorders that are often not identified till secondary deficits such as school refusal, anxiety about attending school, or poor self-esteem have emerged.

Children and adolescents with TSC are at higher risk of academic disorders. Clinicians, educators, and parents should have a low threshold to evaluate those with TSC when they show early signs of not making adequate academic progress.

12.2.5
The Neuropsychological Level

A neuropsychological approach uses various tasks that evaluate different brain skills in order to build up a profile of the strengths and weaknesses of the functional systems distributed throughout the brain. It is sometimes referred to as a brain-referenced model of learning. The field of neuropsychology developed initially through the study of adult patients with specific brain lesions to determine which neuropsychological function was impaired through the absence of a particular part of the brain. More recently, neuropsychology has benefited from advances in functional neuroimaging to confirm old hypotheses and to generate new ones. In addition, the field of developmental neuropsychology has emerged since the 1990s, with the particular interest in understanding how functional brain networks develop and become established in the typically and atypically developing brain.

At a simplistic level, a neuropsychological approach considers brain skills along three axes: the "left–right" axis, the "inside–outside" axis, and the "anterior–posterior" axis. In an adult brain, the left hemisphere is preferentially specialized in language-related functions (such as receptive and expressive language) and local or "detailed" processing, while the right hemisphere is specialized in visuospatial skills (such as pattern recognition, map reading, puzzle, and mental rotation skills) and global or "big picture" processing. The subcortical ("inside") brain structures are primarily regulating (such as the arousal systems of the brain involved in behavioral regulation and emotion recognition) and memory structures. Cortical ("outside") brain structures are associated with the higher order functions. The posterior "processing" functions contrast with the anterior "production and control" functions that include the executive control processes of the brain. In a clinical setting, a neuropsychologist would administer a range of tasks to build up a picture of the range of functional brain skills. After tasks have been performed, the neuropsychologist scores up the individual's performance and converts scores to standardized or percentile scores in relation to their age and gender. By neuropsychological convention, scores that fall below the second percentile (so 98/100 individuals would have performed better) are defined as a neuropsychological "deficit" or "impaired" performance. Some tests use the 5th or 10th percentile as cutoffs to define impaired performance.

12.2.5.1 **Overall Neuropsychological Profiles in TSC**

The first systematic study of neuropsychological skills in TSC was performed by Jambaque *et al.* [35]. The researchers investigated the intellectual (using standardized IQ-type measures) and neuropsychological skills of 23 children with TSC aged 3–16. Fifteen of the 23 (65%) had IQ scores in a normal distribution pattern, while the remaining 35% all had scores below 40, broadly in keeping with the distribution seen in recent epidemiological studies [31]. Among the seven children with normal intellectual ability, the authors observed dyspraxia (developmental coordination disorder), speech delay, visuospatial deficits, memory impairment, and dyscalculia (mathematical disorder). In 1999, Harrison *et al.* described specific deficits of executive skills in a group of seven normally intelligent adults with TSC [36].

The most comprehensive neuropsychological study to date was performed by Prather *et al.* at the Boston TSC clinic. All individuals referred to the clinic between 1999 and 2004 received a detailed neuropsychological evaluation in order to determine the neuropsychological profile of children with TSC [34]. Forty-three children and adolescents were assessed. Of the children with normal global intellectual abilities, 19% had spatial deficits, 24% had language-related deficits, 38% had memory deficits, 54% had complex visuospatial deficits, and 66% had executive control processing deficits. The executive deficits included impairment in planning, attentional set shifting, sequencing, verbal fluency, and monitoring.

12.2.5.2 **Attentional Skills**

In Cambridge, de Vries and colleagues studied the attentional skills of 20 children with TSC and their unaffected siblings [21, 37]. The study addressed the question whether children with TSC had specific neuropsychological deficits in their attentional skills, over and above the expectation for their intellectual abilities. Neuropsychological skills show a strong familial pattern. De Vries therefore wanted to control for the intrafamilial effect on test performance. Using the Test of Everyday Attention for Children (TEA-Ch), results showed that, after controlling for age, gender, and intrafamilial factors, children and adolescents with TSC performed significantly worse than their siblings on tests of selective attention (including visual search tasks), sustained attention (including tasks requiring listening to verbally presented information), attentional switching (including the ability to switch rapidly between different rules), and dual tasking (the ability to perform more than one task at the same time). The TSC group showed very high rates of neuropsychological impairment (performance below the second percentile) on a number of attentional tasks, most particularly on dual tasking where 17/20 children performed in the impaired range on either or both of the dual tasks. Eighteen of the 20 children (90%) showed impairment on one or more of the tasks in the TEA-Ch battery, suggesting that children and young people with TSC are at a very high risk of specific neuropsychological attention deficits, even when they have entirely normal intellectual ability. No systematic study of the neuropsychological attention skills of adults with TSC has been performed to date.

12.2.5.3 **Memory Skills**

Jambaque *et al.* [35] and Prather *et al.* [34] reported specific memory deficits in children with TSC, but no other studies have investigated memory systematically in children. In a study of 25 adults with TSC who had normal intellectual abilities, Ridler measured various aspects of verbal and visuospatial memory and compared their performance to 25 adults without TSC matched for IQ, age, and gender [38]. The most significant group differences were seen on verbal working memory (performance on a digit span backward task), spatial working memory (CANTAB (Cambridge Neuropsychological Test Automated Batteries) spatial span task), a story recall task (difficulty with immediate recall), a paired associates learning task (CANTAB), and a complex self-ordered spatial working memory task (CANTAB SWM task). On many other memory tasks, adults with TSC performed worse than the control group, but results were not statistically highly significantly different. Overall, results confirmed memory deficits and relative memory difficulties in adults with TSC, even when they had normal intellectual abilities.

12.2.5.4 Language Skills

Apart from direct evaluation by Jambaque *et al.* [35] and Prather *et al.* [34] and survey measures suggesting high rates of speech and language delay in TSC [4], there are no published systematic evaluations of language skills in TSC to date.

12.2.5.5 Visuospatial Skills

Jambaque [35] identified specific visuospatial deficits in intellectually able children with TSC, and Prather reported that 54% of the able children in their clinic series showed deficits on a complex visuospatial construction task (the Rey–Osterrieth Complex Figure Drawing) [34]. A number of the memory tasks reported by Ridler as impaired in TSC [38] had strong visuospatial components (spatial span, spatial working memory, paired associates learning). McCartney and de Vries therefore performed a systematic study of 22 adults with TSC who had normal intellectual abilities and compared them to 18 age, gender, and IQ-matched controls (de Vries and McCartney, unpublished data). McCartney included tasks with memory and executive demands and tasks tapping more basic visuospatial processing. Results confirmed the spatial working memory and visuospatial planning and execution difficulties in TSC. Interestingly, McCartney also identified evidence of left unilateral neglect.

Left unilateral neglect is typically seen in patients who have had a right hemisphere cerebrovascular infarct ("stroke"). At its most extreme, patients with left unilateral neglect are unable to perceive any input from the contralateral (left) side. They may not eat food on the left side of the plate, omit to shave on the left side of their face, or may not have awareness of the left side of their bodies. In its mildest form, patients with neglect preferentially attend to stimuli in the right field of vision. So far, the evidence of unilateral neglect in TSC does not seem extreme and there is no evidence that it is specifically associated with tubers or lesions in the right hemisphere of the brain. These findings raise questions about cerebral dominance and hemispheric lateralization in TSC that require further exploration.

12.2.5.6 Executive Control Processes

Executive control processes are also referred to as executive skills. These are the regulatory, control, and production skills of the brain, typically subserved by fronto-subcortical regions. Executive skills include regulatory functions (such as response inhibition and emotional regulation), attentional control functions (such as attentional switching and dual tasking), and goal-directed functions (such as planning, sequencing, monitoring, cognitive flexibility, metacognition and judgment). Impairment in executive skills may have significant adverse effects on an individual's daily living skills, learning, and occupational functioning.

There are various neuropsychological tools that are used to evaluate executive skills. These include the Wisconsin Card Sorting Test (as a measure of cognitive flexibility), the Tower of London Test (as a measure of planning), and verbal fluency tests (as measures of verbal production). A computer-based battery of executive tasks was developed by researchers at the University of Cambridge, UK. The battery, known as CANTAB, captures various aspects of executive and attentional skills, albeit primarily in the visual modality.

Using the CANTAB, Harrison *et al.* first reported specific executive deficits in a handful of adults with TSC [36]. There was no specific pattern of deficits, but almost every adult with TSC had some specific deficit. Subsequent studies using the CANTAB have consistently identified poorer performance in those with TSC on tasks of spatial span and spatial working memory (as reported by Ridler and McCartney), as well as on planning and attentional set-shifting tasks [21, 33]. Taken together, executive skills so far seem to be the most consistently impaired neuropsychological skills in children, adolescents, and adults with TSC. If an individual with TSC is therefore at risk of any specific deficit, the most likely area would be these executive skills.

12.2.5.7 **Is There a Typical Pattern of Neuropsychological Deficits in TSC?**

Even the most able individuals with TSC are at high risk of having some specific neuropsychological deficits. As discussed above, the highest likelihood seems to be for some degree of executive deficit that may include planning, attentional set shifting, self-ordered spatial working performance, or dual tasking. It is clear that studies have typically had small samples and have not always used similar assessment tools, which reduces the certainty with which conclusions can be drawn. To date no particular pattern or subgroups of patterns have been identified in TSC. It seems that the range and inter-individual variability of neuropsychological features are as unique and varied as the genetic mutations seen in TSC.

12.2.6
The Psychosocial Level

The psychosocial level is distinct from the other levels of investigation discussed above. At this level, many psychological and social aspects may play a role to act as vulnerability (negative) or as resilience (positive) factors [39]. It is important to consider the contributions at this level because once a problem is identified at this level, many helpful psychological or social strategies are available to improve the lives of those affected. For instance, having a chronic condition that may be associated with a visible manifestation such as facial angiofibroma or white patches may affect the self-esteem of an individual and thus lead to withdrawal from social interaction, school, or other community activities. By identifying this pattern of "social avoidance" or "safety" behaviors, psychological principles can be used to reduce the individual's isolation and increase exposure to peer support, community support, and active participation in groups or clubs.

An important component of the psychosocial context is the family and family functioning. It is well known that having a family member with a disability, particularly a complex physical and developmental disability such as TSC, can have a significant impact on family stress and family functioning. Where such factors are not identified and supported, families are at high risk of disintegration or of maladaptive patterns of living. It is not uncommon to see one parent withdrawing from support to the affected family member and abdicating responsibility to the other parent (usually the mother). Other maladaptive patterns may include denial of the severity of problems, minimizing of the deficits seen in their child, or an externalizing of anger at professionals or carers outside the family. A particularly problematic maladaptive behavior is adoption of an "overcaring" and overprotective role toward the affected family member. These family factors affect the immediate family including siblings as well as grandparents and others in the wider family network.

Well-adapted family functioning, open communication, and a realistic understanding and acceptance of a child's or a adult's strengths and weaknesses are crucial steps toward accommodating (modifying the environment appropriately to support the individual) and toward skill building (supporting the individual to develop new and better skills). Families and professionals should work collaboratively to identify family and psychosocial stressors, to consider what can be done to reduce stressors, to correct maladaptive behaviors, and to know when additional support may be required from therapists, counselors, or statutory organizations such as social services.

12.2.7
The Biological Level

All the levels discussed above happen in the context of the biological or physical problems associated with TSC. The impact of physical illness (such as renal failure or medications) and of seizures (ictal, interictal, effect of medication on learning and behavior) should be considered in the evaluation of the neurocognitive and neurodevelopmental presentation of an individual with TSC.

In TSC it is particularly important to watch for sudden or unexplained changes in the behavior, mental health, intellectual performance, or neuropsychological skills of any child, adolescent, or adult. In many instances, a change in one or more of these neurocognitive or neurobehavioral levels of investigation may be a sign of a physical disorder (such electrolyte disturbances), a side effect of medication (particularly associated with a recent change in medication), or a growing subependymal giant cell astrocytoma/tumor (SEGA). There have been a number of clinical cases reported where the first sign of a growing SEGA was acute change in a child's behavior (such as onset of aggression or sleep disturbance), or a change in the learning or scholastic abilities or progress of the child (such as gradual or sudden loss of skills, failure to make progress) [40].

Abnormalities and concerns at the biological level have to be investigated as a priority. A child who requires surgical treatment for a SEGA or an adult who has poorly controlled seizures associated with "psychic" phenomena such as auditory, visual, gustatory, or olfactory hallucinations requires surgical and medical workup rather than psychiatric evaluations.

12.3 Assessment and Management of Neurocognitive and Neurobehavioral Difficulties in TSC

It is impossible to provide a comprehensive manual for assessment and management of all individuals with TSC. This section will therefore focus on some general principles of assessment and management.

12.3.1 Assessment

12.3.1.1 Assess the Individual Across all Levels of Investigation (Behavioral, Psychiatric, Intellectual, Academic, Neuropsychological Skills, Psychosocial, Biological)

It should now be clear that individuals with TSC can present with an enormously wide range and combination of difficulties across a number of levels as discussed above. All people with TSC are at an increased risk of this range of difficulties, but not everyone will have everything wrong with them. The only way to determine the profile and therefore the needs of an individual is to perform a comprehensive evaluation that encompasses all the levels of investigation described above. This should be the standard of care.

In 2003, an interdisciplinary panel of experts drew up international consensus clinical guidelines for the assessment of cognitive and behavioral issues in TSC [41]. The guidelines made two main recommendations. The first was that there should be regular, comprehensive assessment of all individuals with TSC, regardless of whether the person has any specific neurodevelopmental or neurocognitive challenges at the time. The goal is to identify the profile and needs of that individual and to treat before any significant impairment or secondary deficits emerge.

The second recommendation of the consensus panel was to perform a comprehensive assessment in response to rapid or unexpected change in cognitive development or behavior in order to identify and treat the underlying cause of the neurobehavioral change.

Table 12.2 shows a summary of the published international consensus clinical guidelines for assessment.

12.3.1.2 Assessment is Likely to Require Multi-agency, Multi-disciplinary Involvement

No single professional or professional group has all the necessary tools or skills to perform a comprehensive assessment as outlined in the international guidelines (Table 12.2). It is likely that a combination of developmental pediatricians, psychiatrists, neurologists, psychologists, neuropsychologists, educational experts, occupational therapists, speech and language therapists, and others may be required,

Table 12.2 The consensus clinical guidelines for the assessment of behavioral, psychiatric, intellectual, academic, and neuropsychological skills in TSC.

Assessment stage	Age range for assessment	General purpose of assessment	General areas to assess	Areas of particular concern in TSC[a]	Behavioral, psychiatric, and academic disorders of particular concern in TSC[a]
At diagnosis		Initial assessment of cognitive and behavioral profile	As listed for chronological age		
Infancy	Birth – 12 months	To perform a baseline assessment for regular monitoring of development	Global standardized assessment of infant development	Impact of seizure onset and treatment on development	
Toddler	1y–2y11m	To identify early developmental delay or developmental disorders	Global intellectual ability and adaptive behaviors		Autism and autism spectrum disorders Severe aggressive outbursts Severe sleep problems
			Specific skills: • Gross and fine motor skills		
			• Social-communication skills	Quality of eye contact, joint attention, reciprocity	
Pre-school	3y to school entry	Evaluation of cognitive and behavioral profile to ensure the provision of appropriate educational programmes	Global intellectual ability	Uneven profile of abilities	Autism and ASD ADHD and related disorders Self-injurious behavior
			Specific neuropsychological skills: • Receptive and expressive language	Poor expressive language	
			• Social-communication skills	Poor reciprocity, peer interaction	
			• Attentional and executive skills	Poor regulation of affect and impulse	
			• Visuospatial skills		
			• Motor skills	Poor bilateral coordination	

(Continued)

Table 12.2 (*Continued*)

Assessment stage	Age range for assessment	General purpose of assessment	General areas to assess	Areas of particular concern in TSC[a]	Behavioral, psychiatric, and academic disorders of particular concern in TSC[a]
Early school years	6y–8y	Monitoring the child's ability to make appropriate educational progress	Global intellectual abilities	Best time to establish baseline to assess whether specific cognitive skills and academic performance are discrepant from global intellectual abilities	Academic difficulties (reading, writing, spelling, mathematics) ADHD and related disorders Peer problems Aggressive behaviors
			Specific neuropsychological skills:		
			• Receptive and expressive language	Poor expressive language and word retrieval	
			• Social-communication skills		
			• Memory	Rote learning difficulties	
			• Attentional and executive skills	Selective attention, sustained attention difficulties	
			• Visuospatial skills		
			• Motor skills		
Middle school years	9y–12y	Comprehensive review of child's abilities, specific learning difficulties, and behavioral problems in preparation for the transition to secondary education	Global intellectual abilities		
			Specific neuropsychological skills:		Asperger's syndrome Peer problems Academic difficulties (reading, writing, spelling, mathematics)
			• Receptive and expressive language		
			• Social-communication skills	Subtle deficits of social communication, unusual interests	
			• Memory skills	Poor working memory, episodic memory	
			• Attentional and executive skills	Planning, organizational abilities, multitasking difficulties	

Adolescence	13y–16y	Determining individual needs and the support required for transition into adult life	Global intellectual abilities Specific neuropsychological skills • Attentional and executive skills Vocational assessment with knowledge of cognitive strengths and weaknesses Adaptive behavior and daily living skills	 Poor judgment, decision-making	Depressive disorders Anxiety disorders Peer problems Epilepsy-related psychotic disorders
Adults	18y +	*Newly diagnosed adults*: assessment of cognitive, behavioral and vocational profile, determining bio-psycho-social needs	Global intellectual abilities Specific neuropsychological skills: • Attentional and executive skills • Memory skills	 Difficulty with integrational skills Working memory, episodic memory problems	Depressive disorders Anxiety disorders Epilepsy-related psychotic disorders
Adults (follow-up)	18y +	Monitoring for emergence of psychiatric problems or changes in existing cognitive and behavioral profile	Dependent adults: • Annual review of social care needs and support Independent adults: • Vocational advice • Genetic counseling as appropriate • Review if problems arise	Pay particular attention to *change* in cognitive abilities or behavior Pay particular attention to *change* in cognitive abilities, vocational performance and behavior	Depressive disorders Anxiety disorders Epilepsy-related psychotic disorders

The table shows the time points recommended for evaluation and the goals of evaluations and lists specific areas of concern for each age group. Table reproduced and modified with permission from Ref. [41].

a) Many features listed in these columns can present at any age, but are listed here at stages most commonly associated with the emergence of such difficulties in TSC.

depending on the needs of the child, adolescent, or adult. Clinicians working with TSC families are encouraged to build a network of professionals who can work collaboratively to provide comprehensive assessment and coordinated intervention.

12.3.1.3 Make Sure You Have an Understanding of the Patient/Individual at Each Level

In order to implement an appropriate management plan, it is essential that the lead clinician should have a good understanding of the challenges and strengths of each person at each of the levels discussed above. Ask yourself whether you have a good overview of the range and nature of behavioral difficulties that the family is concerned about. Do you understand how these behaviors developed, and what might drive or trigger them? If not, you may need to ask a colleague with behavioral expertise for a more detailed behavioral assessment and a functional analysis of behaviors.

Does the individual meet criteria for any psychiatric disorder, particularly one of the disorders that individuals with TSC are at high risk of? If you are not certain, you may need to seek further consultation from an appropriate colleague. Do you know where the intellectual abilities of your patient lie? Are there specific or relative strengths or weaknesses? Might the intellectual profile help to understand the behavioral or academic problems presented? If you are not certain, you may need to ask a psychologist or neuropsychologist colleague for further assessment.

Is the child or adolescent known to have an academic disorder, even if they may have normal intellectual abilities? Has this been assessed in the educational system, for instance, by an educational psychologist? Might this help to explain why the child has school refusal or anxiety about going to school? May the individual have specific neuropsychological deficits, such as memory, attentional, or executive deficits, that can help explain their problems? Children and adults with executive control processing difficulties may often appear stubborn, willful, or emotionally unpredictable. Adults with executive deficits often have difficulties in making and keeping relationships, in finding and keeping jobs, and in taking care of the many aspects of their daily lives such as bills, appointments, and other commitments.

Do you understand the needs of the family and have you considered the impact of TSC on the lives of other family members? Have you considered what the relationship may be between the other organ system involvement of TSC and the individual's learning or behavioral presentation? What is the role of seizures or any medications?

12.3.1.4 Draw Information Together into a "Formulation of Needs"

When people have complex neurodevelopmental and multi-system disorders, simply making a diagnosis is not sufficient. For example, a normally intelligent child with TSC may not meet criteria for a "diagnosis" such as autism or ADHD, but can have a range of disruptive behaviors, an academic disorder, and specific neuropsychological deficits. The combination of these deficits across different levels may have a significant impact on the life of the child and their family and will require interventions. All information available needs to be integrated into a formulation of needs. This formulation can be thought of as the story of the person's life at the time.

12.3.1.5 Discuss the Formulation and a Possible Plan of Action with the Family and the Individual with TSC

Families need to agree with a formulation. It should make sense to them and should sound real and clear to them. At some or all of the levels of investigation, potential interventions or treatments may emerge. These should be discussed with families, advantages and disadvantages discussed, and a joint action plan should be agreed. As a rule of thumb, it is always advisable to deal with one aspect of concern at a time. Prioritization is therefore an important aspect of action planning.

12.3.1.6 Re-assess at Appropriate Intervals as Set Out in the International Clinical Guidelines (Table 12.2)

The clinical guidelines identified key developmental time points for assessment such as infancy, prior to school entry, before secondary school transition, and during adolescence. The goal of regular assessments is to identify emerging difficulties before they have become engrained, before they have led to secondary deficits, and to inform the planning for subsequent life stages.

12.3.1.7 Arrange or Perform an Urgent Reassessment When There is a History of Sudden Change in Learning, Behavior, or Mental Health

Behavioral, psychiatric, learning, and neuropsychological changes can be early and subtle markers of intracranial pathology or poor seizure control including nonconvulsive status epilepticus. Sudden changes in neurocognitive and behavioral profiles require urgent physical workup.

12.3.2 Management Options

12.3.2.1 Psycho-education

Psycho-education refers to the provision and sharing of appropriate and accurate information about the disorders or problems identified. For example, when a child is diagnosed with autism, psycho-educational sessions should provide information about what autism is, characteristics, causes, treatment options, lifetime course, educational needs, and so on. Where assessment identified a specific neuropsychological deficit, psycho-education should explain the deficit in an understandable way, identify how it may cause trouble in daily life, what the individual can do to compensate for some difficulties, and what they can do to develop those skills.

Psycho-education should be provided to parents, grandparents, carers, siblings, and the individual with TSC at a level appropriate to their developmental abilities. It may also be extremely useful to provide psycho-education to schools or other professionals, depending on the situation.

12.3.2.2 Behavioral Interventions

Where the assessment identified specific problem behaviors, focused behavioral interventions can be extremely useful. This is often the case for challenging behaviors such as self-injury, aggression (verbal or physical), sleep problems, or challenges in public places. Behavioral interventions start with a detailed behavioral assessment to collect "baseline" ratings and often include a functional analysis of behavior that aims to understand the antecedents and consequences of behaviors. With expert support, even severely challenging behaviors can be reduced, managed, or shaped into more adaptive behaviors. Behavioral interventions can be particularly powerful when working with individuals who have intellectual disability, even those with limited functional communication skills.

12.3.2.3 Cognitive Behavioral Interventions

Cognitive behavioral interventions (such as cognitive behavioral therapy (CBT)) combine an understanding of maladaptive or unhelpful behaviors with an exploration of the dysfunctional thoughts that typically precede that behavior. CBT has been shown to be of particular benefit in a range of psychiatric disorders including depressive disorder, anxiety disorder, specific phobias, social phobia, OCD, and in psychotic disorders. For instance, in CBT theory of anxiety, a situation (such as walking into a room) triggers a thought that can be neutral ("gosh it's a big room") or maladaptive ("everybody is looking at me"). The thought is associated with "emotional feelings" ("I am not going to be able to do this") and physical feelings (sweating, heart racing, etc.). The combination of the unhelpful thought and the associated feelings leads to a "safety behavior" (typically an avoidance behavior such as leaving the room) that then reinforces the unhelpful thoughts and leads to a vicious circle.

CBT uses a range of strategies to help individuals identify their own unhelpful behaviors and thoughts and then to find ways to break the vicious circle at various points. In the United Kingdom, CBT is recommended as the first-line treatment for depression, anxiety, and OCD in children and adolescents. Medications are only offered if CBT does not lead to sufficient improvement within 3 months or if symptoms are very severe.

The principles of CBT can usefully apply to most people, even if they do not have a particular psychiatric disorder. There are an increasing number of books and Web-based resources for self-help CBT.

12.3.2.4 Coaching Techniques

There is a growing interest in the field of "coaching psychology" that has a potentially wide range of applications from behavioral problems to learning disorders, neuropsychological deficits, and mental health or relationship issues. The principle of coaching psychology is relatively straightforward. First, the "coach" helps the individual to describe or define their target problem behaviors (such as difficulty making relationships or struggling with planning). Next, in a collaborative way, the coach helps to translate the target problem into a "skills language". Skills language refers to the principle that all complex tasks in life require a combination of different skills. All humans have relative strengths in some skills and relative weaknesses in other skills. Turning "problem behaviors" into "skills to develop", the coach then supports the individual to build and establish those skills. Coaching has the advantage of being very practical in nature and can therefore be useful for people with a range of intellectual abilities. Coaching for executive control processes and other academic difficulties is becoming increasingly popular. Similarly, "life coaching" for social and relationship issues is developing a good evidence base [42].

12.3.2.5 Psychodynamic Approaches

The psychodynamic therapies include brief psychodynamic psychotherapy, family therapy, psychodynamic group therapies, and some relationship/marital therapies. The core of psychodynamic therapies is to develop an understanding of the unconscious processes that lead to intrapersonal or interpersonal conflict. Therapy typically focuses on an interpretation of the "transference phenomena" that are experienced by the individual. A simplistic definition of transference is that it is the unconscious emotions, thoughts, and feelings evoked in one person by another. Transferences can be "positive" and pleasant or "negative" and unpleasant. Psychodynamic thinking argues that our behavioral responses to people are governed by positive or negative transferences. Psychotherapy requires a high level of personal insight and willingness for self-reflection and is therefore an appropriate modality only in certain circumstances.

12.3.2.6 Interventions for Autism and Autism Spectrum Disorders

There is a large and growing literature of autism-related interventions. Many therapies make very strong claims of positive change or "cure" of autism. However, the evidence base for most early intervention therapies for ASD is very limited. Experts in autism interventions point out that there is no single treatment programme that has been shown to be effective for every child with ASD. Given the wide spectrum and range of needs, an individualized programme is the key. The most successful programmes are ones that combine behavioral, developmental, and educational strategies, that help to create an "autism-friendly" environment with clear and simple communication, where clear structure and routine and visual schedules and reinforcers are used, where there is a focus on building peer relations and play, where interventions are family focused rather than just child focused, and where interventions start within the first 3 years of life. Importantly, programmes should be reviewed every 3 months or so, and if a programme does not seem to lead to progress, the programme should be changed (see National Autism Plan for Children, www.nas.org.uk, and Ref. [43]).

12.3.2.7 Other Non-pharmacological Approaches

The above examples were given to show that there is a range of options available that could be considered depending on the clinical need of the individual or family. There are also other treatment approaches that may be beneficial for individuals with TSC, such as sensory integration and other occupational therapy programmes.

There have been to date no TSC-specific systematic studies of any non-pharmacological interventions. The choice of therapy should therefore be based on the evidence available in the general literature. Families and clinicians are encouraged to investigate the evidence-base for potential therapies and to weigh up potential risks and benefits before deciding on any intervention strategy.

12.3.2.8 Pharmacological Approaches

Psychopharmacological approaches may have a role to play in the treatment of children, adolescents, and adults with TSC, but should always be incorporated into an overall programme of management appropriate to the disorder treated. Given the multi-system nature of TSC and the possibility that an individual may be on antiepilepsy medication, it is essential to have a cautious approach to pharmacotherapy with due consideration of renal and liver function, cardiac function (e.g., rhythm disturbances), and seizure frequency or intensity.

ADHD When a child or older individual with TSC is diagnosed with ADHD, intervention should always include psycho-education to parents, school, and family about ADHD, teaching of basic behavior management techniques, appropriate provisions for educational support and progress, in addition to consideration of a trial of stimulant or non-stimulant medication [19]. In the United Kingdom, good practice guidelines suggest that medication should be considered only for children with severe ADHD. When medication is prescribed, children should be carefully monitored for sideeffects. In spite of the theoretical risk of increased seizures secondary to stimulant medications such as methylphenidate, there is no systematic evidence to support this concern in TSC. As long as clinicians monitor seizures carefully, stimulants and non-stimulants may have a very positive and beneficial role in the treatment of attentional, hyperactive, and impulsive features of ADHD. No data are available in the TSC population regarding the use of second- and third-line options for the treatment of ADHD such as clonidine, atypical antipsychotics (e.g., risperidone) or tricyclic antidepressants (e.g., imipramine).

Depressive and Anxiety Disorders The mainstay of treatment for mild to moderate depressive and anxiety disorders should include appropriate psycho-education about the features, manifestations, risk management and self-help skills available to individuals. Psychological therapies such as CBT should always be considered. For persistent or moderate to severe depression and anxiety disorders, medication should be considered. No systematic evidence exists specifically for TSC, but SSRI (selective serotonin reuptake inhibitors) such as fluoxetine or citalopram are typical choices as first-line medications. When patients have associated psychotic features, an atypical antipsychotic should be considered.

Sleep Disorders Given that sleep disorders are often associated with seizures, it would be important to consider and review the possibility of nocturnal seizures before pursuing a sleep treatment plan. The treatment of sleep disorders should always include appropriate "sleep hygiene" approaches [44] such as ensuring appropriate bedtime routines, a quiet, organized room, avoidance of overstimulating activities or food/drinks before bed, and so on. There have been parental reports of significant benefit from melatonin in children and adolescents with TSC who had sleep disorders. The systematic evidence base is however fairly small. In a placebo-controlled trial of seven individuals with TSC, O'Callaghan *et al.* showed that melatonin improved total sleep time by about 30 min per night [45]. However, some of the participants in the trial showed significant reductions in their sleep latency (sleep onset time), even though the group findings were not statistically significant. Where the main problem is therefore in initial settling after going to bed, melatonin may be a helpful pharmacological strategy to combine with sleep hygiene approaches in some individuals with TSC.

Challenging Behaviors There is an increasing evidence-base that atypical antipsychotics such as risperidone may be helpful as an adjunctive treatment for challenging behaviors, particularly in people with intellectual disabilities. As discussed earlier, challenging behaviors require a careful functional analysis and should not simply be managed through a psychopharmacological route. There is an increasing awareness of a metabolic syndrome associated with risperidone and other atypical antipsychotic treatment. Good practice guidelines therefore recommend that baseline physical assessments including ECG, full blood count, fasting lipids, liver, and renal function tests should be performed before the initiation of such treatments and that regular physical and biochemical review should be performed. This approach may be particularly important for children and adults with intellectual disabilities.

12.3.2.9 **Educational Interventions**

TSC has an impact on the educational needs of the majority of children with the disorder. A significant group of children may require special educational placements. The process of assessment and provision of such services are different in different countries around the world. The legal framework surrounding educational provision is also very different. In the United States, the two main laws relating to education and special education are the No Child Left Behind (NCLB) Act (2002) and the Individuals with Disabilities Education Improvement Act (IDEIA) of 2004. Many parents, carers, and educators have been frustrated by these and other education-related laws around the world (see, for instance, Ref. [46]).

The hope is that children with very significant global intellectual disabilities will be identified in the first few years of life and that professionals working with the family would support them to determine where the child is at in terms of their developmental needs, and then to identify an appropriate educational setting and programme.

The group of children and adolescents with no intellectual disabilities or mild intellectual disabilities are at particular risk in educational settings. As discussed earlier, children with TSC typically have a pattern of very mixed intellectual strengths and weaknesses. Many have significant specific neuropsychological deficits and are at high risk of academic disorders. The dilemma is that unless a child is comprehensively evaluated, these weaknesses may not be identified. It is not uncommon for children with TSC to have normal or even high "IQ" scores, but have enormous difficulties with attentional or organizational skills, or struggle greatly with production of written work. The executive deficits, in particular, are often "invisible" to the untrained eye and may be interpreted as the signs of an unmotivated, stubborn, or willful child. The most helpful first step is to encourage teachers and educational professionals to understand that even the most able child with TSC is at high risk of such difficulties and that comprehensive evaluations should be performed.

It is often helpful to refer families to advocate groups who could help them negotiate the complex systems of education in their country, state, or province.

12.3.2.10 Social Interventions

Government Social Services departments have a statutory duty to support citizens to ensure their safety and well-being. Unfortunately, it can sometimes be very difficult to access services for children or adults who have disabilities.

Social services departments may be able to support families financially through various forms of disability payments, through access to day care or home care schemes, through holiday or vacation programmes, or through access to respite facilities. A key factor for families would be that social services support packages are planned, tailored to the needs of the individual and that they are predictably provided in partnership with the family. It can be extremely disruptive if an afternoon club or overnight respite stay becomes unavailable (or even available) at short notice for a child or adult with intellectual disability who requires a very clear, predictable programme.

It is often helpful for families to seek support from the TSC parent support organizations around the world (United States, United Kingdom, Germany, Netherlands, Italy, Australia, New Zealand, etc.). These organizations have developed helpful knowledge about appropriate and effective ways to access statutory and non-statutory social support.

12.4 Causes of the Neurocognitive and Neurobehavioral Features of TSC

The behavioral, psychiatric, intellectual, academic, and neuropsychological profiles seen in TSC are wide ranging, variable, and multi-layered. The difficulties and disorders develop over time and the interplay between environmental, genetic, molecular, and other factors therefore requires careful dissection. The challenge for clinically useful causal models would be that the models need to be able to explain or predict the profile or severity of the features in an individual case, not just at a group-based or statistical level. For example, while it is important to know that 50% of children with TSC may have an ASD, it does not help parents much just to know that their child has a 50% chance of developing ASD. A few models have emerged over the past 20 years, all in an attempt to unravel causal aspects of the neurodevelopmental or intellectual manifestations of TSC.

12.4.1 Tuber Models

The main model that emerged in relation to intellectual ability suggests that tubers (number, size, location, or "load") are biomarkers or predictors of the outcome in intellectual ability ("IQ"). Since the first magnetic resonance imaging (MRI) study by Roach *et al.* in 1987 [47], numerous studies have correlated intellectual ability to MRI images of tubers. A meta-analysis by Goodman *et al.* in 1997 proposed that an individual with TSC who had more than seven tubers had a very high likelihood of global intellectual disability [48]. MRI studies until then had used relatively low field strength magnets, and almost no standardized measures of intellectual ability were used in these studies. With the development of improved MRI technologies and computer-based analysis, the use of standardized tools to measure intellectual ability, and the increasing identification of mildly affected individuals with TSC, the evidence base began to change. Ridler *et al.* summarized more recent studies [49] and performed a rigorous investigation of tuber–IQ correlation in a sample of adults. They found no correlation between IQ and any number of tuber features. In an epidemiological sample, O'Callaghan *et al.* showed a positive correlation between IQ and tuber count [50]. However, tuber count accounted for only 33% of the variance observed. O'Callaghan's study included a participant with TSC who had an IQ of 130 but had 27 tubers. In a study by Lyczkowski *et al.* [51], all patients with TSC had seven or more tubers on MRI scan. Included in the study was a pair of identical twins. One of the twins had 65 tubers, but normal intellectual ability. The other twin had 60 tubers, but global intellectual disability. Tuber count and tuber-related correlations are clearly *not sufficient* to explain the neurocognitive phenotype observed. Similarly, there are a number of individuals with severe intellectual disability who have very few tubers, suggesting that tubers are also *not necessary* to explain the phenotype.

Apart from intellectual ability, some studies have formulated hypotheses about autism spectrum disorders and tuber location. In an investigation of 19 children with TSC, Bolton *et al.* reported in 1999 that autism was strongly associated with tubers in the temporal lobes and showed a similar finding in an enriched sample [15]. Similar investigations by Weber and Asano did not replicate the temporal tuber findings. Weber *et al.* found a correlation of cerebellar tubers with ASD, supporting a literature on the role of the cerebellum in ASD [52], while Asano *et al.* proposed a model where functional (not necessarily structural) deficits in the temporal lobes were associated with communication deficits, while functional subcortical abnormalities are associated with the stereotyped and repetitive behaviors and reciprocal social interaction difficulties [53].

At the neuropsychological level, one study has investigated the relationship between tuber location, load, and memory in humans with TSC [38]. Ridler found no correlation between either global intellectual ability or memory and any of the tuber factors. She did, however, find a correlation between long-term memory, verbal and spatial working memory, and subcortical gray matter density, particularly in the thalamus and other basal ganglia structures. No systematic structure–function investigations have been performed on other aspects of neuropsychological skills to date.

12.4.2 Seizure Models

There is no doubt that seizures, particularly infantile spasms, can have an adverse impact on neurodevelopment. This clinical observation has led to "seizure models" suggesting that aspects of seizures (type, age of onset, control) are the determinants of intellectual outcome in TSC. In an epidemiological study of the association of infantile spasms, other seizures, and intellectual outcome, O'Callaghan [50] reported that a history of infantile spasms showed statistically significant correlation with intellectual outcome. However, a history of infantile spasms contributed only 14% of the overall 47% variance in IQ explained by tubers and seizures. That is, more than half of the predicted determinants of IQ were not accounted for by measuring tuber count, a history of infantile spasms, or other seizure disorders. In a similar study, Raznahan *et al.* showed that a history of epilepsy was not correlated with IQ and that a history of infantile spasms explained only 13% of the IQ variance observed [54]. The age of onset of seizures did not show any correlation with intellectual outcome in their study. At least a third of individuals with TSC who had infantile spasms have normal intellectual outcome. Taken together, results suggest that seizure variables, including infantile spasms, are *not sufficient* to explain the intellectual disability. Similarly, clinical data suggest that seizures may *not be necessary* for global intellectual disability or specific neuropsychological deficits to occur.

Bolton *et al.* also investigated the role of seizures in a causal model of autism. They argued that the combination of temporal lobe tubers, temporal lobe epileptiform discharges, and early-onset spasms may render individuals with TSC at high risk of autism spectrum disorders [15]. The findings of their study have not yet been replicated elsewhere.

12.4.3
Genotype–Phenotype Models

Studies comparing groups of individuals with *TSC1* and *TSC2* mutations have suggested that those with *TSC2* are typically more severely affected, and this included intellectual ability [55, 56]. On closer inspection, very few of the early studies used any standardized measures of "IQ" to determine intellectual ability. Some studies divided individuals based on their ability to attend mainstream schools while others used an oversimplistic classification based on communicative language. The first study to use standardized cognitive assessments was performed by Lewis *et al.* [5]. She reported that the presence of a *TSC2* mutation was associated with a higher risk of low IQ, autism spectrum disorders, and infantile spasms than with *TSC1* mutations. However, *TSC1* mutations were not inevitably associated with mild phenotypes and neither were *TSC2* mutations inevitably associated with severe intellectual disability. It is clear from data of Lewis' and others that there are individuals with *TSC1* mutations who have profound intellectual disability. There are also individuals with *TSC2* mutations who have extremely high IQs. As clinical assessments improve and mutation analysis is performed more consistently, it is becoming increasingly clear that simply knowing whether an individual has a *TSC1* or a *TSC2* mutation is unlikely to be a useful marker or predictor of neurocognitive or behavioral outcome.

12.4.4 Molecular Models

Until very recently, the main causal models of neurocognitive and neurobehavioral outcome in TSC were based on tuber, seizure, combined tuber–seizure or gross genotype–phenotype models. In 2007, de Vries and Prather pointed out some of the weaknesses of these models in relation to cognitive/behavioral work, as mentioned above, and suggested that molecular models were required [32].

Later in 2007, de Vries joined forces with Christopher Howe, a biochemist at the University of Cambridge, to formulate the first molecular hypothesis of neurocognitive outcomes in TSC, referred to as the GRIPP hypothesis [57]. For a detailed account of the molecular and signaling mechanisms in TSC, please see Chapter 6. De Vries and Howe pointed out that the TSC1-2 complex functions as a global regulator and integrator of a range of physiological processes (GRIPP) in the PI3K–TSC–mTOR, the AMPK and MAPK pathways and that many of the signaling proteins involved in these pathways had independently been shown to be involved in neurobiological processes such as neuronal development, myelination, forebrain development, cytoskeletal integrity, protein synthesis, and long-term potentiation. The authors therefore proposed that mutations in the *TSC1* or *TSC2* gene would be sufficient to disrupt these crucial pathways and lead directly to neurodevelopmental, cognitive, and behavioral manifestations in TSC. The GRIPP hypothesis had two separate components. The first (GRIPP-I) stated that tubers and seizures were not necessary or sufficient to explain the neurocognitive phenotype in TSC, which instead resulted from disturbances in signaling pathways.

The GRIPP-I hypothesis therefore suggests that the TSC1-2 proteins act as a "molecular switchboard" where haploinsufficiency for TSC1 or TSC2 would be associated with upstream "input" signaling defects and with downstream mTOR signaling defects. The GRIPP-I hypothesis would be supported by evidence of neurocognitive deficits in the absence of structural brain abnormalities and seizures, and by the ability to reverse such deficits through mTOR inhibition. Memory deficits were, for instance, proposed to be improved by the administration of rapamycin, an mTOR inhibitor [57]. De Vries and Howe proposed that experimental animal models may be powerful tools to test such molecular hypotheses.

de Vries and Howe also commented that the biochemical literature was showing increasing evidence that different TSC mutations are associated with different functional consequences both in terms of the ability of the tuberin–hamartin complex to receive upstream signals from targets such as AKT, ERK1/2, and p38MAPK and in terms of signaling to downstream effectors such as mTOR and S6 kinase [57, 58]. Given that different *TSC1* and *TSC2* mutations will have different functional consequences at the molecular level, the authors proposed the second component of the GRIPP hypothesis (GRIPP-II), that the variability of neurocognitive manifestations observed in TSC will be explained, at least in part, by the differential functional deficits of different mutations and consequent signaling impact on input and output pathways.

A prediction of the GRIPP-II hypothesis is that different animal models of TSC would show different neurocognitive and neurobehavioral manifestations and that these differences would be explained by the functional consequences of mutations. Apart from exploring these questions directly in animal models, the hypothesis would also suggest that in humans with TSC, there should be evidence of less intra-mutational variability than inter-mutational variability in neurocognitive and behavioral phenotypes.

To date there are limited data to examine intra- and inter-mutational variability. A recent paper on intrafamilial variability, however, did show interesting results [51]. Lyczkowski and colleagues examined various aspects of phenotypic expression in five multi-generation TSC families. They showed significant variability in peripheral system involvement and in seizure patterns within families. However, global intellectual abilities, tuber counts, and specific neuropsychological skills were remarkably consistent within families. In only one of the five families (family B) were IQ scores significantly different between two identical twins ($IQ = 85$ versus $IQ < 42$; one with occasional seizures and the other with intractable seizures and infantile spasms). All the other families showed within-family clustering of IQ scores around low average, average, or above average scores. Tuber counts were remarkably similar within families but quite different between families. For example, the twins in family B had 60 and 65 tubers, respectively, while family E had 7, 8, 12, 14, and 18 tubers, respectively. At the behavioral level, all members of family A had clinically concerning behavioral rating scores, while all members of family E were in the normal range on behavior ratings. Neuropsychologically, all members of family C showed significant executive deficits while all members of family E showed above normal expressive language skills. These data are consistent with a model in which the nature of the *TSC1* or *TSC2* mutation plays a major part in determining parameters such as intellectual ability, tuber count, and neuropsychological skills, as predicted by the second part of the GRIPP hypothesis. It is likely that modifier genes and other factors such as the effectiveness of seizure control will make additional contributions in a comprehensive causal model.

Clearly, more data will be required to test aspects of any molecular hypotheses of neurocognition in TSC. The strength of the GRIPP hypothesis lies in the fact that it provides an empirically testable model. In parallel, investigations should be performed into genetic modifier effects, gene–environment interactions, perhaps with particular regard to the molecular consequences of seizures.

12.5
Animal Models for Behavioral, Psychiatric, Intellectual, Learning, and Neuropsychological Deficits in TSC

For a detailed account of animal models of TSC, please see Chapter 7. The first study of neurocognition using a TSC animal model was performed in 2006 by Waltereit *et al.* [59]. The authors used the Eker rat, a naturally occurring $Tsc2^{+/-}$ animal, and performed Morris water maze, conditioned taste aversion, light/dark conditioning, and radial maze tasks. The rats were free from seizures and had virtually no structural brain abnormalities. The authors found no differences between $Tsc2^{+/-}$

and wild-type rats on the conditioning tasks but reported an unusual "enhanced episodic memory" ability in the *Tsc2*$^{+/-}$ rats. The authors were not able to explain this behavioral observation. In parallel electrophysiological experiments, they reported impaired synaptic plasticity, both in long-term potentiation (LTP) and long-term depression (LTD) in the *Tsc2*$^{+/-}$ rats [60].

In 2007, Goorden *et al.* in Utrecht studied a *Tsc1*$^{+/-}$ mouse model [61]. The *Tsc1*$^{+/-}$ mice had no structural brain abnormalities and no seizures. However, they showed significant spatial learning deficits on the water maze task as well as conditioning deficits during cued and uncued conditions. Furthermore, the *Tsc1*$^{+/-}$ mice had reduced socialization behaviors in comparison to the wild-type mice on social interaction and nest building tasks [61].

Wong and colleagues created a *Tsc1* conditional knockout mouse where the expression of the Tsc1 protein was selectively knocked out in glia. Behavioral testing before the onset of seizures in their conditional knockout mice revealed significant learning deficits in the water maze and a conditioned fear paradigm task with associated deficits in hippocampal LTP. Results on a battery of sensory, motor, and coordination tests were mostly normal [62].

Using a *Tsc2*$^{+/-}$ mouse model, also free from cerebral lesions and seizures, Silva and colleagues performed similar behavioral experiments [63]. They reported context fear conditioning, water maze, as well as radial maze deficits, but did not observe any social interaction deficits in their *Tsc2*$^{+/-}$ mice. They confirmed abnormal hippocampal LTP and showed that the intraperitoneal administration of rapamycin to adult mice for 5 days was sufficient to reverse the LTP abnormality as well as the contextual fear conditioning and the water maze deficits [63].

These data from mouse models provide empirical evidence that some aspects of developmental disorders (such as socialization deficits in autism) and neuropsychological deficits (such as visuospatial memory) can be studied in the mouse with some insight into the molecular mechanisms involved and the role of Tsc1/Tsc2. In addition, they provide support for the GRIPP-I hypothesis [57] that the intellectual, learning, and neuropsychological deficits seen in TSC can be directly explained by the intracellular signaling abnormalities caused by the TSC mutations, rather than by tubers and seizures. Clearly, further study is required, but there is great potential for mouse models to increase our understanding of the mechanisms that underlie the neurocognitive and behavioral difficulties seen in TSC, and to contribute to the development of novel treatment strategies.

12.6
Future Directions for the Understanding of Behavioral, Psychiatric, Intellectual, Academic, and Neuropsychological Deficits in TSC

To understand the mechanisms underlying cognition and behavior is one of the grand challenges of the twenty-first century. TSC is in an excellent position to integrate the knowledge gained across genetics, molecular biology, clinical observations, and neuroscience studies to provide both an improved understanding of neurocognitive and behavioral processes in general, and their specific deficits and mechanisms that are seen in affected individuals.

However, it is clear that there are significant gaps in knowledge both at the clinical and at the research level. At almost every level of investigation described above, much more work, data collection, and longitudinal monitoring are required. At the behavioral level, standardized rating scale measures should be used in larger populations and followed up over time to determine the developmental trajectory and to identify factors that contribute to behavioral difficulties. The psychiatric level requires standardized research-based tools to reach rigorous diagnostic criteria across disorders such as mood disorders, ADHD, and anxiety disorders. At the level of intellectual ability, it is still uncommon to have comprehensive and regular evaluation across developmental time points in order to determine the emergence, development, and maturation of intellectual abilities in TSC. Learning disorders (reading, writing, mathematical) are likely to be a significant challenge for those with normal intellectual abilities, yet there are no studies in the literature. While there has been a start at neuropsychological studies of memory, visuospatial, and attentional–executive skills, ongoing and larger studies are required. Investigations should also become more interdisciplinary and investigate cross-talk between different levels in order to ensure that studies are not purely descriptive, but also contribute toward causal understanding. The emergence of animal models of learning and memory can become powerful tools, particularly if models can be studied alongside observations and interventions in humans with TSC.

At the clinical level, in spite of published international guidelines, it is very rare to find systematic, multi-disciplinary assessment of the needs of individuals with TSC. Where clinics do perform evaluations, assessment tools are often different across sites and can therefore not easily be compared or grouped. Neurocognitive and behavioral data are not adequately integrated in the overall formulation to identify the needs of individuals seen in clinics. To date there have been no studies that have investigated any neurocognitive, behavioral, or pharmacological intervention specifically in those with TSC.

In order to understand the mechanisms underlying all the neurocognitive and behavioral phenomena outlined earlier, it will be important to identify subtypes or endophenotypic groups and look for markers of specific deficits and for biomarkers of likely treatment response. The GRIPP hypothesis proposed a model to investigate the inter-individual variability in cognition and neurodevelopment. In addition, careful intra-familial studies may identify specific determinants of intra-mutational variability that may be related to modifier genes, developmental cascades, or the consequences of, for instance, a seizure disorder. The likely comprehensive answers will be in a combinatorial model of molecular, electrophysiological, and other factors.

The finding that rapamycin can reverse specific neurocognitive deficits in adult TSC mice is an example of an exciting targeted molecular treatment, which suggests the possibility that mTOR inhibitors may improve some of the neurocognitive and neurobehavioral features of TSC. At the time of the writing of this chapter, a multi-center trial of rapamycin in AML and LAM [64] was investigating the effect of rapamycin on memory skills in adults with TSC and LAM. Preliminary data showed that immediate and delayed recall memory was improved in 8 of the 13 people in the trial between baseline and 4 months. However, in a further two subjects no significant change was seen, while two trial participants showed negative change in immediate and/or delayed recall [65]. Interestingly, many participants showed positive change on tasks of executive skills, a finding not predicted from the animal literature. In addition, in a clinical trial of TSC patients treated with RAD001, caregivers noted subjective improvements in behavior and social interactions in 19 of 21 subjects (David Franz, personal communication; see also Chapter 11). Neuropsychological testing pre- and posttreatment has not shown any evidence of deterioration, but has also been unable to show any clear objective improvement.

These are early, tentative findings that should be viewed with caution and need to be expanded. However, such data do raise the possibility that rapamycin, other mTOR inhibitors, and compounds acting in the TSC signaling pathway may become targeted treatments for some aspects of the behavioral, psychiatric, intellectual, learning, and neuropsychological deficits seen in TSC.

12.7 How to Live a Positive Life with TSC

It is important to remember that individuals with the disorder and their families have to live with the neurocognitive and behavioral challenges of TSC every day. The stories of TSC families remind clinicians and researchers of the many challenges that come with disabilities in learning and behavior. How does one support a family member to live a positive and independent life without being either unrealistic in expectations or too overprotective? How do adults with TSC who also have children with TSC manage and support their families with an awareness of their own strengths and weaknesses? How do you take charge of what you can and let go of those things you have no control over? And, with the advent of possible new treatments, how do you deal with hope for the future without feeling paralyzed now?

There are no easy answers to these questions, and strategies may be different for different families or individuals. The psychological and disability literature has identified a number of protective or "resilience" factors that can help or support individuals on their journey [39]. It is very protective if families have a good understanding of the problems they are dealing with and have a shared commitment to working with these challenges. Good coordination among professionals in the family network and partnership working between families and professionals are strong supportive factors. Clear communication and authoritative parenting are more helpful than over-involved, overprotective, and permissive parenting (allowing one's child to "get away" with inappropriate behaviors because they have TSC). Involvement of fathers and good marital relations are extremely positive factors, often because they allow both parents to reach the same understanding of the expectation for and developmental profile of their child.

A very protective factor is a strong sense of "self-efficacy" [66]. Self-efficacy refers to an individual's sense of what they are able to do or be in charge of. We learn what we are capable of doing through various routes: sometimes through experience and practice, sometimes by doing something wrong. We may even learn and believe we can do something simply by observing others do a task. People are more likely to develop a strong sense of self-efficacy through "doing" rather than "not doing", through helpful peer and social support, and by having a positive attributional style (a "can do" attitude).

Participation in groups, clubs, joining support organizations such as the TSAlliance (TSA) can offer a supportive social network and enhance family resilience.

There are no easy or magical ways for families to live a positive life with TSC. It is essential to know where you (if you have TSC) or your family members with TSC are at in terms of overall neurocognitive and behavioral needs. Unless you know that, you cannot determine what support you need or when you may need that. Understanding your strengths and weaknesses is the first step toward developing a sense of self-efficacy and the belief that whatever TSC may bring, you have a strategy or can find a strategy to deal with that challenge.

References

1 Critchley, M. and Earl, C.J.C. (1932) Tuberous sclerosis and allied conditions. *Brain*, **55**, 311–346.

2 Hunt, A. (1983) Tuberous sclerosis: a survey of 97 cases. III. Family aspects. *Dev. Med. Child Neurol.*, **25**, 353–357.

3 Hunt, A. (1993) Development, behaviour and seizures in 300 cases of tuberous sclerosis. *J. Intellect. Disabil. Res.*, **37**, 41–51.

4 de Vries, P.J., Hunt, A., and Bolton, P.F. (2007) The psychopathologies of children and adolescents with tuberous sclerosis complex (TSC): a postal survey of UK families. *Eur. Child Adolesc. Psychiatry*, **16**, 16–24.

5 Lewis, J.C., Thomas, H.V., Murphy, K.C., and Sampson, J.R. (2004) Genotype and psychological phenotype in tuberous sclerosis. *J. Med. Genet.*, **41**, 203–207.

6 Pulsifer, M.B., Winterkorn, E.B., and Thiele, E.A. (2007) Psychological profile of adults with tuberous sclerosis complex. *Epilepsy Behav.*, **10**, 402–406.

7 World Health Organization (1993) *The ICD-10 Classification of Mental and Behavioural Disorders. Diagnostic criteria for research*, WHO, Geneva.

8 American Psychiatric Association (1994) *Diagnostic and Statistical Manual of Mental Disorders*, 4th edn, APA, Washington, DC.

9 Lord, C., Rutter, M., and Le Couteur, A. (1994) Autism diagnostic interview-revised: a revised version of a diagnostic interview for caregivers of individuals with possible pervasive developmental disorders. *J. Autism Dev. Disord.*, **24**, 659–685.

10 Lord, C., Risi, S., Lambrecht, L., Cook, E.H., Jr., Leventhal, B.L., DiLavore, P.C., Pickles, A., and Rutter, M. (2000) The autism diagnostic observation schedule-generic: a standard measure of social and communication deficits associated with the spectrum of autism. *J. Autism Dev. Disord.*, **30**, 205–223.

11 Endicott, J. and Spitzer, R.L. (1978) A diagnostic interview: the schedule for affective disorders and schizophrenia. *Arch. Gen. Psychiatry*, **35**, 837–844.

12 Schopler, E., Reichler, R.J., DeVellis, R.F., and Daly, K. (1980) Towards objective classification of childhood autism: childhood autism rating scale (CARS). *J. Autism Dev. Disord.*, **10**, 91–103.

13 Krug, D.A., Arick, J., and Almond, P. (1980) Behavior checklist for identifying severely handicapped individuals with high levels of autistic behavior. *J. Child Psychol. Psychiatry*, **21**, 221–229.

14 Baird, G., Simonoff, E., Pickles, A., Chandler, S., Loucas, T., Meldrum, D., and

Charman, T. (2006) Prevalence of disorders of the autism spectrum in a population cohort of children in South Thames: the Special Needs and Autism Project (SNAP). *Lancet*, **368**, 210–215.
15 Bolton, P.F., Park, R.J., Higgins, J.N., Griffiths, P.D., and Pickles, A. (2002) Neuro-epileptic determinants of autism spectrum disorders in tuberous sclerosis complex. *Brain*, **125**, 1247–1255.
16 Hunt, A. and Shepherd, C. (1993) A prevalence study of autism in tuberous sclerosis. *J. Autism Dev. Disord.*, **23**, 323–339.
17 Smalley, S.L., Tanguay, P.E., Smith, M., and Gutierrez, G. (1992) Autism and tuberous sclerosis. *J. Autism Dev. Disord.*, **22**, 339–355.
18 Cass, H., Sekaran, D., and Baird, G. (2006) Medical investigation of children with autistic spectrum disorders. *Child Care Health Dev.*, **32**, 521–533.
19 Taylor, E., Dopfner, M., Sergeant, J., Asherson, P., Banaschewski, T., Buitelaar, J., Coghill, D., Danckarets, M. *et al.* (2004) European clinical guidelines for hyperkinetic disorder – first upgrade. *Eur. Child Adolesc. Psychiatry*, **13** (Suppl 1), 7–30.
20 Gillberg, I.C., Gillberg, C., and Ahlsen, G. (1994) Autistic behaviour and attention deficits in tuberous sclerosis: a population-based study. *Dev. Med. Child Neurol.*, **36**, 50–56.
21 de Vries, P.J. (2002) The Psychopathologies of Attention in Tuberous Sclerosis. PhD Thesis, University of Cambridge, Cambridge, UK.
22 Muzykewicz, D.A., Newberry, P., Danforth, N., Halpern, E.F., and Thiele, E.A. (2007) Psychiatric comorbid conditions in a clinic population of 241 patients with tuberous sclerosis complex. *Epilepsy Behav.*, **11**, 506–513.
23 Nutt, D.J., Fone, K., Asherson, P., Bramble, D., Hill, P., Matthews, K., Morris, K.A., Santosh, P. *et al.* (2007) Evidence-based guidelines for management of attention-deficit/hyperactivity disorder in adolescents in transition to adult services and in adults: recommendations from the British Association for Psychopharmacology. *J. Psychopharmacol.*, **21**, 10–41.
24 Smalley, S.L., Burger, F., and Smith, M. (1994) Phenotypic variation of tuberous sclerosis in a single extended kindred. *J. Med. Genet.*, **31**, 761–765.
25 Raznahan, A., Joinson, C., O'Callaghan, F., Osborne, J., and Bolton, P.F. (2006) Psychopathology of tuberous sclerosis: an overview and findings in a population-based sample of adults with tuberous sclerosis. *J. Intellect. Disabil. Res.*, **50**, 561–569.
26 Goh, S. and Thiele, E.A. (2005) Anorexia nervosa in a child with tuberous sclerosis complex. *J. Child Neurol.*, **20**, 457–460.
27 de Vries, P.J. and Bolton, P.F. (2003) Tuberous sclerosis, in *Outcome in Neurodevelopmental and Genetic Disorders* (eds P. Howlin and O. Udwin), Cambridge University Press, Cambridge.
28 Lishman, W.A. (1997) *Organic Psychiatry: The Psychological Consequences of Cerebral Disorder*, 3rd edn, Blackwell Science, Oxford.
29 Murphy, K.C. and Owen, M.J. (2001) Velo-cardio-facial syndrome: a model for understanding the genetics and pathogenesis of schizophrenia. *Br. J. Psychiatry*, **179**, 397–402.
30 Boer, H., Holland, A., Whittington, J., Butler, J., Webb, T., and Clarke, D. (2002) Psychotic illness in people with Prader–Willi syndrome due to chromosome 15 maternal uniparental disomy. *Lancet*, **359**, 135–136.
31 Joinson, C., O'Callaghan, F.J., Osborne, J.P., Martyn, C., Harris, T., and Bolton, P.F. (2003) Learning disability and epilepsy in an epidemiological sample of individuals with tuberous sclerosis complex. *Psychol. Med.*, **33**, 335–344.
32 de Vries, P.J. and Prather, P. (2007) The tuberous sclerosis complex. *N. Engl. J. Med.*, **356**, 92.
33 Prather, P. and de Vries, P.J. (2004) Behavioral and cognitive aspects of tuberous sclerosis complex. *J. Child Neurol.*, **19**, 666–674.
34 Prather, P., Thiele, E., and de Vries, P.J. (2006) Neuropsychological profiling in tuberous sclerosis complex (TSC): implications for neurobiology and clinical

interventions. *J. Intellect. Disabil. Res.*, **50**, 788.

35 Jambaque, I., Cusmai, R., Curatolo, P., Cortesi, F., Perrot, C., and Dulac, O. (1991) Neuropsychological aspects of tuberous sclerosis in relation to epilepsy and MRI findings. *Dev. Med. Child Neurol.*, **33**, 698–705.

36 Harrison, J.E., O'Callaghan, F.J., Hancock, E., Osborne, J., and Bolton, P.F. (1999) Cognitive deficits in normally intelligent patients with tuberous sclerosis. *Am. J. Med. Genet.*, **88**, 642–646.

37 de Vries, P.J. and Watson, P. (2008) Attention deficits in tuberous sclerosis complex (TSC): rethinking the pathways to the endstate. *J. Intellect. Disabil. Res.*, **52**, 348–357.

38 Ridler, K., Suckling, J., Higgins, N.J., de Vries, P.J., Stephenson, C.M., Bolton, P.F., and Bullmore, E.T. (2007) Neuroanatomical correlates of memory deficits in tuberous sclerosis complex. *Cereb. Cortex*, **17**, 261–271.

39 Carr, A. (1999) *The Handbook of Child and Adolescent Clinical Psychology: A Contextual Approach*, Routledge, London and New York.

40 Goh, S., Butler, W., and Thiele, E.A. (2004) Subependymal giant cell tumors in tuberous sclerosis complex. *Neurology*, **63**, 1457–1461.

41 de Vries, P., Humphrey, A., McCartney, D., Prather, P., Bolton, P., Hunt, A., and the TSC Behaviour Consensus Panel. (2005) Consensus clinical guidelines for the assessment of cognitive and behavioural problems in tuberous sclerosis. *Eur. Child Adolesc. Psychiatry*, **14**, 183–190.

42 Nelson-Jones, R. (2006) *Human Relationship Skills: Coaching and Self-Coaching*, 4th edn, Routledge, New York.

43 Lord, C. and McGee, J.P. (2001) *Educating Children with Autism*, National Research Council, National Academy Press, Washington, DC.

44 Kotagal, S. and Pinosi, P. (2006) Sleep disorders in children and adolescents. *BMJ*, **332**, 828–832.

45 O'Callaghan, F.J., Clarke, A.A., Hancock, E., Hunt, A., and Osborne, J.P. (1999) Use of melatonin to treat sleep disorders in tuberous sclerosis. *Dev. Med. Child Neurol.*, **41**, 123–126.

46 Pino, S. and Hook, D. (2006) *Success with IEPs – Lessons Learned from Implementing the Middle School Model*, Incentive Publications, Inc., Nashville, TN.

47 Roach, E.S., Williams, D.P., and Laster, D.W. (1987) Magnetic resonance imaging in tuberous sclerosis. *Arch. Neurol.*, **44**, 301–303.

48 Goodman, M., Lamm, S.H., Engel, A., Shepherd, C.W., Houser, O.W., and Gomez, M.R. (1997) Cortical tuber count: a biomarker indicating neurologic severity of tuberous sclerosis complex. *J. Child Neurol.*, **12**, 85–90.

49 Ridler, K., Suckling, J., Higgins, N., Bolton, P., and Bullmore, E. (2004) Standardised whole brain mapping of tubers and subependymal nodules in tuberous sclerosis complex. *J. Child Neurol.*, **19**, 658–665.

50 O'Callaghan, F.J., Harris, T., Joinson, C., Bolton, P., Noakes, M., Presdee, D., Renowden, S., Shiell, A. *et al.* (2004) The relation of infantile spasms, tubers, and intelligence in tuberous sclerosis complex. *Arch. Dis. Child.*, **89**, 530–533.

51 Lyczkowski, D.A., Conant, K.D., Pulsifer, M.B., Jarrett, D.Y., Grant, P.E., Kwiatkowski, D.J., and Thiele, E.A. (2007) Intrafamilial phenotypic variability in tuberous sclerosis complex. *J. Child Neurol.*, **22**, 1348–1355.

52 Weber, A.M., Egelhoff, J.C., McKellop, J.M., and Franz, D.N. (2000) Autism and the cerebellum: evidence from tuberous sclerosis. *J. Autism Dev. Disord.*, **30**, 511–517.

53 Asano, E., Chugani, D.C., Muzik, O., Behen, M., Janisse, J., Rothermel, R., Mangner, T.J., Chakraborty, P.K. *et al.* (2001) Autism in tuberous sclerosis complex is related to both cortical and subcortical dysfunction. *Neurology*, **57**, 1269–1277.

54 Raznahan, A., Higgins, N.P., Griffiths, P.D., Humphrey, A., Yates, J.R., and Bolton, P.F. (2007) Biological markers of intellectual disability in tuberous sclerosis. *Psychol. Med.*, **37**, 1293–1304.

55 Jones, A.C., Daniells, C.E., Snell, R.G., Tachataki, M., Idziaszyk, S.A.,

Krawczak, M., Sampson, J.R., and Cheadle, J.P. (1997) Molecular genetic and phenotypic analysis reveals differences between TSC1 and TSC2 associated familial and sporadic tuberous sclerosis. *Hum. Mol. Genet.*, **6**, 2155–2161.

56 Dabora, S.L., Jozwiak, S., Franz, D.N., Roberts, P.S., Nieto, A., Chung, J., Choy, Y.S., Reeve, M.P. *et al.* (2001) Mutational analysis in a cohort of 224 tuberous sclerosis patients indicates increased severity of TSC2, compared with TSC1, disease in multiple organs. *Am. J. Hum. Genet.*, **68**, 64–80.

57 de Vries, P.J. and Howe, C.J. (2007) The tuberous sclerosis complex proteins – a GRIPP on cognition and neurodevelopment. *Trends Mol. Med.*, **13**, 319–326.

58 Nellist, M., Sancak, O., Goedbloed, M.A., Rohe, C., van Netten, D., Mayer, K., Tucker-Williams, A., van den Ouweland, A.M. *et al.* (2005) Distinct effects of single amino-acid changes to tuberin on the function of the tuberin–hamartin complex. *Eur. J. Hum. Genet.*, **13**, 59–68.

59 Waltereit, R., Welzl, H., Dichgans, J., Lipp, H.-P., Schmidt, W.J., and Weller, M. (2006) Enhanced episodic-like memory and kindling epilepsy in a rat model of tuberous sclerosis. *J. Neurochem.*, **96**, 407–413

60 Von der Brelie, C., Waltereit, R., Zhang, L., Beck, H., and Kirschstein, T. (2006) Impaired synaptic plasticity in a rat model of tuberous sclerosis. *Eur. J. Neurosci.*, **23**, 686–692.

61 Goorden, S.M., van Woerden, G.M., van der Weerd, L., Cheadle, J.P., and Elgersma, Y. (2007) Cognitive deficits in Tsc1$^{+/-}$ mice in the absence of cerebral lesions and seizures. *Ann. Neurol.*, **62**, 648–655.

62 Zeng, L.H., Ouyang, Y., Gazit, V., Cirrito, J.R., Jansen, L.A., Ess, K.C., Yamada, K.A., Wazniak, D.F. *et al.* (2007) Abnormal glutamate homeostasis and impaired synaptic plasticity and learning in a mouse model of tuberous sclerosis complex. *Neurobiol. Dis.*, **28**, 184–196.

63 Ehninger, D., Han, S., Shilyansky, C., Shou, Y., Li, W., Kwiatkowski, D.J., Ramesh, V., and Silva, A.J. (2008) Reversal of learning deficits in a Tsc2$^{+/-}$ mouse model of tuberous sclerosis. *Nat. Med.*, **14** (8), 843–848.

64 Davies, D.M., Johnson, S.R., Tattersfield, A.E., Kingswood, J.C., Cox, J.A., McCartney, D.L., Doyle, T., Elmslie, F. *et al.* (2008) Sirolimus therapy in tuberous sclerosis or sporadic lymphangioleiomyomatosis. *N. Engl. J. Med.*, **358**, 200–203.

65 de Vries, P.J. (2008) What is the effect of rapamycin on cognition in humans with TSC and LAM? Conference Presentation, *mTOR Signaling: From Cancer to CNS Function*, January 17–18, 2008, Bethesda, Maryland.

66 Bandura, A. (1997) *Self-Efficacy: The Exercise of Control*, W. H. Freeman and Company, New York.

Part V
Other Organ Systems

Tuberous Sclerosis Complex: Genes, Clinical Features, and Therapeutics
Edited by David J. Kwiatkowski, Vicky Holets Whittemore, and Elizabeth A. Thiele

ISBN: 978-3-527-32201-5

13
Ophthalmic Manifestations

Shivi Agrawal and Anne B. Fulton

13.1
Introduction

This chapter focuses on eye and vision care in individuals with tuberous sclerosis complex (TSC). Ophthalmic assessment and care depends on knowledge of the structure and function of the eye and visual pathways (Figure 13.1). We describe the ophthalmic problems that are encountered in TSC and the management of these problems.

13.2
Adnexa and Anterior Segment

Occasionally, the structures surrounding the eyes, the adnexa, are involved in TSC. For instance, angiofibromas of the eyelids are quite common in TSC [1–4]. These lesions have been observed in 39% of patients with TSC [2], and at times are the presenting sign [5, 6]. Hamartomas of the iris and ciliary body are uncommon (Table 13.1). It has been proposed that colobomas may be secondary manifestations of hamartomas of the iris and ciliary body pigment epithelia [7]. Other uncommon lesions of the adnexa and anterior segment that have been reported in TSC are listed in Table 13.1. These other abnormalities are not specific to TSC.

13.3
Retinal Lesions

13.3.1
Hamartomas

The classic ocular lesions of TSC (Figure 13.2) are astrocytic retinal hamartomas [8] and have similar histological features to the tubers in the brain. While the term tuberous sclerosis was introduced by Bourneville in 1880 [2], the retinal involvement

Tuberous Sclerosis Complex: Genes, Clinical Features, and Therapeutics
Edited by David J. Kwiatkowski, Vicky Holets Whittemore, and Elizabeth A. Thiele

ISBN: 978-3-527-32201-5

(a)

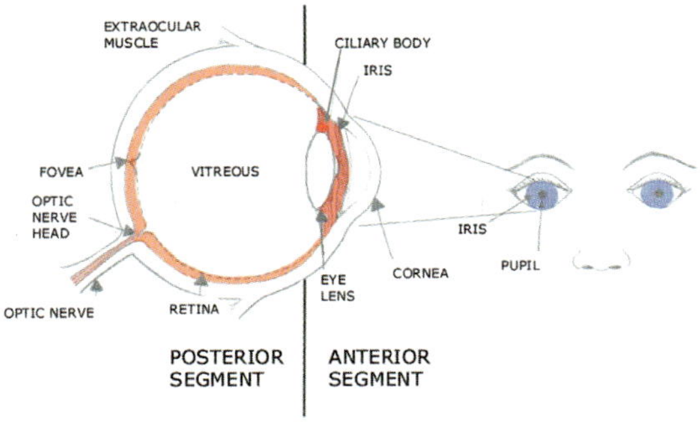

(b)

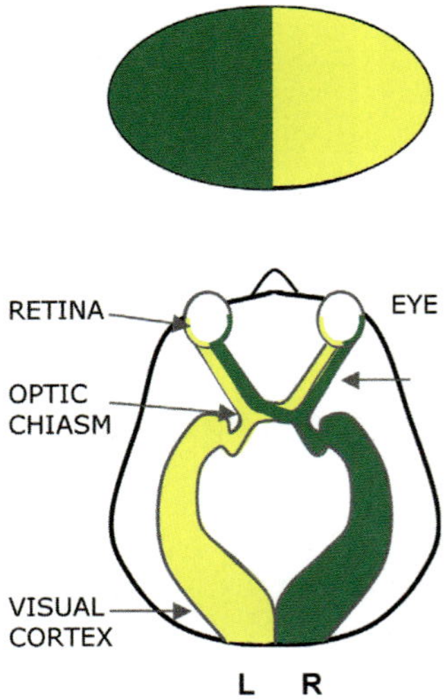

Figure 13.1 Diagram of the eye, visual system, and visual fields. (a) An external view and cross section of the eye. The classic retinal hamartomas of TSC occur in the posterior segment. (b) Schematic of the visual pathway and visual field. The visual pathway passes from the retina, a light-sensitive film of neural tissue lining the posterior segment, along the optic nerves to precise locations in the brain. For example, the pathway originating in left temporal retina (yellow) projects to ipsilateral portions of the brain. The pathway originating from the left nasal retina (green) crosses at the chiasm and projects to the contralateral side of the brain. A lesion at any site along the visual pathway is represented at an exact, corresponding point in the visual field.

of TSC escaped notice until Van der Hoeve's report in 1921 [2, 9]. The detection of retinal hamartomas is of particular importance as it may help in the clinical diagnosis of TSC, particularly in young children who do not yet show classic clinical signs [10, 11]. In these cases, retinal hamartomas can be the first objective sign of TSC [10–12]. Hamartomas may be present prenatally and have been reported in a 7-day-old infant [13]. Hamartomas are composed of a felt-like network of glial astrocytes and typically arise from the retinal nerve fiber layer without disturbance of the deeper retinal lamina [3, 9, 10, 14, 15]. Occasionally, however, the tumors can involve the full thickness of the retina [3, 15] and even the retinal pigment

Table 13.1 Uncommon lesions of adnexa and anterior segment in TSC.

Atypical colobomas of the iris and lens [2, 3, 7, 9, 28, 35]
Poliosis of the eyelashes [2, 4]
Sector depigmentation of the iris [2, 12, 35]
Closed-angle glaucoma [7, 35]
Iris neovascularization [7, 35]
Bony hamartomas of orbital rim [63]

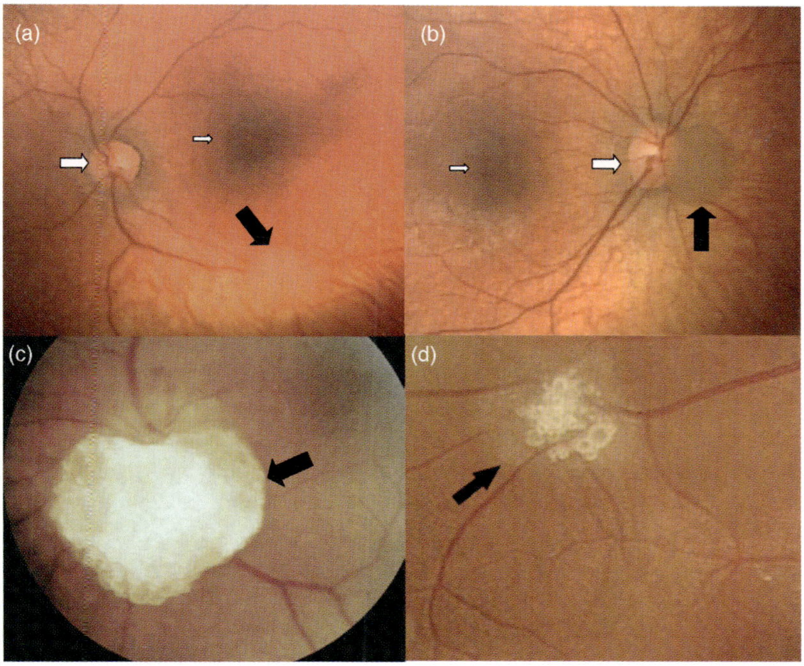

Figure 13.2 Retinal hamartomas in TSC. (a) The hamartoma (black arrow) is approximately of the same diameter as the normal optic nerve head (large white arrow). The fovea (small white arrow) is normal. The hamartoma is located along vessels in the posterior retina. It is a slightly raised, oval, transparent, light gray lesion with indistinct boundaries. (b) This hamartoma (black arrow) is at the optic nerve head (large white arrow) and is raised above the optic nerve. The fovea (small white arrow) is normal and this patient had no visual complaints. (c) This is a calcified, multinodular, mulberry hamartoma (black arrow). This hamartoma obscures much of the underlying optic nerve head. (Reproduced with permission from Figure 2, Ref. [34].) (d) This transitional hamartoma has features of both mulberry and transparent hamartomas (black arrow). It is a multinodular lesion with a mulberry center and a translucent rim. (Reproduced with permission from Figure 3, Ref. [34].)

epithelium [16]. It is thought that TSC signaling dysfunction results in the faulty migration of developing cells of neural crest origin [12, 17] and thus causes retinal disorganization [15] and formation of the hamartoma. Astrocytic hamartomas may also contain giant cells [15, 18, 19] and osseous metaplasia has been reported [20].

Typically, the hamartomas are located near the optic nerve head or other areas of the posterior retina [10]. They may either envelope or remain superficial to the retinal blood vessels [3]. Larger hamartomas may contain drusen-like, calcified material and undergo cystic degeneration with formation of hyaline deposits [3]. At times, the cystic areas contain serous exudates, blood, or calcified areas [3].

It is said that hamartomas are found in approximately 30–50% of individuals with TSC [1–3, 9, 21, 22]. Of those with hamartomas, 34–50% have bilateral lesions [2, 3, 9, 21]. Table 13.2 summarizes the frequency of retinal hamartomas as compiled by Rowley *et al.* [2] and Au *et al.* [22] with addition of their own cases.

Table 13.2 Frequency of retinal hamartomas.

Geographic location	Number of patients	Number with retinal hamartomas
England	100 [2]	44/100 (44%) [2]
United States	139 [2]	69/139 (49%) [2]
	209 [22]	64/209 (30.6%) [22]
Japan	100 [2]	87/100 (87%) [2]

The frequency of retinal hamartomas reported in a population of patients with TSC may depend upon on the expertise of the examiner, the cooperation of the patients, and the technique of examination [23]. The wide-field, three-dimensional view of the fundus through dilated pupils using the indirect ophthalmoscope would, theoretically, identify more hamartomas than a view through an undilated pupil with a direct ophthalmoscope [3, 9]. The pale hamartoma may be more readily visualized against a dark fundus. Fundus photography is also a useful adjunct to clinical examination [24].

Hamartomas are usually well vascularized [3] and blood flow impairment, both primary and secondary to the hamartoma, may induce vascular changes [3]. Although not commonly used in patients with TSC, retinal angiography, both fluorescein and indocyanine green [25], may facilitate identification of hamartomas that are difficult to visualize fundoscopically [26–29]. Optical coherence tomography (OCT), a technique that produces an *in vivo* cross-sectional view of the retina, has become widely available [14, 15, 30, 31]. Potentially, OCT could assist physicians in identifying and characterizing retinal hamartomas; its application in TSC remains to be fully evaluated. The high acoustic density of these lesions can be demonstrated by B-scan ultrasonography [3, 9, 14, 15, 28].

Three types of retinal hamartomas have been described [2, 3, 9, 21]: (i) the noncalcified lesion (Figure 13.2a and b); (ii) the calcified lesion (Figure 13.2c), and (iii) the transitional lesion with both calcified and noncalcified components (Figure 13.2d). Many patients with hamartomas have more than one variety [3, 9].

Generally, retinal hamartomas remain static over time [2, 3, 21, 33]. No correlations between the presence of retinal hamartomas, age, and type of hamartoma have been established [2, 9]. Recent studies, however, show a correlation between retinal hamartomas, male gender [22], and TSC2 gene mutation [22, 32]. Furthermore, some lesions become more elevated, nodular, and calcified [9, 33], and new lesions can also develop from previously normal appearing retina [28, 33]. Elevated, nodular, and calcified lesions are not restricted to adults as they have been identified in infants only a few months old [3, 9].

13.3.1.1 Noncalcified Hamartomas

The noncalcified hamartoma (Figure 13.2a and b), found in approximately 70% of individuals with hamartomas [2, 3, 9], is the most common type [2, 3, 9, 21]. It is raised, transparent, circumscribed, and oval [3, 9]. Their size ranges from a quarter to two disk diameters [2] and they are usually found at the posterior pole [2, 9, 21]. These lesions are often detected by the appearance of a circular light reflex surrounding the tumor [3, 9]. Against the background of a blonde fundus, these lesions may go unnoticed They are frequently located superficial to a retinal vessel [2] and thus when viewed ophthalmoscopically may appear to interrupt a retinal vessel (Figure 13.2a and d). They are best found by searching along the vessels from the disk to the retinal periphery [3, 9].

13.3.1.2 Calcified Hamartomas

The calcified multinodular lesion (Figure 13.2c), found in approximately half of individuals with hamartomas [2, 3, 9], is the second most common type of hamartoma [2, 9]. The configuration of this elevated lesion resembles mulberries [3, 9]. These are frequently located at or near the margin of the disk but may also be seen in the midperipheral retina [3, 9, 21]. Their size ranges from one half to four disk diameters [2, 3] and may be elevated by as much as 7 D [3, 9]. Occasionally, vessels may course through [3, 9] the nodular lesion (Figure 13.2d).

13.3.1.3 Transitional Hamartomas

The transitional hamartoma (Figure 13.2d), found in approximately 10% of individuals with hamartomas, is the least frequently observed lesion and shares characteristics with the calcified and noncalcified hamartomas [2, 3, 9]. These lesions are translucent at the margins but nodular and calcified at the center [2, 3, 9, 27].

13.3.2 Complications and Treatment of Retinal Hamartomas

Fortunately, retinal hamartomas rarely cause loss of vision [1, 3, 21, 34]. Thus, treatment of a hamartoma is seldom indicated [3]. Some patients, however, have had visual loss associated with complications [35] of hamartomas (Table 13.3). The majority of these complications are secondary to enlargement or hemorrhage of a hamartoma [3, 7, 27, 28, 31, 35–38]. These well-vascularized lesions may hemorrhage, increase in size, or become necrotic. These changes may induce visual loss (Table 13.3). Growth of retinal hamartomas tends to occur in early childhood [35]. Lesions located near the optic disk have a greater tendency to proliferate than the more peripheral lesions [35]. Moreover, most symptomatic alterations are found in noncalcified hamartomas [34]. In extremely rare cases, retinal hamartomas are aggressive lesions that fill the entire globe to perforate the cornea [36] or become necrotic [7]. In these rare cases, enucleation is indicated [31, 35]. Ocular complications, other than those resulting from enlarged hamartomas, are not specific to TSC (Table 13.3).

Should complications of retinal hamartomas occur, watchful waiting is the prevailing recommendation [34]. Visual loss due to subretinal fluid, exudates, retinal detachments, and hamartoma enlargement at the optic disk may spontaneously resolve after a few weeks [39, 40]. If, however, edema, detachment, and visual compromise persist, treatment must be considered [27, 34]. In such a situation, fluorescein angiography can be helpful [3, 26, 40, 41]. If angiography identifies the

Table 13.3 Ocular complications associated with visual loss in TSC.

Ocular features	**Literature cited**													
	[2]	**[3]**	**[7]**	**[9]**	**[27]**	**[28]**	**[31]**	**[34]**	**[35]**	**[37]**	**[38]**	**[39]**	**[41]**	**[42]**
Complications of hamartoma growth														
Vitreous hemorrhage		×		×				×	×					×
Vitreous floaters and seeding by a hamartoma						×							×	
Retinal detachment		×	×				×	×	×			×		
Invasion, obstruction, or atrophy of optic nerve						×	×		×					
Compromise of macula		×			×									
Visual field defect – scotomas		×								×	×			
Neovascular glaucoma leading to enucleation		×							×					
Microophthalmia[a)]			×			×								
Cataracts[a)]	×					×	×							×
Coloboma of the choroids[a)]		×	×	×		×			×					

a) Not specific to the TSC disease process.

site of fluid leakage, laser photocoagulation becomes a consideration [3, 27, 34]. While photocoagulation may promote visual stabilization, it is not a completely effective treatment [3]. Moreover, this treatment method for the hamartoma must be weighed against the collateral damage to the neurosensory retina and consequent visual loss [3, 27, 34]. More recently, photodynamic therapy (PDT) has reduced or eliminated vision threatening hamartomas [27]. Favorable anatomical (Figure 13.3) and functional results hold promise that PDT may become the treatment of choice [27, 34]. Also, while long-standing and recurrent vitreous hemorrhages can spontaneously resolve [41, 42], they can also be treated via a pars plana vitrectomy [34].

13.3.3 Chorioretinal Hypopigmented Lesions

Other retinal findings include punched out, hypopigmented lesions (Figure 13.4a and b) found in 39% of TSC patients [1, 2]. They are typically much less than one disk diameter in size, located in the posterior or midperipheral retina [2, 3, 9, 12], and analogous to the hypopigmented macules of the skin [1, 4, 12]. Larger lesions have also been observed [4]. They may resemble isolated patches of paving-stone (Figure 13.4b) degeneration or exhibit a gray-white plaque-like center that obscures the choriodal vasculature [3, 9, 12]. These lesions are deep to the retinal vessels [3] and are seen in both infants and older individuals with TSC. An increase in the number of these chorioretinal lesions with age has not, to our knowledge, been documented [2]. Hyperpigmented, punctate lesions (probably congenital retinal pigment epithelial

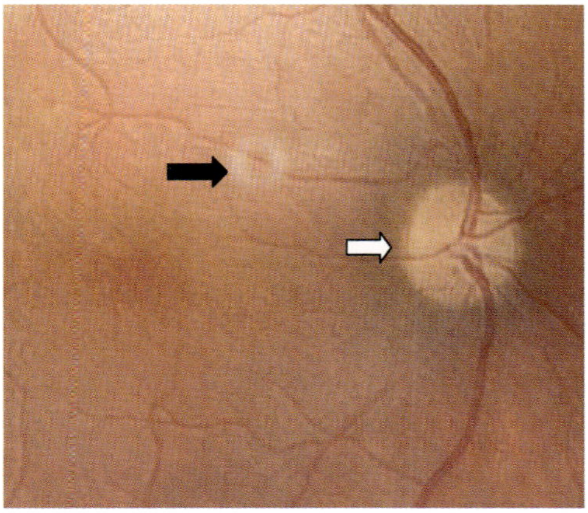

Figure 13.3 Photodynamic therapy for a TSC retina hamartoma. (a) This is a fundus photograph after PDT of a hamartoma (black arrow) with edema and lipid exudates. The hamartoma seen to the left of the optic nerve (white arrow) disappeared 2 weeks after PDT with attendant improvement in visual acuity. No complications were noted during a follow-up of 4 years. (Reproduced with permission from Figure 5, Ref. [34].)

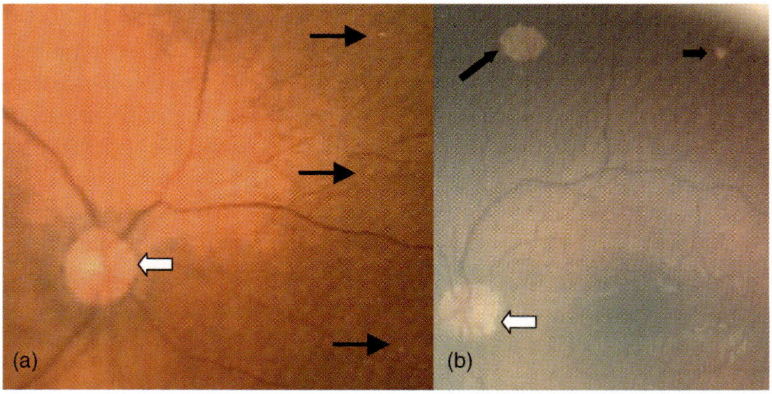

Figure 13.4 Chorioretinal hypopigmented lesions in TSC. (a) This view of the right fundus of a child with TSC shows numerous tiny chorioretinal hypopigmented lesions (black arrows). The chorioretinal hypopigmented lesions are less than one sixteenth of the normal optic nerve head (white arrow). (b) A one half by three quarters disk diameter chorioretinal hypopigmented lesion (black arrow) is shown here along side a much smaller lesion (short black arrow). The optic nerve head (white arrow) is normal.

hypertrophy) have also been reported in TSC but it is unclear as to whether these are specific to the TSC disease process [3, 9].

13.3.4 Differential Diagnosis

The retinal lesions of TSC bear some resemblance to those of neurofibromatosis-1 (NF-1) [43, 44], retinitis pigmentosa, retinoblastoma [28, 35, 44], and other malignant tumors [35, 45, 46]. Although a hamartoma may be the only recognizable feature in a patient with TSC, in practice, the most common differentiator would be the presence of other features of TSC. In the context of adenoma sebaceum, hypomelanotic macules, ungual fibromas, shagreen patches in the skin, or brain lesions typical of TSC, one would be unlikely to confuse hamartomas with the other entities [3, 13]. Retinoblastoma often presents as a white pupil in an infant or a young child; it commonly interferes with vision and produces crossed eyes. In neurofibromatosis-1, retinal lesions are uncommon but axillary freckling and lisch nodules of the irides are present in most patients with NF-1 who are older than 5 years. Imaging of the brain may show neurofibromatosis spots and glioma rather than the lesions characteristic of TSC.

The punched out lesions of TSC have been, at times, mistaken for postinflammatory lesions or even retinal degeneration. For cases in which the underlying diagnosis of TSC has not been recognized, hamartomas may be mistakenly diagnosed as intraocular malignancies such as retinoblastoma in children [35] or uveal melanomas in adults [45]. A complete history, associated physical findings, and genetic testing will help the physician secure the diagnosis of TSC and exclude the others.

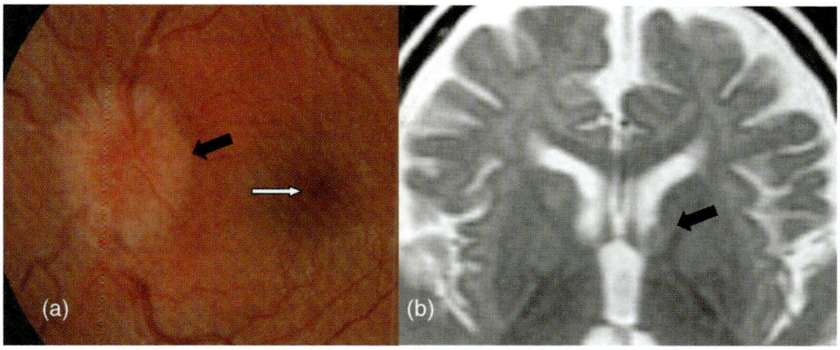

Figure 13.5 Papilledema secondary to an enlarging subependymal giant cell astrocytoma. (a) Papilledema (black arrow) appeared secondary to obstructive hydrocephalus caused by a subependymal giant cell astrocytoma that had been identified as a subependymal nodule on a scan from 3 years earlier. (b) An MRI of a 4-year-old patient shows an enhancing lesion at the left foramen of Munro (arrow) consistent with the subependymal nodule. The lesion appeared solid and did not enhance with gadolinium. At the time of this image, there was no obstructive hydrocephalus and no papilledema.

Intracranial radiologic findings also help establish the correct diagnosis [28]. Other differential diagnoses include choroiditis, Coats' disease, myelinated nerve fibers, and retinal detachment [28].

13.4 Papilledema

Papilledema (Figure 13.5a), which is swelling of the optic nerve head, can result from intracranial hypertension [3, 9] secondary to ventricular obstruction (Figure 13.5b) by a subependymal giant cell astrocytoma (see also Chapter 11) [47, 48]. Papilledema is typically bilateral [3]. While this is a rare occurrence, it is important to remain vigilant for this complication as prompt neurosurgical intervention is indicated. If the papilledema is not relieved, atrophy of the optic nerve and visual loss may ensue [3].

13.5 Visual Field Defects

A field defect can be the consequence of hemispherectomy for treatment of intractable seizures (Figure 13.6). Some medications may also affect visual function. Vigabatrin is an antiepileptic medication, effective in treatment of infantile spasms and partial complex seizures and thus valuable in the management of patients with TSC [49, 50]. Vigabatrin causes significant, irreversible constriction of the visual field [51] in a quarter to a third of those who have taken it, whether or not they have

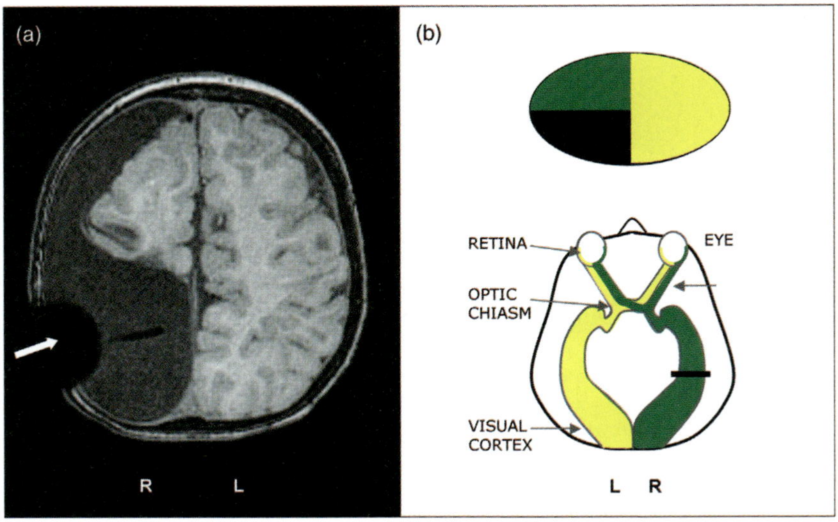

Figure 13.6 Visual field defect secondary to resection of seizure focus involving the occipital lobe. (a) This MRI shows an image of the brain after resection of a large tuber in the right parietooccipital region (white arrow). The white arrow indicates the surgical site and drainage catheter. (b) The resection (black hatch) led to a homonymous inferior visual field defect in each eye. Such a defect is present even with binocular viewing. The visual field defect corresponds to the loss of tissue in the brain (black hatch). (Modified from Ref. [61].)

TSC [51–53]. Vigabatrin-related visual field loss appears to be detectable in young children using procedures developed for pediatric perimetry [53, 54].

Visual field defects may interfere with visuomotor skills such as scanning print and other images and safe ambulation. If visual field defects are identified, orientation and mobility evaluation and instruction are advisable.

13.6 Cerebral Visual Impairment

The central visual processing is often altered in individuals with severe seizures or brain lesions. Difficulties with visual processing may be the consequence of damage to the optic radiations or visual cortex. The frequency of cerebral visual impairment (CVI) in individuals with TSC is not established, perhaps in part because understanding and definitions of CVI have been evolving. A current accepted definition of CVI is visual dysfunction in the presence of otherwise normal eyes and normal pupillary responses [55]. In children with CVI, modest developmental improvement of visual acuity may occur [56–59]. Accordingly, periodic reassessment of the eyes and vision is indicated to characterize vision so as to inform visual support and education services as well as to determine if any ophthalmic interventions are indicated.

13.7 Common Ophthalmic Issues

There are a number of common ophthalmic issues that arise in the course of care of an individual with TSC. These include refractive errors, strabismus, and amblyopia (lazy eye).

13.7.1 Refractive Error

Significant refractive error, whether myopia, hyperopia, anisometropia, or astigmatism, occurs at about the same frequency in individuals with TSC as in the general population [2].

13.7.2 Strabismus and Amblyopia

Strabismus, that is, misalignment of the eyes, is an uncommon finding in individuals with TSC [2, 3, 58]. Exotropia, turning out of the eyes, is slightly more frequent than esotropia, crossed eyes [2, 60]. Strabismus does not necessarily mean that vision is poor. Uncorrected strabismus in a child, however, can lead to poor vision through the development of a fixation preference with the non-preferred eye becoming amblyopic (lazy). Therefore, the presence of strabismus in a child younger than 10 years is an indication for prompt referral to an ophthalmologist [61].

During infancy and childhood, the child is at risk for amblyopia if there is strabismus or anisometropia (unequal refraction in the right versus left eye). Development of the visual system is in progress during these years and it is also during this time that amblyopia treatment must be applied in order to be effective; the most common treatment is patching of the preferred eye. In those with significant refractive errors [62], glasses are recommended to achieve the best possible visual outcome. As the child ages, the plasticity of the visual system decreases and responsiveness to patching becomes gradually less effective.

13.8 Summary and Recommendations

Ophthalmic examinations may contribute to the initial diagnosis of TSC [10]. After the diagnosis is secured, periodic eye examination for surveillance of the common and uncommon ocular and visual problems is essential [1, 10]. Ideally, the initial examination would be close to the time of diagnosis. During infancy, when ocular and visual development is rapid, examination at 6-month intervals is advisable to monitor for departures from the normal course of visual development and to provide intervention should such be indicated. Thereafter, for pediatric patients reassessment on an annual basis is typically appropriate for early identification of ocular and visual issues requiring intervention. Should ophthalmic problems such as those detailed above be identified, the recommendations for care and follow-up must be tailored to that individual. This comprehensive approach assures the best possible visual outcome in individuals with TSC.

References

1 Franz, D.N. (2004) Non-neurologic manifestations of tuberous sclerosis complex. *J. Child Neurol.*, **19**, 690–698.
2 Rowley, S.A., O'Callaghan, F.J., and Osborne, J.P. (2001) Ophthalmic manifestations of tuberous sclerosis: a population based study. *Br. J. Ophthalmol.*, **85**, 420–423.
3 Robertson, D.M. (1991) Ophthalmic manifestations of tuberous sclerosis. *Ann. N.Y. Acad. Sci.*, **615**, 17–25.
4 Awan, K.J. (1982) Leaf-shaped lesions of ocular fundus and white eyelashes in tuberous sclerosis. *South Med. J.*, **75**, 227–228.
5 Mencia-Gutierrez, E. (2004) Eyelid and cutaneous lesions as the sole indicators of tuberous sclerosis. *Arch. Soc. Esp. Oftalmol.*, **79**, 401–404.
6 Mawn, L.A., Jordan, D.R., Nerad, J., Anderson, R., and Jabi, M. (1999) Giant cell angiofibroma of the eyelids: an unusual presentation of tuberous sclerosis. *Ophthalmic. Surg. Lasers*, **30**, 320–322.
7 Eagle, R.C., Jr., Shields, J.A., Shields, C.L., and Wood, M.G. (2000) Hamartomas of the iris and ciliary epithelium in tuberous sclerosis complex. *Arch. Ophthalmol.*, **118**, 711–715.
8 Kwiatkowski, D.J. and Short, M.P. (1994) Tuberous sclerosis. *Arch. Dermatol.*, **130**, 348–354.
9 Nyboer, J.H., Robertson, D.M., and Gomez, M.R. (1976) Retinal lesions in tuberous sclerosis. *Arch. Ophthalmol.*, **94**, 1277–1280.
10 Yassur, Y., Melamed, S., and Ben-Sira, I. (1977) Retinal involvement in tuberous sclerosis. *J. Pediatr. Ophthalmol.*, **14**, 379–381.
11 Grover, W.D. and Harley, R.D. (1969) Early recognition of tuberous sclerosis by funduscopic examination. *J. Pediatr.*, **75**, 991–995.
12 Lucchese, N.J. and Goldberg, M.F. (1981) Iris and fundus pigmentary changes in tuberous sclerosis. *J. Pediatr. Ophthalmol. Strabismus*, **18**, 45–46.
13 Shami, M.J., Benedict, W.L., and Myers, M. (1993) Early manifestation of retinal hamartomas in tuberous sclerosis. *Am. J. Ophthalmol.*, **115**, 539–540.
14 Soliman, W., Larsen, M., Sander, B., Wegener, M., and Milea, D. (2007) Optical coherence tomography of astrocytic hamartomas in tuberous sclerosis. *Acta Ophthalmol. Scand.*, **85**, 454–455.
15 Shields, C.L., Benevides, R., Marterin, M.A., and Shields, J.A. (2006) Optical coherence tomography of retinal astrocytic hamartoma in 15 cases. *Ophthalmology*, **113**, 1553–1557.
16 Paoli, D. (1998) Retinal astrocytoma associated with hypertrophy of the retinal pigment epithelium: clinical report and follow-up. *Ophthalmologica*, **1**, 71–73.
17 Knox, S., Hong, G., Dimitroff, B.D., Ren, Y., Howe, K.A. *et al.* (2007) Mechanism of TSC-mediated control of synapse assembly and axon guidance. *PLoS ONE*, **2**, 375.
18 Jay, V. (1996) Brain and eye pathology in an infant with down syndrome and tuberous sclerosis. *Pediatr. Neurol.*, **15**, 57–59.
19 Kumar, R. and Singh, V. (2004) Subependymal giant cell astrocytoma: a report of five cases. *Neurosurg. Rev.*, **27**, 274–280.
20 Tay, A., Scheithauer, B.W., Cameron, J.D., Myhre, M.J., and Boerner, M.J. (2009) Retinal ependymoma: an immunohistologic and ultrastructural study. *Hum. Pathol.*, **40**, 578–583.
21 Rosser, T., Panigrahy, A., and McClintock, W. (2006) The diverse clinical manifestations of tuberous sclerosis complex: a review. *Semin. Pediatr. Neurol.*, **13**, 27–36.
22 Au, K.S., Williams, A.T., Roach, E.S., Batchelor, L., Sparagana, S.P.,

Delgado, M.R., Wheless, J.W., Baumgartner, J.E., Roa, B.B., Wilson, C.M., Smith-Knuppel, T.K., Cheung, M.Y., Whittemore, V.H., King, T.M., and Northrup, H. (2007) Genotype/phenotype correlation in 325 individuals referred for a diagnosis of tuberous sclerosis complex in the United States. *Genet. Med.*, **9**, 88–100.

23 Roach, E.S. and Sparagana, S.P. (2004) Diagnosis of tuberous sclerosis complex. *J. Child Neurol.*, **19**, 643–649.

24 Baird, H.W., Pileggi, A.J., and Harley, R.D. (1969) Funduscopic photography and fluorescein angioretinography in evaluation of children with neurologic handicaps. *J. Pediatr.*, **74**, 937–945.

25 Koak, N., Saatci, A.O., Kaynak, S., Ergin, M.H., and Ingil, G.C. (2003) Indocyanine green angiography of retinal astrocytomas associated with tuberous sclerosis. *Korean J. Ophthalmol.*, **17**, 145–148.

26 Mennel, S., Meyer, C.H., Eggarter, F., and Peter, S. (2005) Autofluorescence and angiographic findings of retinal astrocytic hamartomas in tuberous sclerosis. *Ophthalmologica*, **219**, 350–356.

27 Mennel, S., Hausmann, N., Meyer, C.H., and Peter, S. (2006) Photodynamic therapy for exudative hamartoma in tuberous sclerosis. *Arch. Ophthalmol.*, **124**, 597–599.

28 Mullaney, P.B., Jacquemin, C., Abboud, E., and Karcioglu, Z.A. (1997) Tuberous sclerosis in infancy. *J. Pediatr. Ophthalmol. Strabismus*, **34**, 372–375.

29 Ahmad, A. (1978) Fluorescein angiography in tuberous sclerosis: case report. *Ann. Ophthalmol.*, **10** (4), 453–455.

30 McKeown, C.A. and Davidson, S.L. (2008) The pediatric eye examination, in *Principles and Practice of Ophthalmology* (eds D.G. Hunter and M.D. Mills), Elsevier, Philadelphia, PA, pp. 4142.

31 Milot, J., Michaud, J., Lemieux, N., Allaire, G., and Gagnon, M.M. (1998) Persistent hyperplastic primary vitreous with retinal tumor in tuberous sclerosis: report of a case including tumoral immunohistochemistry and cytogenetic analyses. *Ophthalmology*, **106** (3), 630–634.

32 Dabora, S.L., Joswiak, S., Franz, D.N., Roberts, P.S., Nieto, A., Chung, J., Choy, Y.S., Reeve, M.P., Thiele, E., Egelhoff, J.C., Kasprzyk-Obara, J., Domanska-Pakiela, D., and Kwiatkowski, D.J. (2001) Mutational analysis in a cohort of 224 tuberous sclerosis patients indicates increased severity of TSC2, compared with TSC1, disease in multiple organs. *Am. J. Hum. Genet.*, **68**, 64–80.

33 Zimmer-Galler, I.E. and Robertson, D.M. (1995) Long-term observation of retinal lesions in tuberous sclerosis. *Am. J. Ophthalmol.*, **119**, 318–324.

34 Mennel, S., Meyer, C., Peter, S., Schmidt, J.C., and Kroll, P. (2007) Current treatment modalities for exudative retinal hamartomas secondary to tuberous sclerosis: review of the literature. *Acta Ophthalmol. Scand*, **85**, 127–132.

35 Shields, J.A., Eagle, R.C., Jr., Shields, C.L., and Marr, B.P. (2004) Aggressive retinal astrocytomas in four patients with tuberous sclerosis complex. *Trans. Am. Ophthalmol. Soc.*, **102**, 139–147, discussion 147–138.

36 Ulbright, T.M., Fulling, K.H., and Helveston, E.M. (1984) Astrocytic tumors of the retina. Differentiation of sporadic tumors from phakomatosis-associated tumors. *Arch. Pathol. Lab. Med.*, **108** (2), 160–163.

37 Prompitak, A., Maberley, A.L., and Shea, M. (1973) An abortive case of tuberous sclerosis without mental deficiency of epilepsy in an adult. *Am. J. Ophthalmol.*, **76** (2), 255–259.

38 van Dorp, D.B. and Kwee, M.L. (1990) Tuberous sclerosis. Diagnostic problems in a family. *Ophthalmic Paediatr Genet*, **11** (2), 95–101.

39 Lee, S.J., Kim, Y.H., Lee, J.H., Rah, S.H., Shin, H.W., and Han, B.K. (2002) Development of parafoveal exudates and serous retinal detachment in a pregnant woman with tuberous sclerosis. *Gynecol. Obstet. Invest.*, **53**, 188–190.

40 Brodsky, M.C. and Safar, A.N. (2007) Optic disc tuber. *Arch. Ophthalmol.*, **125**, 710–712.

41 Jost, B.F. and Olk, R.J. (1986) Atypical retinitis proliferans, retinal telangiectasis,

and vitreous hemorrhage in a patient with tuberous sclerosis. *Retina*, **6** (1), 53–56.

42 Atkinson, A., Sanders, M.D., and Wong, V. (1973) Vitreous haemorrhage in tuberous sclerosis. Report of two cases. *Br. J. Ophthalmol.*, **57** (10), 773–779.

43 Turek, M., Raistrick, E.R., and Hart, C.D. (1977) Retinal tumours in neurofibromatosis. *Can. J. Ophthalmol.*, **12** (1), 68–70.

44 Font, R.L. and Ferry, A.P. (1977) Retinal tumours in neurofibromatosis. *Can. J. Ophthalmol.*, **12** (1), 68–70.

45 Margo, C.E., Barletta, J.P., and Staman, J.A. (1993) Giant cell astrocytoma of the retina in tuberous sclerosis. *Retina*, **13**, 155–159.

46 Ganley, J.P. and Streeten, B.W. (1971) Glial nodules of the inner retina. *Am. J. Ophthalmol.*, **71** (5), 1099–1103.

47 Goh, S., Butler, W., and Thiele, E.A. (2004) Subependymal giant cell tumors in tuberous sclerosis complex. *Neurology*, **63**, 1457–1461.

48 Chong, D.Y., Hirunwiwatkul, P., McKeever, P.E., and Trobe, J.D. (2007) Papilledema in obstructive hydrocephalus caused by giant cell astrocytoma of tuberous sclerosis. *J. Neuroophthalmol.*, **27**, 50–54.

49 Chiron, C., Dulac, O., Luna, D., Palacios, L., Mondragon, S., Beaumont, D., and Mumford, J.P. (1990) Vigabatrin in infantile spasms. *Lancet*, **335**, 363–364.

50 Hancock, E.C., Osborne, J.P., and Edwards, S.W. (2008) Treatment of infantile spasms. *Cochrane Database Syst. Rev.*, Issue 4, Art. No.: CD001770. DOI: 10.1002/14651858. CD001770. pub2.

51 Johnson, M.A., Krauss, G.L., Miller, N.R., Medura, M., and Paul, S.R. (2000) Visual function loss from vigabatrin: effect of stopping the drug. *Neurology*, **55**, 40–45.

52 Krauss, G.L., Johnson, M.A., Sheth, S., and Miller, N.R. (2003) A controlled study comparing visual function in patients treated with vigabatrin and tiagabine. *J. Neurol. Neurosurg. Psychiatry*, **74**, 339–343.

53 Agrawal, S., Mayer, D.L., Hansen, R.M., and Fulton, A.B. (2009) Visual fields in young children treated with vigabatrin. *Optom. Vis. Sci.*, **86**, 767–773.

54 Wilson, M., Quinn, G., Dobson, V., and Breton, M. (1991) Normative values for visual fields in 4- to 12-year-old children using kinetic perimetry. *J. Pediatr. Ophthalmol. Strabismus*, **28**, 151–241.

55 Hoyt, C. and Fredrick, D. (1998) Cortically visually impaired children: a need for more study. *Br. J Ophthalmol.*, **82**, 1225–1227.

56 Lim, M., Soul, J.S., Hansen, R.M., Mayer, D.L., Moskowitz, A., and Fulton, A.B. (2005) Development of visual acuity in children with cerebral visual impairment. *Arch. Ophthalmol.*, **123**, 1215–1220.

57 Mayer, D.L. and Fulton, A.B. (2004) Visual fields, in *Pediatric Ophthalmology and Strabismus* (eds D. Taylor and C.S. Hoyt), Elsevier Saunders, New York, pp. 78–86.

58 Mayer, D.L. and Fulton, A.B. (2005) Perspectives on cerebral visual impairment (CVI), in *Summit on Cerebral/Cortical Visual Impairment: Education, Family and Medical Perspectives* (eds E. Dennison and A.H. Lueck), AFB Press, San Francisco, CA, pp. 65–75.

59 Roman-Lantzy, C.A. (2007) *Cortical Visual Impairment: An Approach to Assessment and Intervention*, ABF Press.

60 Shelton, R.W. (1975) The incidence of ocular lesions in tuberous sclerosis. *Ann. Ophthalmol.*, **7** (6), 771–774.

61 Fulton, A.B., Mayer, D.L., Miller, K.B., and Hansen, R.M. (2006) Eye and vision care, in *Medical Care for Children & Adults with Developmental Disabilities* (eds I.L. Rubin and A.C. Crocker), Paul H. Brookes Publishing Co., Inc., Baltimore, MD, pp. 343–352.

62 Mayer, D.L., Hansen, R.M., Moore, B.D., Kim, S., and Fulton, A.B. (2001) Cycloplegic refractions in healthy children aged 1 through 48 months. *Arch. Ophthalmol.*, **119** (11), 1625–1628.

63 Abel, A., Brockbank, D.T., Farber, M., and Meyer, D.R. (2004) Bony hamartoma of the inferior orbital rim in a patient with tuberous sclerosis. *Arch. Ophthalmol.*, **122**, 780–782.

14
Dermatologic Manifestations of Tuberous Sclerosis Complex (TSC)

Thomas N. Darling, Joel Moss, and Mark Mausner

14.1
Introduction

Skin lesions were part of the initial descriptions of tuberous sclerosis complex (TSC) over 100 years ago, and they continue to be important for recognition and diagnosis. They occur in nearly all individuals with TSC, can be a major concern for patients, and may be a presenting sign. TSC skin lesions are benign, but may be painful, bleed spontaneously, distort normal skin structures, or compromise normal functions of nails, eyes, or nasal passages.

TSC skin lesions include hypomelanotic macules, "confetti" macules, facial angiofibromas, forehead plaques, shagreen patches, and ungual fibromas. Each has a characteristic appearance and age of onset that helps in the differential diagnosis of TSC, since they comprise many of the major and minor features in the diagnostic criteria (see Chapter 3).

14.2
Types of TSC Skin Lesions

14.2.1
Hypomelanotic Macules

Hypomelanotic macules (white spots) are observed in nearly all TSC patients [1–8]. They are evident at birth or during early infancy [1, 2, 5], and were a presenting sign in 6 out of 41 infants diagnosed with TSC before age of 1 year [9]. Hypomelanotic macules, recently termed "Fitzpatrick patches" in honor of their describer Professor Thomas Fitzpatrick [10], are off-white in color and usually polygonal or oval in shape (Figure 14.1a–c). Some hypomelanotic macules are oval at one end and pointed at the other. These lesions are called "ash leaf spots" because they resemble the leaf of the European mountain ash [2]. Hypomelanotic macules may number from 1 to over 20, and can be located anywhere on the body. Most hypopigmented macules measure about 0.5–3 cm in diameter, but large segmental lesions can also occur. Hypopig-

Tuberous Sclerosis Complex: Genes, Clinical Features, and Therapeutics
Edited by David J. Kwiatkowski, Vicky Holets Whittemore, and Elizabeth A. Thiele

ISBN: 978-3-527-32201-5

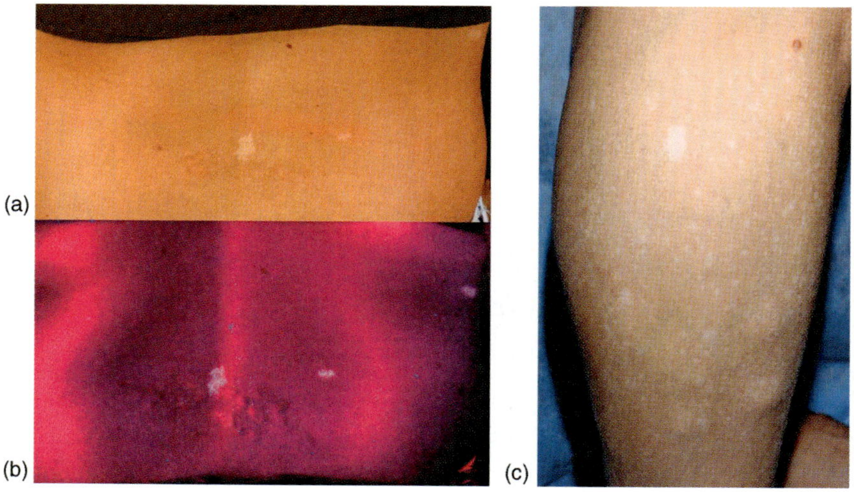

Figure 14.1 Hypomelanotic macules in TSC. (a) Hypomelanotic macules (and a shagreen patch) on the back. (b) Accentuation of the same lesions by Wood lamp illumination. (c) Hypomelanotic macule and confetti macules.

mented macules on the scalp, eyebrows, or eyelashes cause poliosis (a patch of white hair) [1–8].

Some patients have numerous 1–3 mm diameter hypopigmented macules scattered over regions of the body, often the arms or legs. This type of hypopigmentation, termed "confetti," is observed in about 30% of patients [3]. It may occur alone or together with larger hypopigmented macules [2, 4, 5].

Hypomelanotic macules are usually stable for many years. In adulthood, they may slowly fade or even disappear [3]. They may be prone to sunburn due to the lack of pigment, and the patient should use good sun protection (e.g., a broad-spectrum sunscreen protecting against both UVA and UVB, with a sun protection factor of greater than 15). Sunscreen should be applied to all exposed skin, since reduced tanning of normal skin will decrease the contrast with hypopigmented skin. Cosmetic products (e.g., Dermablend® and Covermark®) can be used to temporarily conceal hypomelanotic macules. Topical stains and self-tanning products may be used on the hypopigmented macules, but these can be difficult to match to the surrounding skin color [11].

Hypomelanotic macules are usually diagnosed clinically. To improve their detection in lightly pigmented individuals, one may use the ultraviolet light of a Wood lamp to accentuate pigment contrast. The user should use the lamp in a dark room and allow his or her eyes to adjust. A biopsy is rarely needed for diagnosis. Hypomelanotic macules in TSC have normal number of melanocytes but they have poorly developed dendritic processes, and melanosomes are decreased in number, size, and melanization [4].

The differential diagnosis of TSC hypomelanotic macules includes nevus depigmentosus, nevus anemicus, piebaldism, and vitiligo [4, 12]. Nevus depigmentosus is clinically and histologically similar to TSC hypomelanotic macules except that they are solitary lesions and occur in the general population. Although named nevus depigmentosus, lesions are hypopigmented, not depigmented. Nevus anemicus, which may appear similar to nevus depigmentosus, is caused by altered vascularity, not pigmentation. It is distinguished from a hypopigmented lesion by applying pressure with a glass slide. In nevus anemicus, the border of the nevus disappears, while hypopigmented macules are still apparent. Piebaldism is a congenital disorder with depigmented patches on the ventral trunk and central forehead, often accompanied by a white forelock. In piebaldism, the depigmented patches usually contain small islands of pigmented skin within. Vitiligo is acquired and usually progressive, with expanding and new lesions appearing over time. Lesions in piebaldism and vitiligo are completely depigmented (white) rather than hypopigmented (off-white) as in TSC [12]. Histologically, vitiligo and piebaldism lack melanocytes [4].

The differential diagnosis of confetti macules includes postinflammatory, postinfectious, and posttraumatic hypopigmentation, sarcoidosis, and mycosis fungoides. In adults, confetti macules must be distinguished from idiopathic guttate hypomelanosis, which is limited to sun-exposed areas and appears in adulthood rather than infancy [4].

14.2.2
Facial Angiofibromas

Angiofibromas are slightly less common than hypomelanotic macules, but they are most emblematic of TSC because of their prominence on the face. Angiofibromas had been called adenoma sebaceum, a misnomer since they are neither adenomas nor derived from sebaceous glands [13]. Angiofibromas generally begin to appear at about 2–5 years of age and eventually affect about 75–90% of patients [1, 3, 5, 6]. They are typically 1–4 mm diameter papules with a smooth surface and a color that varies from skin color to red to reddish brown. Their number ranges from 1 to over 100. They predominate on the central face, especially involving the nasolabial folds and extending symmetrically onto the cheeks, nose, and chin (Figure 14.2a–c). They are usually dome-shaped but may be pedunculated or form larger nodules or clusters in the nasolabial folds [3, 5]. Angiofibromas may occur on the forehead, scalp, or eyelids, with relative sparing of the upper lip and lateral face. Some angiofibromas have telangiectasias, or patients may have a ruddy complexion with multiple telangiectasias scattered among papules. In unusual cases, angiofibromas are unilateral, suggesting a segmental or mosaic defect [14].

The development of angiofibromas may be preceded by mild erythema, which is intensified by emotion or heat [1, 3]. Angiofibromas tend to grow progressively until adulthood, with reports of increased growth during puberty [5, 15]. During adulthood, they tend to be stable in size and redness may gradually diminish. Angiofibromas may present psychosocial problems, and can bleed spontaneously or in response to mild trauma. In rare instances, their size may obstruct vision or nasal passages [16]. As detailed later, angiofibromas are usually treated surgically. Measures should be taken to avoid traumatizing the lesions, such as not shaving in areas with angiofibromas.

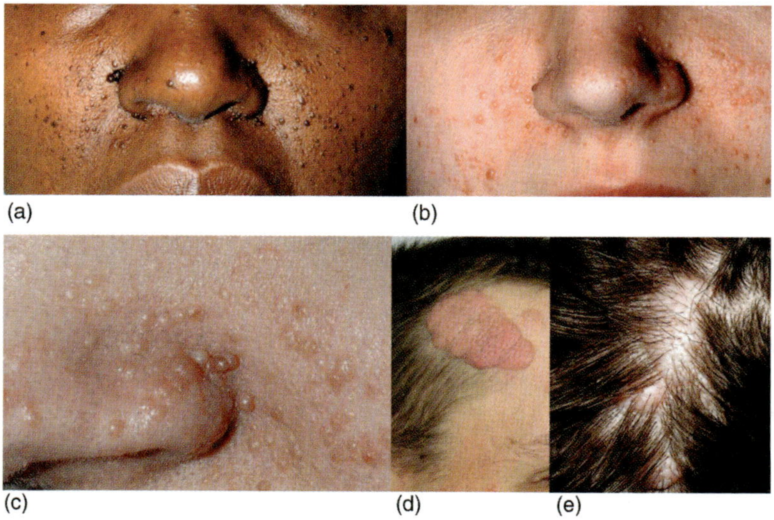

Figure 14.2 Angiofibromas and forehead plaque in TSC. (a, b) Typical distribution of angiofibromas. (c) Clustering of angiofibromas in nasolabial folds. (d) Forehead plaque. (e) Scalp plaque with slight decrease in hair density.

When the clinical diagnosis is in doubt, a shave biopsy will show plump, spindle-shaped or stellate fibroblastic cells in the dermis among increased number of dilated vessels. Collagen fibers are oriented in onion-skin pattern around follicles and vessels [13, 15, 17].

Angiofibromas were once considered pathognomonic for TSC. It is now known that angiofibromas occur in other syndromes, including multiple endocrine neoplasia type 1 (MEN1) [18–21] and as an unusual feature of Birt–Hogg–Dube syndrome [22]. Also, angiofibromas occur sporadically in the general population, but these "fibrous papules" are solitary lesions [23].

The differential diagnosis of angiofibromas includes benign skin tumors that are observed in several rare syndromes. Skin lesions in multiple familial trichoepithelioma appear as small skin-colored papules on the face. Lesions may be located in the nasolabial folds, but they are also found on the upper lip, forehead, eyelids, scalp, and ears [24]. They lack the redness and telangiectasias typical of angiofibromas. In Birt–Hogg–Dube syndrome, patients have multiple gray-white fibrofolliculomas and trichodiscomas that tend to occur on the nose, cheeks, neck, and upper trunk [25, 26]. Patients with Cowden syndrome have multiple facial tricholemmomas (also called trichilemmomas) but these skin-colored papules have a rough surface compared to the smooth surface of angiofibromas [25]. Multiple familial trichoepithelioma, Birt–Hogg–Dube syndrome, and Cowden syndrome may be distinguished from TSC by these differences in clinical appearance, a tendency for a later age of onset of skin lesions, and characteristic histological findings. Moreover, they lack constellation of other TSC-associated skin and internal findings.

Clinicians may mistake angiofibromas for common skin lesions (e.g., acne, warts, moles). Patients with acne vulgaris or acne rosacea have red inflammatory papules and these are transient and accompanied by comedones or pustules. Small angiofibromas may appear similar to flat warts, but these pink flat-topped papules are often on the lateral face without a predilection for the nasolabial folds, and may occur in a linear array. Dermal melanocytic nevi have the permanence of angiofibromas but lack telangiectasias and do not tend to cluster in the nasolabial folds. Syringomas are multiple small skin-colored papules that tend to form on or around the lower eyelids in adults. Lupus pernio is a papular form of sarcoidosis that involves the nose, but these lesions are more violaceous and onset is in adulthood.

14.2.3 Forehead Plaques

Forehead plaques, also referred to as fibrous facial plaques, may be congenital or show gradual development over the first 10 years of life [5, 15, 27]. They are present in 20–40% of TSC patients [3, 5, 6] and may be a presenting feature [5, 27]. Forehead plaques are irregular, rubbery to firm, connective tissue nevi with variable color, ranging from skin colored to pink to yellowish brown (Figure 14.2d), or dark brown in darkly pigmented individuals. Although forehead plaques are commonly located unilaterally on the forehead, they can also be found on the scalp (Figure 14.2e), cheeks, and elsewhere on the face. Fibrous facial plaques can be solitary or multiple. They arise as plaques and not from confluence of papules. Histologically, they resemble angiofibromas [13, 15]. Larger and more protuberant scalp plaques can be more fibromatous than forehead plaques and are referred to as scalp fibromas.

The forehead plaque is generally stable in area but becomes thicker over time [5]. It may be treated surgically, or patients may cover forehead plaques with their hair style.

The differential diagnosis for a forehead plaque includes nevus sebaceous, a congenital plaque that can occur on the face and/or scalp. Nevus sebaceous is distinguished from fibrous facial plaques because of its yellowish color, but confusion with TSC may arise since patients with linear nevus sebaceous syndrome have extracutaneous findings including seizures and cognitive disability [28]. Another feature that discriminates nevus sebaceous from the fibrous plaque is that scalp lesions of nevus sebaceous cause balding within the lesion, whereas TSC scalp plaques show a variable decrease in hair density [3]. Other skin lesions occurring on the forehead include lipomas and cysts, both of which are deeper in the skin, without the mammillated surface or follicular accentuation of the TSC forehead plaque. Granuloma faciale, angiosarcoma, and cutaneous metastasis can occur on the forehead, but these are more violaceous in color and have later ages of onset.

14.2.4 Shagreen Patch

The shagreen patch is observed in about 50% of patients [1, 3, 5, 6]. It can be present at birth but usually becomes apparent during the first 10 years of life [1, 3, 5]. It is a firm or

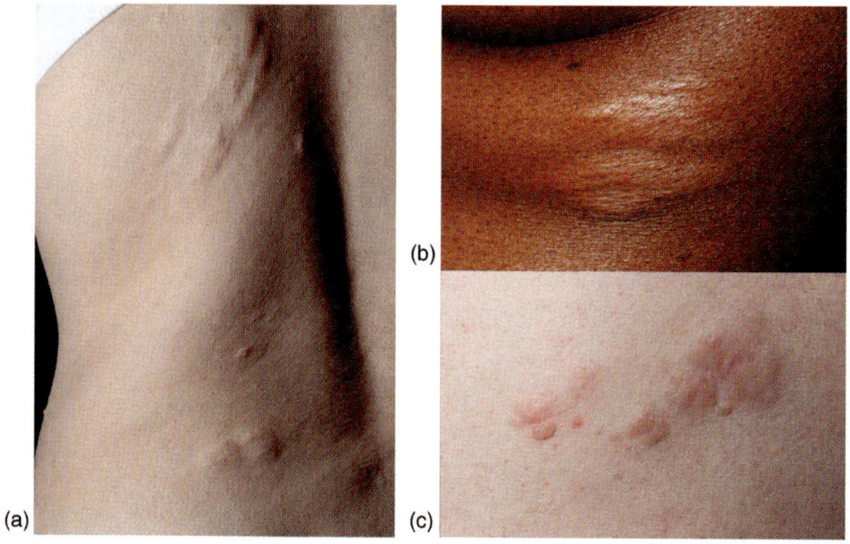

Figure 14.3 Shagreen patch in TSC. (a) Shagreen patch accentuated with side lighting. (b) Shagreen patch with prominence of follicular openings. (c) Shagreen patch borders may vary from sharply demarcated to indistinct.

rubbery irregular plaque ranging in size from 1 to 10 cm (Figure 14.3a–c). The surface appears bumpy with coalescing papules and nodules, and it may have the surface appearance of an orange peel. The color may be that of the surrounding skin, or it may be slightly pink or brown. It can be single or multiple, sometimes with one large lesion and scattered smaller lesions, or there may be scattered smaller oval papules without a larger plaque. Shagreen patches are most commonly distributed asymmetrically on the lower back and, less frequently, on the upper back, buttocks, or thighs [3, 5, 15].

The significance of the shagreen patch is mostly diagnostic, since it is generally asymptomatic and of less cosmetic concern than angiofibromas. It is usually diagnosed clinically. Histologically, the shagreen patch shows sclerotic bundles of collagen in the reticular dermis with reduced elastic fibers. The microscopic appearance may be mistaken for normal skin [15].

The plaques of Buschke–Ollendorff syndrome (dermatofibrosis lenticularis disseminate) can appear similar to the shagreen patch. The lesions of Buschke–Ollendorff syndrome can occur anywhere on the body and usually contain increased elastic fibers, so a skin biopsy may be diagnostic [29]. Connective tissue nevi with increased collagen are observed in other syndromes, such as MEN1, Birt–Hogg–Dube syndrome, and Cowden syndrome. Collagenomas in MEN1 appear as multiple skin-colored to whitish, dome-shaped papules or nodules on the trunk and proximal extremities, without a propensity for the lower back [18, 19]. Collagenomas in Birt–Hogg–Dube syndrome can appear similar to the collagenomas in MEN1 [26] or rarely may appear similar to a shagreen patch, as reported in one patient [30]. The multiple storiform collagenomas (sclerotic fibromas) observed in Cowden syndrome are histologically distinctive because they are well circumscribed and have prominent clefts between collagen bundles, which are arranged in a storiform or whorled pattern [31]. Other conditions with connective tissue nevi include familial cutaneous collagenoma and eruptive collagenoma. These plaques are typically oval in shape and distributed on the trunk or extremities without a predilection for the lower back as in TSC [32].

14.2.5 Ungual Fibromas

Ungual fibromas, also called Koenen tumors, usually appear later than other TSC-associated skin lesions, typically after the first decade and even as late as the fifth decade [3, 5]. Though infrequent in childhood, ungual fibromas eventually affect up to 88% of adults with TSC [3, 5]. The size of ungual fibromas ranges from barely detectable to about 1 cm. They are more common on the toes than the fingers [3]. When they arise from under the proximal nail fold, they are termed periungual fibromas. Those located under the nail plate are called subungual fibromas, although there is variability in the literature regarding the use of these terms. Periungual fibromas are pink to red papules and nodules that may be soft and rounded or have a pointed hyperkeratotic tip (Figure 14.4a and b). They press on the nail matrix to cause a longitudinal groove in the nail [15] (Figure 14.4c), and sometimes a groove forms in

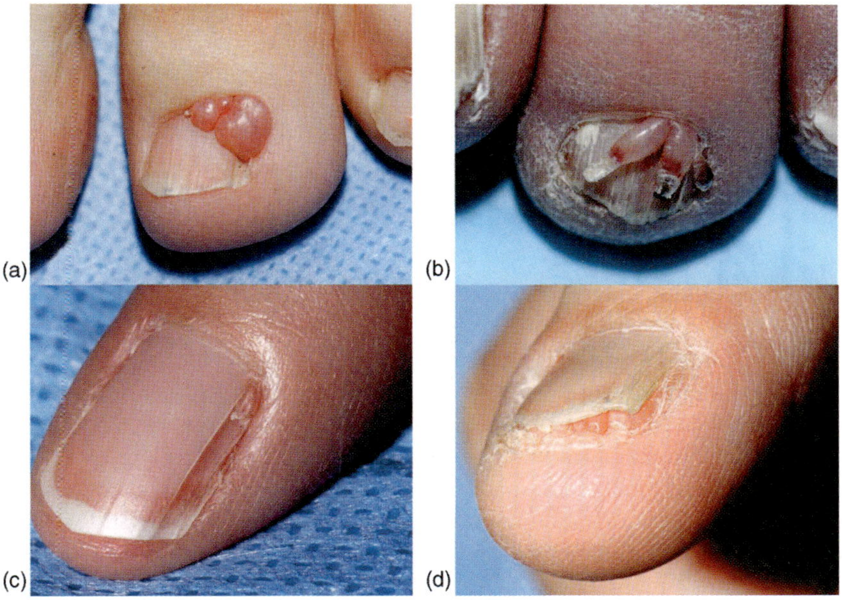

Figure 14.4 Periungual and subungual fibromas in TSC. (a) Periungual fibromas. (b) Periungual fibromas with hyperkeratotic tips. (c) Longitudinal nail groove from periungual fibroma. (d) Subungual fibromas.

the nail without an evident papule [3]. Subungual fibromas can be seen through the nail plate as red or white oval lesions or as red papules protruding from the distal nail plate (Figure 14.4d). Ungual fibromas tend to catch on clothing, bleed when traumatized, and can be painful [15]. They disfigure the nail, and distortion of the nail anatomy may predispose to secondary fungal infections. Histologically, ungual fibromas are similar to angiofibromas, but with more extensive hyperkeratosis and a variable increase in vascularity [15, 33].

The differential diagnosis of ungual fibromas includes several similar-appearing but solitary papules that occur sporadically, usually in response to trauma. These are called by a variety of names including periungual fibroma, garlic-clove fibroma, acral fibrokeratoma, and acquired periungual fibrokeratoma [34]. Other lesions that may occur in the periungual location include epidermoid cyst and pseudomyxoid cyst. Lesions occurring in the subungual location may include subungual exostosis and subungual horn. Other lesions in the differential diagnosis include Bowen disease, squamous cell carcinoma, superficial acral fibromyxoma, warts, pyogenic granuloma, juvenile xanthograuloma, onychomycosis, and psoriasis. Multiple acral fibromas with a myxoid stroma were reported in one patient with familial retinoblastoma [35].

14.2.6
Other Skin Lesions

Molluscum fibrosum pendulum is the name given for multiple skin-colored or hyperpigmented skin tags in TSC. These range from soft pedunculated papules to larger firm pedunculated nodules located on the neck, axillae, trunk, and flexures [5]. Since skin tags are common in the general population, molluscum fibrosum pendulum is not included as a diagnostic criterion for TSC. *Miliary fibromas* are patches of multiple minute papules usually on the neck or trunk that appear like "gooseflesh." *Pachydermodactyly* is a benign thickening of the proximal fingers that has been observed in a few patients with TSC [36, 37].

14.2.7
Significance of Skin Lesions for Diagnosis of TSC

Three or more hypopigmented macules constitute a major feature for diagnosis of TSC. This number is based on the observation of one or two (and in rare individuals up to three) hypomelanotic macules in 4.7% of the general population [38]. Angiofibromas must be multiple to be counted as a major feature. A solitary angiofibroma is clinically and histologically indistinguishable from the fibrous papule that occurs sporadically as a single lesion in the general population [23]. The forehead plaque is grouped together with angiofibromas as a major feature for diagnosis [39]. A nontraumatic ungual fibroma is a major feature for diagnosis of TSC [39]. Solitary lesions (also termed acral or acquired digital fibrokeratomas) are also observed in the general population, especially following nail trauma [40, 41].

14.3
Pathogenesis of TSC Skin Lesions

TSC skin tumors are hamartomas, overgrowths containing cells typically found in the skin but abnormal in amounts and organization [42]. TSC skin tumors show increased cellularity reflecting increased number of fibroblast-like cells, vessels, and dermal mononuclear phagocytes (Figure 14.5a–f) [43]. These tumors are highly angiogenic [44].

The propensity for skin tumors in TSC is the result of a mutation in either *TSC1* or *TSC2* (see Chapter 4). Similar to many other tumors in TSC, skin tumors show loss of the second, normal allele for either *TSC1* or *TSC2*, in accord with the Knudson two-hit model of tumor development [45] (see also Chapter 4). The cells sustaining this two-hit inactivation are fibroblast-like cells as derived from both facial angiofibromas and ungual fibromas, and show loss of hamartin (TSC1) or tuberin (TSC2) expression [46], biallelic mutation of *TSC2* [45], and activation of signaling through mTORC1 [45]. These findings suggest that TSC skin tumors may respond to treatment with inhibitors of mTORC1, such as rapamycin. Indeed, improvement in angiofibromas was observed in a TSC patient treated with rapamycin for renal transplantation [47]. However, rapamycin has many potential significant side effects when given orally. It may be possible to decrease these side effects through topical administration. Topical rapamycin decreased tumor growth in a mouse model of tuberous sclerosis [48], suggesting that this approach should also be tested in humans.

Several other distinctive findings have been made in the fibroblast-like cells derived from TSC skin tumors, suggesting important mechanisms of how these lesions develop, as well potential therapeutic targets. The fibroblast-like cells show increased release of paracrine factors that can stimulate angiogenesis [49], recruit mononuclear phagocytes through secretion of MCP-1 [50], and appear to stimulate epidermal proliferation through secretion of epiregulin [45]. In addition, fibroblast-like cells produce greater amounts of several extracellular matrix molecules [51–53] and matrix degrading enzymes [54]. Although these findings are of considerable interest, they have not yet been extended to clinical application in either a clinical trial or routine clinical use, and further study is needed. Currently, surgery is the major approach to treatment of TSC skin lesions.

14.4
Considerations for Surgical Treatment of TSC Skin Lesions

14.4.1
Patient Evaluation

In treating children and adults for the skin manifestations of TSC, it is vitally important to understand that we are treating the whole patient. The complete patient

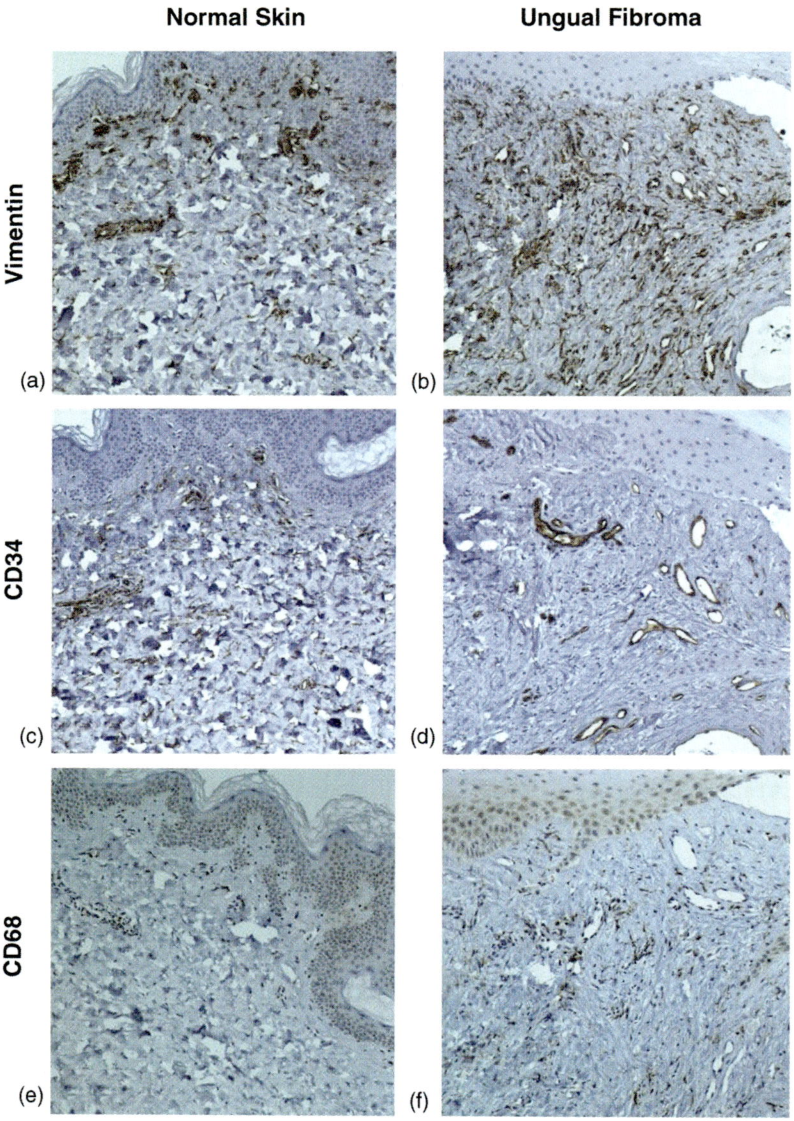

Figure 14.5 Altered cellular composition of TSC skin tumors. Immunohistochemical staining for fibroblasts and other mesenchymal cells (vimentin), blood vessels (CD34), and mononuclear phagocytes (CD68) shows increased number of these cells, stained brown, in an ungual fibroma (b, d, f) compared to normal-appearing skin (a, c, e). Each image of tumor or normal skin shows the epidermis overlying the tumor stroma or normal dermis, respectively.

assessment determines whether the patient is a good candidate for treatment from both the physical and emotional standpoints.

A full-body skin evaluation should be performed by the treating surgeon. All lesions, whether treatable or not, should be documented. The color, number, size, texture, and distribution of the lesions should be recorded. History should be obtained regarding age of onset, growth, and symptoms of skin lesions, as well as medications and other health problems. Photographs can be obtained before and after surgery to document the extent of involvement and evaluate postoperative results.

14.4.2
Indications for Treatment and Preoperative Considerations

A young child is usually brought by the family requesting early treatment of flat "red spots" on the face before they become more raised. Older children tend to seek treatment for growing angiofibromas, and adults also desire removal of ungual fibromas. Indications for treatment may include bleeding, chronic irritation, pain, infection, or functional compromise (e.g., obstruction of nasal airways or vision, problems wearing shoes). The most common reason to seek treatment is the appearance of facial angiofibromas. The positive effects of surgical treatment of angiofibromas on the emotional and psychological well-being of the patient cannot be overemphasized. Although patients with TSC may suffer with seizures, kidney tumors, and respiratory problems, their major concern may be the negative effects of angiofibromas on their social interactions and self-esteem [55].

Factors to consider before surgically treating an adult or a mature child with TSC include the indications for surgery, anticipated treatment outcome, patient motivation, and developmental status. With young children or developmentally delayed patients, the surgeon must understand the patient–parent dynamic and answer the question "Am I treating the patient or am I treating the parents?" Additional patient-related issues include

1) **Expectations**: Does the patient and family understand current limitations of treatments? Will they be accepting of the outcome?
2) **Associated medical conditions**: Are there preexisting medical problems that would preclude surgical treatment (e.g., respiratory compromise, deteriorating kidney function, symptomatic rhabdomyoma, active facial skin infection)? Allergies to anesthetics are a relative contraindication to treatment.
3) **Current medications**: Will there be any adverse interactions between current antiseizure and other medications and postoperative pain medications or perioperative antibiotics? Is the patient on steroid medication that needs to be managed during the immediate perioperative period?
4) **Appropriate postoperative care**: Will the patient, parents, or caregivers be able to follow the postoperative care regimen and be available to return for appropriate postoperative follow-up?

14.4.3
Patient, Family, and Caregiver Education

The surgeon should discuss treatment approaches, potential risks including anesthesia risks, and complications. Images from previous surgeries are helpful for the patient/family to understand potential outcomes. It is preferable to refer to TSC skin lesions as "growths" and avoid terms such as "abnormal," "disfigured," or "deformed," to allay the fears of already anxious patients/parents. In addressing patient/family expectations, the following points should be stressed:

1) Rarely is one treatment sufficient; multiple-stage procedures may be indicated, at least until adulthood, because of recurrences or new lesions.
2) Although early outcomes are usually described by the patient/family as either good or excellent, long-term results are variable.
3) The future growth of lesions cannot be predicted.
4) All surgical treatments have complications and risks.
5) Facial skin will never be completely smooth; cobblestoning is common and background redness (exacerbated by stress, fever, exercise, and during menstrual periods) will remain despite adequate treatment.

Patients or their families may be hesitant regarding laser surgery because of one or several unsuccessful treatments, or they may know someone who had a negative experience. There can be many explanations for previous failed surgery, including inappropriate laser selection or settings, surgeon inexperience, and inadequate number of laser treatments. It is therefore critical for the treating surgeon to a priori avoid excluding a patient from surgery based on previous outcomes. At the other end of the spectrum, the surgeon needs to counsel the potential patient/family about realistic prospects and limitations of laser treatment.

14.4.4 Insurance Issues

Because TSC is not a well-known disease and laser treatment is not widely available, insurance companies need to be educated about the problems caused by the skin manifestations of the disease and the indications for treatment. To that end, it is helpful to provide the insurance company with

1) a letter explaining TSC, such as one available from the Tuberous Sclerosis Alliance; and
2) a cover letter from the treating physician explaining the need for lesion excision and proposed treatment, including International Classification of Diseases-9 (ICD-9) and Current Procedural Terminology (CPT) codes. It is helpful for the family to develop a personal contact at the insurance company to resolve any coverage issues that might arise. Since the patient seeking care of the skin manifestations has most likely filed previous insurance claims for other medical issues arising from TSC, the insurance company has a track record of the patient's needs and this can be used to get the precertification/claim approved.

14.5 Treatment of Angiofibromas

14.5.1 Approaches

Facial angiofibromas have been treated using multiple different surgical approaches [56], including laser surgery, excision, curettage, cryotherapy, electrodessication, timed electrosurgery [57], and shave excision and dermabrasion [58–62]. Several different types of lasers have been used, including argon [63, 64], copper vapor [65], erbium:YAG [66], "hot" KTP [67], carbon dioxide (CO_2) [68], scanning CO_2 [69, 70], and vascular lasers [71, 72]. It is difficult to draw firm conclusions about the optimal treatment based on this literature, because of variability in the ages of patients, extent of skin lesions, methods of assessing treatment response, number of treatments, and duration of follow-up. Treatment recommendations are based on the literature and surgeon experience [56, 71–73]. One of the authors (M.M.) has treated over 250 TSC patients and has concluded the following:

1) Flat and red lesions in young children (<4 years old) can be treated with a vascular laser.
2) Raised and red lesions in older children and adults require the use of both the CO_2 and vascular lasers either sequentially or simultaneously (Figure 14.6a and b). Recurrence is likely until adulthood.
3) Raised and normopigmented lesions are frequently seen both in the growing child and in the adult. The CO_2 laser is used to treat these fibrous tumors (Figure 14.7a and b).

14.5.2 Timing of Treatment

In the majority of cases, the treatment of angiofibromas is an elective procedure. The timing is based on symptoms, age of the patient, and anticipated growth of the

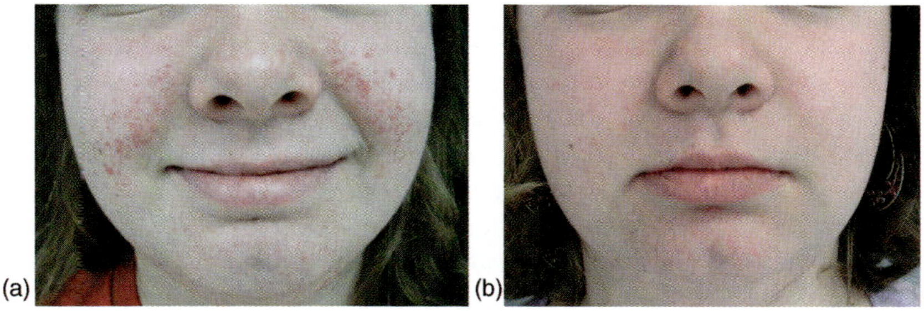

Figure 14.6 Treatment of facial angiofibromas with CO_2 and vascular lasers. (a) Preoperative appearance of multiple angiofibromas. (b) Postoperative appearance 9 months after simultaneous CO_2 and vascular laser treatment.

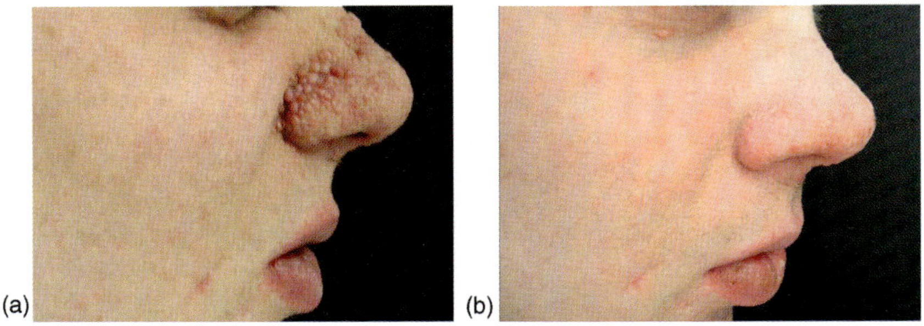

Figure 14.7 Treatment of facial angiofibromas with CO_2 laser. (a) Preoperative appearance of alar groove angiofibroma clusters. (b) Postoperative appearance 1 year following CO_2 laser ablation.

lesions. Our impression is that children should be treated before starting school. By treating at this early age, it allows the surgeon to ablate the flat red/pink lesion using only the vascular laser, perhaps thwarting future growth. This may also spare the child from embarrassing comments made by others. Another developmental time point for treatment is prior to puberty. Angiofibromas may grow more rapidly during puberty, and early treatment may require less aggressive and fewer procedures than treatment performed later in life, and may also result in a better final outcome. Timing of treatment in adults is less critical than childhood since the number and size of angiofibromas are fairly stable. Adults may be treated at any time without significantly impacting final results.

14.5.3
Patient Preparation

Surgical preparation for all patients scheduled for CO_2 laser treatment, regardless of age, should include good control of seizure activity when possible, discontinuance of all medications with anticoagulant properties at least 2 weeks in advance (e.g., aspirin, Vitamin E, herbal preparations), and a complete history and physical performed by his/her primary care physician or neurologist to exclude medical contraindications to surgery and identify drug allergies. Kidney and lung function should be normal or stable. For patients over the age of 16, perioperative administration of antiviral prophylaxis (Valtrex 500 mg BID for 10 days) is advisable and a pregnancy test is warranted for all females.

The day before surgery, the child's seizure medications should be carefully managed. If possible, the child should have nothing by mouth (NPO) from midnight and the procedure performed early in the morning. The parents should be asked to bring the child's medication to the perianesthesia care unit (PACU) for use upon awakening. Because general anesthesia may raise the threshold for seizures, the risk of seizure activity is usually confined to the PACU and even then the risk is small.

14.5.4 Operating Room

Inadequate sedation will lead to inadequate treatment. Hence, it is suggested that the procedure be performed in an outpatient setting under general anesthesia administered via either an endotracheal or laryngeal mask airway (LMA) tube. This allows for proper patient monitoring and increased safety. For children who are unwilling or unable to cooperate, it is helpful to have either two anesthesiologists or an additional anesthesia assistant present to start the intravenous line as the patient is being induced. It is comforting to both child and family to have one parent accompany the child into the operating room.

14.6 Laser Treatments of Angiofibromas

14.6.1 CO_2 Laser

The CO_2 laser has several practical advantages. It is readily available and many surgeons have extensive experience in its use. The depth of treatment can be controlled using this laser, and recovery is fairly rapid. The patient can have repeated treatments if there is regrowth.

In preparation for treatment using the CO_2 laser, the patient's face is cleansed with betadine. The patient's eyes are irrigated with balanced salt solution and a sterile eye lubricant on polished stainless steel laser eye shields is applied. To prevent fires, the endotracheal tube is wrapped with tin foil (dull side out) and the face is draped in moist sterile towels. The surgical lights are adjusted to provide tangential lighting, thereby highlighting the textures of the facial skin. The room is kept warm (72 °F) or a heating blanket is used to avoid postoperative hypothermia.

The surgeon should use 3.5 power loupe magnification for accurate assessment of each raised angiofibroma. The CO_2 laser (M.M. uses the Coherent Ultrapulse 5000 laser) is set to 3–4 W power in continuous mode. With the skin stretched, each individual lesion is treated using the 0.2 mm cutting handpiece in a *defocused mode*. If "blistering" of the treated tissue is noted, then the surgeon must place the handpiece closer to the lesion for an adequate depth of treatment. Conversely, to avoid cutting too deep into the surrounding tissue, the handpiece needs to be farther from the skin than the focal length metal adapter. For each angiofibroma, multiple passes are made and the char removed with moist gauze pads between each pass. When larger clusters of angiofibromas are encountered, commonly in the alar groove, it is best to "debulk" them first with the laser directed tangentially to the base, with care taken to avoid damage or injury to surrounding unaffected tissue. The operative field is kept dry to minimize dispersion of the beam. The end point of treatment is a smooth skin surface with a chamois color at the base of the treated tissue, usually after three passes in teenage children and adults and two passes in younger children. The skin is rinsed with saline and Aquaphor is applied liberally to the treated areas. Nonadherent gauze (e.g., Telfa) pads are applied. The laser eye shields are removed and the eyes are once again irrigated with balanced salt solution. The young patient should leave the operating room on a stretcher with bumpers to avoid injury upon awakening and the parents should be present in the recovery room to reorient and reassure their child.

14.6.2 CO_2 Laser Postoperative Care

All treated areas are covered with Aquaphor, applied continuously during the 9–12 days of healing (until complete reepithelialization). This "open" care technique allows visualization of the healing tissue, increases patient comfort, and is easy to perform. The head is kept elevated and showers are allowed on the second day. If itching causes discomfort, cool water/dilute vinegar soaks with a soft washcloth are used as needed to soothe the face. After complete epithelialization, use of topical 1% hydrocortisone cream twice daily for 2–3 weeks reduces redness. The addition of bleaching cream (4% hydroquinone) to the regimen in more darkly pigmented individuals (Fitzpatrick types III and IV) can help minimize hyperpigmentation. An appropriate schedule for follow-up is at least twice during the first 10 days, at 2-week intervals until the sixth week, and then every 3 months for the first year. After the first year, it is valuable to see the patient yearly for 5–10 years, both to assess the patient's needs and to gain clinical knowledge of the outcomes of the various treatments.

14.6.3 Complications and Risks of CO_2 Laser Treatment

Depending upon the experience of the surgeon, the laser chosen, the energy levels used, the patient's skin type, the facial location of the lesion being treated, and the attentiveness to postoperative care, complications can occur with use of the CO_2 laser.

1) *Infection* is rare. Patients are placed on broad-spectrum antibiotics perioperatively to further decrease the likelihood of infection.
2) *Pain* is well tolerated and treatable. Pain is minimized by using Aquaphor as an interface between the freshly treated areas and the ambient air. Acetaminophen or, rarely, oral opioids can be used for pain control.
3) *Bleeding* is minimal. The CO_2 laser has the effect of cauterizing the tissue, so there is minimal blood loss intraoperatively. Postoperative blood loss may occur with trauma, so the patient should avoid touching the treated areas.
4) *Hypopigmentation* can result from the use of high energy levels or repeated treatments within a short time period. If treatment is being performed for purely cosmetic reasons, one must ask "Is a flat white spot better than a raised red one?" It is important to achieve a uniform pigmentation of each facial aesthetic unit.

   To avoid spotty hypopigmentation, it might be necessary to use the CPG (computer pattern generator) handpiece during the same operative session. After treating the individual angiofibromas with the cutting handpiece, the CPG handpiece can be used to treat entire aesthetic units of the face (including affected and unaffected skin). This is typically reserved for adults or *older teenage* children when there is extensive skin involvement. When using the CPG handpiece, an appropriate size and shape is selected for the anatomical area to be treated and the laser is set to 300 mj (150 mj can be used for "half passes" in areas of thinner skin such as the upper and lower eyelids).

5) *Hyperpigmentation* is most often a response to healing in Fitzpatrick III–V skin types. It is usually self-limited and resolves with time. Bleaching creams with hydroquinone can speed resolution.
6) *Scarring* can be a response to the use of high energy levels and/or multiple deep passes with the CO_2 laser. The treated skin should demonstrate a chamois color as the end point of treatment. This will indicate that the surgeon has reached the interface between the papillary and reticular dermis. Proceeding into deeper tissue increases the risk of scarring. In our experience, the areas at greatest risk for hypertrophic scarring include the chin and the alar grooves. Further, one would expect the possibility of keloid scarring in certain skin types. Although hypertrophic scarring has been encountered in four children (in the perialar and chin regions), keloid scars have not been seen in our clinical experience with over 250 patients.
7) *Allergic conjunctivitis* can occur secondary to the laser eye shields if they contain even trace amounts of nickel. This is usually self-limiting but can be treated with ophthalmic steroid ointment or drops.
8) *Milia* are commonly seen in the paranasal region from the 12th day through the sixth week of healing. They are easy to treat by unroofing the individual milia with a 30-gauge needle and expressing the contents or, more gradually by the patient, by using facial puff pads (nonmedicated coarse texture found in most pharmacies) when washing the face.
9) *Prolonged erythema* can last 3–6 months but has been seen to last for up to 1 year. Topical steroids and an SPF 45 sunblock (for UVA and UVB) can help resolve the erythema.
10) *Anesthesia risks.*

14.6.4
Limitations of CO_2 Laser Treatment

1) **Recurrence/regrowth**: Because treatment using the carbon dioxide laser only affects the lesion to the mid-dermal (papillary) level, recurrence rates should theoretically approach 100%. Clinical experience, though, has revealed a much lower recurrence rate. It is believed that deep dermal fibrosis and perivascular fibrosis can prevent or slow regrowth of the angiofibroma.
2) "**Background**" **red/pink pigmentation**: Despite adequately treating the fibrous portion of the angiofibroma, unless CO_2 laser treatment reaches the deep dermis, vascular pigmentation may remain and will require a further treatment of the skin with the vascular laser.

3) **Skin texture retains its “cobblestone” appearance**: Although the CO_2 laser can ablate individual angiofibromas, even facial skin that is resurfaced in aesthetic units will still exhibit a “cobblestone” appearance. This rough texture of the skin is due to residual, subclinical, enlarging angiofibromas interspersed within unaffected skin and thickened skin secondary to laser treatment.

14.6.5
Vascular Laser

There are a plethora of vascular lasers on the market including but not limited to Nd: YAG, Alexandrite, pulsed dye, KTP, intense pulsed light (IPL), and diode. Although most vascular lasers can be used, our preference is to choose a laser that is easy to use, efficient, portable, reliable, and does not cause posttreatment purpura. No matter which laser is selected in the hemoglobin wavelength, the end point of treatment is blanching of the vessel.

If not sure as to the proper energy setting for the laser, the surgeon can perform a few “test spots” at various energy levels using only a topical anesthetic (e.g., EMLA cream, BLT, Topicaine) or an ice pack. The immediate skin reaction and subsequent vascular response at 6 weeks can be recorded, thereby allowing appropriate treatment at a later date.

14.6.6
Vascular Laser Postoperative Care

Because there is little in the way of tissue damage from treatment with the vascular laser, pain is negligible and no local wound care is necessary in the postoperative period. In the event that there is mild discomfort, oral acetaminophen is effective. No antibiotics are needed and topical care with Aquaphor or Vaseline helps soothe the skin. If blistering of the skin occurs, limited topical care with an antibacterial ointment (e.g., Bacitracin) is recommended. When treating patients with Fitzpatrick III–IV skin types, pretreatment once a day with 1% cortisone cream and 4% hydroquinone 2 weeks prior to surgery can reduce hyperpigmentation. This regimen can be reinstituted 1 week after treatment and can be continued for 2–4 weeks as needed.

14.6.7
Complications and Risks of Vascular Laser Treatment

Although uncommon, complications can occur with the use of vascular lasers.

1) *Infection* is rare because the epidermis remains intact. In the event that blistering occurs, topical antibacterial ointments can be used.
2) *Pain* is minimized by using ice for a few minutes immediately after treatment. This can be continued at home as needed during the first 24 h. Rarely is pain medication needed.
3) *Hypopigmentation* can result from the use of high energy levels or repeated treatments within a short time period.

4) *Hyperpigmentation*, when it occurs, is most commonly seen in Fitzpatrick III–V skin types due to a postinflammatory response. Often it will resolve spontaneously but it can be treated with topical steroids, hydroquinone, and sunblock.
5) *Scarring* is rare but can be a response, after blistering resolves, to the use of high energy levels on a patient of any skin type.
6) *Allergic conjunctivitis* secondary to eye shields.
7) *Anesthesia risks.*

14.6.8 Limitations of Vascular Laser Treatment

Skin always retains its "background" redness. Cheek skin will "flush" (vascular dilation) secondary to extreme emotion, fever, strenuous activity, and even during menses. Despite the possible drawbacks to the combined use of the CO_2 and vascular lasers, patients and family routinely rate the results as "good to excellent." Because of the possibility of lesion regrowth and/or the growth of new angiofibromas, both the patient and the surgeon must understand that repeated treatments may be required during the patient's lifetime.

14.7 Treatment of other TSC Skin Lesions

14.7.1 Facial and Scalp Plaques

Surgical excision is an option for scalp plaques, and small forehead and cheek plaques when in a favorable anatomic location. Scalp excisions may be closed with direct approximation or rotation-advancement flaps. Small forehead plaques may be closed in a horizontal orientation with excellent cosmetic results. A small plaque on the cheek may also be excised if the resultant scar can be placed into a facial fold or following the natural lines of facial skin relaxation. For larger forehead plaques, limited ablation using the CO_2 laser in a defocused mode (as described above) can achieve acceptable, albeit temporary, results. Recurrences are common after debulking and flattening plaques with the CO_2 laser. Ablating the full thickness of the dermis in an attempt to eradicate plaques will leave the patient with an unacceptable scar. The pink/red color of facial plaques cannot be cleared with a vascular laser even with multiple treatments. The carbon dioxide laser can be used for cheek plaques but, in our experience, the final results are poor.

14.7.2 Ungual Fibromas

Ungual fibromas can be excised or ablated using a different approach depending on whether the lesion is periungual or subungual (Figure 14.8a and b). The fingers or toes are prepped and draped in the usual sterile fashion. The affected digit(s) is

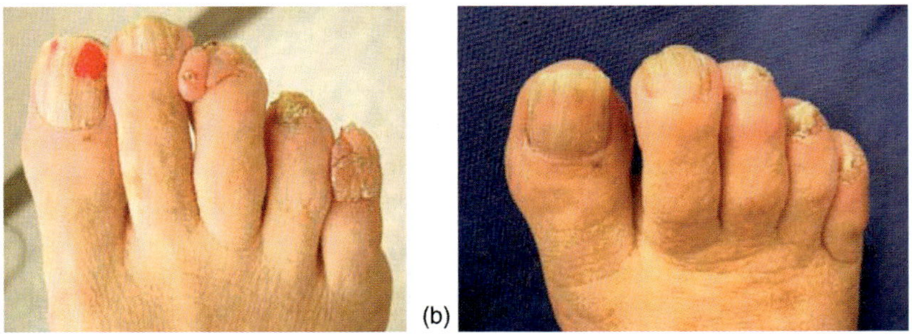

(a) (b)

Figure 14.8 Treatment of periungual fibromas. (a) Periungual fibromas on third and fifth digits. (b) Postoperative appearance 1 year following surgical excision and CO_2 laser ablation.

blocked with 0.5% marcaine and a digital tourniquet is applied. Using loupe magnification will allow the surgeon to accurately differentiate the fibroma from the nail matrix at its base.

For periungual fibromas, whether located in the central portion of the eponychium or along the paronychium, a longitudinal incision over the lesion to its base will allow adequate exposure. Reflect the skin flaps on either side, dissecting proximally to the base of the fibroma. The lesion can be excised with a scalpel or ablated with the CO_2 laser, using a 0.2 mm handpiece set at 4 W power in cutting mode. Additionally, curettage of the lesion base helps to assure complete removal and prevent recurrence. The skin flaps are reapproximated with 4-0 or 5-0 Monocryl sutures. The tourniquet is then released, an antibacterial ointment applied, followed by a dry sterile finger/hand (or toe/foot) dressing.

For subungual fibromas, the lesion is accessed by carefully removing the nail plate by blunt dissection, separating it from the surrounding soft tissue. If the nail plate is healthy, clean it of all soft tissue, create a hole in its center with electrocautery and place it in betadine-soaked gauze. The eponychium/paronychium is then incised, the fibroma excised and the skin approximated (as described above). After applying an antibiotic ointment to the nail bed, loosely secure the previously prepared nail plate under the eponychium and to the skin of the fingertip with 5-0 Monocryl or Vicryl sutures. If the nail is dystrophic, then it should be discarded. In its place, a thin piece of silicone sheeting can be used to protect the nail bed, decrease postoperative pain secondary to exposure, and prevent the eponychium from adhering to the nail bed. The tourniquet is then released, an antibacterial ointment applied, and a dry sterile finger/hand (or toe/foot) dressing applied.

It is important to keep dressings dry, maintain elevation, and use ice packs on the surgical site for the first 48 h. If the fibroma was on the toe, minimize ambulation for 3 days. Daily dressings should consist of an antibiotic ointment covered by a bandage for 10 days. Sutures are removed at 14 days and steri-strips applied for additional support. If the nail has been reapplied as a protective "splint," remove the proximal and distal sutures in 2 weeks. The new growing nail will displace the old nail plate within the first month.

14.7.3 Shagreen Patch

Because these lesions typically do not cause any symptoms, the surgeon's role is limited to addressing the cosmetic appearance of the affected skin. Carbon dioxide laser resurfacing has not proven to be effective in ablating these thick, fibrous lesions and can cause significant and unsightly scarring. Others have tried shaving the lesion with a dermatome or combined wide excision with skin grafting but sequelae and complications include additional scarring (donor site), change in pigmentation, and possible loss of the graft and lesion recurrence.

14.8 Future of Medical/Surgical Treatment of TSC Skin Lesions

The treatment of TSC skin lesions continues to evolve. Outcomes have improved over the years with increased surgical experience and improved technologies. Patients benefit from improved appearance and function, but still experience recurrences and unpredictable long-term outcomes. The ideal treatment – one that is quick, painless, permanent, and leaves no scarring – is still being sought. Research into the cause of TSC has led to the identification of new targets for therapy, such as the mTOR pathway and angiogenesis. Drugs targeting, these pathways may have the potential for serious side effects when used systemically, limiting their use for skin lesions as the sole indication. Drugs with fewer adverse effects, or drugs applied topically, may one day augment or replace surgical therapies. Until then, surgery will continue to be the mainstay of treatment, lessening the impact of skin lesions on the lives of those with TSC.

Acknowledgments

We thank John Crawford and Mary King at NIH Medical Arts for patient photography. Research cited in this chapter was supported in part by the National Institutes of Health (NIH) grant R01 CA100907 (T.N.D.) and the Intramural Research Program, NIH, NHLBI (J.M.).

References

1 Hunt, A. (1983) Tuberous sclerosis: a survey of 97 cases. II. Physical findings. *Dev. Med. Child Neurol.*, **25**, 350–352.
2 Fitzpatrick, T.B. (1991) History and significance of white macules, earliest visible sign of tuberous sclerosis. *Ann. N.Y. Acad. Sci.*, **615**, 26–35.
3 Webb, D.W., Clarke, A., Fryer, A., and Osborne, J.P. (1996) The cutaneous features of tuberous sclerosis: a population study. *Br. J. Dermatol.*, **135**, 1–5.
4 Jimbow, K. (1997) Tuberous sclerosis and guttate leukodermas. *Semin. Cutan. Med. Surg.*, **16**, 30–35.

5 Jozwiak, S., Schwartz, R.A., Janniger, C.K., Michalowicz, R., and Chmielik, J. (1998) Skin lesions in children with tuberous sclerosis complex: their prevalence, natural course, and diagnostic significance. *Int. J. Dermatol.*, **37**, 911–917.

6 Dabora, S.L., Jozwiak, S., Franz, D.N., Roberts, P.S., Nieto, A., Chung, J., Choy, Y.S., Reeve, M.P. *et al.* (2001) Mutational analysis in a cohort of 224 tuberous sclerosis patients indicates increased severity of TSC2, compared with TSC1, disease in multiple organs. *Am. J. Hum. Genet.*, **68**, 64–80.

7 Sogut, A., Ozmen, M., Sencer, S., Caliskan, M., Aydinli, N., Ertugrul, T., and Peksayar, G. (2002) Clinical features of tuberous sclerosis cases. *Turk. J. Pediatr.*, **44**, 98–101.

8 Sun, X.F., Yan, C.L., Fang, L., Shen, F.M., and Liao, K.H. (2005) Cutaneous lesions and visceral involvement of tuberous sclerosis. *Chin. Med. J. (Engl.)*, **118**, 215–219.

9 Datta, A.N., Hahn, C.D., and Sahin, M. (2008) Clinical presentation and diagnosis of tuberous sclerosis complex in infancy. *J. Child Neurol.*, **23**, 268–273.

10 Schwartz, R.A., Fernandez, G., Kotulska, K., and Jozwiak, S. (2007) Tuberous sclerosis complex: advances in diagnosis, genetics, and management. *J. Am. Acad. Dermatol.*, **57**, 189–202.

11 Schaffer, J.V. and Bolognia, J.L. (2003) The treatment of hypopigmentation in children. *Clin. Dermatol.*, **21**, 296–310.

12 Bolognia, J.L. (1999) A clinical approach to leukoderma. *Int. J. Dermatol.*, **38**, 568–572.

13 Sanchez, N.P., Wick, M.R., and Perry, H.O. (1981) Adenoma sebaceum of Pringle: a clinicopathologic review, with a discussion of related pathologic entities. *J. Cutan. Pathol.*, **8**, 395–403.

14 Hall, M.R., Kovach, B.T., and Miller, J.L. (2007) Unilateral facial angiofibromas without other evidence of tuberous sclerosis: case report and review of the literature. *Cutis*, **80**, 284–288.

15 Nickel, W.R. and Reed, W.B. (1962) Tuberous sclerosis. Special reference to the microscopic alterations in the cutaneous hamartomas. *Arch. Dermatol.*, **85**, 209–226.

16 Earnest, L., Byrne, P., and Califano, J. (2003) Massive facial angiofibroma in a patient with tuberous sclerosis. *Otolaryngol. Head Neck Surg.*, **128**, 151–153.

17 Reed, R.J. and Ackerman, A.B. (1973) Pathology of the adventitial dermis. Anatomic observations and biologic speculations. *Hum. Pathol.*, **4**, 207–217.

18 Asgharian, B., Turner, M.L., Gibril, F., Entsuah, L.K., Serrano, J., and Jensen, R.T. (2004) Cutaneous tumors in patients with multiple endocrine neoplasm type 1 (MEN1) and gastrinomas: prospective study of frequency and development of criteria with high sensitivity and specificity for MEN1. *J. Clin. Endocrinol. Metab.*, **89**, 5328–5336.

19 Darling, T.N., Skarulis, M.C., Steinberg, S.M., Marx, S.J., Spiegel, A.M., and Turner, M. (1997) Multiple facial angiofibromas and collagenomas in patients with multiple endocrine neoplasia type 1. *Arch. Dermatol.*, **133**, 853–857.

20 Hoang-Xuan, T. and Steger, J.W. (1999) Adult-onset angiofibroma and multiple endocrine neoplasia type I. *J. Am. Acad. Dermatol.*, **41**, 890–892.

21 Sakurai, A., Matsumoto, K., Ikeo, Y., Nishio, S.I., Kakizawa, T., Arakura, F., Ishihara, Y., Saida, T. *et al.* (2000) Frequency of facial angiofibromas in Japanese patients with multiple endocrine neoplasia type 1. *Endocr. J.*, **47**, 569–573.

22 Schaffer, J.V., Gohara, M.A., McNiff, J.M., Aasi, S.Z., and Dvoretzky, I. (2005) Multiple facial angiofibromas: a cutaneous manifestation of Birt–Hogg–Dube syndrome. *J. Am. Acad. Dermatol.*, **53**, S108–S111.

23 Bansal, C., Stewart, D., Li, A., and Cockerell, C.J. (2005) Histologic variants of fibrous papule. *J. Cutan. Pathol.*, **32**, 424–428.

24 Lee, D.A., Grossman, M.E., Schneiderman, P., and Celebi, J.T. (2005) Genetics of skin appendage neoplasms

and related syndromes. *J. Med. Genet.*, **42**, 811–819.

25 Vincent, A., Farley, M., Chan, E., and James, W.D. (2003) Birt–Hogg–Dube syndrome: a review of the literature and the differential diagnosis of firm facial papules. *J. Am. Acad. Dermatol.*, **49**, 698–705.

26 Toro, J.R., Glenn, G., Duray, P., Darling, T., Weirich, G., Zbar, B., Linehan, M., and Turner, M.L. (1999) Birt–Hogg–Dube syndrome: a novel marker of kidney neoplasia. *Arch. Dermatol.*, **135**, 1195–1202.

27 Fryer, A.E., Osborne, J.P., and Schutt, W. (1987) Forehead plaque: a presenting skin sign in tuberous sclerosis. *Arch. Dis. Child.*, **62**, 292–293.

28 Menascu, S. and Donner, E.J. (2008) Linear nevus sebaceous syndrome: case reports and review of the literature. *Pediatr. Neurol.*, **38**, 207–210.

29 Kawamura, A., Ochiai, T., Tan-Kinoshita, M., and Suzuki, H. (2005) Buschke–Ollendorff syndrome: three generations in a Japanese family. *Pediatr. Dermatol.*, **22**, 133–137.

30 Weintraub, R. and Pinkus, H. (1977) Multiple fibrofolliculomas (Birt–Hogg–Dube) associated with a large connective tissue nevus. *J. Cutan. Pathol.*, **4**, 289–299.

31 Requena, L., Gutierrez, J., and Sanchez Yus, E. (1992) Multiple sclerotic fibromas of the skin. A cutaneous marker of Cowden's disease. *J. Cutan. Pathol.*, **19**, 346–351.

32 Uitto, J., Santa Cruz, D.J., and Eisen, A.Z. (1980) Connective tissue nevi of the skin. clinical, genetic, and histopathologic classification of hamartomas of the collagen, elastin, and proteoglycan type. *J. Am. Acad. Dermatol.*, **3**, 441–461.

33 Kint, A. and Baran, R. (1988) Histopathologic study of Koenen tumors. Are they different from acquired digital fibrokeratoma? *J. Am. Acad. Dermatol.*, **18**, 369–372.

34 Carlson, R.M., Lloyd, K.M., and Campbell, T.E. (2007) Acquired periungual fibrokeratoma: a case report. *Cutis*, **80**, 137–140.

35 Dereure, O., Savoy, D., Doz, F., Junien, C., and Guilhou, J.J. (2000) Multiple acral fibromas in a patient with familial retinoblastoma: a cutaneous marker of tumour-suppressor gene germline mutation? *Br. J. Dermatol.*, **143**, 856–859.

36 Bardazzi, F., Neri, I., Fanti, P.A., and Patrizi, A. (1996) Pachydermodactyly in two young girls. *Pediatr. Dermatol.*, **13**, 288–291.

37 Lo, W.L. and Wong, C.K. (1993) Localized pachydermodactyly in tuberous sclerosis. *Clin. Exp. Dermatol.*, **18**, 146–147.

38 Vanderhooft, S.L., Francis, J.S., Pagon, R.A., Smith, L.T., and Sybert, V.P. (1996) Prevalence of hypopigmented macules in a healthy population. *J. Pediatr.*, **129**, 355–361.

39 Roach, E.S., Gomez, M.R., and Northrup, H. (1998) Tuberous sclerosis complex consensus conference: revised clinical diagnostic criteria. *J. Child Neurol.*, **13**, 624–628.

40 Zeller, J., Friedmann, D., Clerici, T., and Revuz, J. (1995) The significance of a single periungual fibroma: report of seven cases. *Arch. Dermatol.*, **131**, 1465–1466.

41 Kojima, T., Nagano, T., and Uchida, M. (1987) Periungual fibroma. *J. Hand. Surg. [Am.]*, **12**, 465–470.

42 Darling, T.N. (2006) Hitting the mark in hamartoma syndromes. *Adv. Dermatol.*, **22**, 181–200.

43 Benjamin, D.R. (1996) Cellular composition of the angiofibromas in tuberous sclerosis. *Pediatr. Pathol. Lab. Med.*, **16**, 893–899.

44 Arbiser, J.L., Brat, D., Hunter, S., D'Armiento, J., Henske, E.P., Arbiser, Z.K., Bai, X., Goldberg, G. *et al.* (2002) Tuberous sclerosis-associated lesions of the kidney, brain, and skin are angiogenic neoplasms. *J. Am. Acad. Dermatol.*, **46**, 376–380.

45 Li, S., Takeuchi, F., Wang, J.A., Fan, Q., Komurasaki, T., Billings, E.M., Pacheco-Rodriguez, G., Moss, J. *et al.* (2008) Mesenchymal–epithelial interactions involving epiregulin in tuberous sclerosis complex hamartomas. *Proc. Natl. Acad. Sci. USA*, **105**, 3539–3544.

46 Fackler, I., DeClue, J.E., Rust, H., Vu, P.A., Kutzner, H., Rutten, A., Kaddu, S., Sander, C.A. *et al.* (2003) Loss of expression of tuberin and hamartin in tuberous sclerosis complex-associated but not in sporadic angiofibromas. *J. Cutan. Pathol.*, **30**, 174–177.

47 Hofbauer, G.F., Marcollo-Pini, A., Corsenca, A., Kistler, A.D., French, L.E., Wuthrich, R.P., and Serra, A.L. (2008) The mTOR inhibitor rapamycin significantly improves facial angiofibroma lesions in a patient with tuberous sclerosis. *Br. J. Dermatol.*, **159**, 473–475.

48 Rauktys, A., Lee, N., Lee, L., and Dabora, S.L. (2008) Topical rapamycin inhibits tuberous sclerosis tumor growth in a nude mouse model. *BMC Dermatol.*, **8**, 1.

49 Nguyen-Vu, P.A., Fackler, I., Rust, A., DeClue, J.E., Sander, C.A., Volkenandt, M., Flaig, M., Yeung, R.S. *et al.* (2001) Loss of tuberin, the tuberous-sclerosis-complex-2 gene product is associated with angiogenesis. *J. Cutan. Pathol.*, **28**, 470–475.

50 Li, S., Takeuchi, F., Wang, J.A., Fuller, C., Pacheco-Rodriguez, G., Moss, J., and Darling, T.N. (2005) MCP-1 overexpressed in tuberous sclerosis lesions acts as a paracrine factor for tumor development. *J. Exp. Med.*, **202**, 617–624.

51 Uysal, H. and Hemming, F.W. (1999) Changes in the expression and distribution of fibronectin, laminin and tenascin by cultured fibroblasts from skin lesions of patients with tuberous sclerosis. *Br. J. Dermatol.*, **141**, 658–666.

52 Uysal, H., Saxton, J., and Hemming, F.W. (1997) Changes in the secretion and glycosylation of fibronectin by human skin fibroblasts associated with tuberous sclerosis. *Glycoconj. J.*, **14**, 439–447.

53 Papakonstantinou, E., Dionyssopoulos, A., Pesintzaki, C., Minas, A., and Karakiulakis, G. (2003) Expression of proteoglycans and glycosaminoglycans in angiofibroma and fibrous plaque skin lesions from patients with tuberous sclerosis. *Arch. Dermatol. Res.*, **295**, 138–145.

54 Papakonstantinou, E., Dionyssopoulos, A., Aletras, A.J., Pesintzaki, C., Minas, A., and Karakiulakis, G. (2004) Expression of matrix metalloproteinases and their endogenous tissue inhibitors in skin lesions from patients with tuberous sclerosis. *J. Am. Acad. Dermatol.*, **51**, 526–533.

55 Kane, Y. (2004) The "bumps" on my face. *J. Am. Acad. Dermatol.*, **51**, S11–S12.

56 Sweeney, S.M. (2004) Pediatric dermatologic surgery: a surgical approach to the cutaneous features of tuberous sclerosis complex. *Adv. Dermatol.*, **20**, 117–135.

57 Capurro, S. and Fiallo, P. (2001) Timed surgery for treatment of angiofibromas in tuberous sclerosis. *Dermatol. Surg.*, **27**, 486–488.

58 Menon, P.A. (1982) Dermabrasion for the management of angiofibromas in tuberous sclerosis. *J. Dermatol. Surg. Oncol.*, **8**, 984–985.

59 Drake, D.B., Morgan, R.F., and Cooper, P.H. (1992) Shave excision and dermabrasion for facial angiofibroma in tuberous sclerosis. *Ann. Plast. Surg.*, **28**, 377–380.

60 Fischer, K., Blain, B., Zhang, F., Richards, L., and Lineaweaver, W.C. (2001) Treatment of facial angiofibromas of tuberous sclerosis by shave excision and dermabrasion in a dark-skinned patient. *Ann. Plast. Surg.*, **46**, 332–335.

61 Widgerow, A.D. (1989) Shaving and dermabrasion of the facial lesions in tuberous sclerosis. A case report. *S. Afr. Med. J.*, **76**, 169–170.

62 Verheyden, C.N. (1996) Treatment of the facial angiofibromas of tuberous sclerosis. *Plast. Reconstr. Surg.*, **98**, 777–783.

63 Arndt, K.A. (1982) Adenoma sebaceum: successful treatment with the argon laser. *Plast. Reconstr. Surg.*, **70**, 91–93.

64 Pasyk, K.A. and Argenta, L.C. (1988) Argon laser surgery of skin lesions in tuberous sclerosis. *Ann. Plast. Surg.*, **20**, 426–433.

65 Kaufman, A.J., Grekin, R.C., Geisse, J.K., and Frieden, I.J. (1995) Treatment of adenoma sebaceum with the copper vapor laser. *J. Am. Acad. Dermatol.*, **33**, 770–774.

66 Cole, R.P., Widdowson, D., and Moore, J.C. (2008) Outcome of erbium:yttrium aluminium garnet laser resurfacing treatments. *Lasers Med. Sci.*, **23**, 427–433.

67 Tope, W.D. and Kageyama, N. (2001) "Hot" KTP-laser treatment of facial angiofibromata. *Lasers Surg. Med.*, **29**, 78–81.

68 Bellack, G.S. and Shapshay, S.M. (1986) Management of facial angiofibromas in tuberous sclerosis: use of the carbon dioxide laser. *Otolaryngol. Head. Neck. Surg.* **94**, 37–40.

69 Bittencourt, R.C., Huilgol, S.C., Seed, P.T., Calonje, E., Markey, A.C., and Barlow, R.J. (2001) Treatment of angiofibromas with a scanning carbon dioxide laser: a clinicopathologic study with long-term follow-up. *J. Am. Acad. Dermatol.*, **45**, 731–735.

70 Song, M.G., Park, K.B., and Lee, E.S. (1999) Resurfacing of facial angiofibromas in tuberous sclerosis patients using CO_2 laser with flashscanner. *Dermatol. Surg.*, **25**, 970–973.

71 Boixeda, P., Sanchez-Miralles, E., Azana, J.M., Arrazola, J.M., Moreno, R., and Ledo, A. (1994) CO_2, argon, and pulsed dye laser treatment of angiofibromas. *J. Dermatol. Surg. Oncol.*, **20**, 808–812.

72 Papadavid, E., Markey, A., Bellaney, G., and Walker, N.P. (2002) Carbon dioxide and pulsed dye laser treatment of angiofibromas in 29 patients with tuberous sclerosis. *Br. J. Dermatol.*, **147**, 337–342.

73 Janniger, C.K. and Goldberg, D.J. (1990) Angiofibromas in tuberous sclerosis: comparison of treatment by carbon dioxide and argon laser. *J. Dermatol. Surg. Oncol.*, **16**, 317–320.

15
Renal Manifestations of Tuberous Sclerosis Complex

John J. Bissler and Elizabeth P. Henske

15.1
Introduction

The renal manifestations of tuberous sclerosis complex (TSC) include angiomyolipomata, polycystic kidney disease, oncocytomas, and rarely carcinomas. All of these manifestations can occur in both children and adults with TSC. It is notable that these lesions are pathologically diverse, representing both epithelium (cysts, oncocytomas, carcinomas) and mesenchyme (angiomyolipomata). Renal involvement in TSC most often progresses insidiously and can result in significant morbidity, including retroperitoneal hemorrhage and renal insufficiency, and even mortality. The precise incidence of end-stage renal disease in the tuberous sclerosis complex population has not been well defined. European surveys suggest that approximately 1% of the TSC patient population with normal intellect is receiving dialytic renal replacement therapy [1, 2]. Shepherd *et al.* reported that 11 patients (27%) died from renal failure in a series of 40 TSC patients at the Mayo Clinic, identifying renal involvement as the leading cause of death [3]. These studies provide some information on the scope of renal morbidity in TSC, though more quantitative information on the frequency of end-stage renal failure due to polycystic kidney disease, multiple interventions for hemorrhage, and replacement of renal tissue by angiomyolipomata is needed. An ongoing natural history study should provide this information.

15.2
Angiomyolipomata

Angio324myolipomata, once classified as hamartomata [4] or choristoma [5], are now classified as PEComas (tumors showing perivascular epithelioid cell differentiation). The World Health Organization defines PEComas as mesenchymal tumors containing distinctive perivascular epithelioid cells. PEComas include angiomyolipoma, lymphangioleiomyomatosis (LAM), clear cell "sugar" tumor of the lung, and a group of rare, morphologically and immunophenotypically similar lesions arising from a variety of soft tissues and visceral sites. Nearly all PEComas exhibit immunoreactivity

Tuberous Sclerosis Complex: Genes, Clinical Features, and Therapeutics
Edited by David J. Kwiatkowski, Vicky Holets Whittemore, and Elizabeth A. Thiele

ISBN: 978-3-527-32201-5

for both melanocytic (HMB-45 and/or melan-A) and smooth muscle (actin and/or desmin) markers (Figure 15.1). The normal tissue counterpart and the origin of the PEC cell are unknown. Because PEComas express melanocytic proteins, and melanocytes are neural crest derived, it has been postulated that neural crest lineage cells give rise to angiomyolipomata [6]. Neural crest lineage cells were recently found to contribute to renal development [7]. Alternatively, since angiomyolipomata are mesenchymal tumors and the epithelial elements of the kidney arise from embryonic mesenchyme, it has been proposed that angiomyolipomata arise from a renal mesenchymal precursor cell [8]. Supporting this proposal is the fact that rtPCR can

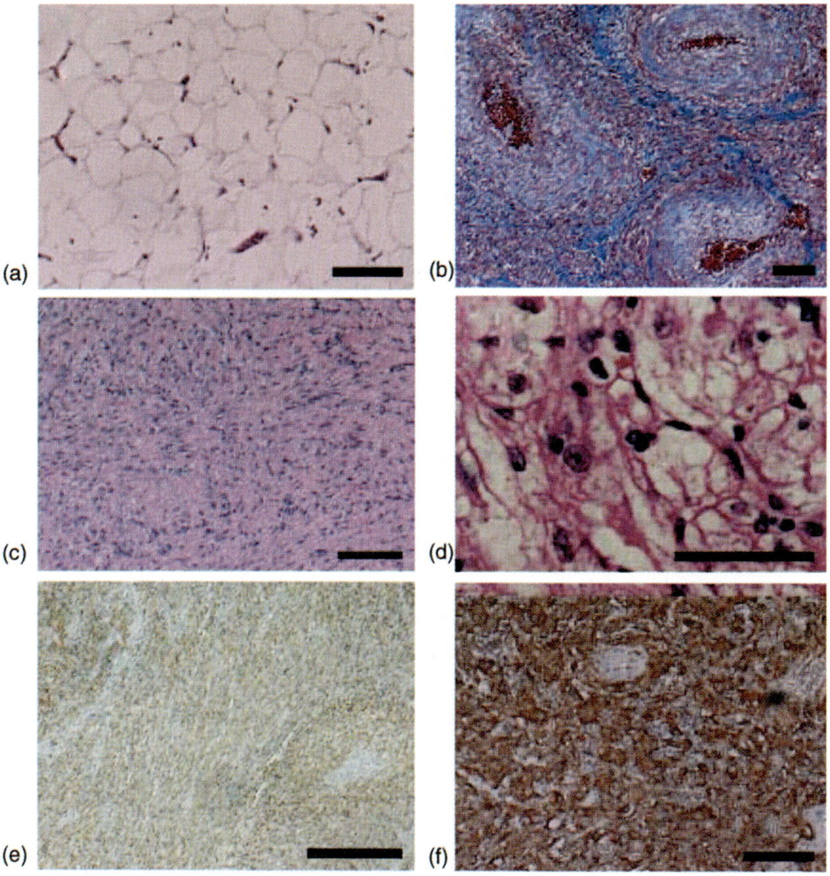

Figure 15.1 Histology of renal angiomyolipomata. (a) Hematoxylin and eosin stain of a typical lesion with predominantly adipose tissue. (b) Masson's trichrome stain of a fat-poor angiomyolipoma with many vascular structures. (c) Hematoxylin and eosin stain of a fat-poor angiomyolipoma consisting of smooth muscle-like cells. (d) Hematoxylin and eosin stain of a fat-poor angiomyolipoma consisting primarily of epithelioid cells. (e) Fat-poor angiomyolipoma consisting of spindle-shaped cells stained an antibody for smooth muscle actin. (f) HMB-45 staining in an angiomyolipoma. Black bars in each panel represent 100 μm in length.

detect mRNA for gp100, the antigen target of HMB-45 in both the proximal and distal tubules, although the protein product is below the level of detection with HMB-45 [9]. Furthermore, all three components of angiomyolipomata (vascular cells, immature smooth muscle cells, and fat cells) contain "second-hit" mutations and are therefore derived from a common progenitor cell [10, 11]. Cells lacking TSC2, therefore, appear to have unusual mesenchymal differentiation plasticity. Since kidney epithelium is derived from embryonic mesoderm, it is possible that angiomyolipomata reflect an aberrant mesenchymal–epithelial transition during kidney development.

While the classic form of angiomyolipoma contains vascular, smooth muscle, and adipose tissue, the contribution of each component can vary from lesion to lesion in the same kidney. For example, the vascular or the smooth muscle component may be scant or dominate the lesion [12]. Five different types of vessels within angiomyolipomata have been recognized [13], some of which are deficient in elastic fibers [14]. The smooth muscle component can be limited to a small amount cuffing the vasculature [15], or may be so extensive that imaging detects a solid mass devoid of fat [16]. In classic angiomyolipomata, there can be patches of smooth muscle cells that exhibit nuclear atypia with mitotic figures; despite these findings, such lesions typically do not behave as a malignant lesion. Like the smooth muscle and vascular component, the adipose element of angiomyolipomata can be variable. Most often the fat component is made up of mature fat tissue. Very rarely, renal angiomyolipomata can infiltrate surrounding tissue [17]. While most often described in the kidneys, angiomyolipomata have been identified in many other sites, most often in the liver but also in the ovary, fallopian tube, spermatic cord, palate, and colon.

Fat poor angiomyolipomata may consist of smooth muscle cells, predominantly vascular elements, or epithelial cells in the epithelioid variant [18, 19]. The epithelioid cells may be polygonal with a slight degree of nuclear atypia, or atypical and variable in size; these lesions may exhibit significant mitotic activity. There are few or no abnormal vessels or fat cells in this latter lesion.

The frequency of angiomyolipomata in patients with TSC varies with the age of the population being studied. A longitudinal study found that 80% of children with TSC had renal lesions by age of 10.5 years [20]. Seventy-five percent of the lesions were angiomyolipomata and 17% were renal cysts. Angiomyolipomata increased in size and/or number in about 60% of the children, highlighting the need for close surveillance of the kidneys even if an initial renal ultrasound is normal. Two boys in this study had tumors that grew by 4 cm in diameter in 1 year. An autopsy study found that 67% of TSC patients have angiomyolipomata [21] and a recent TSC clinic survey revealed that 85% of patients had renal angiomyolipomata [22]. Growth of angiomyolipomata in patients with TSC is often first detected during childhood [23] and continues into adulthood [24–26]. While no clear gender predominance has been defined, angiomyolipomata often express receptors for both estrogen and progesterone [27, 28].

There are two significant morbidities associated with renal angiomyolipomata. The first is retroperitoneal hemorrhage originating in the angiomyolipoma [29]. Angiomyolipoma, as they enlarge, frequently develop both micro- and macroaneurysms that can rupture [30–32]. The size of the aneurysm appears to be proportional to the risk of hemorrhage, and lesions larger than 5 mm are at a higher risk [33]. Patients with such a hemorrhage most often experience a sudden, painful, and possibly life-threatening event and are most often first seen in the emergency room. An important symptom to help differentiate if a hemorrhage has occurred is significant nausea and vomiting. Up to 20% of patients with such hemorrhages present in shock [34]. Volume resuscitation and embolization are the preferred therapies. However, with such an emergent presentation, the treatment in centers without knowledge about TSC may be a total nephrectomy. The risk of hemorrhage from renal angiomyolipomata has been reported to be between 25 and 50% [35, 36]. A population study suggests that the cumulative risk of a hemorrhage is 20% for females and 10% for males [37]. The size of an angiomyolipoma roughly correlates with the risk of hemorrhage [25, 38–41] but the risk has been more firmly linked to the presence of an aneurysm within the angiomyolipoma [33]. The likely scenario is that larger lesions, with a larger blood supply, are more likely to have larger aneurysms that rupture and bleed into the retroperitoneal space or renal pelvis.

The second morbidity of renal angiomyolipomata is the effects of infiltration and/or coalescence of angiomyolipomata leading to chronic kidney disease and even end-stage renal failure [1, 2]. European surveys suggest that approximately 1% of the TSC patient population with normal intelligence is receiving dialytic renal replacement therapy [1, 2]. However, renal function is often normal even in patients with apparently bilateral confluent angiomyolipomata and this function should be protected. Kidneys with a large angiomyolipoma burden can still contribute significantly to total renal function and the decision to undergo a nephrectomy should be carefully balanced against the loss of renal function.

15.3 Epithelioid and Malignant Angiomyolipomata

The term "epithelioid angiomyolipoma" is used to refer to one type of fat-poor angiomyolipoma, which can sometimes exhibit an aggressive phenotype [42], recur after resection, and even be fatal [43–45]. The pathologic distinction between epithelioid angiomyolipoma and the similar malignant angiomyolipoma is unclear [46, 47]. There are several reports of aggressive epithelioid or malignant angiomyolipomata in TSC patients. A recent review, based on uterine lesions, stratified PEComa aggressiveness and suggested that criteria for malignancy should include a size greater than 8.0 cm, greater than one mitotic figure per 50 high-power fields, and necrosis [48]. Similar data for renal epithelioid lesions have not yet been compiled.

15.4 Renal Cystic Disease

Benign epithelial cysts occur in both *TSC1* and *TSC2* patients [49]. The *TSC2* gene is adjacent to the *PKD1* gene on chromosome 16p13, and large genomic mutations

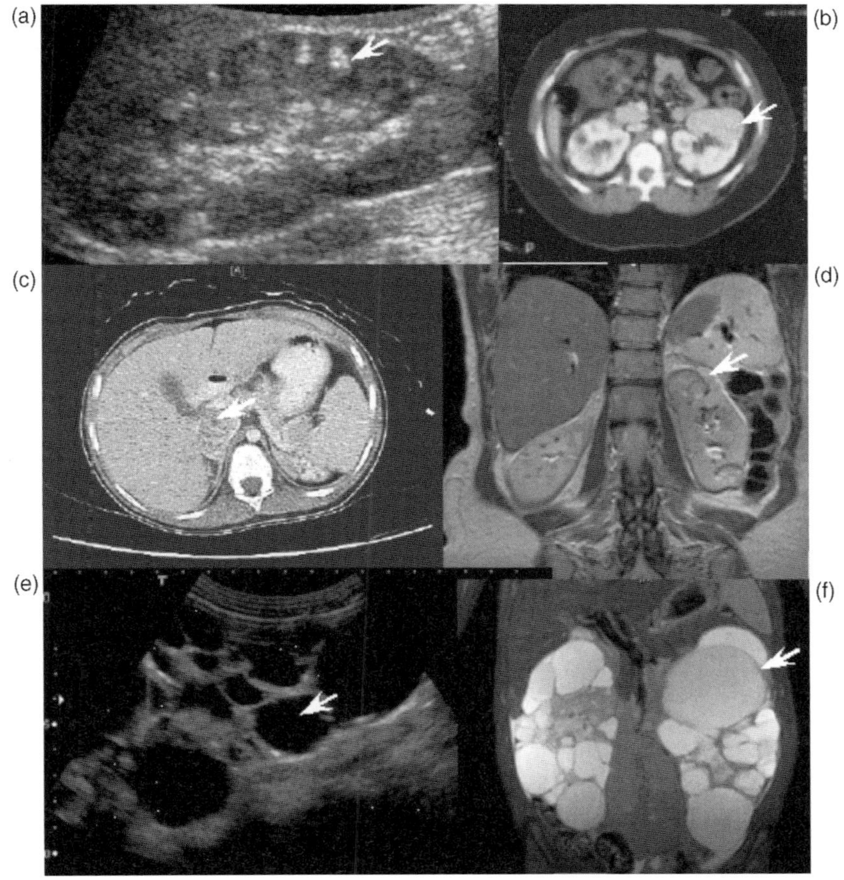

Figure 15.2 Renal angiomyolipoma imaging. (a) Renal ultrasound demonstrates multiple small angiomyolipomata (arrow). (b) CT scan reveals a solid angiomyolipoma on the left kidney (arrow). (c) CT scan with intracaval angiomyolipoma (arrow). (d) MRI with angiomyolipomata in the left kidney upper pole; lesion is marked with an arrow. (e) Ultrasound of a patient with the polycystic variety of TSC. (f) MRI of a patient with the polycystic variety of TSC. (e and f) Arrows highlight individual cysts.

(deletions) that involve portions or all of both genes typically result in a severe PKD phenotype that presents during childhood, or even at birth [50] (Figure 15.2e and f). These children represent about 2% of all those with TSC, are usually due to new mutations, and often exhibit mosaicism for these genomic deletions [51, 52]. Such patients have been reported to often develop significant renal insufficiency as teenagers [51]. However, it is quite possible that when such patients are identified today, they experience improved renal function and survival due to the use of longer acting angiotensin converting enzyme inhibitors and angiotensin receptor blockers. In addition, individuals with mosaic deletion of these two genes appear to have a milder phenotype, at least in some cases [51]. In addition, other TSC patients without these genomic deletions in *TSC2–PKD1* can also develop cystic kidney disease resembling autosomal dominant polycystic kidney disease (ADPKD), with later onset of cysts and a slower rate of progression [51]. It is not yet clear if the *PKD1* gene is involved directly in these patients in some way, or whether this is an atypical manifestation of mutation within *TSC1* or *TSC2* alone.

The mechanism by which deletion of *TSC2–PKD1* accelerates polycystic kidney disease is not clear. It seems likely that second-hit events in the renal epithelium lead to complete loss of function of both *TSC2* and *PKD1*, and this may be important in accelerating cyst development. Mechanisms of cyst pathogenesis in TSC and functional and biochemical links to other cystic diseases are areas of active investigation [53].

15.5 Oncocytoma

There are multiple case reports of oncocytomas in TSC patients [54]. Radiologically, oncocytomas are solid fat-poor renal lesions. Histologically, they contain dense eosinophilic cytoplasm, which may be exclusively or predominantly granular, generally uniform nuclei, and abundant mitochondria [55]. The cell of origin is thought to be the intercalated cell of the cortical portion of the collecting tubule [56, 57]. Renal oncocytomas are considered to be benign tumors and are uncommon in the general population, accounting for only 3–5% of renal parenchymal tumors [58].

15.6 Renal Cell Carcinoma

The exact incidence of renal cell carcinoma (RCC) in TSC is not known, but they are rare affecting perhaps fewer than 2% of TSC patients. In contrast, the incidence of angiomyolipomata is at least 80%. However, it is important to note that renal carcinomas in TSC patients occur at an average age of 28 years [59], which is 25 years younger than the average age of RCC in the general population, and there are many reports in children with TSC [59, 60] and one report in an infant girl [61].

The reported frequency of RCC in TSC is difficult to interpret at the present time for a variety of reasons. The epithelioid variant of angiomyolipoma histologically resembles renal cell carcinoma [62], and not all the reported renal cell carcinomas have been studied in sufficient detail to exclude misclassification of an atypical angiomyolipoma [63]. Second, HMB-45 is a diagnostic marker for angiomyolipoma, which for a time had performance flaws, so that misdiagnoses may have been made [64]. Renal tumors in TSC patients that are suspected to be renal cell carcinomas need to be carefully analyzed pathologically by an experienced renal pathologist, including immunostains for cytokeratins, HMB-45, and other markers that can help distinguish a fat-poor angiomyolipoma from a carcinoma [63]. Cytogenetics may be necessary to differentiate fat-poor angiomyolipoma from the TFEB (t(6;11)) HMB45-positive pediatric RCC (translocation morphology RCC) [65]. This latter RCC is known to occur in the pediatric patient but is not associated with TSC.

15.7
Monitoring Renal Lesions

Infrequent and/or erratic renal follow-up can put TSC patients at unnecessary risk for renal complications. Significant growth of angiomyolipomata can occur in a single year or insidiously develop over 5–10 years. Even if initial imaging does not reveal a lesion, follow-up is warranted because new angiomyolipomata lesions can develop in both childhood and adulthood. At the current time, there are no proven noninvasive methods to distinguish fat-poor angiomyolipomata from carcinomas; biopsy may be indicated. Growth rate has recently been introduced as a possible method to help distinguish renal cell carcinomas from fat-poor lesions [66]. Ongoing natural history studies will better define many of the issues about lesion growth.

Yearly renal imaging is generally advised for most TSC patients, although this has not been prospectively studied and longer intervals may be appropriate for patients with very stable lesions (Figure 15.2). Previously, ultrasonography was considered the renal imaging method of choice in TSC because of its sensitivity in the detection of both the adipose component of angiomyolipomata and the fluid in cysts (Figure 15.2a and e), as well as speed, convenience, and avoidance of radiation exposure. However, ultrasound may not detect solid lesions until significant tissue distortion occurs and is inherently imprecise for measurements.

Computed tomography (CT) gives significantly more detail, is extremely fast, does not have user bias, and can be used to analyze the radiographic density of lesions quantitatively (Hounsfield units) (Figure 15.2b and c). However, it has the risk of contrast nephrotoxicity, if contrast is given, as well as radiation exposure.

Magnetic resonance imaging (MRI) is becoming the major approach for renal imaging in TSC patients (Figure 15.2d and f). It lacks the risks of CT and can give excellent detail even without contrast. A drawback of MRI is that it is slow and some patients require sedation or general anesthesia for the procedure. At the University of Cincinnati, we most often use MRI and combine the abdominal study with central nervous system imaging, so that the child is sedated once. MRI and CT imaging files can be used to view the images in a variety of planes, and software can be used to measure actual volume and even estimate density. The strengths and weaknesses of the different imaging modalities must be born in mind when making comparisons between studies using different techniques to avoid an erroneous interpretation of sudden growth. For example, an ultrasound may not detect the solid component of a lesion, whereas the MRI will.

15.8
Treatment

Specific treatment plans vary tremendously based on the age of the patient, the size of the lesion(s), the radiographic appearance of the lesions, and renal function, but there are critical core concepts regardless of the renal phenotype. First and most critically, the majority of TSC patients have multiple, bilateral angiomyolipomata and may require intervention at multiple points during their lives. Therefore, renal parenchyma must be protected and no surgical approach that involves the removal of normal parenchyma should be performed without extensive consideration of alternative approaches. One of the most frequent reasons for surgical removal of renal lesions in TSC is a radiographically detected solid lesion. As discussed earlier, the growth rate and/or a biopsy may prevent renal surgery for what proves pathologically to be a fat-poor angiomyolipoma (Figure 15.3). Management of fat-poor lesions requires careful imaging follow-up and possibly biopsy if the lesion is growing at a rate greater than 0.5 cm per year [66]. A multidisciplinary team that includes a nephrologist, radiologist, and urologist who are all well versed in TSC is absolutely necessary in diagnostically confusing cases.

Humans must lose more than half of their kidney function to manifest an elevation in serum creatinine, so that a normal creatinine does not indicate normal kidney function. Nonsteroidal pain relievers and other potentially nephrotoxic agents, including IV contrast, should be avoided if possible. Uncontrolled hypertension hastens renal failure. Patients must have their blood pressures measured at every visit and intervention initiated if hypertension is detected. For children with significant renal involvement, intravascular volume status is also a key issue. Alterations of the renal architecture as well as disease state can limit renal urine concentrating capacity. While under normal circumstances this is compensated by increased thirst, infants and neurologically impaired patients are at significant risk for dehydration, especially

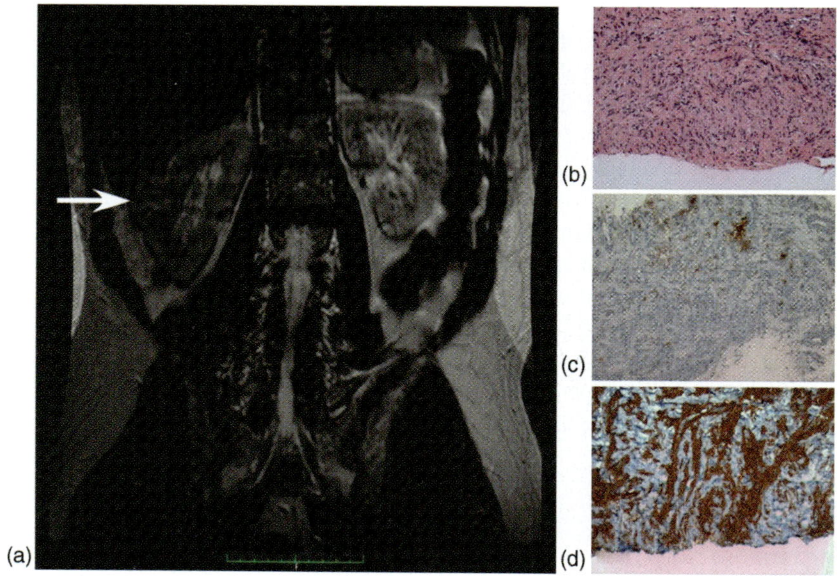

Figure 15.3 A fat-poor lesion can be a smooth muscle predominant angiomyolipoma. (a) Fast spin echo T2 MRI revealing a fat-poor lesion, possible carcinoma (arrow). (b–d) Biopsy of the lesion stained with hematoxylin and eosin (b), HMB-45 (c), and smooth muscle actin (d) demonstrates that it is a fat-poor angiomyolipoma that is focally positive for both HMB-45 and smooth muscle actin.

in the setting of vomiting or diarrhea. In order to avoid serious dehydration caused by increased losses and defective renal concentrating capacity, a simple scale can be tremendously helpful. Specifically, the family should weigh the patient when they first detect an illness, and at regular intervals during the illness, in order to estimate fluid losses and adjust hydration as necessary.

The current standard of care for large angiomyolipomata includes embolization and renal sparing surgery. While a size of 4–6 cm in diameter has been used as a trigger for further investigation and possible embolization, aneurysms are the major risk factor for bleeding (Figure 15.4). There is a correlation, though imperfect, between angiomyolipoma size and presence and size of aneurysms [67]. Some TSC

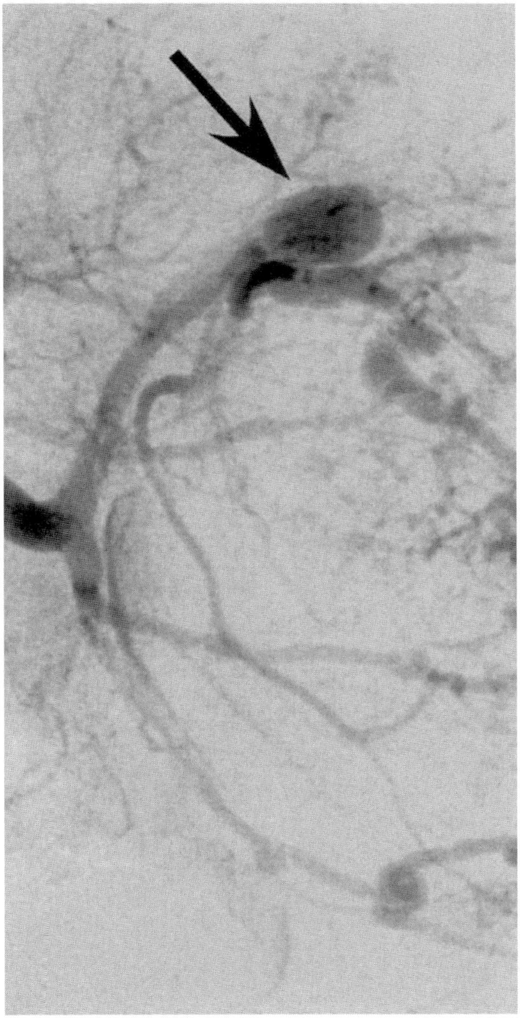

Figure 15.4 Subselective angiography revealing distorted vessels and a large aneurysm (arrow).

centers have developed prophylactic embolization protocols that are designed to take a proactive approach to reduce the risk of hemorrhage. Prior to embolization procedures, a nuclear cortical scan can be very helpful in mapping areas of the kidney that can be embolized without significant effects on renal function [68]. In some cases, CT angiography can be very helpful in defining the aneurismal load. Sometimes, the vascular supply for such lesions is very poor and embolization offers more risk than benefit. In these cases, either radiofrequency or cryoablation may be excellent choices. Some urologists can do these procedures endoscopically. Exophytic lesions can be also treated by a surgical approach, although as emphasized earlier surgical removal of normal renal parenchyma should be avoided if at all possible.

The embolization procedure is aimed at reducing the risk of hemorrhage, but the lesions most often shrink as well [68]. The patient's response to the ischemic and necrotic tissue following embolization can be dramatic. About 90% of patients that undergo renal angiomyolipomata embolization experience postembolization syndrome. This consists of severe pain and fever that can last for several weeks depending on the mass of tissue involved in the embolization. These unfortunate side effects can be mitigated by using a steroid treatment in the form of a Solumedrol bolus during the procedure followed by a tapering schedule of prednisone [32]. Other rare consequences of embolization include renin-mediated hypertension. The presumed etiology involves the reduction of blood flow to functional renal tissue by the embolization procedure. Treatment with angiotensin-converting enzyme inhibitors or angiotensin receptor blockers may be required temporarily, or in very rare cases be required for an extended duration.

It is also gratifying to note that progress toward drug treatment of angiomyolipomata is advancing rapidly. This development is due to clinical translation of basic research findings in TSC, though caution and additional study are required. As discussed in detail in Chapter 6, the TSC1/TSC2 proteins have a major role in the regulation of the state of activation of mTOR, such that this pathway is aberrantly activated in TSC angiomyolipoma. In addition, there is a class of drugs, termed rapalogues after the first drug identified, rapamycin, that specifically target and block activated mTOR. We have recently published a clinical proof-of-principle trial that suggests that rapamycin and analogues may have benefit in the control of the growth of TSC angiomyolipoma [69]. Twenty patients were enrolled on this trial that completed the 12-month evaluation, and 18 have been followed through the 24-month evaluation. After taking rapamycin for 12 months, the mean ($\pm$SD) angiomyolipoma volume had decreased to $53 \pm 27\%$ of the baseline volume ($P < 0.001$). However, at 24 months (12 months after discontinuation of rapamycin) the mean volume had increased back to $86 \pm 28\%$ of the baseline volume ($P = 0.005$). Nonetheless, at 24 months, 5 of 18 patients had a persistent reduction in angiomyolipoma volume of 30% or more. In addition, 11 patients on this trial also had pulmonary LAM. At the end of the year of rapamycin therapy, the mean forced expiratory volume in 1 s (FEV1) increased by 118 ± 330 ml ($P = 0.06$), the forced vital capacity (FVC) increased by 390 ± 570 ml ($P < 0.001$), and the residual volume decreased by 439 ± 493 ml ($P = 0.02$) in these patients. At 24 months of follow-up, their average FEV1 was increased 62 ± 411 ml, FVC was increased 346 ± 712 ml, and residual volume was 333 ± 570 ml below the starting value. Five patients had six serious adverse events while receiving sirolimus, including pyelonephritis, stomatitis, diarrhea, and respiratory infections. A similar angiomyolipoma response to rapamycin was also reported by Davies *et al.* in seven TSC patients [70]. In addition, preclinical studies with autosomal dominant polycystic kidney disease have also shown positive results with rapamycin and led to studies of mTOR inhibitors to limit the cystic disease progression in that condition as well (ClinicalTrials.gov identifiers NCT00346918 and NCT00491517).

Despite these positive studies, the number of patients studied overall is small, and significant toxicity has been seen. Additional evaluation is recommended before rapamycin, or more generally other mTOR inhibitors, can be recommended in the routine management of angiomyolipomata.

15.9
Conclusions and Future Directions

The presence of both mesenchymal and epithelial renal tumors sets TSC apart from all other genetic syndromes associated with renal tumors. While the vast majority of renal lesions in TSC patients are benign angiomyolipomata, malignant lesions (both malignant angiomyolipomata and renal cell carcinomas) occur at low frequency, and necessitating careful review of imaging studies and wisdom in management. Loss of functional renal parenchyma occurs with each surgical procedure, and therefore they should be avoided if possible. Therefore, biopsy is usually considered for confirmation prior to surgical resection when malignancy is suspected. In most TSC patients with renal lesions, yearly imaging of the kidneys is recommended unless there is good evidence of stable disease over many years. Currently, many angiomyolipomata are treated with embolization. At this time, targeted therapeutic approaches are not recommended for the routine management of angiomyolipomata, but this is an area of intense clinical investigation and cautious optimism.

References

1 Schillinger, F. and Montagnac, R. (1996) Chronic renal failure and its treatment in tuberous sclerosis. *Nephrol. Dial. Transplant.*, **11**, 481–485.

2 Clarke, A., Hancock, E., Kingswood, C., and Osborne, J.P. (1999) End-stage renal failure in adults with the tuberous sclerosis complex. *Nephrol. Dial. Transplant.*, **14**, 988–991.

3 Shepherd, C.W., Gomez, M.R., Lie, J.T., and Crowson, C.S. (1991) Causes of death in patients with tuberous sclerosis. *Mayo. Clin. Proc.*, **66**, 792–796.

4 Fischer, W. (1911) Die niereutemoren bi der tuberosen hirsklerose. *Beitr z path Anat u z allg Path*, **1**, 235.

5 Dickinson, M., Ruckle, H., Beaghler, M., and Hadley, H.R. (1998) Renal angiomyolipoma: optimal treatment based on size and symptoms. *Clin. Nephrol.*, **49**, 281–286.

6 Sarnat, H.B. and Flores-Sarnat, L. (2005) Embryology of the neural crest: its inductive role in the neurocutaneous syndromes. *J. Child Neurol.*, **20**, 637–643.

7 Hao, Z., Guo, C., Jiang, X., Krueger, S., Pietri, T., Dufour, S., Cone, R.E., and O'Rourke, J. (2006) New transgenic evidence for a system of sympathetic axons able to express tissue plasminogen activator (t-PA) within arterial/arteriolar walls. *Blood*, **108**, 200–202.

8 Henske, E.P. (2005) Tuberous sclerosis and the kidney: from mesenchyme to epithelium, and beyond. *Pediatr. Nephrol.*, **20**, 854–857.

9 Lin, F., Yang, W., Betten, M., Teh, B.T., and Yang, X.J. (2006) Expression of S-100 protein in renal cell neoplasms. *Hum. Pathol.*, **37**, 462–470.

10 El-Hashemite, N., Walker, V., Zhang, H., and Kwiatkowski, D.J. (2003) Loss of Tsc1 or Tsc2 induces vascular endothelial growth factor production through mammalian target of rapamycin. *Cancer Res.*, **63**, 5173–5177.

11 Karbowniczek, M., Yu, J., and Henske, E.P. (2003) Renal angiomyolipomas from patients with sporadic lymphangiomyomatosis contain both neoplastic and non-neoplastic vascular structures. *Am. J. Pathol.*, **162**, 491–500.

12 Lin, C.N., Chiang, H.S., Hsu, S.I., Huang, A.H., and Chuang, S.S. (1994) Renal angiomyolipoma with a prominent angiomatous component and extramedullary hematopoiesis: a case report. *Chung Hua i Hsueh Tsa Chih–Chinese Medical Journal*, **53**, 185–187.

13 Karbowniczek, M., Yu, J., and Henske, E.P. (2003) Renal angiomyolipomas from patients with sporadic lymphangiomyomatosis contain both neoplastic and non-neoplastic vascular structures. *Am. J. Pathol.*, **162**, 491–500.

14 Tweeddale, D.N., Dawe, C.J., and McDonald, J.R. (1955) Angiolipo-leiomyoma of the kidney: report of a case with observations on histogenesis. *Cancer*, **8**, 764–770.

15 Wong, A.L., McGeorge, A., and Clark, A.H. (1981) Renal angiomyolipoma: a review of the literature and a report of 4 cases. *Brit. J. Urol.*, **53**, 406–411.

16 Obuz, F., Karabay, N., Secil, M., Igci, E., Kovanlikaya, A., and Yorukoglu, K. (2000) Various radiological appearances of angiomyolipomas in the same kidney. *Eur. Radiol.*, **10**, 897–899.

17 Farrow, G.M., Harrison, E.G., Jr., Utz, D.C., and Jones, D.R. (1968) Renal angiomyolipoma: a clinicopathologic study of 32 cases. *Cancer*, **22**, 564–570.

18 Mai, K.T., Perkins, D.G., and Collins, J.P. (1996) Epithelioid cell variant of renal angiomyolipoma. *Histopathology*, **28**, 277–280 (see comments).

19 Eble, J.N., Amin, M.B., and Young, R.H. (1997) Epithelioid angiomyolipoma of the kidney: a report of five cases with a prominent and diagnostically confusing epithelioid smooth muscle component. *Am. J. Surg. Pathol.*, **21**, 1123–1130.

20 Ewalt, D.H., Sheffield, E., Sparagana, S.P., Delgado, M.R., and Roach, E.S. (1998) Renal lesion growth in children with tuberous sclerosis complex. *J. Urol.*, **160**, 141–145.

21 Bernstein, J. and Robbins, T.O. (1991) Renal involvement in tuberous sclerosis. *Ann. N.Y. Acad. Sci.*, **615**, 36–49.

22 Rakowski, S.K., Winterkorn, E.B., Paul, E., Steele, D.J., Halpern, E.F., and Thiele, E.A. (2006) Renal manifestations of tuberous sclerosis complex: Incidence, prognosis, and predictive factors. *Kidney Int.*, **70**, 1777–1782.

23 Ewalt, D.H., Sheffield, E., Sparagana, S.P., Delgado, M.R., and Roach, E.S. (1998) Renal lesion growth in children with tuberous sclerosis complex. *J. Urol.*, **160**, 141–145.

24 Lemaitre, L., Robert, Y., Dubrulle, F., Claudon, M., Duhamel, A., Danjou, P., and Mazeman, E. (1995) Renal angiomyolipoma: growth followed up with CT and/or US. *Radiology*, **197**, 598–602 (see comments).

25 Steiner, M.S., Goldman, S.M., Fishman, E.K., and Marshall, F.F. (1993) The natural history of renal angiomyolipoma. *J. Urol.*, **150**, 1782–1786.

26 Kennelly, M.J., Grossman, H.B., and Cho, K.J. (1994) Outcome analysis of 42 cases of renal angiomyolipoma. *J. Urol.*, **152**, 1988–1991 (see comments).

27 Henske, E.P., Ao, X., Short, M.P., Greenberg, R., Neumann, H.P., Kwiatkowski, D.J., and Russo, I. (1998) Frequent progesterone receptor

immunoreactivity in tuberous sclerosis-associated renal angiomyolipomas. *Mod. Pathol.*, **11**, 665–668.
28 Logginidou, H., Ao, X., Russo, I., and Henske, E.P. (2000) Frequent estrogen and progesterone receptor immunoreactivity in renal angiomyolipomas from women with pulmonary lymphangioleiomyomatosis. *Chest*, **117**, 25–30.
29 Chesa Ponce, N., Artiles Hernandez, J.L., Ponce Socorro, J.M., del Rosario Medina, J., Castro Lopez-Torrella, V., and Betancort de Leon, R. (1995) Wunderlich's syndrome as the first manifestation of a renal angiomyolipoma. *Arch. Esp. Urol.*, **48**, 305–308.
30 Ou, Y.C., Wu, H.C., Yang, C.R., Chang, C.L., Hwang, T.I., and Chang, C.H. (1991) Renal angiomyolipoma: experience of 23 patients. *Chung Hua i Hsueh Tsa Chih – Chinese Medical Journal*, **48**, 217–223.
31 Adler, J., Greweldinger, J., and Litzky, G. (1984) "Macro" aneurysm in renal angiomyolipoma: two cases, with therapeutic embolization in one patient. *U. Radiol.*, **6**, 201–203.
32 Bissler, J.J., Racadio, J., Donnelly, L.F., and Johnson, N.D. (2002) Reduction of postembolization syndrome after ablation of renal angiomyolipoma. *Am. J. Kidney Dis.*, **39**, 966–971.
33 Yamakado, K., Tanaka, N., Nakagawa, T., Kobayashi, S., Yanagawa, M., and Takeda, K. (2002) Renal angiomyolipoma: relationships between tumor size, aneurysm formation, and rupture. *Radiology*, **225**, 78–82.
34 Pode, D., Meretik, S., Shapiro, A., and Caine, M. (1985) Diagnosis and management of renal angiomyolipoma. *Urology*, **25**, 461–467.
35 Mouded, I.M., Tolia, B.M., Bernie, J.E., and Newman, H.R. (1978) Symptomatic renal angiomyolipoma: report of 8 cases, 2 with spontaneous rupture. *J. Urol.*, **119**, 684–688.
36 Kessler, O.J., Gillon, G., Neuman, M., Engelstein, D., Winkler, H., and Baniel, J. (1998) Management of renal angiomyolipoma: analysis of 15 cases. *Eur. Urol.*, **33**, 572–575.
37 Webb, D.W., Kabala, J., and Osborne, J.P. (1994) A population study of renal disease in patients with tuberous sclerosis. *Br. J. Urol.*, **74**, 151–154.
38 Zagoria, R.J., Dyer, R.B., Assimos, D.G., Scharling, E.S., and Quinn, S.F. (1991) Spontaneous perinephric hemorrhage: imaging and management. *J. Urol.*, **145**, 468–471.
39 van Baal, J.G., Smits, N.J., Keeman, J.N., Lindhout, D., and Verhoef, S. (1994) The evolution of renal angiomyolipomas in patients with tuberous sclerosis. *J. Urol.*, **152**, 35–38 (see comments).
40 Koike, H., Muller, S.C., and Hohenfellner, R. (1994) Management of renal angiomyolipoma: a report of 14 cases and review of the literature. Is nonsurgical treatment adequate for this tumor? *Eur. Urol.*, **25**, 183–188.
41 Dickinson, M., Ruckle, H., Beaghler, M., and Hadley, H.R. (1998) Renal angiomyolipoma: optimal treatment based on size and symptoms. *Clin. Nephrol.*, **49**, 281–286.
42 Eble, J.N. (1998) Angiomyolipoma of kidney. *Semin. Diagn. Pathol.*, **15**, 21–40.
43 Bjornsson, J., Short, M.P., Kwiatkowski, D.J., and Henske, E.P. (1996) Tuberous sclerosis-associated renal cell carcinoma. Clinical, pathological, and genetic features. *Am. J. Pathol.*, **149**, 1201–1208.
44 Hardman, J.A., McNicholas, T.A., Kirkham, N., and Fletcher, M.S. (1993) Recurrent renal angiomyolipoma associated with renal carcinoma in a patient with tuberous sclerosis. *Brit. J. Urol.*, **72**, 983–984.
45 Eble, J.N., Amin, M.B., and Young, R.H. (1997) Epithelioid angiomyolipoma of the kidney: a report of five cases with a prominent and diagnostically confusing epithelioid smooth muscle component. *Am. J. Surg. Pathol.*, **21**, 1123–1130.
46 Ferry, J.A., Malt, R.A., and Young, R.H. (1991) Renal angiomyolipoma with sarcomatous transformation and pulmonary metastases. *Am. J. Surg. Pathol.*, **15**, 1083–1088.
47 Lowe, B.A., Brewer, J., Houghton, D.C., Jacobson, E., and Pitre, T. (1992) Malignant transformation of

angiomyolipoma. *J. Urol.*, **147**, 1356–1358.
48 Folpe, A.L., Mentzel, T., Lehr, H.A., Fisher, C., Balzer, B.L., and Weiss, S.W. (2005) Perivascular epithelioid cell neoplasms of soft tissue and gynecologic origin: a clinicopathologic study of 26 cases and review of the literature. *Am. J. Surg. Pathol.*, **29**, 1558–1575.
49 Dabora, S.L., Jozwiak, S., Franz, D.N., Roberts, P.S., Nieto, A., Chung, J., Choy, Y.S., Reeve, M.P., Thiele, E., Egelhoff, J.C., Kasprzyk-Obara, J., Domanska-Pakiela, D., and Kwiatkowski, D.J. (2001) Mutational analysis in a cohort of 224 tuberous sclerosis patients indicates increased severity of TSC2, compared with TSC1, disease in multiple organs. *Am. J. Hum. Genet.*, **68**, 64–80.
50 Brook-Carter, P.T., Peral, B., Ward, C.J., Thompson, P., Hughes, J., Maheshwar, M.M., Nellist, M., Gamble, V., Harris, P.C., and Sampson, J.R. (1994) Deletion of the TSC2 and PKD1 genes associated with severe infantile polycystic kidney disease – a contiguous gene syndrome. *Nat. Genet.*, **8**, 328–332.
51 Sampson, J.R., Maheshwar, M.M., Aspinwall, R., Thompson, P., Cheadle, J.P., Ravine, D., Roy, S., Haan, E., Berstein, J., and Harris, P.C. (1997) Renal cystic disease in tuberous sclerosis: role of the polycystic kidney disease 1 gene. *Am. J. Hum. Genet.*, **61**, 843–851.
52 Kozlowski, P., Roberts, P., Dabora, S., Franz, D., Bissler, J., Northrup, H., Au, K.S., Lazarus, R., Domanska-Pakiela, D., Kotulska, K., Jozwiak, S., and Kwiatkowski, D.J. (2007) Identification of 54 large deletions/duplications in TSC1 and TSC2 using MLPA, and genotype–phenotype correlations. *Hum. Genet.*, **121**, 389–400.
53 Siroky, B., Czyzyk-Krzeska, M.F., and Bissler, J.J. (2009) Renal involvement in tuberous sclerosis complex and von Hippel Lindau disease: shared disease mechanism? *Nat. Clin. Pract. Nephrol.*, **5**, 143–156.
54 Jimenez, R.E., Eble, J.N., Reuter, V.E., Epstein, J.I., Folpe, A.L., de Peralta-Venturina, M., Tamboli, P., Ansell, I.D., Grignon, D.J., Young, R.H., and Amin, M.B. (2001) Concurrent angiomyolipoma and renal cell neoplasia: a study of 36 cases. *Mod. Pathol.*, **14**, 157–163.
55 Amin, M.B., Crotty, T.B., Tickoo, S.K., and Farrow, G.M. (1997) Renal oncocytoma: a reappraisal of morphologic features with clinicopathologic findings in 80 cases. *Am. J. Surg. Pathol.*, **21**, 1–12.
56 Eble, J.N. and Hull, M.T. (1984) Morphologic features of renal oncocytoma: a light and electron microscopic study. *Hum. Pathol.*, **15**, 1054–1061.
57 Zerban, H., Nogueira, E., Riedasch, G., and Bannasch, P. (1987) Renal oncocytoma: origin from the collecting duct. *Virchows Arch B Cell Pathol. Incl. Mol. Pathol.*, **52**, 375–387.
58 Siracusano, S., Zanon, M., D'Aloia, G., Plaino, F., Trombetta, C., and Bussani, R. (1998) Rare association of renal angiomyolipoma and oncocytoma. *Urology*, **51**, 837–839.
59 Al-Saleem, T., Wessner, L.L., Scheithauer, B.W., Patterson, K., Roach, E.S., Dreyer, S.J., Fujikawa, K., Bjornsson, J., Bernstein, J., and Henske, E.P. (1998) Malignant tumors of the kidney, brain, and soft tissues in children and young adults with the tuberous sclerosis complex. *Cancer*, **83**, 2208–2216.
60 Robertson, F.M., Cendron, M., Klauber, G.T., and Harris, B.H. (1996) Renal cell carcinoma in association with tuberous sclerosis in children. *J. Pediatr. Surg.*, **31**, 729–730.
61 Breysem, L., Nijs, E., Proesmans, W., and Smet, M.H. (2002) Tuberous sclerosis with cystic renal disease and multifocal renal cell carcinoma in a baby girl. *Pediatr. Radiol.*, **32**, 677–680.
62 Saito, K., Fujii, Y., Kasahara, I., Kobayashi, N., Kasuga, T., and Kihara, K. (2002) Malignant clear cell "sugar" tumor of the kidney: clear cell variant of epithelioid angiomyolipoma. *J. Urol.*, **168**, 2533–2534.
63 Pea, M., Bonetti, F., Martignoni, G., Henske, E.P., Manfrin, E., Colato, C., and Bernstein, J. (1998) Apparent renal cell carcinomas in tuberous sclerosis are heterogeneous: the identification of malignant epithelioid angiomyolipoma.

Am. J. Surg. Pathol., **22**, 180–187 (see comments).

64 Bonetti, F., Pea, M., Martignoni, G., Mombello, A., Colombari, R., Zamboni, G., Scarpa, A., Piubello, Q., Bacchi, C.E., and Gown, A.M. (1991) False-positive immunostaining of normal epithelia and carcinomas with ascites fluid preparations of antimelanoma monoclonal antibody HMB45. *Am. J. Clin. Pathol.*, **95**, 454–459.

65 Argani, P. and Ladanyi, M. (2005) Translocation carcinomas of the kidney. *Clin. Lab. Med.*, **25**, 363–378.

66 Patel, U., Simpson, E., Kingswood, J.C., and Saggar-Malik, A.K. (2005) Tuberose sclerosis complex: analysis of growth rates aids differentiation of renal cell carcinoma from atypical or minimal-fat-containing angiomyolipoma. *Clin. Radiol.*, **60**, 665–673; discussion 663–664.

67 Yamakado, K., Tanaka, N., Nakagawa, T., Kobayashi, S., Yanagawa, M., and Takeda, K. (2002) Renal angiomyolipoma: relationships between tumor size, aneurysm formation, and rupture. *Radiology*, **225**, 78–82.

68 Williams, J.M., Racadio, J.M., Johnson, N.D., Donnelly, L.F., and Bissler, J.J. (2006) Embolization of renal angiomyolipomata in patients with tuberous sclerosis complex. *Am. J. Kidney Dis.*, **47**, 95–102.

69 Bissler, J.J., McCormack, F.X., Young, L.R., Elwing, J.M., Chuck, G., Leonard, J.M., Schmithorst, V.J., Laor, T., Brody, A.S., Bean, J., Salisbury, S., and Franz, D.N. (2008) Sirolimus for angiomyolipoma in tuberous sclerosis complex or lymphangioleiomyomatosis. *N. Engl. J. Med.*, **358**, 140–151.

70 Davies, D.M., Johnson, S.R., Tattersfield, A.E., Kingswood, J.C., Cox, J.A., McCartney, D.L., Doyle, T., Elmslie, F., Saggar, A., de Vries, P.J., and Sampson, J.R. (2008) Sirolimus therapy in tuberous sclerosis or sporadic lymphangioleiomyomatosis. *N. Engl. J. Med.*, **358**, 200–203.

16
Cardiac and Vascular Manifestations

Sergiusz Jóźwiak and Maria Respondek-Liberska

16.1
Introduction

The frequent appearance of cardiac tumors in patients with tuberous sclerosis complex (TSC) has been recognized for many years. In 1862, von Recklinghausen [1] first described the association of cardiac masses with intracerebral sclerotic areas in a newborn who died soon after birth. Since then many similar cases have been reported.

Cardiac tumors, usually in the form of cardiac rhabdomyomas (CRs), represent the earliest detectable hamartoma in TSC and are the only lesion in TSC that may regress with age. Their incidence in patients with TSC had been underestimated until noninvasive detection of the tumors by pre- and postnatal ultrasonography became possible.

Although often asymptomatic, both in prenatal life and in postnatal life, they may result in arrhythmias and cardiac failure, and in some series have been the most frequent cause of death among TSC children below 10 years of age [2].

16.2
Prevalence and Natural History of Cardiac Rhabdomyomas

16.2.1
Prevalence of Cardiac Rhabdomyomas

Primary cardiac tumors are diagnosed in 0.20% of children presenting to pediatric cardiac referral centers [3] and in 0.27% of cases of pediatric autopsies [4]. Rhabdomyoma is the most common primary cardiac tumor in infancy and childhood, representing 36–42% tumors in autopsy series [4, 5] and 79% in clinical series [3]. Cardiac tumors are the most common tumors detected in fetuses.

Tuberous Sclerosis Complex: Genes, Clinical Features, and Therapeutics
Edited by David J. Kwiatkowski, Vicky Holets Whittemore, and Elizabeth A. Thiele

ISBN: 978-3-527-32201-5

16.2.2
Association Between Cardiac Rhabdomyomas and Tuberous Sclerosis Complex

It is currently thought that the majority of children with CRs have TSC, though an initial study concerning the association between TSC and CRs by Fenoglio *et al.* [6] reported 11 TSC confirmed autopsies out of 36 (31%) patients who died due to CRs. In a review of all cases of CRs published up until 1990, Harding and Pagon [7] found that at least 172 of 335 patients (51%) had TSC. Data regarding TSC manifestations were inadequate to confirm the diagnosis of TSC in an additional 117 cases. If these additional cases are considered possible TSC, then up to 86% of CRs would be related to TSC in this series. The highest proportion of TSC among CRs was reported by Bader *et al.* [8], who confirmed a TSC diagnosis on follow-up in 25 of 26 patients (96%) with postnatally diagnosed cardiac tumors.

It should be stressed that the true prevalence of TSC in patients with pediatric CRs may be close to 100%. Since many clinical manifestations of TSC are either difficult to detect or completely absent in early childhood, and radiographic evaluation requires sedation, a definite diagnosis of TSC is often impossible at this age. In addition, genetic mosaicism may lead to limited clinical features. For example, we reported a child in whom multiple CRs were identified on routine fetal ultrasound (US) at 22 weeks of gestation [9]. Postnatally, the child showed normal development. Molecular genetic evaluation was performed leading to identification of a TSC2 mutation. Later, brain CT performed at 6 years of age showed periventricular calcifications consistent with subependymal nodules, providing a definite diagnosis of TSC. At that age, the patient had no skin, neurological, or other obvious signs of TSC. This case represents an example of a patient with CRs who had no other signs of TSC at birth and would have been classified as non-TSC-related CRs.

16.2.3
Natural History of Cardiac Rhabdomyomas in TSC Patients

Conversely, the prevalence of CRs in TSC patients is highly dependent upon the age at which the cardiac ultrasound examination is performed and is estimated to be 47–67% [10–12]. Smith *et al.* [10] noted cardiac tumors in 64% (9/14) of patients under the age of 6 years, 47% (8/17) in the group of children aged 6–12, 67% (8/12) aged 12–18, and in 18% (3/17) of adult patients.

In our series, we found CRs on initial ultrasound examination in 74 of 154 (48%) TSC patients [11]. CRs were seen at highest frequency in those of age <2 years (66%) and were significantly less common in those of age 2–11 years (26%, $p < 0.0001$) (Figure 16.1). CRs were found to be more common in the age 12–15, 54%, consistent with a growth in these lesions during puberty. Interestingly, the difference in prevalence of CRs was greater in girls, as in girls the difference between younger children, ages 2–11 years, and older children, ages 12–15 (6/32; 19% versus 4/6; 67%, respectively), was statistically significant ($p = 0.015$).

Serial echocardiographic studies were performed in 55 patients, including 38 patients in whom the first examination revealed at least one CR. Of these 38 children,

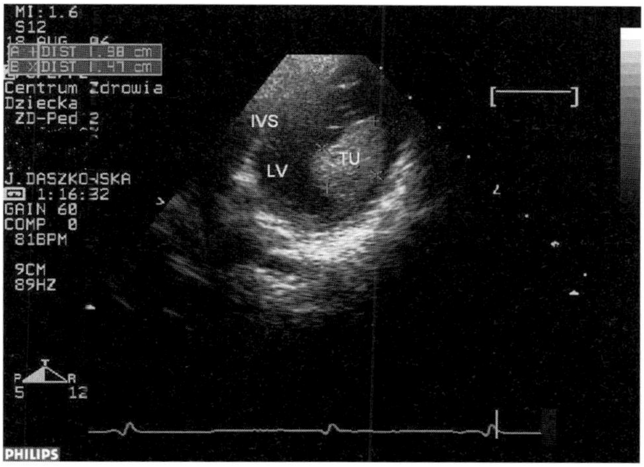

Figure 16.1 Large cardiac rhabdomyoma (2 cm in the largest diameter) in the left ventricle in a newborn with TSC. TU, tumor; LV, left ventricle; IVS, interventricular septum. Reproduced with permission of Professor Wanda Kawalec and Dr Jadwiga Daszkowska, The Children's Memorial Health Institute, Warsaw.

tumors disappeared in 7 (18%) and decreased in 19 (50%) patients [11]. The mechanism of the regression of CRs is completely unknown. It seems possible that there is a relationship between the exposure of the fetus to high levels of maternal progestational hormones followed by abrupt withdrawal at birth, and the typical growth pattern of CRs. Though we have no evidence that this is the mechanism, the occurrence of another lesion in TSC that is highly related to female hormones (pulmonary lymphangioleiomyoma) supports this possible mechanism.

It is also important to note that cardiac tumors may grow and/or appear *de novo* in occasional adolescents with TSC. In a group of 55 TSC patients followed with serial USG, we found 6 (10%) patients in whom CRs appeared *de novo* or grew in size. The ages of the TSC patients in whom new tumors were identified were 10, 12, and 14. In addition, the three children in whom CR tumor growth was seen in serial follow-up studies had ages 11, 12, and 15. Thus, all TSC patients in whom CR growth was observed by echocardiography were in the pubertal age range, consistent with some growth-promoting effect of that process [11]. Despite this increase in size during puberty, none of these lesions were so large as to cause clinical symptoms or require intervention. Thus, routine periodic screening for CR growth during puberty does not seem necessary.

One should also be aware that the natural history of CRs may be influenced by ACTH therapy, which is often used in the treatment of infantile spasms. Hishitani *et al.* [13] reported the rapid enlargement of cardiac rhabdomyomas in two TSC infants receiving corticotropin therapy.

Data concerning a sex difference in the incidence of CRs are inconsistent. In a review of the literature on cardiac rhabdomyomas, Harding and Pagon [7] reported a higher incidence in males than females – 1.55 : 1 (161/104). However, in our experience with 154 TSC patients, the frequency of cardiac rhabomyomas is very similar in males and females (45% versus 44%, respectively) [11].

16.3 Clinical Manifestations

The majority of cardiac tumors in TSC patients have no clinical manifestations. In our series of 74 children with TSC and documented CRs, clinical symptoms were seen in only 29 (39%) patients. When they occurred, symptoms were usually related to tumor size or its location within the heart [11].

Cardiac rhabdomyomas can cause clinical manifestations through at least three different mechanisms: (1) obstruction of an inflow or outflow tract due to a large intracavitary tumor, (2) myocardial involvement extensive enough to impair ventricular function, and (3) cardiac rhythm abnormalities [14].

Although apparently rare, CRs may first cause significant physiologic compromise and even fetal loss during intrauterine development. Obstructive CRs and/or cardiac arrhythmia may lead to development of hydrops fetalis and spontaneous abortion [15–17].

Since postnatal regression of CRs is common, the neonatal or early infantile period is the most common period during which cardiac symptoms may become evident, including sudden death due to large obstructing intracavitary tumors [15, 18, 19]. In patients with very large, hemodynamically significant lesions, early surgical intervention is appropriate [20, 21].

Congestive heart failure develops in 2–5% of TSC children with CRs [11, 22]. CRs arising in the interventricular septum or located in the vicinity of cardiac valves may result in valvular dysfunction and congestive heart failure [23]. These children may demonstrate tachypnea, cyanosis, tachycardia, and cardiomegaly (Figure 16.2). In our series of 154 children with TSC and CRs, only a single child died due to rhabdomyoma. This child had multiple tumors in all heart cavities and presented with cardiac failure at birth. She died at the age of 4 months [11].

In addition to congestive heart failure, infants with TSC may present with cardiomyopathy due to replacement of ventricular muscle by noncontractile tumor tissue. In our series of children with TSC, we had six (4%) patients in whom the diagnosis of cardiomyopathy was established in the first months of life [11]. In follow-up studies, the cardiomyopathy improved/disappeared due to spontaneous regression of the tumor. The prognosis for this group of patients is clearly better than for patients with intracavitary tumors that may result in cardiac failure and early death due to obstruction.

Interestingly, a cardiac murmur is heard in a relatively small proportion of patients with CRs. Only 11 out of 74 (15%) children with TSC and CRs had a detectable cardiac murmur on routine examination [11].

The size of CRs is usually 5–15 mm in diameter. In the vast majority of patients with CRs, the tumors are multiple. In the Fenoglio *et al.* [6] series of 36 patients with

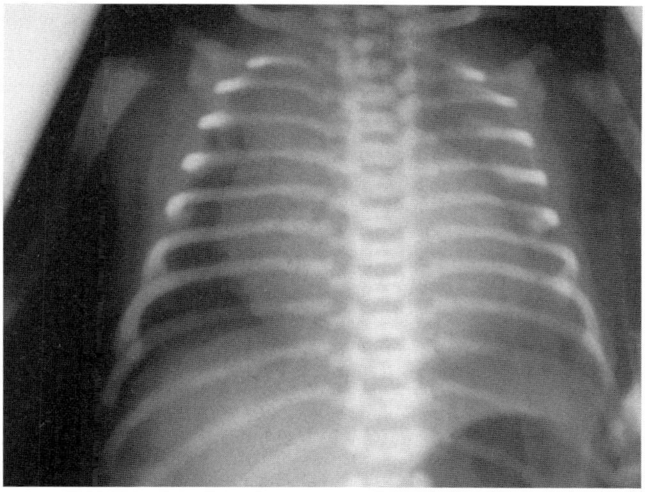

Figure 16.2 Cardiomegaly in a newborn with prenatally diagnosed multiple cardiac rhabdomyomas.

rhabdomyomas only three (8%) had solitary tumors. Multiple CRs are more frequent in children under 2 years of age [11].

CRs may be located in any of the cardiac chambers. There are contradictory data about their appearance in the left or right side of the heart, but all reports confirm a higher incidence of tumors in the ventricles than in the atria [6, 11, 24–26].

There are only a few studies on the incidence of cardiac rhythm abnormalities in children with TSC. The possibility that cardiac tumors might interfere with normal conduction pathways was first raised in 1945 when Duras [27] observed frequent extrasystoles in a patient with severe mental retardation and "adenoma sebaceum." Since then many different types of cardiac rhythm abnormalities have been described in patients with TSC: atrial or ventricular tachycardia, second- or third-degree atrioventricular block, sinus nodal dysfunction, and Wolff–Parkinson–White (WPW) syndrome [28–31].

These rhythm abnormalities in TSC patients are all caused by intramural rhabdomyomas that interrupt normal conduction pathways, leading to ectopic electrical foci or an accessory electrical circuit producing preexcitation (Wolff–Parkinson–White syndrome). WPW syndrome may be diagnosed at all ages [17, 32], leading some to recommend that an electrocardiogram be performed in all newly diagnosed TSC patients [33]. However, the frequency of WPW in TSC is relatively low (9%) and a routine electrocardiogram is a relatively insensitive technique for detection of many arrhythmias and is very poorly sensitive in the detection of cardiac rhabdomyomas [11, 12, 22].

Special attention should be paid to the children with CRs treated with carbamazepine and possibly other antiepileptic drugs that affect cardiac repolarization. Weig and Pollack [34] reported a 13-month-old boy with a single 2–3 mm mass in the intrarventricular septum who developed second-degree atrioventricular block while on carbamazepine. After cessation of this treatment, the rhythm abnormalities resolved. EKG monitoring is recommended in all TSC patients receiving carbamazepine for epileptic seizures.

16.4 Pathology and Molecular Biology of Cardiac Tumors

Cardiac rhabdomyomas are the most frequent cardiac tumors seen in children overall. All cardiac tumors found in TSC patients are rhabdomyomas on pathological examination. However, one case of a pericardial mesothelioma has been reported in a TSC patient [35].

On gross examination, CRs are unencapsulated, have a gray or yellowish-white color, usually measure 5–15 mm in diameter, and are well demarcated from the surrounding myocardium. In autopsy series, the percentage of multiple tumors reaches 92% [6] and is much higher than in echocardiographic studies [11]. Most tumors are intramural sometimes protruding into the cardiac chambers.

The tumor cells range in size from 80 to 200 μm in diameter and are larger than the surrounding myocardial cells. Rhabdomyoma cells appear to be aberrant glycogen-filled myocytes. On routine histologic processing, loss of intracellular glycogen leads to a distinctive appearance referred to as "spider" cells, due to the radial arrangement of residual sarcoplasm extending out from the nucleus [14] (Figure 16.3a and b).

The observation that rhabdomyomas do not undergo malignant transformation and tend to regress on follow-up may support the hypothesis that CR represents a fetal hamartoma rather than a true neoplasm [36].

Following the Knudson model of tumorigenesis in TSC, tumor formation should be accompanied by two independent mutations on both alleles of either TSC1 or TSC2. There is limited support for this model of rhabdomyoma development, as only 4 of 15 (27%) rhabdomyomas have shown loss of heterozygosity (LOH) for markers near TSC1 or TSC2 [37–41]. However, this may reflect technical limitations in the ability to perform this study. Nonetheless, in contrast LOH is seen in the majority of renal angiomyolipomas from TSC patients (see Chapter 4 for additional details) although it is also rare in TSC cortical tubers.

In order to understand the process of formation of CRs in TSC, an animal model of CR has been recently developed [42]. Meikle *et al.* [42] used a conditional, floxed allele of *Tsc1* and a modified myosin light chain 2v allele in which cre recombinase expression occurs in ventricular myocytes. Mice with ventricular loss of Tsc1 developed cardiomyopathy with enlarged myocytes containing excessive glycogen and expressing elevated levels of phospho-S6, similar to findings in human rhabdomyoma cells. The authors concluded that the results confirm that rhabdomyomas occur through a two-hit mechanism of pathogenesis. However, there were distinct differences between the pathology seen in these mice and in human rhabdomyomas.

In addition, the lack of consistent finding of TSC1 or TSC2 LOH in CRs has led to recent studies to investigate potential alternative mechanisms of pathogenesis.

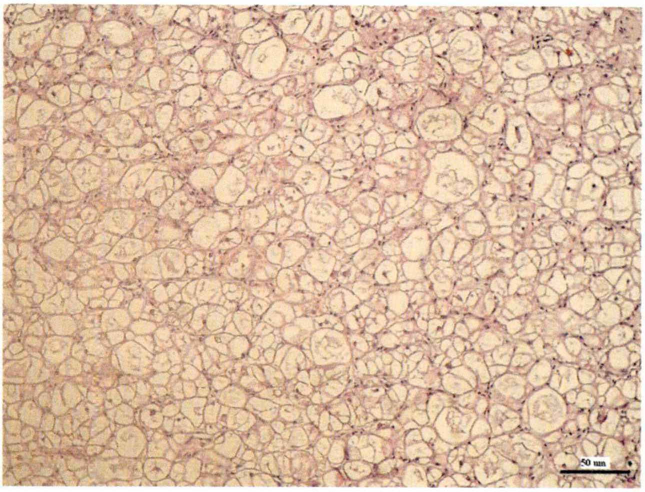

(a)

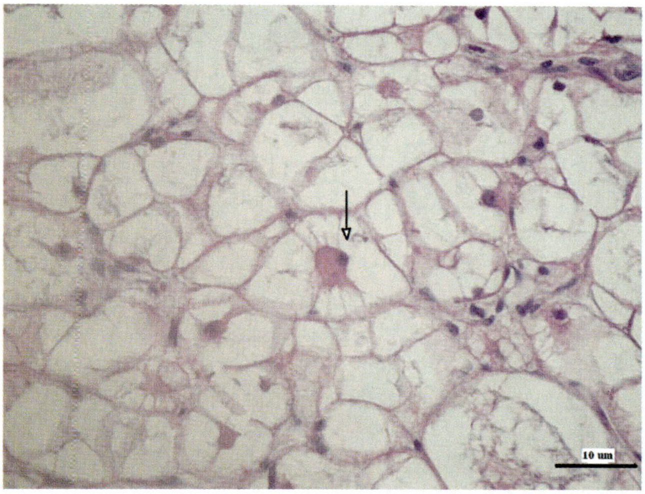

(b)

Figure 16.3 (a) Cardiac rhabdomyoma. Well-differentiated, large, rounded, and polygonal cells with abundant acidophilic cytoplasm containing variable amounts of lipid and glycogen (hematoxylin and eosin, 100×). Courtesy of Dr Wiesława Grajkowska, The Children's Memorial Health Institute, Warsaw. (b) Typical "spider cells" of rhabdomyoma (in the middle of the picture) with centrally located nucleus and peripheral threads of sarcoplasm (hematoxylin and eosin, 400×). Courtesy of Dr Wiesława Grajkowska, The Children's Memorial Health Institute, Warsaw.

We studied several TSC patients in whom cardiac rhabdomyomas were excised. One tumor was heterozygous for a F1510 deletion in exon 34 of *TSC2*, arguing against an LOH event. The mTOR, Akt, and Erk pathways were studied in this tumor, with findings that Akt was not upregulated, while mTOR and Erk and their substrates Elk-1 and S6K1 were hyperactive [43].

Increased expression of mTOR, pS6K, pErk, and 4E-BP1 has been seen consistently in several CR samples [44]. However, TSC1 and TSC2 expression was also decreased in these samples, suggesting the possibility of occult LOH, or silencing of the remaining normal allele of TSC1 or TSC2 by other means.

As discussed above, the peculiar natural history of rhabdomyomas – that is, growth prenatally followed by regression postnatally – is very poorly understood. We have previously suggested that high-level progestational hormone exposure during embryonic development contributes to transient cardiac rhabdomyoma growth. However, this is an hypothesis that is difficult to prove. Some differences in expression of Bax in rhabdomyomas in comparison to normal heart tissue have been seen by Kotulska *et al.* [44], suggesting a possible defect in apoptosis regulation. However, much more work needs to be done to have a detailed understanding of the pathogenesis of these intriguing neoplasms [45].

16.5 Diagnosis

As the vast majority of TSC patients with rhabdomyomas are completely asymptomatic, cardiovascular symptoms and signs lead to a diagnosis of CRs in a very small proportion of patients with TSC. Major progress in the diagnosis of CRs occurred with the introduction of sector echocardiography, allowing noninvasive detection of tumors including the use of prenatal screening ultrasounds. This modality became and remains today the diagnostic procedure of choice for CRs. Echocardiography can assess the tumor size, shape, attachment, and mobility and can efficiently screen all four chambers. In our series of 74 children in whom CRs were revealed echocardiographically, 45 patients (61%) remained asymptomatic [11].

According to current recommendations of the Tuberous Sclerosis Consensus Conference, an echocardiographic examination should be carried out in TSC patients of any age with symptoms of a CR [46]. However, due to the potential risk of congestive heart failure or sudden death in patients with multiple CRs, as well as the diagnostic value of this noninvasive procedure, our practice is to carry out at least one cardiac ultrasound examination in all patients with TSC, as soon as the diagnosis is considered. In addition, there are several reports of sudden death in asymptomatic children who had multiple CRs at autopsy [19]. Echocardiography, together with electrocardiographic examination, enables identification of children with the highest risk of sudden death.

However, as for all noninvasive diagnostic procedures, there are limits to the resolution of echocardiography. Nir *et al.* [22] reported a child who had a normal echocardiography at 11 months of age, and then died suddenly 1.5 years later. Multiple CRs were found at autopsy.

Even if echocardiography remains a gold standard in the diagnosis of CRs in TSC patients, there is an increasing number of publications indicating a role of cardiac MRI in detecting the tumors [47]. MRI seems to be particularly useful in children with cardiac tumors in whom the diagnosis of rhabdomyomas is considered unlikely. In such cases, MRI may help to differentiate tumor tissue type and determine the prognosis. In general, tumor location, size, and borders are best determined by T1-weighted standard spin echo, or more recently by fast spin echo with double inversion recovery sequences obtained in orthogonal planes. A T2-weighted spin echo sequence is helpful in distinguishing vascular from avascular tumors [47]. For further evaluation, electrocardiogram-gated MRI may be used.

According to the latest revision of the diagnostic criteria for TSC, single or multiple CRs are considered a major diagnostic criterion of this entity [48]. However, our experience has been that multiple CRs hold a greater diagnostic significance than single lesions. As described above, there is reasonable evidence that nearly all infants with multiple CRs have a diagnosis of TSC. In our experience, 29 of 30 patients with multiple CRs had a diagnosis of TSC and the single remaining patient continues in follow up (unpublished data). Another large series of 94 pediatric patients with cardiac tumors also found a much higher risk of TSC in patients with multiple (61 of 64; 95%) versus single cardiac tumors (7 of 30; 23%) ($p < 0.001$)[49].

The diagnostic significance of CRs is especially high in newborns and small infants, in whom other features of TSC may be absent. Echocardiography appears to be the most useful single diagnostic test for TSC in this age group. However, the absence of detectable cardiac lesions in a newborn does not exclude a diagnosis of TSC.

It should be noted that most patients from the literature with "idiopathic" CRs are small infants with very short follow-up. In such cases, the diagnosis of TSC cannot be excluded based upon negative skin examination and neuroimaging studies, or any other criteria. In patients who have negative results by clinical examination for signs of TSC, molecular genetic analysis may be considered for diagnostic confirmation.

It is worth stressing that according to consensus diagnostic criteria, a TSC diagnosis may be established by the presence of single or multiple CRs and three hypomelanotic macules [48]. Therefore, a very useful approach in infants and young children is careful and repeated skin examinations, as signs of TSC may become apparent and give a definite diagnosis [50].

16.6 Fetal Cardiac Rhabdomyomas and Diagnosis of TSC

The introduction of high-resolution prenatal ultrasounds during pregnancy has led to increased recognition of fetal CRs [51, 52]. The largest meta-analysis of prenatal CRs was published in 2008 and includes 266 fetuses [53]. The earliest diagnosis of a CR was made during the 15th week of gestation [49], but most cases are detected after 24 weeks of gestational age. Cardiac rhabdomyoma may be the earliest sign of TSC

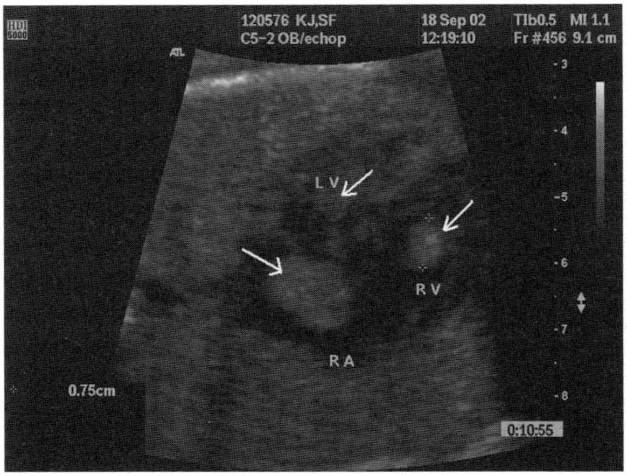

Figure 16.4 Multiple fetal heart tumors at 35 weeks of gestation: tumors (arrows) in both ventricles and the right atrium. RV, right ventricle; LV, left ventricle; RA, right atrium.

in utero, and definitely are the easiest to detect antenatally. Their appearance generally precedes the detection of brain or kidney lesions. Findings of fetal CRs clearly provide a major clue toward the diagnosis of TSC.

The size of the fetal heart tumors varies from 3 to 52 mm in diameter. Large tumors (>20 mm) have a higher risk of prenatal death [53]. Small fetal heart tumors generally have little effect, similar to those in the newborn, allowing normal fetal growth. They are typically seen in the left or right ventricular outflow tracts. However, fetal CRs may cause cardiac arrhythmias and result in fetal or neonatal sudden death. Fetal CRs diagnosis is established using 2D ultrasound imaging and fetal Doppler (Figure 16.4). In recent years, 3D echocardiography has also been used to characterize these lesions [54]. To suppress fetal motion artifact and obtain a better visualization of the tumors, fetal ultrafast sequences have been recently reported [55–57].

This rapidly growing literature on the prenatal diagnosis of cardiac tumors has raised the same question posed after birth: how many of these tumors are due to TSC [8, 49, 53, 58–64]? As discussed above, in some cases CRs will be the sole manifestation of TSC in the newborn, leading to an underestimate of the prevalence of TSC. Thus, with short follow-up, the overall prevalence of TSC in these infants is 68%, but it is likely that the true prevalence is much higher.

Therefore, in our opinion all children with prenatally diagnosed multiple CRs should be considered to have probable (if not confirmed) TSC. Prenatal neurosonography or brain MRI can confirm the diagnosis in about 50% of fetuses [57, 65] (Figure 16.5). The earliest fetal brain MRI diagnosis of TSC has been reported in a 23-week fetus [62]. After birth, infants with multiple CRs should be examined thoroughly for signs of TSC, and molecular genetic analysis for germline mutations in TSC1 and TSC2 should be considered to enable early implementation of further diagnostic evaluation and genetic counseling [66].

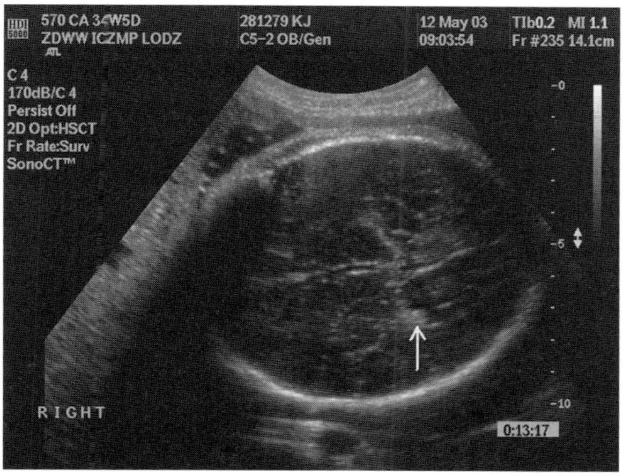

Figure 16.5 Periventricular calcification (arrow) on prenatal ultrasonography of the brain at 35 weeks of gestation in a child with multiple cardiac tumors. The examination confirms the diagnosis of TSC.

16.7 Treatment

Prenatal echocardiography (with positive brain MRI) allows the antenatal diagnosis of TSC. This noninvasive procedure should be offered to every family in which one member (parent or child) is affected with TSC. To be most comprehensive, we recommend two fetal echocardiography examinations: the first one at the 13–14th week of pregnancy, and the second at the 18–20th week. Early detection of cardiac tumors enables early treatment planning if the child is going to be carried to term. Alternatively, this approach allows consideration of therapeutic abortion, if desired.

However, the decision about method of delivery (vaginal versus cesarean section) of a fetus with cardiac tumors is complex. Under optimal circumstances (a second or third delivery, good fetal condition, normal heart anatomy, and normal heart study by fetal targeted echocardiography), spontaneous vaginal delivery can proceed. However, in the case of fetal deterioration, functional fetal heart abnormalities (evidence of critical aortic or pulmonary stenosis, or atrioventricular valve insufficiency), or signs of congestive heart failure, elective cesarean section is appropriate.

In the case of a neonate with a prenatal diagnosis of multiple cardiac rhabdomyoma, prostaglandin intravenous infusion is administered for hours or days to maintain a fetal type of circulation [67]. Neonatal echocardiography may be repeated weekly during the first month of life and later on monthly to provide information about progression or regression of the tumor. After birth all patients with CRs with suspicion or diagnosis of TSC should be referred to a pediatric neurologist. Careful neurological assessment is important to enhance detection of infantile spasms and other seizures, to facilitate early treatment, and to improve cognitive outcome [68, 69].

Almost all CRs regress spontaneously after birth, so a conservative approach in the majority of patients is recommended. More intensive follow-up echocardiography evaluations may be required if the patient is considered to be at relative risk of lesion growth (e.g., due to ACTH treatment) or cardiac failure appears.

Some patients with congestive heart failure may require treatment with digitalis, diuretics, and salt restriction. Cardiac arrhythmia should be treated with appropriate antiarrhythmic drugs. In patients with rhythm abnormalities resistant to drug medication, a cardiac pacemaker or other intervention should be considered.

Surgical excision of a cardiac rhabdomyoma in infancy is appropriate only in the presence of life-threatening hemodynamic compromise. In these patients, the operative risk is high and depends on the number, size, and location of the tumors. However, the prognosis for survival without surgical intervention in these critically ill children is very poor. In such cases, limited tumor resection, in order to limit perioperative morbidity, may be sufficient as the residual tumor mass will probably involute. In infants with large multiple CRs producing severe heart failure, cardiac transplantation may be the best approach.

Recently, several clinical trials of rapamycin in TSC have been initiated, and a few have been reported [70]. Thus far, the potential benefit of rapamycin or related drugs for treatment of rhabdomyomas has not been reported. Although these drugs are immunosuppressants that have significant risk in the infant, they may also be considered for therapy in this situation.

16.8 Genotype–Phenotype Correlations with Rhabdomyomas

A single report has examined the frequency of rhabdomyomas according to mutation genotype in TSC patients [11]. Cardiac tumors were seen in 50 (54%) of 93 patients with TSC2 mutations, in comparison to 3 (20%) of 15 of those with TSC1 mutations ($p < 0.0001$). In addition, cardiac failure has been reported in four TSC patients with TSC2 mutations, but none with TSC1 mutations. Thus, similar to other manifestations of TSC, cardiac rhabdomyomas are more common and more severe in patients with TSC2 mutations (see Chapter 5 for more details).

16.9 Vascular Abnormalities in TSC

Arterial aneurysms have been reported in multiple TSC patients, at an early age, leading Davidson [71] to suggest that they might be caused by a congenital defect of the arterial wall in TSC. In this respect, there is similarity between TSC and polycystic kidney disease, as both conditions are associated with a relatively high incidence of intracranial aneurysms and arteriopathy.

At present, there are about 20 patients with TSC and aortic aneurysms described in the literature. In a major review, Jost found that the aneurysm was localized to the abdominal aorta in 75% of cases and in the descending thoracic aorta in the remainder [72]. The mean age at diagnosis of abdominal aortic aneurysm was

5 years, and the mean age at the diagnosis of thoracic aortic aneurysm was 11.7 years. The youngest TSC patient reported with an aneurysm was 4.5 months old [73]. There appears to be a high risk of rupture, with 40% patients dying due to aneurysmal disease. This high vascular mortality has lead some authors to suggest careful screening for aortic aneurysms, particularly in young TSC children at diagnosis and at regular 2–3-year intervals thereafter [72].

Pathological examination of resected aneurysms usually reveals dysplastic features in addition to loss of elastin fibers, medial atrophy, and focal medial disruption. Dysplasia and medial atrophy contribute to the loss of aortic wall strength and aneurysm formation [72]. These findings suggest the possibility that there is focal loss of the second allele of TSC1/TSC2 in these lesions, leading to aberrant smooth muscle cell function.

Aortic aneurysms are accompanied, at times, by vascular lesions in other arteries, including the renal arteries [74, 75]. In addition, aneurysms of other large arteries have also been reported in TSC, without apparent involvement of the aorta [76, 77]. Aneurysms are relatively common in kidney angiomyolipomas in TSC patients, but appear to be intrinsic to those lesions [78] (Chapter 15). Nonetheless, the two-hit mechanism that has been confirmed in angiomyolipomas provides further support for this model of development of aortic and other aneurysms.

Carotid artery aneurysms have also been reported in multiple TSC patients (Figure 16.6) [79, 80]. Occasionally, these aneurysms extend to involvement of the intracerebral circulation. Strokes have been rarely reported in TSC children, and it seems likely that these are due to aneurismal involvement of the cerebral artery leading to dissection and thrombotic occlusion [81].

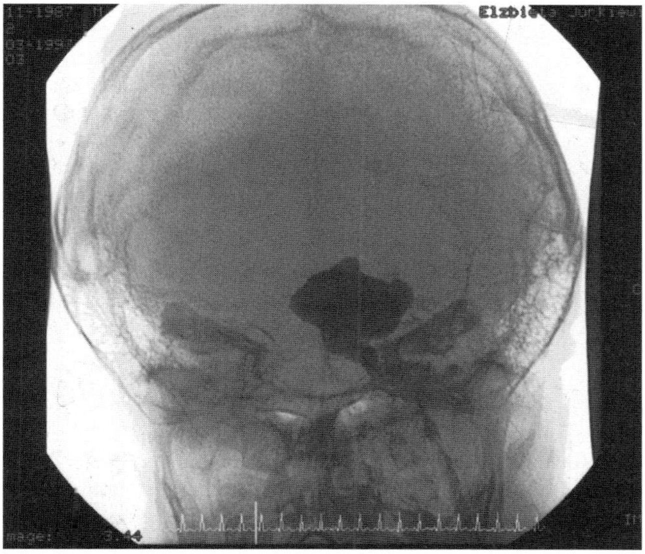

Figure 16.6 Digital conventional angiography in a 9-year-old boy with TSC. Partially thrombosed, giant saccular aneurysm of the left supraclinoid internal carotid artery. Reproduced by courtesy of Prof. Elzbieta Jurkiewicz, The Children's Memorial Health Institute, Warsaw.

References

1 Von Recklinghausen, F. (1862) Ein Herz von einem Neugeborenen, welches mehrere theils nach aussen, theils nach den Höhlen prominirende tumouren (myomen) trug. *Monatschr. Geburtsk. Frauenkr.*, **20**, 1–2.

2 Shepherd, C.W., Gomez, M.R., and Crowson, C.S. (1991) Causes of death in patients with tuberous sclerosis. *Mayo Clin. Proc.*, **66**, 792–796.

3 Freedom, R.M., Lee, K.J., MacDonald, C., and Taylor, G. (2000) Selected aspects of cardiac tumors in infancy and childhood. *Pediatr. Cardiol.*, **21**, 299–316.

4 Nadas, A.S. and Ellison, R.C. (1968) Cardiac tumors in infancy. *Am. J. Cardiol.*, **21**, 363–366.

5 Butany, J., Nair, V., Naseemuddin, A., Nair, G.M., Catton, C., and Yau, T. (2005) Cardiac tumors: diagnosis and management. *Lancet Oncol.*, **6**, 219–228.

6 Fenoglio, J.J., McAllister, H.A., and Ferrans, V.J. (1976) Cardiac rhabdomyoma: a clinicopathologic and electron microscopic study. *Am. J. Cardiol.*, **38**, 241–251.

7 Harding, C.O. and Pagon, R.A. (1990) Incidence of tuberous sclerosis in patients with cardiac rhabdomyoma. *Am. J. Med. Genet.*, **37**, 443–446.

8 Bader, R.S., Chitayat, D., Kelly, E., Ryan, G., Smallhorn, J.F., Toi, A., and Hornberger, L.K. (2003) Fetal rhabdomyomas: prenatal diagnosis, clinical outcome, and incidence of associated tuberous sclerosis complex. *J. Pediatr.*, **143**, 620–624.

9 Jóźwiak, S., Domańska-Pakieła, D., Kwiatkowski, D., and Kotulska, K. (2005) Multiple cardiac rhabdomyomas as a sole symptom of tuberous sclerosis complex: case report with molecular confirmation. *J. Child Neurol.*, **20**, 988–989.

10 Smith, H.C., Watson, G.H., Patel, R.G., and Super, M. (1989) Cardiac rhabdomyomata in tuberous sclerosis: their course and diagnostic value. *Arch. Dis. Child.*, **64**, 196–200.

11 Jóźwiak, S., Kotulska, K., Kasprzyk-Obara, J., Domańska-Pakieła, D., Tomyn-Drabik, M., Roberts, P., and Kwiatkowski, D. (2006) Clinical and genotype studies of cardiac tumors in 154 patients with tuberous sclerosis complex. *Pediatrics*, **118** (4), e1146–e1151.

12 Muhler, E.G., Turniski-Harder, V., Engelhardt, W., and von Bernuth, G. (1994) Cardiac involvement in tuberous sclerosis. *Br. Heart J.*, **72**, 584–590.

13 Hishitani, T., Hoshino, K., Ogawa, K. *et al.* (1997) Rapid enlargement of cardiac rhabdomyoma during corticotropin therapy for infantile spasms. *Can. J. Cardiol.*, **13**, 72–74.

14 Mair, D.D., Edwards, W.D., and Seward, J.B. (1999) Cardiac manifestations, in *Tuberous Sclerosis Complex* (eds M.R. Gomez, J.R. Sampson, and V.H. Whittemore), Oxford University Press, New York, pp. 194–206.

15 De Leon, G.A., Zaeri, N., and Foley, C.M. (1988) Olfactory hamartomas in tuberous sclerosis. *J. Neurol. Sci.*, **87**, 187–194.

16 Guereta, L.G., Burgueros, M., Elorza, M.D., Alix, A.G., Benito, F., and Gamallo, C. (1986) Cardiac rhabdomyoma presenting as fetal hydrops. *Pediatr. Cardiol.*, **7**, 171–174.

17 Masoumi, H., Kinney, H.C., Chadwick, A.E., Rubio, A., and Krous, H.F. (2007) Sudden unexpected death in childhood associated with cardiac rhabdomyomas, involuting adrenal ganglioneuroma, and megalencephaly: another expression of tuberous sclerosis. *Pediatr. Dev. Pathol.*, **10**, 129–133.

18 Bordarier, C., Lellouch-Tubiana, A., and Robain, O. (1994) Cardiac rhabdomyoma and tuberous sclerosis in three fetuses: a neuropathological study. *Brain Dev.*, **16**, 467–471.

19 Grellner, W. and Henssge, C. (1996) Multiple cardiac rhabdomyoma with exclusively histological manifestation. *Forensic Sci. Int.*, **78**, 1–5.

20 Ibrahim, C.P.H., Thakker, P., Miller, P.A., and Barron, D. (2003) Cardiac rhabdomyomas presenting as left ventricular outflow tract obstruction in a neonate. *Interact. Cardiovasc. Thorac. Surg.*, **2**, 572–574.

21 Karnak, I., Alehan, D., Ekinci, S., and Buyukpamukcu, N. (2007) Cardiac rhabdomyoma as an unusual mediastinal mass in a newborn. *Pediatr. Surg. Int.*, **23** (8) 811–814.

22 Nir, A., Tajik, A.J., Freeman, W.K., Seward, J.B., Offord, K.P., Edwards, W.D., Mair, D.D., and Gomez, M.R. (1995) Tuberous sclerosis and cardiac rhabdomyoma. *Am. J. Cardiol.*, **6**, 419–421.

23 Ilina, M.V., Jaeggi, E.T., and Lee, K.J. (2007) Neonatal rhabdomyoma causing right ventricular inflow obstruction with duct-dependent pulmonary blood flow: successful stenting of PDA. *Catheter. Cardiovasc. Interv.*, **69** (6), 881–885.

24 Bass, J.L., Breningstall, G.N., and Swaiman, K.F. (1985) Echocardiographic incidence of cardiac rhabdomyoma in tuberous sclerosis. *Am. J. Cardiol.*, **55**, 1379–1382.

25 Bosi, G., Lintermans, J.P., Pellegrino, P.A., Svaluto-Moreolo, G., and Vliers, A. (1996) The natural history of cardiac rhabdomyoma with and without tuberous sclerosis. *Acta Paediatr.*, **85**, 928–931.

26 Jóźwiak, S., Kawalec, W., Dłużewska, J., Daszkowska, J., Mirkowicz-Małek, M., and Michałowicz, R. (1994) Cardiac tumours in tuberous sclerosis: their incidence and course. *Eur. J. Pediatr.*, **153**, 155–157.

27 Duras, F.P. (1945) Cardiac manifestations in a case of tuberous sclerosis. *Br. Heart J.*, **7**, 37–40.

28 Cowley, C.G., Tani, L.Y., Judd, V.E., and Shaddy, R.E. (1996) Sinus node dysfunction in tuberous sclerosis. *Pediatr. Cardiol.*, **17**, 51–52.

29 O'Callaghan, F.J., Clarke, A.C., Joffe, H., Keeton, B., Martin, R., Salmon, A., Thomas, R.D., and Osborne, J.P. (1998) Tuberous sclerosis complex and Wolff–Parkinson–White syndrome. *Arch. Dis. Child.*, **78**, 159–162.

30 Kagan, K.O., Schmidt, M., Kuhn, U., and Kimmig, R. (2004) Ventricular outflow obstruction, valve aplasia, bradyarrhythmia, pulmonary hypoplasia and non-immune fetal hydrops because of a large rhabdomyomas in a case of unknown tuberous sclerosis: a prenatal diagnosed cardiac rhabdomyomas with multiple symptoms. *Int. J. Obstet. Gynaecol.*, **111**, 1478–1480.

31 Shiono, J., Horigome, H., Yasui, S., Miyamoto, T., Takahashi-Igari, M., Iwasaki, N., and Matsui, A. (2003) Electrocardiographic changes in patients with cardiac rhabdomyomas associated with tuberous sclerosis. *Cardiol. Young*, **13**, 258–263.

32 Bar-Cohen, Y., Silka, M.J., and Sklansky, M.S. (2007) Neonatal tuberous sclerosis and multiple cardiac arrhythmias. *Circulation*, **115**, e395–e397.

33 Mohan, P., Raghavan, K., and Sagili, H. (2008) Wolff–Parkinson–White syndrome in association with tuberous sclerosis. *N. Z. Med. J.*, **121**, 71–74.

34 Weig, S.G. and Pollack, P. (1993) Carbamazepine-induced heart block in a child with tuberous sclerosis and cardiac rhabdomyoma: implications for evaluation and follow up. *Ann. Neurol.*, **34**, 617–619.

35 Naramoto, A., Itoh, N., Nakano, M., and Shigematsu, H. (1989) An autopsy case of tuberous sclerosis associated with primary pericardial mesothelioma. *Acta Pathol. Jpn.*, **39**, 400–406.

36 Curatolo, P., Bombardieri, R., and Jozwiak, S. (2008) Tuberous sclerosis. *Lancet*, **372** (9639), 657–668.

37 Carbonara, C., Longa, L., Grosso, E., Mazzucco, G., Borrone, C., Garrè, M.L., Brisigotti, M., Filippi, G., Scabar, A., Giannotti, A., Falzoni, P., Monga, G., Garini, G., Gabrielli, M., Riegler, P., Danesino, C., Ruggieri, M., Magro, G., and Migone, N. (1996) Apparent preferential loss of heterozygosity at TSC2 over TSC1 chromosomal region in tuberous sclerosis hamartomas. *Genes Chromosomes Cancer*, **15**, 18–25.

38 Henske, E.P., Scheithauer, B.W., Short, M.P., Wollmann, R., Nahmias, J., Hornigold, N., van Slegtenhorst, M., Welsh, C.T., and Kwiatkowski, D.J. (1996) Allelic loss is frequent in tuberous sclerosis kidney lesions but rare in brain lesions. *Am. J. Hum. Genet.*, **59**, 400–406.

39 Niida, Y., Stemmer-Rachamimov, A.O., Logrip, M., Tapon, D., Perez, R., Kwiatkowski, D.J., Sims, K., MacCollin, M., Louis, D.N., and Ramesh, V. (2001)

Survey of somatic mutations in tuberous sclerosis complex hamartomas suggests different genetic mechanisms for pathogenesis of TSC lesions. *Am. J. Hum. Genet.*, **69**, 493–503.

40 Green, A.J., Smith, M., and Yates, J.R. (1994) Loss of heterozygosity on chromosome 16p13.3 in hamartomas from tuberous sclerosis patients. *Nat. Genet.*, **6** (2), 193.

41 Sepp, T., Yates, J.R., and Green, A.J. (1996) Loss of heterozygosity in tuberous sclerosis hamartomas. *J. Med. Genet.*, **33**, 962–964.

42 Meikle, L., McMullen, J.R., Sherwood, M.C., Lader, A.S., Walker, V., Chan, J.A., and Kwiatkowksi, D.J. (2005) A mouse model of cardiac rhabdomyomas generated by loss of TSC1 in ventricular myocytes. *Hum. Mol. Genet.*, **14**, 429–435.

43 Jóźwiak, J., Sahin, M., Jóźwiak, S., Kotulska, K., Ploski, R., Szperl, M., and Wlodarski, P.K. (2009) Cardiac rhabdomyoma in tuberous sclerosis: hyperactive Erk signaling. *Int. J. Cardiol.*, **132** (1), 145–147.

44 Kotulska, K., Larysz-Brysz, M., Grajkowska, W., Jóźwiak, J., Włodarski, P., Sahin, M., Lewin-Kowalik, J., Domańska-Pakieła, D., and Jóźwiak, S. (2009) Cardiac rhabdomyomas in tuberous sclerosis complex show apoptosis regulation and mTOR pathway abnormalities. *Pediatr. Dev. Pathol.*, **12** (2), 89–95.

45 Jóźwiak, J., Jóźwiak, S., and Wlodarski, P. (2008) Possible mechanisms of disease development in tuberous sclerosis. *Lancet Oncol.*, **9** (1), 73–79.

46 Roach, E.S., DiMario, F.J., Kandt, R.S., and Northrup, H. (1999) Tuberous sclerosis complex consensus conference: recommendations for diagnostic evaluation. *J. Child Neurol.*, **14**, 401–407.

47 Kiaffas, M.G., Powell, A.J., and Geva, T. (2002) Magnetic resonance imaging evaluation of cardiac tumor characteristics in infants and children. *Am. J. Cardiol.*, **89**, 1229–1233.

48 Roach, E.S., Gomez, M.R., and Northrup, H. (1998) Tuberous sclerosis complex consensus conference: revised clinical diagnostic criteria. *J. Child Neurol.*, **13**, 624–628.

49 Tworetzky, W., McElhinney, D.B., Margossian, R., Moon-Grady, A.J., Sallee, D., Goldmuntz, E., van der Velde, M.E., Silverman, N.H., and Allan, L.D. (2003) Association between cardiac tumors and tuberous sclerosis in the fetus and neonate. *Am. J. Cardiol.*, **92**, 487–489.

50 Jóźwiak, S., Schwartz, R.A., Krysicka-Janniger, C., and Bielicka-Cymerman, J. (2000) Usefulness of diagnostic criteria of tuberous sclerosis complex in pediatric patients. *J. Child Neurol.*, **15**, 652–659.

51 Czechowski, J., Langille, E.L., and Varady, E. (2000) Intracardiac tumour and brain lesions in tuberous sclerosis. A case report of antenatal diagnosis by ultrasonography. *Acta Radiol.*, **41**, 371–374.

52 Gushiken, B.J., Callen, P.W., and Silverman, N.H. (1999) Prenatal diagnosis of tuberous sclerosis in monozygotic twins with cardiac masses. *J. Ultrasound Med.*, **18**, 165–168.

53 Chao, A.S., Chao, A., Wang, T.H., Chang, Y.C., Chang, Y.L., Hsieh, C.C., Lien, R., and Su, W.J. (2008) Outcome of antenatally diagnosed cardiac rhabdomyomas: case series and a meta-analysis. *Ultrasound Obstet. Gynecol.*, **31**, 289–295.

54 Hata, T., Yan, F., Dai, S.Y., Kanenishi, K., and Yanagihara, T. (2007) Real-time 3-dimensional echocardiographic features of fetal cardiac tumor. *J. Clin. Ultrasound*, **35** (6), 338–340.

55 Chen, C.P., Liu, Y.P., Huang, J.K., Chang, T.Y., Chen, M.R., Chiu, N.C., and Wang, W. (2005) Contribution of ultrafast magnetic resonance imaging in prenatal diagnosis of sonographically undetected cerebral tuberous sclerosis associated with cardiac rhabdomyomas. *Prenat. Diagn.*, **25** (6), 523–524.

56 Kivelitz, D.E., Mühler, M., Rake, A., Scheer, I., and Chaoui, R. (2004) MRI of cardiac rhabdomyoma in the fetus. *Eur. Radiol.*, **14** (8), 1513–1516.

57 Mühler, M.R., Rake, A., Schwabe, M., Schmidt, S., Kivelitz, D., Chaoui, R., and Hamm, B. (2007) Value of fetal cerebral MRI in sonographically proven cardiac

rhabdomyoma. *Pediatr. Radiol.*, **37** (5), 467–474.

58 Geipel, A., Krapp, M., Germer, U., Becker, R., and Gembruch, U. (2001) Perinatal diagnosis of cardiac tumors. *Ultrasound Obstet. Gynecol.*, **17**, 17–21.

59 Gamzu, R., Achiron, R., Hegesh, J., Weiner, E., Tepper, R., Nir, A., Rabinowitz, R., Auslander, R., Yagel, S., Zalel, Y., and Zimmer, E. (2002) Evaluating the risk of tuberous sclerosis in cases with prenatal diagnosis of cardiac rhabdomyoma. *Prenat. Diagn.*, **22**, 1044–1047.

60 Pipitone, S., Mongiovi, M., Grillo, R., Gagliano, S., and Sperandeo, V. (2002) Cardiac rhabdomyoma in intrauterine life: clinical features and natural history. A case series and review of published reports. *Ital. Heart J.*, **3**, 48–52.

61 Das, B.B. and Sharma, J. (2003) Cardiac rhabdomyoma and tuberous sclerosis: prenatal diagnosis and follow-up. *Indian J. Pediatr.*, **70**, 87–89.

62 Fesslova, V., Villa, L., Rizzuti, T., Mastrangelo, M., and Mosca, F. (2004) Natural history and long-term outcome of cardiac rhabdomyomas detected prenatally. *Prenat. Diagn.*, **24**, 241–248.

63 Zhou, Q.C., Fan, P., Peng, Q.H., Zhang, M. Fu, Z., and Wang, C.H. (2004) Prenatal echocardiographic differential diagnosis of fetal cardiac tumors. *Ultrasound Obstet. Gynecol.*, **23**, 165–171.

64 Jóźwiak, S. and Kotulska, K. (2006) Are all prenatally diagnosed multiple cardiac rhabdomyomas a sign of tuberous sclerosis? *Prenat. Diagn.*, **26**, 867–869.

65 Wortmann, S.B., Reimer, A., Creemers, J.W.T., and Mullaart, R.A. (2008) Prenatal diagnosis of cerebral lesions in tuberous sclerosis complex. Case report and review of the literature. *Eur. J. Paediatr. Neurol.*, **12**, 123–126.

66 Milunsky, A., Shim, S.H., Ito, M., Jaekle, R.K., Bassett, L.L., Brumund, M.R., and Milunsky, J.M. (2005) Precise prenatal diagnosis of tuberous sclerosis by sequencing the *TSC2* gene. *Prenat. Diagn.*, **25**, 582–585.

67 Lacey, S.R. and Donofrio, M.T. (2007) Fetal cardiac tumors: prenatal diagnosis and outcome. *Pediatr. Cardiol.*, **28**, 61–67.

68 Jóźwiak, S., Domańska-Pakiela, D., Kotulska, K., and Kaczorowska, M. (2007) Treatment before seizures: new indications for antiepileptic therapy in children with tuberous sclerosis. *Epilepsia*, **48** (8), 632.

69 Schwartz, R.A., Fernandez, G., Kotulska, K., and Jóźwiak, S. (2007) Tuberous sclerosis complex: advances in diagnosis, genetics and management. *J. Am. Acad. Dermatol.*, **57**, 189–202.

70 Bissler, J.J., McCormack, F.X., Young, L.R., Elwing, J.M., Chuck, G., Leonard, J.M., Schmithorst, V.J., Laor, T., Brody, A.S., Bean, J., Salisbury, S., and Franz, D.N. (2008) Sirolimus for angiomyolipoma in tuberous sclerosis complex or lymphangioleiomyomatosis. *N. Engl. J. Med.*, **358** (2), 140–145.

71 Davidson, S. (1974) Tuberous sclerosis with fusiform aneurysms of both internal carotid arteries manifested by unilateral visual loss and papilledema. *Bull. Los Angeles Neurol. Soc.* **39**, 128–132.

72 Jost, C.J., Gloviczki, P., Edwards, W.D., Stanson, A.W., Joyce, J.W., and Pairolero, P.C. (2001) Aortic aneurysms in children and young adults with tuberous sclerosis: report of two cases and review of the literature. *J. Vasc. Surg.*, **33**, 639–642.

73 Lavocat, M.P., Teyssier, G., Allard, D., Tronchet, M., and Freycon, F. (1992) Abdominal aortic aneurysm and Bourneville's tuberous sclerosis. *Pediatrie [French]*, **47**, 517–519.

74 Paraf, F. and Bruneval, P. (1996) Dysplasie fibromusculaire arteérielle et sclérose tubéreuse de Bourneville. *Ann. Pathol.*, **16**, 203–206.

75 Ng, S.H., Ng, K.K., Pai, S.C., and Tsai, C.C. (1988) Tuberous sclerosis with aortic aneurysm and rib changes: CT demonstration. *J. Comput. Assist. Tomogr.*, **12**, 666–668.

76 Calgani, G., Gesualdo, F., Tamisier, D., Brunelle, F., Sidi, D., and Ou, P. (2008) Arterial aneurysms and tuberous sclerosis: a classic but little known association. *Pediatr. Radiol.*, **38**, 795–797.

77 Carette, M.F., Antoine, M., Bazelly, B., Cadranel, J., and Khalil, A. (2006) Primary pulmonary artery aneurysm in tuberous sclerosis: CT, angiography and pathological study. *Eur. Radiol.*, **16**, 2369–2370.

78 Martinez-Barcina, M.J., Quintian Schwieters, C., Bonnin-Sanchez, D., and Alberola Bou, J.M. (2008) Intratumoral pseudoaneurysm: atypical radiologic manifestation of bleeding from an angiomyolipoma. *Radiologia*, **50**, 79–81.

79 Beltramello, A., Puppini, G., Bricolo, A., Andreis, I.A., el-Dalati, G., Longa, L., Polidoro, S., Zavarise, G., and Marradi, P. (1999) Does the tuberous sclerosis complex include intracranial aneurysms? A case report with a review of the literature. *Pediatr. Radiol.*, **29**, 206–211.

80 Jurkiewicz, E. and Jóźwiak, S. (2006) Giant intracranial aneurysm in a 9-year-old boy with tuberous sclerosis. *Pediatr. Radiol.*, **36**, 463.

81 Gomez, M.R. (1989) Strokes in tuberous sclerosis: are rhabdomyomas a cause? *Brain Dev.*, **11**, 14–19.

17
Lymphangioleiomyomatosis and Pulmonary Disease in TSC

Francis X. McCormack and Elizabeth P. Henske

17.1
Introduction

Lymphangioleiomyomatosis (LAM) is the primary pulmonary manifestation of tuberous sclerosis complex (TSC). LAM occurs almost exclusively in females, for reasons that are not understood, although there are a few biopsy-documented cases of LAM in men with TSC. The histopathologic hallmark of LAM in the lung is interstitial expansion with benign appearing smooth muscle cells, which infiltrate all lung structures, including alveolar septa, airways, blood vessels, lymphatics, and pleura. Immunohistochemical stains reveal that LAM lesions express markers of lymphatic and melanocytic differentiation. LAM most commonly presents with progressive dyspnea on exertion and recurrent pneumothorax in the third or fourth decade of life [1]. About 30–40% of women who have TSC have cystic changes on CT scan consistent with LAM, although these women are frequently asymptomatic. LAM also occurs in women who do not have TSC, a disease that is referred to as sporadic LAM (S-LAM). Interestingly, about one half to two-thirds of women with sporadic LAM also have renal angiomyolipomas, but other manifestations of TSC such as skin and brain lesions are not found. In some women with sporadic LAM, mutations in both alleles of the TSC2 gene are present in LAM cells and angiomyolipoma cells but not in normal lung or kidney or in the circulating lymphocytes. These genetic data have led to the current model that pulmonary LAM cells reach the lungs via a metastatic process. The origin of the invading cells is unknown, but available evidence suggests an extrapulmonary source. The prognosis in LAM depends on mode of presentation and is more favorable in patients who are ascertained through screening, pneumothorax, or incidental findings on studies obtained for other purposes than through shortness of breath. There are currently no treatments that are known to be effective. Advances in our understanding of the molecular pathogenesis of LAM have led to ongoing clinical trials with targeted agents.

Tuberous Sclerosis Complex: Genes, Clinical Features, and Therapeutics
Edited by David J. Kwiatkowski, Vicky Holets Whittemore, and Elizabeth A. Thiele

ISBN: 978-3-527-32201-5

17.2 Historical Features of LAM

Lymphangioleiomyomatosis was first described by Lutembacher in 1918, in a TSC patient who presented with bilateral pneumothoraces [2]. The first report of a case of sporadic LAM appeared in Germany in 1937 [3]. LAM has been reported under many different pathological labels, including lymphangiopericytoma, lymphangiomyoma, leiomyomatosis, lymphangiomatous malformation, and intrathoracic angiomyomatous hyperplasia. Clinical, radiologic, and pathologic descriptions of LAM by Cornog and Enterline [4] and Corrin, Leibow, and Friedman [5] in the 1960s and 1970s refined the nomenclature for LAM and highlighted the lymphatic origins of LAM, the close association between LAM and TSC, and the nonmalignant, neoplastic nature of the LAM lesion.

17.3 Epidemiology

The prevalence of LAM is difficult to estimate. As of this writing, the LAM Foundation has registered over 1400 patients, 65% from the United States and 35% from foreign countries (personal communication, Leslie Sullivan-Stacey, CEO, The LAM Foundation). However, only about 11–15% of the LAM patients registered with the LAM Foundation report that they have TSC. This is surprising, since we know from several studies that cystic change consistent with LAM is present in 30–40% of women with TSC [6–8] and that the estimated prevalence of TSC in the population is between 1/6000 and 1/12 500 [9, 10]. These data indicate that TSC-LAM must affect approximately 200 000–250 000 women worldwide and about 15 000 in the United States. These values are much greater than the 2000–2500 or so LAM patients who are known to be registered with LAM organizations around the world. In addition, in most of the large case series and pulmonary clinics, S-LAM patients outnumber TSC-LAM patients 5 or 6 to 1. Yet, at a prevalence of roughly 1–2 per million, S-LAM is predicted to be more than 10-fold less common than TSC-LAM [9, 11]. The reasons that the TSC-LAM patients do not come to medical attention and/or register with patient organizations are not clear. Data from screening studies indicate that LAM in patients with TSC is often subclinical and mild, and it is possible that LAM may be less of a health priority for patients who are suffering with other manifestations of TSC than for women with the sporadic form of LAM [6–8]. The concept that TSC-LAM may be a different disease than S-LAM is important to consider; Table 17.1 compares the manifestations of these two disorders. There is no evidence for a racial, geographic, or ethnic predilection for LAM, but access to health care and information appear to play a major role in ascertainment. Further studies clearly need to be done in this area.

The National Heart Lung and Blood Institute established a registry for LAM in 1997. The baseline data from that study was published in 2006 [12]. All 243 of the NHLBI (National Heart, Lung and Blood Institute) Registry registrants were

Table 17.1 Comparison of S-LAM and TSC-LAM.

	S-LAM	TSC-LAM	Reference
Demographics			
LAM Foundation database	89%	11%	[1]
NHLBI Registry	85%	15%	[12]
Estimated number worldwide	15 000	250 000	[1]
Described in males	Yes (1)	Yes (3)	[104]
Average disease duration at diagnosis	1.5 years	1.1 years	[12]
Average age at diagnosis	41 years	39 years	[12]
TSC manifestations			
Neurocutaneous lesions	No	Yes	[105]
Cognitive impairment	No	Yes	[105]
Pulmonary manifestations			
Breathlessness	74%	71%	[12]
Pneumothorax	57%	47%	[12]
Chylothorax	24%	6%	[12]
Chylous ascites	10%	6%	[37]
Thoracic duct dilation	4%	0%	[37]
Multifocal nodules/MMPH	1%	12%	[37]
FEV1	69% predicted	79% predicted	[12]
DLCO	66% predicted	77% predicted	[12]
Obstructive physiology	61%	38%	[12]
Bronchodilator response	17%	17%	[12]
Abdominal manifestations			
Cystic lymphangioma	29%	9%	[37]
Renal angiomyolipoma	32%	93%	[37]
Hepatic angiomyolipoma	2%	33%	[37]
Nephrectomy	11%	41%	[37]
Embolization	2%	9%	[37]

women, and 13 patients were excluded based on prior lung transplant. The average age at onset of symptoms was 38.9 years and the average age at LAM diagnosis was 41.0 years, which is similar to that reported from France (36.3 years/39.3 years [11]), but much later than in the United Kingdom (31 years/35 years [13]) or Japan (31.6 years/34 years [14]. Earlier recognition in Japan is likely related to routine annual chest X-ray screening for tuberculosis of all working people and the convention that first pneumothoraces in patients without apparent lung disease are frequently evaluated with chest CT scanning in Japan (but not in the United States, Europe, or Australia).

There are four cases of biopsy-documented LAM in men in the literature, three in men who had definite or probable TSC [15–17] and one in a man who had no evidence of TSC [18]. Other (non-biopsy-documented) cases of cystic lung changes in male TSC patients have been reported, but subclinical LAM in men with TSC is almost certainly a rare occurrence. Screening of approximately 20 male TSC patients at the NIH [8] and in Cincinnati (unpublished observation) did not reveal a single case. In our Cincinnati-based clinic of 450 patients with TSC, there is one male with LAM.

17.4 Clinical Presentation

The average interval between the onset of symptoms and diagnosis in LAM is 2.4 years in Japan [14], 3.0 years in France [11], and 3.5 years in the United States [12], and is longer in patients with S-LAM than in those with TSC-LAM (Table 17.1). This delay is frequently related to the failure of the physician consulted to consider the diagnosis, and in many cases patients are first told that they have asthma or chronic obstructive lung disease. The most common initial manifestations of LAM are pneumothorax and progressive dyspnea on exertion [1]. Pulmonary symptoms were the presenting features of the disease in 86.5% of patients in the NHLBI Registry, including pneumothorax in 35% [12]. Over the course of illness, pneumothorax eventually occurred in about 55% of Registry patients, lower than the average of 65% from other series (France [11], United Kingdom [13, 19], Japan [20], and Korea [21]), perhaps because the Registry selected for patients who were well enough to travel by air to enrolling sites. Even in patients with pneumothorax as the presenting sign, the diagnosis is often delayed. Almoosa reported that the average number of pneumothoraces prior to the diagnosis of LAM is 2.2 [22]. This was true even in patients with underlying TSC. Other symptoms and signs reported by the NHLBI registrants included cough (31%), wheezing (46.5%), angiomyolipoma (38%), hemoptysis (30%), chylous effusion (21%), and chylous ascites (4.3%). Chest pain has been reported in 32–50% of patients in other series, but was not mentioned in the NHLBI series. Only about 15% of Registry patients had TSC. Comparisons between the patients with TSC-LAM and S-LAM revealed that the patients with TSC-LAM were about 2 years younger at the time of diagnosis and, in general, had less severe pulmonary physiologic and gas exchange impairment (Table 17.1). It is likely that earlier identification through screening played a role in identification of a population with more mild LAM manifestations.

17.4.1 Physical Examination

The physical examination in LAM can yield useful clues [1]. Wheezes and rhonchi may be heard on examination of the chest, but crackles are unusual. Dullness to percussion and lack of transmission of spoken sounds to the chest wall suggest pleural effusion. Elevated neck veins, a right ventricular heave, or a tricuspid regurgitant murmur may suggest pulmonary hypertension and should trigger an evaluation including echocardiogram and possibly right heart catheterization. Large angiomyolipomas and cystic lymphangiomyomas may be palpable on abdominal exam. Chylous ascites can produce abdominal fullness, tension, or shifting dullness. For patients without known TSC, a detailed skin and eye exam should be performed by clinicians familiar with the cutaneous and ocular manifestations of TSC.

17.5 Diagnosis

The most common pitfall in making the diagnosis of LAM is failure to consider the disease in young women who present with progressive, unexplained dyspnea on exertion or a sentinel pneumothorax. All clinicians who care for women with TSC should recognize the implications of these symptoms and signs in their patients and should obtain a high-resolution CT scan (HRCT) of the chest. In a nonsmoking woman with TSC, the diagnosis of pulmonary LAM is considered definite in the presence of typical cystic changes on HRCT and biopsy is not necessary. In contrast, in a woman without TSC, the diagnosis requires corroborative clinical features in addition to a typical HRCT, such as a known angiomyolipoma or chylothorax, or a tissue biopsy that reveals LAM. The diagnosis is most commonly made by video-assisted thoracoscopic lung biopsy but occasionally transbronchial biopsy with HMB-45 staining can be sufficient in a patient with an otherwise typical presentation for LAM. A recent study suggested that the diagnosis of LAM can be made by immunohistochemical staining of LAM cell clusters found in chylous effusions and ascites [81].

17.6 Pathology and Laboratory Studies

On gross examination at the time of transplant or death, when disease is advanced, the lungs are enlarged and diffusely cystic, with dilated airspaces ranging in size from a few mm to 2.0 cm in diameter [5, 23]. Microscopic examination of the lung reveals foci of smooth muscle cell infiltration of the lung parenchyma, airways, lymphatics, pleura, and blood vessels, associated with areas of thin-walled cystic change (Figure 17.1). Many of the cysts are lined with hyperplastic alveolar type II cells. Focal collections of cells termed LAM nodules are composed of actin-positive, spindle-shaped cells that stain with proliferative markers such as proliferating cellular nuclear antigen (PCNA) and less abundant cuboidal epithelioid cells that stain with a monoclonal antibody called HMB-45 [24]. This immunohistochemical study is very useful diagnostically, since other smooth muscle predominant lesions in the lung do not react with the antibody [25]. Estrogen and progesterone receptors may also be present in some LAM lesions [26, 27], but not in normal lung tissue [28]. In 2004, Kumasaka *et al.* reported abundant lymphangiogenesis in the lymphatic systems of patients with LAM [29] associated with marked elevation of serum vascular endothelial growth factor D (VEGF-D) [30]. Serum VEGF-D levels do not appear to be elevated in the serum of patients with other chylous and cystic lung diseases that can mimic LAM, such as emphysema, Sjogren's syndrome, or

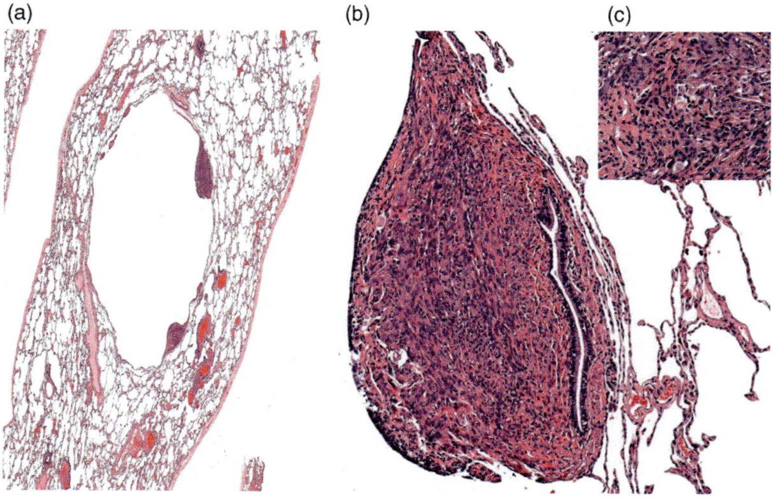

Figure 17.1 LAM histopathology. Cystic spaces with clusters of smooth muscle cells are evident at (a) low power, (b) medium power, and (c) high power (inset). LAM nodules are composed of haphazardly arranged spindle-shaped and epithelioid cells with abundant eosinophilic cytoplasm.

Langerhans cell histiocytosis [31]. Thus, VEGF-D may be useful as a diagnostic tool as well as a marker of disease progression and response to therapy in LAM.

TSC patients can also develop multinodular multifocal pneumocyte hyperplasia (MMPH), a diffuse nodular proliferation of alveolar epithelial cells that occurs in both men and women with TSC, and that appears to arise separately from LAM cells [7, 32]. The histological features are papillary or tubular proliferation of cytokeratin and surfactant protein B positive type II cells without nuclear atypia. HMB-45 staining, alpha smooth muscle actin, p53, CEA, and hormonal specimens are typically negative. In women with TSC, MMPH can occur in the presence or absence of LAM. MMPH does not have any known physiologic or prognostic consequences, but can mimic atypical adenomatous hyperplasia, metastatic malignancy, or miliary tuberculosis [33, 34]. There is no evidence that it is a precursor lesion to lung cancer.

17.7 Physiology

The most typical features of LAM on pulmonary function tests are airflow obstruction, gas trapping, and reduced diffusing capacity. Quality controlled lung function data were collected prospectively by the NHLBI Registry [12]. Spirometry revealed obstructive changes in about 57% of patients, restrictive changes in about 21%, and normal spirometric results in about 34% [12]. Hyperinflation was present in only about 6% of patients. The average residual volume was 125% of predicted when measured by plethysmography, but was 103% of predicted determined with gas dilution methods. These data suggest that a significant proportion of gas trapped in the chest is not in communication with the airway. Reduction in diffusing capacity for carbon monoxide (DLCO) and increase in residual volume are generally considered to be the earliest physiologic manifestations of LAM, and it is not unusual for DLCO to be reduced out of proportion of FEV1 (forced expiratory volume in 1 s) [35]. Cardiopulmonary exercise testing in patients with LAM reveals that exercise-induced desaturation is often present, even in patients with normal or near-normal DLCO and FEV1 [36]. Six-minute walk testing is a useful clinical tool that can be used to follow the effectiveness of clinical interventions.

17.8 Radiology

The chest radiograph is often normal early in the disease. Bilateral and symmetric reticulonodular infiltrates, cysts and bullae, or a honeycomb appearance may evolve over time but are virtually never specific enough to suggest LAM in the absence of other features. Pleural effusions and pneumothoraces may be apparent in patients who present with dyspnea, and are sometimes discovered incidentally in asymptomatic patients. Chylous effusions appear to be less common in TSC-LAM than in S-LAM. HRCT scan of the chest is the most useful and most sensitive radiographic test. The HRCT reveals thin-walled cysts of sizes varying from a few mm to several cm in all lung distributions (Figure 17.2). The morphology of the cysts is useful in differentiating LAM from other cystic lung diseases. The presence of internal septa or a "centrilobular dot" consistent with a vessel is frequently seen in emphysema, but never in LAM. The number of cysts varies in LAM from a few to near-complete replacement of the normal lung tissue. Mediastinal and hilar adenopathy are not uncommon. The abdominal CT may reveal enlarged axial lymph nodes, cystic lymphangiomyomas, or angiomyolipomas in the kidney, liver, spleen, or adrenal

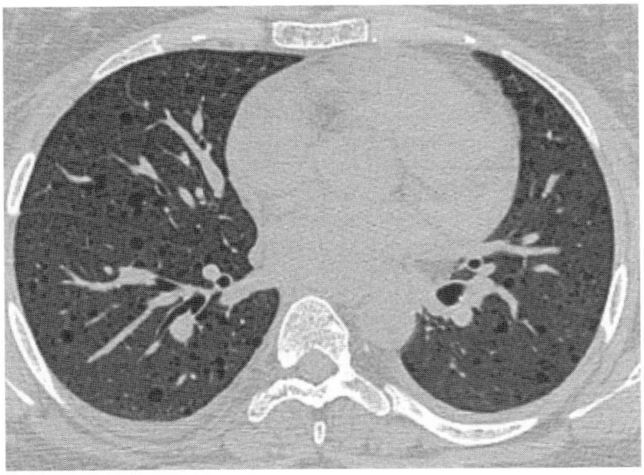

Figure 17.2 HRCT scan of the lung in a patient with LAM. HRCT reveals scattered cysts ranging in size from a few mm to 2 cm, some of which abut the pleura.

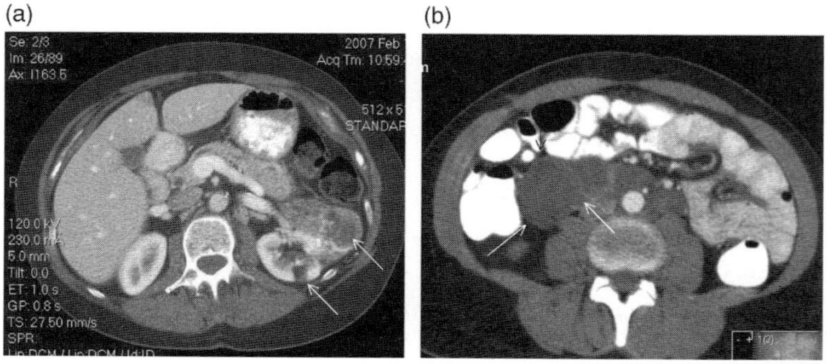

Figure 17.3 Abdominal tumors in a patient with LAM. (a) The abdominal CT in a TSC-LAM patient reveals a large angiomyolipoma that distorts the normal renal architecture as well as a second small angiomyolipoma (arrows). (b) A cystic lymphangiomyoma is also seen in this patient (arrows), though these lesions are less common in patients with TSC-LAM than those with S-LAM.

(Figure 17.3). As shown in Table 17.1, patients with TSC-LAM in the NHLBI program, compared to patients with S-LAM, had a lower frequency of abdominal lymphangioleiomyomas (9% versus 29%), thoracic duct dilation (0% versus 4%), pleural effusion (6% versus 12%), and ascites (6% versus 10%) [37]. In contrast, patients with TSC-LAM had a higher frequency of noncalcified pulmonary nodules consistent with MMPH, and hepatic and renal angiomyolipomas, and prior nephrectomy [37].

17.9 Clinical Course and Management

17.9.1 Pulmonary Function

Patients with LAM lose lung function at an accelerated rate. The average annual rate of decline in FEV1 and DLCO in 275 patients studied in a single lab at the NHLBI was 75 cc $\pm$ 9 ml and 0.69 $\pm$ 0.07 ml/min/mmHg, respectively [38]. In other series from Europe, the rate of decline in FEV1 was considerably higher, estimated at approximately 100–120 cc per year [11, 39, 40]. There was some evidence in these studies that rate of decline in lung function correlates with initial DLCO, with menopausal status, and with progesterone treatment. It is not clear whether the rate of decline in lung function is different in TSC-LAM compared to S-LAM.

17.9.2 Pleural Complications

Pleural complications are extremely common in LAM. About 60–70% of patients with LAM will have a pneumothorax at some point in their course. Of those NHLBI

Registry participants who had a history of an initial pneumothorax, the average number of recurrences was 3.4 in the United States [12] and 2.0 in Japan [14]. Given >70% chance of recurrence, the LAM Foundation Pleural Consensus Group advocated the use of a pleurodesis procedure with the initial pneumothorax [22]. Chemical sclerosis, mechanical abrasion, talc poudrage, and pleurectomy have all been effective in patients with LAM. It is important to note, however, that the failure rate with chemical and surgical pleurodesis is approximately 35%, for reasons that are not understood. Although prior pleural procedures can increase preoperative bleeding in transplant patients, they do not appear to affect candidacy for transplant or posttransplant survival [41]. Chylous fluid does not generally cause pleural inflammation or fibrosis, and small pleural effusions often require no intervention. Shortness of breath may mandate drainage, however, and pleural symphysis may be required to prevent nutritional and lymphocyte deficiencies that can result from repeated taps or persistent drainage. Chemical pleurodesis is generally an effective therapy for chylothorax, as is mechanical abrasion and talc poudrage [42]. Mechanical abrasion and doxycycline pleurodesis produce less intense pleural fusion than talc and pleurectomy, and are generally preferred by transplant surgeons.

17.9.3 Screening and Follow Up

Several screening studies have revealed that 30–40% of patients with TSC have cystic changes in their lung consistent with LAM [6–8], and the Tuberous Sclerosis Alliance recommends that women with TSC be screened by HRCT at least once after reaching the age of 18 [43]. It is reasonable to consider screening asymptomatic women with TSC with pulmonary function tests, including spirometry, lung volumes, and diffusing capacity for carbon monoxide, every 1–3 years. The wisdom and appropriate interval of periodic screening with HRCT beyond the initial scan is debated because of the lifetime radiation risk. In our clinic, women with TSC and no known cystic change are rescanned with HRCT at an interval of every 3–5 years. In patients with sporadic LAM or symptomatic TSC-LAM, the interval for follow-up testing varies with clinical context but in general pulmonary function tests are obtained every 6–12 months and HRCTs are repeated every 1–5 years.

17.9.4 Medical Treatment

Most of the current treatment strategies for LAM are empiric and unproven. The results of small series of patients treated with progestins [11, 40, 44], GnRh agonists [45–47], and oophorectomy [48] are inconclusive and conflicting. A large retrospective study of the effect of progestin therapy on the rate of decline in pulmonary function revealed no effect on FEV1, and perhaps an acceleration in the rate of decline in DLCO [44]. The only completed controlled trial involving patients with LAM was the Cincinnati Angiomyolipoma Sirolimus Trial (NCT00457808), which included lung function measures as secondary end points [49]. Twenty-three patients with angiomyolipomas and either tuberous sclerosis or LAM or both were treated for 1 year with escalating doses of sirolimus. By the fourth month, all patients were receiving doses of the drug that produced serum levels of 10–15 ng/ml. Renal tumor volume measured by MRI revealed a 50% reduction in tumor size at the end of the first year, but the kidney tumor size returned to 85% of the original volume over the course of the following year. Average FEV1 and FVC improved by 118 and 394 cc on drug, and the residual volume fell by 400 cc. Although FEV1 and FVC began to decline again off drug, these values remained significantly above baseline at 1 year. The reduction in residual volume was also durable through the 2-year point. The total lung capacity, diffusing capacity, and, most significantly, the six-minute walk test distance did not change on sirolimus. There were a number of side effects, including six hospitalizations, while patients were on the drug. A preliminary, interim report from a similar trial in England (NCT00490789) revealed similar effects with regard to angiomyolipoma size, but no improvement in lung function [50]. To explore the possibility that sirolimus has a beneficial effect on lung function, a larger placebo-controlled study, called the Multicenter International LAM Efficacy of Sirolimus Trial (MILES), was launched in December 2006 (NCT000414648). Other mTOR inhibitor trials that include lung function end points include one using RAD001 (NCT00457964) and the other sirolimus (NCT0012667). As of this writing, the efficacy and safety of mTOR inhibitor therapy for LAM remains unclear.

17.9.5
Transplantation

The United Network for Organ Sharing has recorded 126 transplants for LAM from 1989 through 2007, including 77 double lung transplants and 49 single lung transplants. The 1-, 3-, and 5-year survivals for single and double lung transplants were 87, 73, and 61 and 92, 83, and 77%, respectively.[1)] These survival rates are equal to or better than those for other disease groups transplanted in the same time frame. Although the question of bilateral versus unilateral transplantation has not been directly studied in LAM, bilateral lung transplantation produces slightly better functional outcomes in other obstructive lung diseases such as emphysema [51]. There have been four case reports of recurrence of LAM in the donor allograft [52–55]. The recurrences did not appear to contribute to death in any of these patients, and at present we do not feel that recurrence should be considered in judging the candidacy of patients for transplant. More than half of LAM patients who have undergone lung transplantation have had a prior history of a pleural fusion procedure, and although postoperative bleeding risk is increased, the operative mortality and long-term survival do not appear to be affected [22].

1) Based on OPTN data as of January 31, 2008. This work was supported in part by Health Resources and Services Administration contract 234-2005-370011C. The content is the responsibility of the authors alone and does not necessarily reflect the views or policies of the Department of Health and Human Services, nor does mention of trade names, commercial products, or organizations imply endorsement by the US Government.

17.9.6 Lifestyle and Miscellaneous Issues

Patients with LAM are often advised to avoid air travel because of the theoretical risk of cyst rupture and pneumothorax during flight. In a survey of 308 women registered with the LAM Foundation, 35% of subjects had been advised by their doctor to avoid air travel [56]. Of the 276 patients who had ever traveled by plane, 22% had reported anxiety, 14% had experienced shortness of breath, 12% chest pain, 8% oxygen desaturation, and 2% pneumothorax associated with flight. Five of the 10 patients with pneumothoraces had symptoms that began before they boarded the flight. The conclusion was that air travel can be associated with adverse events, but that most patients tolerate flying without difficulty. Since 1995, the NHLBI has conducted an intramural protocol for LAM that included obtaining a chest roentgenogram upon arrival. Taveira-DaSilva *et al.* reported that 7 of the 281 LAM patients arrived at the NIH with a new pneumothorax [57]. They concluded that the incidence of pneumothorax in patients was about 1/100 flights and 1/200 ground transportation trips (car or train). Since patients did not obtain a CXR prior to the drive, train trip, or flight to the NIH, it is impossible to be sure that the pneumothoraces detected were precipitated by travel. Nonetheless, the authors concluded that the association of pneumothorax with air travel was more likely to be related to the high incidence of pneumothorax in LAM than to the mode of travel. Patients with LAM should be advised that there have been multiple case reports of exacerbation of LAM associated with pregnancy and exogenous estrogen use [58–60]. Patients contemplating pregnancy or the use of estrogen-containing products including birth control pills and topical creams should consult with their physicians. Close attention should be paid to bone health, vaccinations, and oxygen requirements in this susceptible population.

17.10 Genetic Basis and Molecular Pathology

17.10.1 Tuberous Sclerosis Complex-Associated LAM

Radiographic evidence of LAM is present in about one-third of women with TSC, although only a fraction of these women have clinically significant pulmonary symptoms [6–8]. Germline mutations in both TSC1 and TSC2 are associated with LAM in TSC [7, 61–65]. The mutations in women with TSC and LAM are found throughout these genes and include the two most frequent TSC2 mutations (R611Q and an 18-base pair in-frame deletion in exon 40) and a missense mutation in the last exon (exon 41) of TSC2 [61]. Therefore, there is no evidence for genotype–phenotype correlation, though the number of patients available for this possible correlation is less than a dozen.

TSC1 and TSC2 are tumor suppressor genes, which in the classic "two-hit" tumor suppressor gene model are associated with disease when a germline mutation inactivates one allele and a second inactivating mutation occurs in somatic tissues [66]. Often the somatic, "second hit" mutation involves loss of the chromosomal region containing the entire wild-type copy of TSC1 or TSC2. This chromosomal loss is detected when DNA markers within or near the genes are found to be heterozygous in the patient's normal DNA and homozygous in tumor DNA, which is referred to as loss of heterozygosity (LOH). LOH for each of TSC1 and TSC2 has been detected in LAM cells from different TSC patients [67].

17.10.2
Sporadic LAM

About 30–60% of women with sporadic LAM have renal angiomyolipomas [68, 69]. One of the first clues to the pathogenesis of sporadic LAM was the finding of TSC2 LOH in angiomyolipomas from women with the sporadic form of LAM [70]. Subsequently, inactivating mutations in the remaining allele were detected, implicating TSC2 inactivation in the pathogenesis of the angiomyolipomas. Identical TSC2 mutations in the angiomyolipomas were also present in microdissected pulmonary LAM cells of five sporadic LAM patients, but not in normal DNA from the kidney, lung, or peripheral blood mononuclear cells from these patients [63]. TSC2 LOH was found in the sporadic LAM cells proving that these cells, like other tumor cells in TSC, fit the two-hit tumor suppressor gene model. These data, which were confirmed in Japanese sporadic LAM patients [64], indicated that somatic TSC2 mutations are a cause of sporadic LAM. The mutational pattern – present in the LAM and angiomyolipoma cells, but not in normal cell types – suggested that LAM cells may spread or metastasize to the lungs from the another site [63]. Strong additional support for the "benign metastasis" model of LAM pathogenesis arose from studies of women with the sporadic form of LAM who had recurrent LAM after lung transplantation: the same TSC2 mutations were present in LAM cells before and after transplantation, and in cases in which the donor lung was male, the recurrent LAM cells lacked a Y chromosome, indicating that the LAM cells spread to the transplanted lung [54, 55]. LAM cells carrying TSC2 mutations have also been detected in mediastinal lymph nodes and circulating in the blood of women with LAM [71]. Taken together, these data support a model in which the pathogenesis of LAM involves the metastasis of benign cells; the fact that LAM occurs primarily in women suggests that the metastasis is estrogen-driven.

17.10.3
LAM Cells Have Evidence of mTOR Activation

The pace and trajectory of LAM research has been dramatically accelerated by the recognition of the genetic relationship of LAM to TSC. The precise mechanisms through which loss of TSC2 leads to the proliferation of LAM cells are not fully understood (see Chapter 6 for additional discussion). It is known that expression of TSC2 in LAM-derived cells inhibits their growth, migration, and invasion [72, 73].

The fraction of sporadic LAM cases that are associated with TSC1 or TSC2 mutations is not known. However, the majority of LAM and angiomyolipoma specimens that have been studied show immunohistochemical evidence of hyperactivation of the mTOR pathway, suggesting that mTOR activation is a universal event in LAM pathogenesis [74]. The TSC1/TSC2 protein complex has an important role in regulating the state of activation of Rheb (GTP-loading), and Rheb-GTP activates the mTORC1 complex, (Figure 17.4; see also Chapter 6 for more details). Activating mutations in Rheb and Rheb-like protein (RLP) were not detected in angiomyolipomas from women with LAM [75], suggesting that this possible mechanism of mTORC1 activation independent of TSC1/TSC2 probably does not occur in LAM.

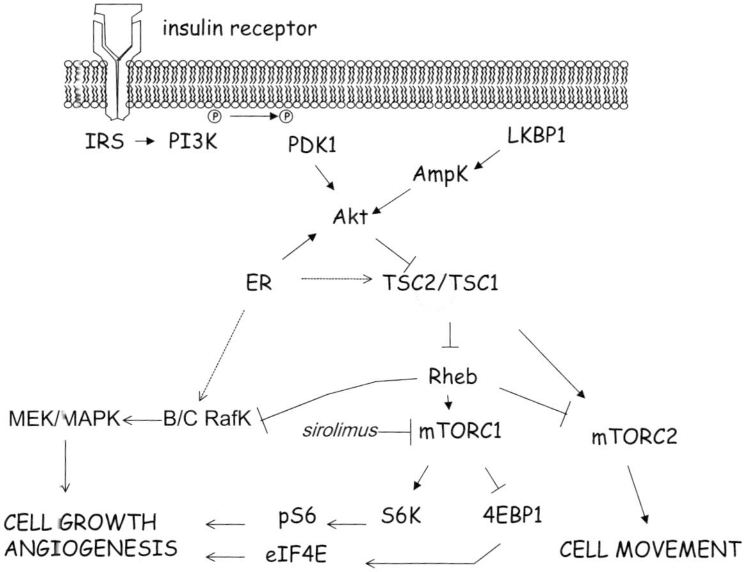

Figure 17.4 Tuberous sclerosis proteins TSC1 and TSC2 regulate signaling through the Akt pathway. The binding of a ligand to a cell surface receptor (such as the insulin receptor) activates PI3K and PDK1, followed by Akt. Akt phosphorylates TSC2, which blocks its ability to maintain Rheb in a GTP-depleted state and stimulates activation of mTORC2 and downstream targets associated with cell movement and cytoskeletal dynamics such as Rho. Activated (GTP-loaded) Rheb is abundant when TSC1 or TSC2 are missing or defective, and through phosphorylation of mTOR, activates downstream targets S6K and 4EBP1. These kinases phosphorylate the ribosomal protein S6 and liberate the translation initiation factor, eIF4E, respectively, and activate the cellular translational machinery to promote cell growth. Activated Rheb also inhibits the Raf kinases, which signal through MEK and MAPK to promote cell growth. This may be a mechanism by which cells protect against constitutive activation of Rheb. In the presence of the ligand, estrogen, however, estrogen receptor activation can stimulate Akt activation and/or "reactivate" Raf kinase and MEK/MAPK signaling in TSC1/TSC2-deficient cells. Sirolimus (rapamycin) blocks mTORC1, but not MTORC2 activation. See also Chapter 6.

A key area of uncertainty is whether either TSC1/TSC2 or Rheb have mTORC1-independent targets that are relevant to LAM pathogenesis. Rheb is known to inhibit the activity of B-Raf kinase and C-Raf kinase, resulting in downregulation of the Raf/MEK/MAPK signaling cascade [76, 77]. This activity of Rheb is unaffected by rapamycin, and therefore TORC1-independent. The role of B-Raf/C-Raf inhibition in LAM pathogenesis is not yet known. One hypothesis is that reactivation of C-Raf by estrogen contributes to the female predominance of LAM. Other investigators have studied rapamycin-independent functions of TSC2, including Finlay, who found that RhoA is activated in Tsc2-null cells in a rapamycin-independent manner [78].

17.10.4
The Cell-of-Origin of LAM Is Unknown

The genetic data discussed above indicate that LAM cells spread to the lung through a metastatic mechanism. If LAM cells arise outside the lung, where do they originate? LAM cells and angiomyolipoma cells histologically resemble immature smooth muscle cells, yet their distinctive expression of melanocyte-associated proteins, including the melanocytic transcription factor, MITF [79, 80], suggests that their origin is not from a simple smooth muscle cell precursor. It has been recognized for decades that LAM cells are immunoreactive to HMB-45, a monoclonal antibody to the melanoma-associated surface antigen, gp-100. In fact, HMB-45 immunoreactivity is widely used to diagnose LAM, since few other human tumors or diseases are HMB-45 positive: melanoma, angiomyolipomas, sugar cell tumors. Electron microscope studies have confirmed the presence of premelanosomes in the cytoplasm of LAM cells, consistent with expression of MITF as well gp-100 and other "melanocyte-specific" proteins [24]. The expression of melanocytic and other neural crest lineage makers has led to speculation that LAM cells may be of neural crest origin [80].

Lymphangiogenesis is also thought to play a role in LAM pathogenesis, and serum VEGF-D is elevated up to 30-fold in patients with LAM [30]. LAM cell clusters enveloped by lymphatic endothelial cells can be identified in the chylous pleural and ascites fluid from LAM patients and in lymphatic channels in lymph nodes [29, 81]. These observations suggest the possibility that LAM cells may have vascular or lymphatic origins. The expression of HMB-45 by this group of tumor types has led LAM to be classified by some groups with "perivascular epithelioid cell tumors" or "PEComas" [82]. Yet the origin of the putative perivascular epithelioid cell remains unknown. PEComas of the uterus and soft tissues have been reported [83, 84] more frequently in women and in patients with TSC, as well as in patients with sporadic LAM and TSC-LAM [84].

17.10.5
Estrogen May Promote LAM Pathogenesis

The fact that LAM affects primarily young women has led to the hypothesis that estrogen plays a critical role in LAM pathogenesis. The carboxy-terminus of tuberin was shown to interact with the estrogen receptor (ER) [85] and tuberin was found to function *in vitro* as a transcriptional corepressor of the estrogen receptor [86], resulting in a twofold decrease in ERE-luciferase reporter response. Finlay *et al.* confirmed the interaction between ER alpha and tuberin and showed that reexpression of tuberin in tuberin-null ELT-3 cells (from rat uterine leiomyoma) abrogated estradiol (E_2)-induced growth *in vitro* [87]. York *et al.* recently showed that tuberin and ER alpha interact at endogenous expression levels in multiple cell types [88]. Additional work is clearly needed to understand whether and how these data are related to LAM pathogenesis.

In primary cells derived from a sporadic LAM angiomyolipoma proven to carry biallelic TSC2 inactivation, estrogen stimulated cell growth. This proliferation was associated with increased phosphorylation of p42/44 MAPK at 5 min and increased expression of c-myc at 4 h. These findings, which need to be confirmed in other patient-derived cell lines, are consistent with activation of both genomic and non-genomic signaling pathways [89].

The TSC/Rheb/mTOR pathway plays a critical role in the regulation of estrogen-induced proliferation signals. In MCF-7 cells, 17-beta estradiol (E_2) rapidly increased the phosphorylation of downstream targets of mTOR: p70 ribosomal protein S6 kinase (S6K), ribosomal protein S6, and eukaryotic initiation factor 4E-binding protein 1 (4E-BP1). siRNA downregulation of endogenous Rheb blocked the E_2-stimulated proliferation of MCF-7 cells, demonstrating that Rheb is a key determinant of E_2-dependent cell growth [90]. Whether these effects of estrogen have relevance to LAM is unknown.

17.10.5
Cystic Lung Disease in LAM

The pathogenesis of cystic lung disease in LAM is incompletely understood, in part because of the lack of a LAM animal model. It is known that LAM cells express matrix metalloproteinases (MMPs), including MMP2 and MT1-MMP, and that the expression of MMPs may be lower after treatment with antihormonal agents [91]. Increased expression of serum response factor (SRF) has been demonstrated in LAM cells, which may lead to MMP activation [92] and downregulation of tissue inhibitor of metalloproteinase (TIMP)-3 in LAM cells [93], and thereby contribute to the tissue destruction in LAM. A simple explanation, therefore, is that expression of MMPs is entirely responsible for cyst formation. However, while many malignant tumor cells that metastasize to the lungs express MMPs, cystic degeneration of the surrounding lung parenchyma is rarely observed in cancer, suggesting that other mechanisms may contribute to cyst formation. Identifying these mechanisms will be critical to LAM therapy, especially at early stages, in order to prevent the loss of lung parenchyma, which is a major contributor to irreversible pulmonary function decline.

Studies of Birt–Hogg–Dubé (BHD) syndrome may yield clues to the pathogenesis of cyst formation in LAM. BHD is an autosomal dominant disorder characterized by hamartomas of skin follicles, lung cysts, spontaneous pneumothorax, and renal cell carcinoma [94–96]. The BHD gene was cloned in 2002 and encodes folliculin, which has no significant homology to other human proteins [97]. It has recently been discovered that BHD functions in the TOR pathway in *Schizosaccharomyces pombe* [98] and in mammalian cells [99, 100]. Surprisingly, in *S. pombe* the BHD homologue functions as an activator of Tor2 (one of the two homologues of TOR), in contrast to the TSC1/TSC2 homologues, which function as inhibitors of Tor2. The precise relationship between BHD and mTOR in mammalian cells is not yet clear, but it is tempting to speculate that in BHD, inappropriate mTOR inhibition leads to lung cysts, while in TSC, inappropriate mTOR activation leads to LAM cell proliferation and cysts. One possible mechanism for these apparent contradictory results involves the balance between mTOR's two distinct complexes in mammalian cells, mTORC1 (mTOR and raptor) and mTORC2 (mTOR, rictor, and SIN1). Inhibition of mTORC1 with rapamycin alters the stoichiometry between mTORC1 and mTORC2 in a cell-type specific manner, with loss of mTOR–raptor binding at early time points and loss of mTOR–rictor binding at later time points [101], indicating that the balance between mTOR activation and inhibition is tightly regulated.

17.11
Challenges and Future Directions

Progress in LAM research has been slowed by several challenges, including the lack of an animal model that recapitulates LAM and by the difficulties in growing LAM cells in culture. The problems with culturing and validating LAM cells are complex: First, LAM-derived cells in culture include a mixture of cell types, and the LAM cells appear to undergo senescence after several passages (which is not surprising given that they are histologically benign and usually slow growing). Second, since most tissue specimens are acquired at the time of lung transplantation and therefore represent end-stage disease, the proportion of LAM cells varies greatly between cultures and between passages of a given culture. Third, while the detection of TSC1 or TSC2 mutations in cultured LAM cells has been proposed as a "gold standard," these mutations can be challenging and expensive to detect.

Despite these challenges, remarkable advances in our understanding of the pathogenesis of LAM have occurred since the year 2000, when TSC2 mutations were first found in LAM cells, and studies of sporadic LAM continue to elucidate TSC-associated LAM. It is hoped that continued basic, translational, and clinical research will lead to highly effective, targeted therapies for women with LAM. The development of well-validated cell culture and animal models of LAM would speed this progress. Finally, the efficient design of clinical trials is critical, since the number of available patients is small, and many different potential therapeutic approaches have been already proposed, including statins [78], estrogen antagonists, interferon gamma [102, 103], and matrix metalloproteinase inhibitors, as well as rapamycin and related drugs.

Glossary

DLCO	Diffusing capacity for carbon monoxide. This test measures the removal of a trace amount of carbon monoxide from an inhaled mixture of oxygen, nitrogen, and carbon monoxide, as an estimate of the ease with which oxygen is taken up into the bloodstream by the lung. In essence, it measures the integrity of the pulmonary capillary bed.
FEV1	Forced expiratory volume in 1 s. The volume of air that is expelled over the first second during a maximal forced exhalation. Reduced out of proportion to total volume expelled over time in the obstructive lung diseases.
FVC	Forced vital capacity. The volume of air that is forcibly expelled over the entire period of a maximal forced exhalation.
HMB-45	The monoclonal human melanoma black antibody raised against melanoma antigens. HMB-45 recognizes the gp-100 epitope that is expressed in LAM cells for unknown reasons.
LAM cell clusters	Spherical collections of smooth muscle cells, enveloped by a single layer of lymphatic endothelial cells and found in the lumen of lymphatic channels or chylous fluid collections in the chest or abdomen.
Lymphangiomyoma	A tumor of lymphatic vessels characterized by smooth muscle infiltration, cystic dilation, and luminal obstruction.
Pleurodesis	A variety of techniques that are used to induce fusion of the parietal and visceral pleura, to obliterate the pleural space and prevent recurrence of pneumothorax or pleural effusion.
Poudrage	Technique of instilling talc into the chest cavity as a powder through a bulb syringe.
Sporadic LAM (S-LAM)	LAM that occurs in patients who do not have germline mutations in tuberous sclerosis genes. S-LAM is not heritable. S-LAM is often associated with unilateral or small renal angiomyolipomas and lymphangiomyomas, but not with brain or skin lesions.
Tuberous sclerosis-associated LAM (TSC-LAM)	LAM that occurs in patients who meet diagnostic criteria for tuberous sclerosis and/or have germline mutations in tuberous sclerosis genes.
Vascular endothelial growth factor D	A lymphangiogenic growth factor that is known to be expressed by LAM cells and to be elevated in the serum of LAM patients.

References

1 McCormack, F.X. (2008) Lymphangioleiomyomatosis: a clinical update. *Chest*, **133**, 507–516.

2 Lutembacher, R. (1918) Dysembryomes metatypiques des reins; carcinose submiliaire aigue du poumon avec emphyseme generalise et double pneumothorax. *Ann. Med.*, **5**, 435–450.

3 von Stossel, E. (1937) Uber muskulare Cirrhose der Lunge (Muscular cirrhosis of the lung). *Beitr. Klin. Tuberk.*, **90**, 432–442.

4 Cornog, J.L., Jr. and Enterline, H.T. (1966) Lymphangiomyoma, a benign lesion of chyliferous lymphatics synonymous with lymphangiopericytoma. *Cancer*, **19**, 1909–1930.

5 Corrin, B., Leibow, A.A., and Friedman, P.J. (1975) Pulmonary lymphangiomyomatosis: a review. *Am. J. Pathol.*, **79**, 348–382.

6 Costello, L.C., Hartman, T.E., and Ryu, J.H. (2000) High frequency of pulmonary lymphangioleiomyomatosis in women with tuberous sclerosis complex. *Mayo Clin. Proc.*, **75**, 591–594.

7 Franz, D.N., Brody, A., Meyer, C., Leonard, J., Chuck, G., Dabora, S., Sethuraman, G., Colby, T.V., Kwiatkowski, D.J., and McCormack, F.X. (2001) Mutational and radiographic analysis of pulmonary disease consistent with lymphangioleiomyomatosis and micronodular pneumocyte hyperplasia in women with tuberous sclerosis. *Am. J. Respir. Crit. Care Med.*, **164**, 661–668.

8 Moss, J., Avila, N.A., Barnes, P.M., Litzenberger, R.A., Bechtle, J., Brooks, P.G., Hedin, C.J., Hunsberger, S., and Kristof, A.S. (2001) Prevalence and clinical characteristics of lymphangioleiomyomatosis (LAM) in patients with tuberous sclerosis complex. *Am. J. Respir. Crit. Care Med.*, **164**, 669–671.

9 Osborne, J.P., Fryer, A., and Webb, D. (1991) Epidemiology of tuberous sclerosis. *Ann. N. Y. Acad. Sci.*, **615**, 125–127.

10 O'Callaghan, F.J., Shiell, A.W., Osborne, J.P., and Martyn, C.N. (1998) Prevalence of tuberous sclerosis estimated by capture–recapture analysis. *Lancet*, **351**, 1490.

11 Urban, T., Lazor, R., Lacronique, J., Murris, M., Labrune, S., Valeyre, D., and Cordier, J.F. (1999) Pulmonary lymphangioleiomyomatosis. A study of 69 patients. groupe d'Etudes et de recherche sur les Maladies "Orphelines" Pulmonaires (GERM"O"P). *Medicine*, **78**, 321–337.

12 Ryu, J.H., Moss, J., Beck, G.J., Lee, J.C., Brown, K.K., Chapman, J.T., Finlay, G.A., Olson, E.J., Ruoss, S.J., Maurer, J.R., Raffin, T.A., Peavy, H.H., McCarthy, K., Taveira-Dasilva, A., McCormack, F.X., Avila, N.A., Decastro, R.M., Jacobs, S.S., Stylianou, M., and Fanburg, B.L. (2006) The NHLBI Lymphangioleiomyomatosis Registry: characteristics of 230 patients at enrollment. *Am. J. Respir. Crit. Care Med.*, **173**, 105–111.

13 Johnson, S.R. and Tattersfield, A.E. (2000) Clinical experience of lymphangioleiomyomatosis in the UK. *Thorax*, **55**, 1052–1057.

14 Hayashida, M., Seyama, K., Inoue, Y., Fujimoto, K., and Kubo, K. (2007) The epidemiology of lymphangioleiomyomatosis in Japan: a nationwide cross-sectional study of presenting features and prognostic factors. *Respirology*, **12**, 523–530.

15 Aubry, M.C., Myers, J.L., Ryu, J.H., Henske, E.P., Logginidou, H., Jalal, S.M., and Tazelaar, H.D. (2000) Pulmonary lymphangioleiomyomatosis in a man. *Am. J. Respir. Crit. Care Med.*, **162**, 749–752.

16 Miyake, M., Tateishi, U., Maeda, T., Kusumoto, M., Satake, M., Arai, Y., and Sugimura, K. (2005) Pulmonary lymphangioleiomyomatosis in a male patient with tuberous sclerosis complex. *Radiat. Med.*, **23**, 525–527.

17 Kim, N.R., Chung, M.P., Park, C.K., Lee, K.S., and Han, J. (2003) Pulmonary

lymphangioleiomyomatosis and multiple hepatic angiomyolipomas in a man. *Pathol. Int.*, **53**, 231–235.

18 Schiavina, M., Di Scioscio, V., Contini, P., Cavazza, A., Fabiani, A., Barberis, M., Bini, A., Altimari, A., Cooke, R.M., Grigioni, W.F., and D'Errico-Grigioni, A. (2007) Pulmonary lymphangioleiomyomatosis in a karyotypically normal man without tuberous sclerosis complex. *Am. J. Respir. Crit. Care Med.*, **176**, 96–98.

19 Johnson, S.R., Whale, C.I., Hubbard, R.B., Lewis, S.A., and Tattersfield, A.E. (2004) Survival and disease progression in UK patients with lymphangioleiomyomatosis. *Thorax*, **59**, 800–803.

20 Kitaichi, M., Nishimura, K., Itoh, H., and Izumi, T. (1995) Pulmonary lymphangioleiomyomatosis: a report of 46 patients including a clinicopathologic study of prognostic factors. *Am. J. Respir. Crit. Care Med.*, **151**, 527–533.

21 Oh, Y.M., Mo, E.K., Jang, S.H., Yoo, C.G., Kim, Y.W., Seo, J.W., Han, S.K., Im, J.G., and Shim, Y.S. (1999) Pulmonary lymphangioleiomyomatosis in Korea. *Thorax*, **54**, 618–621.

22 Almoosa, K.F., Ryu, J.H., Mendez, J., Huggins, J.T., Young, L.R., Sullivan, E.J., Maurer, J., McCormack, F.X., and Sahn, S.A. (2006) Management of pneumothorax in lymphangioleiomyomatosis: effects on recurrence and lung transplantation complications. *Chest*, **129**, 1274–1281.

23 Carrington, C.B., Cugell, D.W., Gaensler, E.A., Marks, A., Redding, R.A., Schaaf, J.T., and Tomasian, A. (1977) Lymphangioleiomyomatosis. Physiologic–pathologic–radiologic correlations. *Am. Rev. Respir. Dis.*, **116**, 977–995.

24 Matsumoto, Y., Horiba, K., Usuki, J., Chu, S.C., Ferrans, V.J., and Moss, J. (1999) Markers of cell proliferation and expression of melanosomal antigen in lymphangioleiomyomatosis. *Am. J. Respir. Cell Mol. Biol.*, **21**, 327–336.

25 Hoon, V., Thung, S.N., Kaneko, M., and Unger, P.D. (1994) HMB-45 reactivity in renal angiomyolipoma and lymphangioleiomyomatosis. *Arch. Pathol. Lab. Med.*, **118**, 732–734.

26 McCarty, K.S., Jr., Mossler, J.A., McLelland, R., and Sieker, H.O. (1980) Pulmonary lymphangiomyomatosis responsive to progesterone. *N. Engl. J. Med.*, **303**, 1461–1465.

27 Colley, M.H., Geppert, E., and Franklin, W.A. (1989) Immunohistochemical detection of steroid receptors in a case of pulmonary lymphangioleiomyomatosis. *Am. J. Surg. Pathol.*, **13**, 803–807.

28 Berger, U., Khaghani, A., Pomerance, A., Yacoub, M.H., and Coombes, R.C. (1990) Pulmonary lymphangioleiomyomatosis and steroid receptors. An immunocytochemical study. *Am. J. Clin. Pathol.*, **93**, 609–614.

29 Kumasaka, T., Seyama, K., Mitani, K., Sato, T., Souma, S., Kondo, T., Hayashi, S., Minami, M., Uekusa, T., Fukuchi, Y., and Suda, K. (2004) Lymphangiogenesis in lymphangioleiomyomatosis: its implication in the progression of lymphangioleiomyomatosis. *Am. J. Surg. Pathol.*, **28**, 1007–1016.

30 Seyama, K., Kumasaka, T., Souma, S., Sato, T., Kurihara, M., Mitani, K., Tominaga, S., and Fukuchi, Y. (2006) Vascular endothelial growth factor-D is increased in serum of patients with lymphangioleiomyomatosis. *Lymphat. Res. Biol.*, **4**, 143–152.

31 Young, L.R., Inoue, Y., and McCormack, F.X. (2008) Diagnostic potential of serum VEGF-D for lymphangioleiomyomatosis. *N. Engl. J. Med.*, **358**, 199–200.

32 Muir, T.E., Leslie, K.O., Popper, H., Kitaichi, M., Gagne, E., Emelin, J.K., Vinters, H.V., and Colby, T.V. (1998) Micronodular pneumocyte hyperplasia. *Am. J. Surg. Pathol.*, **22**, 465–472.

33 Kobashi, Y., Sugiu, T., Mouri, K., Irei, T., Nakata, M., and Oka, M. (2008) Clinicopathological analysis of multifocal micronodular pneumocyte hyperplasia associated with tuberous sclerosis in Japan. *Respirology*, **13**, 1076–1081.

34 Kobashi, Y., Sugiu, T., Mouri, K., Irei, T., Nakata, M., and Oka, M. (2008)

Multifocal micronodular pneumocyte hyperplasia associated with tuberous sclerosis: differentiation from multiple atypical adenomatous hyperplasia. *Jpn. J. Clin. Oncol.*, **38**, 451–454.
35 Taveira-DaSilva, A.M., Hedin, C., Stylianou, M.P., Travis, W.D., Matsui, K., Ferrans, V.J., and Moss, J. (2001) Reversible airflow obstruction, proliferation of abnormal smooth muscle cells, and impairment of gas exchange as predictors of outcome in lymphangioleiomyomatosis. *Am. J. Respir. Crit. Care Med.*, **164**, 1072–1076.
36 Taveira-DaSilva, A.M., Stylianou, M.P., Hedin, C.J., Kristof, A.S., Avila, N.A., Rabel, A., Travis, W.D., and Moss, J. (2003) Maximal oxygen uptake and severity of disease in lymphangioleiomyomatosis. *Am. J. Respir. Crit. Care Med.*, **168**, 1427–1431.
37 Avila, N.A., Dwyer, A.J., Rabel, A., and Moss, J. (2007) Sporadic lymphangioleiomyomatosis and tuberous sclerosis complex with lymphangioleiomyomatosis: comparison of CT features. *Radiology*, **242**, 277–285.
38 Taveira-DaSilva, A.M., Steagall, W.K., and Moss, J. (2006) Lymphangioleiomyomatosis. *Cancer Control*, **13**, 276–285.
39 Lazor, R., Valeyre, D., Lacronique, J., Wallaert, B., Urbane, T., and Cordier, J.F. (2004) Low initial KCO predicts rapid FEV1 decline in pulmonary lymphangioleiomyomatosis. *Respir. Med.*, **98**, 536–541.
40 Johnson, S.R. and Tattersfield, A.E. (1999) Decline in lung function in lymphangioleiomyomatosis: relation to menopause and progesterone treatment. *Am. J. Respir. Crit. Care Med.*, **160**, 628–633.
41 Boehler, A., Speich, R., Russi, E.W., and Weder, W. (1996) Lung transplantation for lymphangioleiomyomatosis. *N. Engl. J. Med.*, **335**, 1275–1280.
42 Ryu, J.H., Doerr, C.H., Fisher, S.D., Olson, E.J., and Sahn, S.A. (2003) Chylothorax in lymphangioleiomyomatosis. *Chest*, **123**, 623–627.
43 Roach, E.S., Gomez, M.R., and Northrup, H. (1998) Tuberous sclerosis complex consensus conference: revised clinical diagnostic criteria. *J. Child Neurol.*, **13**, 624–628.
44 Taveira-DaSilva, A.M., Stylianou, M.P., Hedin, C.J., Hathaway, O., and Moss, J. (2004) Decline in lung function in patients with lymphangioleiomyomatosis treated with or without progesterone. *Chest*, **126**, 1867–1874.
45 Harari, S., Cassandro, R., Chiodini, J., Taveira-DaSilva, A.M., and Moss, J. (2008) Effect of a gonadotrophin-releasing hormone analogue on lung function in lymphangioleiomyomatosis. *Chest*, **133**, 448–454.
46 Seyama, K., Kira, S., Takahashi, H., Ohnishi, M., Kodama, Y., Dambara, T., Kobayashi, J., Kitamura, S., and Fukuchi, Y. (2001) Longitudinal follow-up study of 11 patients with pulmonary lymphangioleiomyomatosis: diverse clinical courses of LAM allow some patients to be treated without anti-hormone therapy. *Respirology*, **6**, 331–340.
47 Schiavina, M., Contini, P., Fabiani, A., Cinelli, F., Di Scioscio, V., Zompatori, M., Campidelli, C., and Pileri, S.A. (2007) Efficacy of hormonal manipulation in lymphangioleiomyomatosis. A 20-year-experience in 36 patients. *Sarcoidosis Vasc. Diffuse Lung Dis.*, **24**, 39–50.
48 Banner, A.S., Carrington, C.B., Emory, W.B., Kittle, F., Leonard, G., Ringus, J., Taylor, P., and Addington, W.W. (1981) Efficacy of oophorectomy in lymphangioleiomyomatosis and benign metastasizing leiomyoma. *N. Engl. J. Med.*, **305**, 204–209.
49 Bissler, J.J., McCormack, F.X., Young, L.R., Elwing, J.M., Chuck, G., Leonard, J.M., Schmithorst, V.J., Laor, T., Brody, A.S., Bean, J., Salisbury, S., and Franz, D.N. (2008) Sirolimus for angiomyolipoma in tuberous sclerosis complex or lymphangioleiomyomatosis. *N. Engl. J. Med.*, **358**, 140–151.
50 Davies, D.M., Johnson, S.R., Tattersfield, A.E., Kingswood, J.C., Cox, J.A., McCartney, D.L., Doyle, T., Elmslie, F., Saggar, A., de Vries, P.J., and Sampson, J.R. (2008) Sirolimus therapy in tuberous sclerosis or sporadic

lymphangioleiomyomatosis. *N. Engl. J. Med.*, **358**, 200–203.

51 Bando, K., Paradis, I.L., Keenan, R.J., Yousem, S.A., Komatsu, K., Konishi, H., Guilinger, R.A., Masciangelo, T.N., Pham, S.M., Armitage, J.M. *et al.* (1995) Comparison of outcomes after single and bilateral lung transplantation for obstructive lung disease. *J. Heart Lung Transplant.*, **14**, 692–698.

52 Nine, J.S., Yousem, S.A., Paradis, I.L., Keenan, R., and Griffith, B.P. (1994) Lymphangioleiomyomatosis: recurrence after lung transplantation. *J. Heart Lung Transplant.*, **13**, 714–719.

53 O'Brien, J.D., Lium, J.H., Parosa, J.F., Deyoung, B.R., Wick, M.R., and Trulock, E.P. (1995) Lymphangioleiomyomatosis recurrence in the allograft after single lung transplantation. *Am. J. Respir. Crit. Care*, **151**, 2033–2036.

54 Karbowniczek, M., Astrinidis, A., Balsara, B.R., Testa, J.R., Lium, J.H., Colby, T.V., McCormack, F.X., and Henske, E.P. (2003) Recurrent lymphangiomyomatosis after transplantation: genetic analyses reveal a metastatic mechanism. *Am. J. Respir. Crit. Care Med.*, **167**, 976–982.

55 Bittmann, I., Rolf, B., Amann, G., and Lohrs, U. (2003) Recurrence of lymphangioleiomyomatosis after single lung transplantation: new insights into pathogenesis. *Hum. Pathol.*, **34**, 95–98.

56 Pollock-BarZiv, S., Cohen, M.M., Downey, G.P., Johnson, S.R., Sullivan, E., and McCormack, F.X. (2007) Air travel in women with lymphangioleiomyomatosis. *Thorax*, **62**, 176–180.

57 Taveira-DaSilva, A.M., Burstein, D., Hathaway, O.M., Fontana, J.R., Gochuico, B.R., Avila, N.A., and Moss, J. (2009) Pneumothorax after air travel in lymphangioleiomyomatosis, idiopathic pulmonary fibrosis, and sarcoidosis. *Chest*, **136**, 665–670.

58 Hughes, E. and Hodder, R.V. (1987) Pulmonary lymphangiomyomatosis complicating pregnancy. A case report. *J. Reprod. Med.*, **32**, 553–557.

59 Shen, A., Iseman, M.D., Waldron, J.A., and King, T.E. (1987) Exacerbation of pulmonary lymphangioleiomyomatosis by exogenous estrogens. *Chest*, **91**, 782–785.

60 Wilson, A.M., Slack, H.L., Soosay, S.A., Taylor, T., Carey, F.A., Grove, A., Brown, P.H., and Winter, J.H. (2001) Lymphangioleiomyomatosis. A series of three case reports illustrating the link with high oestrogen states. *Scott. Med. J.*, **46**, 150–152.

61 Strizheva, G.D., Carsillo, T., Kruger, W.D., Sullivan, E.J., Ryu, J.H., and Henske, E.P. (2001) The spectrum of mutations in TSC1 and TSC2 in women with tuberous sclerosis and lymphangiomyomatosis. *Am. J. Respir. Crit. Care Med.*, **163**, 253–258.

62 Moss, J., DeCastro, R., Patronas, N.J., and Taveira-DaSilva, A. (2001) Meningiomas in lymphangioleiomyomatosis. *J. Am. Med. Assoc.*, **286**, 1879–1881.

63 Carsillo, T., Astrinidis, A., and Henske, E.P. (2000) Mutations in the tuberous sclerosis complex gene TSC2 are a cause of sporadic pulmonary lymphangioleiomyomatosis. *Proc. Natl. Acad. Sci. USA*, **97**, 6085–6090.

64 Sato, T., Seyama, K., Fujii, H., Maruyama, H., Setoguchi, Y., Iwakami, S., Fukuchi, Y., and Hino, O. (2002) Mutation analysis of the TSC1 and TSC2 genes in Japanese patients with pulmonary lymphangioleiomyomatosis. *J. Hum. Genet.*, **47**, 20–28.

65 Sato, T., Seyama, K., Kumasaka, T., Fujii, H., Setoguchi, Y., Shirai, T., Tomino, Y., Hino, O., and Fukuchi, Y. (2004) A patient with TSC1 germline mutation whose clinical phenotype was limited to lymphangioleiomyomatosis. *J. Intern. Med.*, **256**, 166–173.

66 Knudson, A.G. (2001) Two genetic hits (more or less) to cancer. *Nat. Rev. Cancer*, **1**, 157–162.

67 Yu, J., Astrinidis, A., and Henske, E.P. (2001) Chromosome 16 loss of heterozygosity in tuberous sclerosis and sporadic lymphangiomyomatosis. *Am. J. Respir. Crit. Care Med.*, **164**, 1537–1540.

68 Avila, N.A., Kelly, J.A., Chu, S.C., Dwyer, A.J., and Moss, J. (2000) Lymphangioleiomyomatosis:

abdominopelvic CT and US findings. *Radiology*, **216**, 147–153.

69 Bernstein, S.M., Newell, J.D., Jr., Adamczyk, D., Mortensen, R., King, T.E., Jr., and Lynch, D.A. (1995) How common are renal angiomyolipomas in patients with pulmonary lymphangiomyomatosis? *Am. J. Respir. Crit. Care Med.*, **152**, 2138–2143.

70 Smolarek, T.A., Wessner, L.L., McCormack, F.X., Mylet, J.C., Menon, A.G., and Henske, E.P. (1998) Evidence that lymphangiomyomatosis is caused by TSC2 mutations: chromosome 16p13 loss of heterozygosity in angiomyolipomas and lymph nodes from women with lymphangiomyomatosis. *Am. J. Hum. Genet.*, **62**, 810–815.

71 Crooks, D.M., Pacheco-Rodriguez, G., DeCastro, R.M., McCoy, J.P., Jr., Wang, J.A., Kumaki, F., Darling, T., and Moss, J. (2004) Molecular and genetic analysis of disseminated neoplastic cells in lymphangioleiomyomatosis. *Proc. Natl. Acad. Sci. USA*, **101**, 17462–17467.

72 Goncharova, E.A., Goncharov, D.A., Spaits, M., Noonan, D.J., Talovskaya, E., Eszterhas, A., and Krymskaya, V.P. (2006) Abnormal growth of smooth muscle-like cells in lymphangioleiomyomatosis: role for tumor suppressor TSC2. *Am. J. Respir. Cell Mol. Biol.*, **34**, 561–572.

73 Goncharova, E.A., Goncharov, D.A., Lim, P.N., Noonan, D., and Krymskaya, V.P. (2006) Modulation of cell migration and invasiveness by tumor suppressor TSC2 in lymphangioleiomyomatosis. *Am. J. Respir. Cell Mol. Biol.*, **34**, 473–480.

74 El-Hashemite, N., Zhang, H., Henske, E.P., and Kwiatkowski, D.J. (2003) Mutation in TSC2 and activation of mammalian target of rapamycin signalling pathway in renal angiomyolipoma. *Lancet*, **361**, 1348–1349.

75 Robb, V.A., Astrinidis, A., and Henske, E.P. (2006) Frequent [corrected] hyperphosphorylation of ribosomal protein S6 [corrected] in lymphangioleiomyomatosis-associated angiomyolipomas. *Mod. Pathol.*, **19**, 839–846.

76 Karbowniczek, M., Cash, T., Cheung, M., Robertson, G.P., Astrinidis, A., and Henske, E.P. (2004) Regulation of B-Raf kinase activity by tuberin and Rheb is mammalian target of rapamycin (mTOR)-independent. *J. Biol. Chem.*, **279**, 29930–29937.

77 Karbowniczek, M., Robertson, G.P., and Henske, E.P. (2006) Rheb inhibits C-raf activity and B-raf/C-raf heterodimerization. *J. Biol. Chem.*, **281**, 25447–25456.

78 Finlay, G.A., Malhowski, A.J., Liu, Y., Fanburg, B.L., Kwiatkowski, D.J., and Toksoz, D. (2007) Selective inhibition of growth of tuberous sclerosis complex 2 null cells by atorvastatin is associated with impaired Rheb and Rho GTPase function and reduced mTOR/S6 kinase activity. *Cancer Res.*, **67**, 9878–9886.

79 Juvet, S.C., McCormack, F.X., Kwiatkowski, D.J., and Downey, G.P. (2007) Molecular pathogenesis of lymphangioleiomyomatosis: lessons learned from orphans. *Am. J. Respir. Cell Mol. Biol.*, **36**, 398–408.

80 Lim, S.D., Stallcup, W., Lefkove, B., Govindarajan, B., Au, K.S., Northrup, H., Lang, D., Fisher, D.E., Patel, A., Amin, M.B., and Arbiser, J.L. (2007) Expression of the neural stem cell markers NG2 and L1 in human angiomyolipoma: are angiomyolipomas neoplasms of stem cells? *Mol. Med.*, **13**, 160–165.

81 Kumasaka, T., Seyama, K., Mitani, K., Souma, S., Kashiwagi, S., Hebisawa, A., Sato, T., Kubo, H., Gomi, K., Shibuya, K., Fukuchi, Y., and Suda, K. (2005) Lymphangiogenesis-mediated shedding of LAM cell clusters as a mechanism for dissemination in lymphangioleiomyomatosis. *Am. J. Surg. Pathol.*, **29**, 1356–1366.

82 Bonetti, F., Pea, M., Martignoni, G., Doglioni, C., Zamboni, G., Capelli, P., Rimondi, P., and Andrion, A. (1994) Clear cell ("sugar") tumor of the lung is a lesion strictly related to angiomyolipoma: the concept of a family of lesions characterized by the presence of the perivascular epithelioid cells (PEC). *Pathology*, **26**, 230–236.

83 Folpe, A.L., Mentzel, T., Lehr, H.A., Fisher, C., Balzer, B.L., and Weiss, S.W. (2005) Perivascular epithelioid cell neoplasms of soft tissue and gynecologic origin: a clinicopathologic study of 26 cases and review of the literature. *Am. J. Surg. Pathol.*, **29**, 1558–1575.

84 Hornick, J.L. and Fletcher, C.D. (2006) PEComa: what do we know so far? *Histopathology*, **48**, 75–82.

85 Henry, K.W., Yuan, X., Koszewski, N.J., Onda, H., Kwiatkowski, D.J., and Noonan, D.J. (1998) Tuberous sclerosis gene 2 product modulates transcription mediated by steroid hormone receptor family members. *J. Biol. Chem.*, **273**, 20535–20539.

86 Noonan, D.J., Lou, D., Griffith, N., and Vanaman, T.C. (2002) A calmodulin binding site in the tuberous sclerosis 2 gene product is essential for regulation of transcription events and is altered by mutations linked to tuberous sclerosis and lymphangioleiomyomatosis. *Arch. Biochem. Biophys.*, **398**, 132–140.

87 Finlay, G.A., York, B., Karas, R.H., Fanburg, B.L., Zhang, H., Kwiatkowski, D.J., and Noonan, D.J. (2004) Estrogen-induced smooth muscle cell growth is regulated by tuberin and associated with altered activation of platelet-derived growth factor receptor-beta and ERK-1/2. *J. Biol. Chem.*, **279**, 23114–23122.

88 York, B., Lou, D., Panettieri, R.A., Jr., Krymskaya, V.P., Vanaman, T.C., and Noonan, D.J. (2005) Cross-talk between tuberin, calmodulin, and estrogen signaling pathways. *FASEB J.*, **19**, 1202–1204.

89 Yu, J., Astrinidis, A., Howard, S., and Henske, E.P. (2004) Estradiol and tamoxifen stimulate LAM-associated angiomyolipoma cell growth and activate both genomic and nongenomic signaling pathways. *Am. J. Physiol. Lung Cell. Mol. Physiol.*, **286**, L694–L700.

90 Yu, J. and Henske, E.P. (2006) Estrogen-induced activation of mammalian target of rapamycin is mediated via tuberin and the small GTPase Ras homologue enriched in brain. *Cancer Res.*, **66**, 9461–9466.

91 Matsui, K., Takeda, K., Yu, Z.X., Travis, W.D., Moss, J., and Ferrans, V.J. (2000) Role for activation of matrix metalloproteinases in the pathogenesis of pulmonary lymphangioleiomyomatosis. *Arch. Pathol. Lab. Med.*, **124**, 267–275.

92 Zhe, X., Yang, Y., and Schuger, L. (2005) Imbalanced plasminogen system in lymphangioleiomyomatosis: potential role of serum response factor. *Am. J. Respir. Cell Mol. Biol.*, **32**, 28–34.

93 Zhe, X., Yang, Y., Jakkaraju, S., and Schuger, L. (2003) Tissue inhibitor of metalloproteinase-3 downregulation in lymphangioleiomyomatosis: potential consequence of abnormal serum response factor expression. *Am. J. Respir. Cell Mol. Biol.*, **28**, 504–511.

94 Birt, A.R., Hogg, G.R., and Dube, W.J. (1977) Hereditary multiple fibrofolliculomas with trichodiscomas and acrochordons. *Arch. Dermatol.*, **113**, 1674–1677.

95 Toro, J.R., Glenn, G., Duray, P., Darling, T., Weirich, G., Zbar, B., Linehan, M., and Turner, M.L. (1999) Birt–Hogg–Dube syndrome: a novel marker of kidney neoplasia. *Arch. Dermatol.*, **135**, 1195–1202.

96 Zbar, B., Alvord, W.G., Glenn, G., Turner, M., Pavlovich, C.P., Schmidt, L., Walther, M., Choyke, P., Weirich, G., Hewitt, S.M., Duray, P., Gabril, F., Greenberg, C., Merino, M.J., Toro, J., and Linehan, W.M. (2002) Risk of renal and colonic neoplasms and spontaneous pneumothorax in the Birt–Hogg–Dube syndrome. *Cancer Epidemiol. Biomarkers Prev.*, **11**, 393–400.

97 Nickerson, M.L., Warren, M.B., Toro, J.R., Matrosova, V., Glenn, G., Turner, M.L., Duray, P., Merino, M., Choyke, P., Pavlovich, C.P., Sharma, N., Walther, M., Munroe, D., Hill, R., Maher, E., Greenberg, C., Lerman, M.I., Linehan, W.M., Zbar, B., and Schmidt, L.S. (2002) Mutations in a novel gene lead to kidney tumors, lung wall defects, and benign tumors of the hair follicle in patients with the Birt–Hogg–Dube syndrome. *Cancer Cell*, **2**, 157–164.

98 van Slegtenhorst, M., Khabibullin, D., Hartman, T.R., Nicolas, E., Kruger, W.D., and Henske, E.P. (2007) The Birt–Hogg–Dube and tuberous sclerosis complex homologs have opposing roles in amino acid homeostasis in *Schizosaccharomyces pombe*. *J. Biol. Chem.*, **282**, 24583–24590.

99 Baba, M., Hong, S.B., Sharma, N., Warren, M.B., Nickerson, M.L., Iwamatsu, A., Esposito, D., Gillette, W.K., Hopkins, R.F., 3rd, Hartley, J.L., Furihata, M., Oishi, S., Zhen, W., Burke, T.R., Jr., Linehan, W.M., Schmidt, L.S., and Zbar, B. (2006) Folliculin encoded by the BHD gene interacts with a binding protein, FNIP1, and AMPK, and is involved in AMPK and mTOR signaling. *Proc. Natl. Acad. Sci. USA*, **103**, 15552–15557.

100 Baba, M., Furihata, M., Hong, S.B., Tessarollo, L., Haines, D.C., Southon, E., Patel, V., Igarashi, P., Alvord, W.G., Leighty, R., Yao, M., Bernardo, M., Ileva, L., Choyke, P., Warren, M.B., Zbar, B., Linehan, W.M., and Schmidt, L.S. (2008) Kidney-targeted Birt–Hogg–Dube gene inactivation in a mouse model: Erk1/2 and Akt-mTOR activation, cell hyperproliferation, and polycystic kidneys. *J. Natl. Cancer Inst.*, **100**, 140–154.

101 Sarbassov, D.D., Ali, S.M., Kim, D.H., Guertin, D.A., Latek, R.R., Erdjument-Bromage, H., Tempst, P., and Sabatini, D.M. (2004) Rictor, a novel binding partner of mTOR, defines a rapamycin-insensitive and raptor-independent pathway that regulates the cytoskeleton. *Curr. Biol.*, **14**, 1296–1302.

102 Lee, L., Sudentas, P., and Dabora, S.L. (2006) Combination of a rapamycin analog (CCI-779) and interferon-gamma is more effective than single agents in treating a mouse model of tuberous sclerosis complex. *Genes Chromosomes Cancer*, **45**, 933–944.

103 Lee, L., Sudentas, P., Donohue, B., Asrican, K., Worku, A., Walker, V., Sun, Y., Schmidt, K., Albert, M.S., El-Hashemite, N., Lader, A.S., Onda, H., Zhang, H., Kwiatkowski, D.J., and Dabora, S.L. (2005) Efficacy of a rapamycin analog (CCI-779) and IFN-gamma in tuberous sclerosis mouse models. *Genes Chromosomes Cancer*, **42**, 213–227.

104 McCormack, F.X. and Moss, J. (2007) S-LAM in a man? *Am. J. Respir. Crit. Care Med.*, **176**, 3–5.

105 Crino, P.B., Nathanson, K.L., and Henske, E.P. (2006) The tuberous sclerosis complex. *N. Engl. J. Med.*, **355**, 1345–1356.

18
Endocrine, Gastrointestinal, Hepatic, and Lymphatic Manifestations of Tuberous Sclerosis Complex

Finbar J. O'Callaghan and John P. Osborne

18.1
Introduction and Summary

Tuberous sclerosis lesions infrequently cause symptoms in the endocrine, gastro-intestinal (GI), and lymphatic systems. It is unclear, however, why these organ systems should be less prone to hamartoma formation than the central nervous system, kidneys, or skin. Indeed, it is not entirely clear that hamartomas occur at lower frequency in the endocrine, GI, and lymphatic systems, but perhaps they rather remain small, clinically insignificant, and unrecognized in most cases. There is a growing body of evidence, though anecdotal, that TSC patients develop hormone-secreting tumors involving the neuroendocrine system at higher frequency than the general population. Cushing's disease [1, 2], hypoglycemia secondary to insulinomas [3–7], precocious puberty [8, 9], thyrotoxicosis [10], hypercalcemia secondary to parathyroid adenomas [10–12], hyperprolactinemia [13], and acromegaly [14] have all been reported in TSC patients. In addition, in at least one TSC patient, multiple endocrine abnormalities have been noted [10]. These and other clinical observations have led to speculation that there is an overlap between TSC and multiple endocrine neoplasia type 1 (MEN type 1). However, the brain, skin, and renal involvement by TSC is quite distinct from anything seen in the MEN syndromes, apart from facial angiofibromas that can be seen in 88% of patients with MEN type 1 [15]. Recently, there has been evidence linking neuroendocrine tumors to the AKT/mTOR/S6 kinase pathway that is regulated by the hamartin/tuberin (TSC1/TSC2) complex. Thus, it is beginning to appear that the occurrence of these neuroendocrine tumors (NET) in TSC is more than random coincidence.

Tuberous Sclerosis Complex: Genes, Clinical Features, and Therapeutics
Edited by David J. Kwiatkowski, Vicky Holets Whittemore, and Elizabeth A. Thiele

ISBN: 978-3-527-32201-5

18.2
Endocrine Manifestations of TSC

18.2.1
Theoretical Relationship Between TSC and Neuroendocrine Tumors

Tumors of the neuroendocrine axis are well known to occur in autosomal dominant familial syndromes such as Carney Complex and MEN types 1 and 2. It is less well recognized that neuroendocrine tumors occur in the phakomatoses such as TSC, von Hippel–Lindau disease, and neurofibromatosis. Recent evidence suggests that the same AKT/mTOR/S6 kinase pathway that is involved in TSC may also be involved in some NETs. For example, overexpression of the proto-oncogene AKT/PKB has been demonstrated in certain NETs and AKT activates mTOR [16]. Similarly, upregulation of the phosphatidylinositol kinase/AKT pathway has been described in some sporadic pituitary tumors [17]. The rapamycin analogue RAD 001 that inhibits mTOR activity has been shown to inhibit NET cell line proliferation [18]. Interestingly, the Eker rat, which has a germline mutation of the TSC2 gene, is prone to form pituitary adenomas and these adenomas show loss of heterozygosity for the TSC2 gene [19]. In addition, treatment of these rats with rapamycin improved their clinical state and prolonged their survival [20]. Thus, there is much circumstantial evidence supporting the concept that TSC1/TSC2 has an important role in growth regulation in neuroendocrine tissues.

18.2.2
Pituitary

Pituitary adenomas causing oversecretion of growth hormone, prolactin, or adrenocorticotrophin hormone (ACTH) have all been described in TSC.

Hoffman *et al.* described a case of acromegaly in an 11-year-old prepubertal child [14]. He presented with headache, enlarging hands, and growth of 10 cm in the year prior to admission. A skull X-ray revealed an expanded sella turcica and intracranial calcification. On investigation it was found that he had elevated growth hormone, prolactin, and IGF-1 levels and he underwent surgical removal of the pituitary tumor. Histopathology of the tumor revealed that it was an eosinophilic adenoma.

In addition to the patient described by Hoffman, there have been two other case reports of hyperprolactinemia in TSC patients. Bloomgarden *et al.* described a young woman who having been diagnosed with TSC at 16 years of age presented with amenorrhea and galactorrhea at the age of 25 and after delivery of her third child [13]. Her serum prolactin level did not increase with the administration of dopamine antagonists such as chlorpromazine or haloperidol and did not decrease with the administration of L-Dopa or bromocriptine. Because manipulation with dopamine agonists and antagonists usually produces predictable effects in pituitary prolactin secretion, the authors speculated that the prolactin might have been ectopically produced by a hamartoma.

The other case report of hyperprolactinemia in TSC occurred in a patient who also had growth hormone deficiency and diabetes insipidus. No abnormality of the sellar or suprasellar regions was demonstrable in this patient on CT scanning, but this patient presented the same before the days of routine magnetic resonance imaging [21].

There have been two reports of Cushing's disease in TSC patients, one in a child and one in an adult. In the first case, a 13.5-year-old boy, TSC had been diagnosed at 5 years of age [2]. His mother and maternal grandfather both had confirmed TSC. At the age of 9.5, the patient underwent investigation for short stature. He had normal IGF-1, IGF-BP3, and growth hormone levels. At 13.5 years of age, he had a bone age of 11 years and he began to exhibit clinical features of Cushing's disease. He had a round face with plethora and acne, a hump at the back of his neck and central obesity. He had elevated urinary free cortisol (458.3 μg/24 h) and an elevated morning serum cortisol of 858 nmol/l. An MRI scan of his pituitary suggested the presence of a microadenoma and petrosal sinus sampling confirmed the presence of an ACTH-secreting tumor. He underwent transsphenoidal removal of a cystic microadenoma and histopathology showed that the adenoma contained regressive changes, including cholesterol clefts, lymphocytic infiltration, and siderophages possibly secondary to hemorrhage.

The adult case was of a 32-year-old man who had a history of epilepsy since childhood [1]. The diagnosis of TSC was not made until adulthood and was based on the presence of periungual fibromas, shagreen patches, facial angiofibromatosis, retinal astrocytoma, and subependymal nodules. One year after the diagnosis he was referred to an endocrine clinic because he had noted a change in his facial appearance, increasing central obesity, increased bruising, and a fatty lump at the back of his neck. He was also suffering from depression. Clinical examination confirmed the classical Cushingoid features and in addition noted proximal myopathy and hypertension. He was hypokalemic and had elevated early morning serum cortisol (1018 nmol/l) and ACTH (50 ng/l) levels and increased urinary cortisol excretion (797 μmol/24 h). A sudden clinical deterioration necessitated emergency bilateral adrenalectomy before petrosal sinus sampling could confirm Cushing's disease. Ten years later he presented with symptoms and signs of raised intracranial pressure due to a subependymal giant cell astrocytoma (SEGA), which was treated with the insertion of a ventriculoperitoneal shunt. Six months later an MRI revealed a pituitary microadenoma that was resected via the transphenoidal route.

18.2.3 Parathyroid

There are at least three reports describing hypercalcemia and parathyroid adenomas in TSC patients. Mortensen and Rungby described the case of a 15-year-old boy, previously undiagnosed with TSC, being admitted to hospital with acute abdominal pain [11]. He was diagnosed with acute pancreatitis and hypercalcemia. Further investigation revealed an eightfold increase in parathyroid hormone activity. Thallium–technetium scintigrams and ultrasonography revealed a parathyroid adenoma in the right side of the neck. The adenoma was removed and the diagnosis was confirmed by histopathology. The boy had a positive family history of TSC in his sister and mother. On close examination he was found to have facial angiofibromatosis, hypomelanic macules on his thighs, and calcified subependymal nodules on cranial CT.

Ilgren and Westmoreland described the case of a 20-year-old female TSC patient presenting with hypercalcemia [10]. She had been diagnosed with TSC as a child and had a history of facial angiofibromatosis, bilateral polycystic kidneys, epilepsy, and pneumothorax. Her hypercalcemia did not respond to hydrocortisone suppression and she underwent a parathyroidectomy that revealed four hyperplastic parathyroid glands. Although surgery rendered her normocalcemic, she unfortunately subsequently developed thyrotoxicosis, heart failure, and hemoptysis. She died of sepsis. Postmortem examination revealed adenomas in multiple endocrine organs in addition to the parathyroid glands, namely, in the pituitary, adrenals, and islet cells of the pancreas.

Yin *et al.* described the case of a 14-year-old girl with TSC who presented with bony pain in her feet, hips, legs, and lumbar region and who consequently had progressive difficulty with walking [12]. She also had anorexia, nausea, constipation, polydipsia, and polyuria. Her blood calcium level, however, was normal. Skeletal X-rays revealed generalized osteoporosis. At surgery she was discovered to have a parathyroid adenoma. Postoperatively she developed tetany but recovered and was discharged on calcium and vitamin D supplementation.

18.2.4 Thyroid

Papillary adenomas of the thyroid gland, which are uncommon in the general population, are the most common thyroid lesions found in TSC patients. They were first described by Perou [22] and Verma [23] in the 1960s. They both described non-encapsulated papillary adenomas within the thyroid gland. Gomez described the presence of a papillary adenoma of the thyroid in an 8-year-old boy who died of obstructive hydrocephalus secondary to a large subependymal giant cell astrocytoma [24]. However, papillary adenomas do not appear to cause thyroid dysfunction.

There are three case reports of abnormal thyroid function in TSC patients. The first is the 20-year-old female patient described above who presented with hypercalcemia but who went on to develop congestive cardiac failure associated with thyrotoxicosis [10]. Sareen *et al.* described three patients in the 1970s who were hypothyroid [25]. Two of the patients had palpable goiters and one had evidence of circulating antithyroid antibodies. Bereket and Wilson described a child with congenital hypothyroidism and TSC who had a dysgenetic thyroid gland [26].

Dicorato *et al.* have recently reported a unique case of a medullary thyroid carcinoma arising in a patient with TSC [27].

18.2.5 Pancreas

There have been several reports of insulinomas in TSC patients. All these patients developed symptomatic hypoglycemia secondary to pathological insulin secretion from insulinomas. In the first case, a woman with TSC developed seizures at the age of 18. The seizures all occurred early in the morning and were associated with excessive agitation. It was also noticed that she put copious amounts of sugar in her tea [3]. In another case, a patient also had an increased appetite for sweet food and was excessively tired and sleepy after exertion and developed seizures after a 15-year period of seizure freedom and despite adequate doses of two anticonvulsants [4]. Kim *et al.* described a 28-year-old man with TSC and learning difficulties who underwent a significant behavioral change and who exhibited new symptoms of agitation and, on other occasions, profound lethargy [5]. Eledrisi *et al.* related the story of a 43-year-old man with TSC who presented with confusion, unusual behavior and slurred speech. There was also a history of episodes of sweating and dizziness developing after fasting that were terminated by drinking fruit juice [7].

The relatively frequent association of insulinomas, hypoglycemia, and TSC probably warrants screening for these lesions in any TSC patient who undergoes significant behavioral change with agitation and lethargy, who presents with a recurrence of seizure activity after a long seizure-free interval, or who presents with seizures for the first time in late childhood or adulthood.

There has been one case report of a patient with TSC developing a gastrinoma and Zollinger–Ellison syndrome. This patient had been diagnosed with TSC in childhood but presented with esophagitis, stomach ulceration, and weight loss at 34 years of age. A biopsy of a pancreatic tumor revealed a gastrinoma. Unfortunately, he already had liver metastases at the time of diagnosis and these spread to his spine, lungs, and lymphatic system and he died soon after diagnosis [28].

A nonfunctioning somatostatinoma has also been described in a 24-year-old TSC patient [29]. Somatostatinomas have also been described in neurofibromatosis and von Hippel–Lindau syndrome.

There are two case reports of malignant non-secretory endocrine pancreatic tumors. Both patients presented in the first decade of life and both had highly vascular retroperitoneal tumor arising from the pancreas. Both patients had TSC2 disease and there was 16p13 loss of heterozygosity in the pancreatic tumor but not in the normal pancreas [30, 31].

18.2.6 Adrenal

The adrenal gland may be the endocrine gland most commonly affected by TSC. The most commonly found adrenal abnormality in TSC is the adrenal angiomyolipoma (AML) [32, 33]. These angiomyolipomas are morphologically similar to those found in the kidney and liver. However, unlike angiomyolipomas in the kidney, they rarely, if ever, hemorrhage. One estimate is that angiomyolipomas will be present in the adrenal glands of a quarter of TSC patients [34].

Abnormalities of adrenal function have been described in TSC. Clearly, there is excess cortisol production when the gland is subject to excess ACTH production by the pituitary. However, there are also examples of abnormality of function intrinsic to the adrenal gland. Holtzman *et al.* reported on an 8-year-old girl with TSC and masculinization of the external genitalia secondary to deficiency of 11-β-hydroxylase [35].

There is one case report of a pheochromocytoma arising in a 29-year-old female TSC patient. She presented with recurrent fever and abdominal pain in the right upper quadrant. She was anemic and had a leukocytosis. Ultrasonography, angiography, and CT scanning revealed a right-sided adrenal lesion that was shown to be a pleomorphic pheochromocytoma on histology. Unfortunately, the tumor recurred less than a year postexcision and because it was found to involve the spinal cord, it was inoperable and the patient died [36].

18.2.7 Gonads

Both angiomyolipomas and fibroadenomas occur in the testis and may grow to a significant size and require surgical removal (Figure 18.1). The angiomyolipomas are histologically similar to the lesions that arise in the kidney. As with angiomyolipomas in the liver and adrenal gland and in comparison to the lesions in the kidney, angiomyolipomas in the testis do not seem to have a tendency to hemorrhage.

Martin *et al.* reported on the case of a benign Leydig cell tumor of the testis in a patient with TSC. This patient also had an eruptive seborrheic keratosis associated with human papillomavirus type 33 infection [37].

In 2002, Anderson *et al.* [38] reported a case of an ovarian angiomyolipoma in a 39-year-old female with a history of TSC and bilateral renal angiomyolipomas. The ovarian angiomyolipoma reached 4.5 cm in diameter. Histology of the lesion revealed a mixture of epithelioid cells, smooth muscle, large thick-walled blood vessels, and adipose tissue. The epithelioid cells were strongly immunoreactive for HMB-45.

There has also been a case report of a juvenile granulosa cell tumor arising in the ovary of an 8-year-old girl with TSC. There was a family history of TSC in both her sister and father. Diagnosis of TSC was based on the family history, calcified subependymal nodules on CT scan, and hypomelanic macules. She presented at the age of seven with increasing abdominal girth and vaginal bleeding. On examination she exhibited signs of precocious puberty with Tanner stage IV breast development and pubic hair growth. A pelvic ultrasound revealed a 20 cm × 22 cm mass. She had a right salpingo-oophorectomy and histology revealed a juvenile granulosa cell tumor. Three months after the initial surgery she suffered a recurrence on the left ovary and she underwent left salpingo-oophorectomy with removal of lymph nodes and the omentum [39].

Bonetti *et al.* described the occurrence of perivascular epithelial cell tumors (PEComas) arising in a 41-year-old woman with TSC [40]. Tumors were present in the uterus and on the right ovary. The patient underwent successful surgical resection of the tumors [40]. The concept of PEComa has been developed in the past 10 years, initially by Bonetti [40, 69]. The cell of origin for these tumors is uncertain, but it may be a variant of the perivascular epithelioid cell, which normally lines small arteries. PEComas occur in a variety of sites, including the lungs and female pelvic organs, and

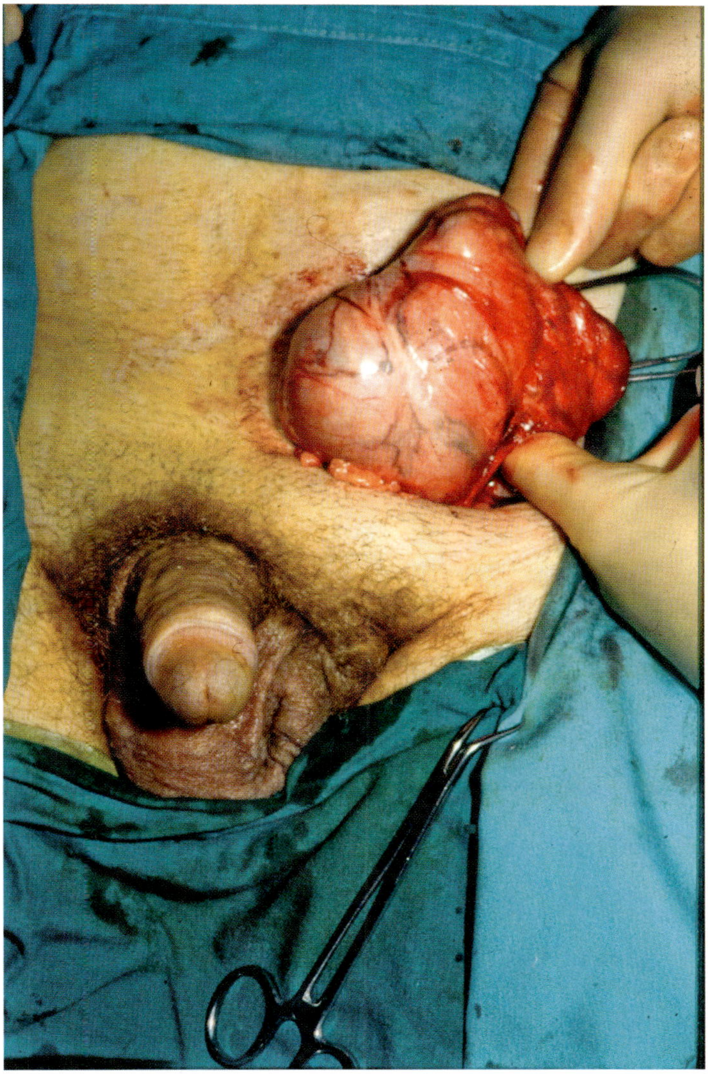

Figure 18.1 Angiomyolipoma of the testis. Photograph of the surgical removal of a large testicular angiomyolipoma in a TSC patient.

include renal angiomyolipomas and pulmonary lymphangioleiomyomatosis (LAM) (see Chapters 15 and 17). Although AMLs and LAM are seen at much higher frequency in TSC individuals than in non-TSC subjects, other types of PEComa do not share this marked predisposition in TSC individuals. The epithelioid cells in all types of PEComas express the melanocytic marker HMB-45. PEComas can display a malignant natural history in some cases, with distant metastases or aggressive local invasion [69]. Anecdotal reports indicate that rapamycin and related drugs may be helpful for malignant PEComa management [70].

18.2.8
Precocious Puberty and TSC

There are many cases in the literature of precocious puberty in individuals with TSC. Often the mechanism is unclear. Some patients develop early puberty as a result of activation of the hypothalamic-pituitary-gonadal axis. Cummings *et al.* document a TSC patient with central precocious puberty, gynecomastia, and hyperprolactinemia who did not have any demonstrable lesion on imaging in the region of the hypothalamus or pituitary [41]. De Cornulier [9] described a case where the development of precocious puberty revealed underlying TSC. Precocious puberty developed at 13 months with enlargement of the penis and testes, appearance of pubic hair, acne, and deepening of the voice. Linear growth had recently accelerated and bone age was advanced. Plasma testosterone was elevated and a LHRH injection induced a raid rise in plasma LH and FSH. Brain imaging revealed a hypothalamic hamartoma and calcified subependymal nodules. Administration of a LHRH analogue provided effective treatment of the precocious puberty [9].

Precocious puberty has also been described in the context of an ovarian granulosa cell tumor in a 7-year-old child as described above [39]. There is also at least one case of gonadotrophin-independent precocious puberty. In this male with TSC, his basal gonadotrophin levels were virtually undetectable and his gonadotrophin response to administration of releasing hormone was negligible [8].

18.3
Gastrointestinal Manifestations of TSC

18.3.1
Mouth

Dental pits have been known to occur in TSC for some time and they are probably present in almost all patients with TSC (Figure 18.2). In a recent study of 58 adult patients with TSC, Sparling *et al.* identified multiple dental enamel pits in 56 (97%) [42]. In 1975, Hoff *et al.* described the dental radiograph findings of six individuals with TSC demonstrating that all the tooth surfaces had pit-shaped enamel defects [43]. There was no clear pattern of distribution. The number of pits varied from 1 to 11 and averaged 3 per tooth surface. They were able to demonstrate pits that were not visible to the naked eye and found these to be up to 60 μm in diameter: pits visible with the naked eye were approximately 100 μm in diameter. They are believed to be areas of deficient enamel, but it is not known how they arise. Hoff *et al.* suggested that the enamel hypoplasia is caused by a malfunction of the ameloblasts, and because the amorphous material found in the pits was close to the dentin, the first phase of amelogenesis might be defective. We do not know if they are hamartomas or are in some way related to hypomelanic macules. They are not always easily seen until or unless they become stained. Examination with a light, magnifying glass, and dental disclosing solution is necessary to have the best chance of detection clinically.

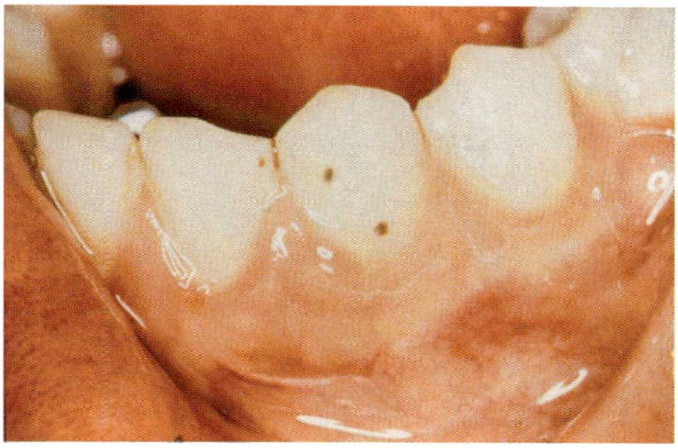

Figure 18 2 Dental pits. Photograph of dental pits in a TSC patient that have been stained with a dental disclosing solution in order to ease identification. Pits can be found in both secondary and deciduous teeth in TSC patients.

They are most likely to be visible on the premolar teeth and are therefore in areas easy to examine. They may be less common in milk teeth than in permanent teeth but they arise in all teeth.

Are dental pits specific to TSC? Investigators have attempted to examine this. Mlynarczyk looked at 50 patients with TSC and 250 controls [44]. He found controls to have dental pits but not as often as individuals with TSC – who were more likely to have them as they aged. A subsequent study of deciduous teeth by Russell *et al.* failed to find dental pits in controls with cerebral palsy, phenylketonuria, Down's syndrome, and normal individuals [45]. In conclusion, they are not a reliable diagnostic sign of TSC, but their presence can be used to suggest that further investigation for the possibility of TSC is needed.

Treatment for pits is not required unless for cosmetic reasons. They are not associated with increased rates of dental decay.

It is, however, very important to pursue good dental hygiene in TSC. Many TSC individuals find it difficult to tolerate dental brushing and the resulting problem can be increased by the effects of antiepileptic drugs. Phenytoin (Epanutin), in particular, can cause gum hypertrophy. Thus, it is good advice to get the TSC infant used to dental brushing as soon as teeth erupt. Use an ultrasonic toothbrush or similar. Take the individual to the dentist and let them get used to sitting in a dental chair. Make sure that fluoride is either in the toothpaste or the drinking water, or both, to enhance resistance of enamel to decay.

In addition to dental pits, oral fibromas are another common manifestation of TSC. Sparling *et al.* found oral fibromas in 40 of his 58 patients (69%); they were seen mostly commonly on the gingival mucosa (Figure 18.3) [42]. Other sites of involvement included the buccal and labial mucosa, the superior labial frenulum, palate, and tongue (Figure 18.4). They rarely cause symptoms even when large, but if necessary, fibroma resection can be performed. In our experience, significant regrowth is rare. Bleeding from oral fibromas is also rare, unless general periodontal disease affects the whole mouth.

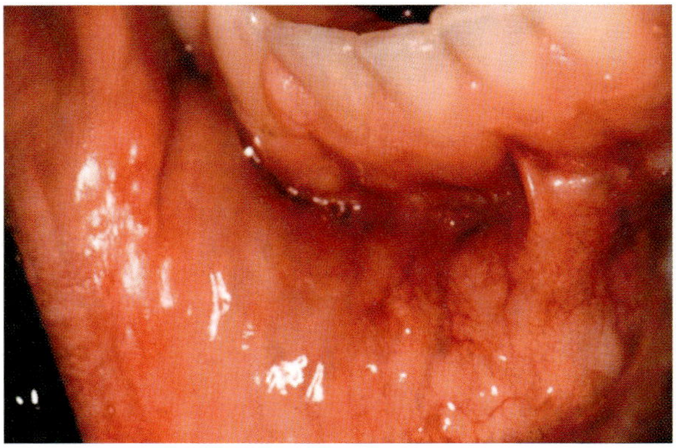

Figure 18.3 Gingival fibroma. Photograph of gingival fibroma in a TSC patient.

Oral fibromas most commonly arise in the late first and early second decade of life, similar to facial angiofibromas, with which they also share pathologic features.

18.3.2 Esophagus and Stomach

The whole of the gastrointestinal tract can be affected by TSC. The most common lesions are fibroadenomatous polyps or angiomyolipomas. These lesions are less commonly seen in the esophagus and stomach. However, Hizawa *et al.* reported two cases of TSC who had multiple polyps throughout the gastrointestinal tract and comment that in one, a 21-year-old male, there were multiple tiny protrusions in the esophagus [46]. Kim *et al.* described a 42-year-old female patient who had multiple polyps in the fundus of her stomach [47]. Gastric angiomyolipoma has been described in the literature but not in the context of TSC. However, there is no reason to believe that angiomyolipomas could not arise in the stomach in TSC.

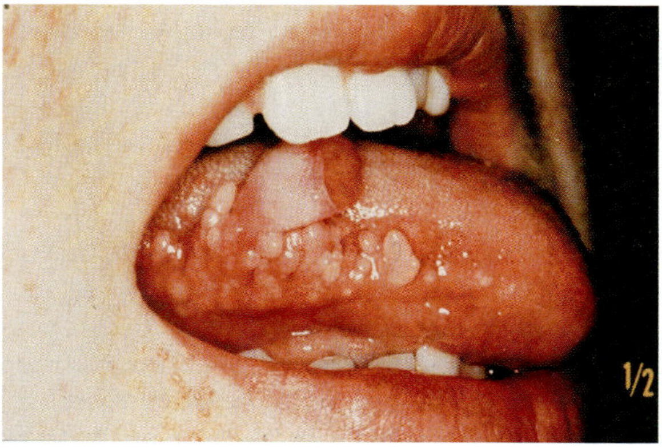

Figure 18.4 Glossal hamartomata. Large disfiguring glossal hamartomata in a female teenage TSC patient of normal intellect.

18.3.3
Small Bowel

Polyps have also been seen in the small bowel in TSC. However, they do not cause symptoms and do not lead to bleeding. Theoretically, there is no reason that angiomyolipomas or other PEComas could not arise in the ileum or jejunum in the context of TSC, but there are no case reports as yet in the literature.

18.3.4
Large Bowel and Rectum

The lower bowel is often involved with polyps that are histologically characteristic of TSC, but their presence is usually overlooked because they do not cause symptoms and do not lead to occult blood loss (Figure 18.5). In fact, in one study, rectal polyps were identified in 14 out of 18 TSC patients [48]. The polyps are not believed to have malignant potential. It is, however, very interesting to note that the LKB1 tumor suppressor gene that is the genetic cause of Peutz–Jegher syndrome encodes a protein that also acts in the mTOR pathway through regulation of AMP-activated protein kinase. Peutz–Jegher syndrome is characterized by widespread GI polyps that can progress to carcinoma, in contrast to the GI polyps of TSC.

PEComas of the large bowel have been described. Goh *et al.* [49] reported the case of a 30-year-old Chinese female who presented with intractable constipation for which a colectomy was performed. Pathological examination of the resected specimen revealed diffuse thickening of the bowel wall and multiple confluent nodular CD34 and CD117 negative smooth muscle proliferation. The epithelioid smooth muscle cells were HMB-45 positive.

Even in the absence of obvious explanatory lesions, constipation is very common in TSC probably because of the increased risk of hypotonia following early-onset seizures leading to weaker abdominal wall muscles. Added to this is the common

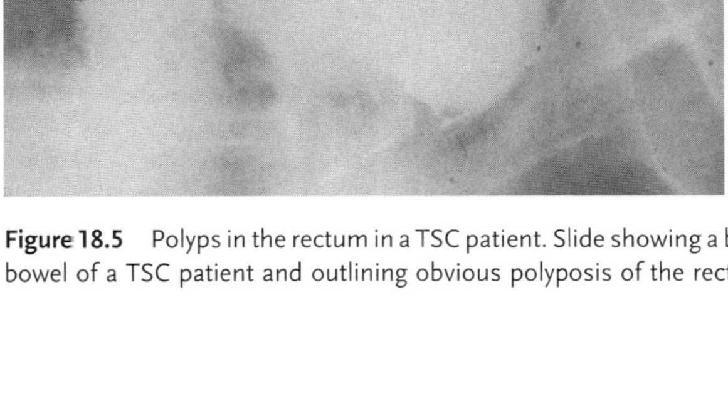

Figure 18.5 Polyps in the rectum in a TSC patient. Slide showing a barium examination of the large bowel of a TSC patient and outlining obvious polyposis of the rectal wall.

difficulty in potty training and toileting. Treatment with behavioral modification and laxatives is important – usually without the use of laxatives in childhood and for a long period (years as often as months), persistent problems will become entrenched because of the dilated large bowel that results from persistent constipation. Diarrhea is a frequent symptom of constipation because of the overflow of loose stool that squeezes past the constipated stool, and leaks out of the anus to cause soiling. However, encopresis (passing a normal stool in an abnormal place) is fortunately not common in TSC.

18.4 Hepatic Manifestations of TSC

Liver lesions are frequently seen during ultrasound scanning of the abdomen and are thought to be angiomyolipomas based on their imaging characteristics. However, histological confirmation is rare because clinically these lesions do not cause major problems, and therefore biopsy is rarely performed. Multiple types of liver neoplasms were described in TSC in the past, including racemose angioma [50], liver adenomas [51], and lipomyomas of fatty mesenchymatous tumors [52]. It is likely that these were all angiomyolipomas. On ultrasound, the lesions display the characteristics of angiomyolipomas seen in the kidney and on CT it is often possible to demonstrate the presence of fat. They are often seen in association with renal angiomyolipomas but can occur in isolation.

The frequency of hepatic angiomyolipomas in TSC is approximately 16–24% [53, 54]. However, the prevalence may be higher in adults as these numbers are derived primarily from pediatric populations, and we know that these lesions become more common with age. Patients without TSC can also present with liver angiomyolipomas (as well as renal angiomyolipoma – commonly seen in association with pulmonary lymphangioleiomyomatosis): such isolated liver angiomyolipomas may arise through a spontaneous sporadic double hit on one of the TSC genes. Jozwiak *et al.* reported on a systematic examination of liver angiomyolipomas in children with TSC and found that females were more likely than males to have liver lesions with a ratio of 5 : 1 [54]. This female preponderance was also confirmed by Fricke *et al.* [55] in his study of 62 TSC patients. In the seven patients in his study that had multiple hepatic angiomyolipomas, five were females.

There is only one report in the literature of spontaneous hemorrhage from a liver AML, presumably because of the low blood pressure in the portal system [56]. We have seen one female not known to have TSC who presented in her forties with abdominal pain and was thought to have bled from a hepatic angiomyolipoma. She subsequently had a partial hepatectomy and histology confirmed that the lesion was an angiomyolipoma. Subsequent investigation confirmed the presence of subependymal nodules and facial angiofibromatosis, confirming the diagnosis of TSC. She represented 10 years later with a second episode of mild abdominal pain and was found to have a lesion with the imaging characteristics of an AML with a small area of possible hemorrhage.

Lenci *et al.* reported on the case of 26-year-old female patient who was admitted with symptoms of anorexia, abdominal pain, and severe malnutrition [57]. Examination revealed a huge hepatic angiomyolipoma compressing the stomach. Prior to proceeding to partial hepatectomy, the authors instigated a trial of tamoxifen therapy that was associated with a significant symptomatic improvement, weight gain, and diminution of the size of the lesion on follow-up magnetic resonance imaging. The apparent therapeutic success of tamoxifen is interesting in view of El-Hashemite *et al.* demonstrating the importance of estrogen signaling *in vivo* for the growth of TSC lesions in $TSC1^{+/-}$ mice [58].

There is one report of hepatocellular carcinoma arising in a TSC patient in conjunction with multiple hepatic angiomyolipomas [59]. The lesion was discovered in a 51-year-old female patient with known hepatic angiomyolipomas. Although most lesions had been hyperechoic on ultrasound, indicating presence of fat, one lesion was markedly hypoechoic. This lesion was subsequently resected and histology confirmed a hepatocellular carcinoma that stained negatively for HMB-45. This may be a coincidence.

18.5 Splenic Manifestations of TSC

The spleen is also occasionally found to have lesions with the imaging characteristics of angiomyolipomas on ultrasound or CT, but these are not common. There are no reports of hemorrhage, but one report of pain associated with a rapidly enlarging splenic lesion [60]. Splenic hamartomas are not common in the absence of TSC. Reports of histology of splenic lesions in TSC are from many years ago and again one suspects that these lesions would be interpreted as angiomyolipoma today. Although arterio–venous shunting has been reported on angiography, symptoms from such lesions have not been reported. No intervention is advised if a lesion suggestive of a splenic AML is found – although there is one report of a man with TSC who had a splenic sarcoma making it difficult to know if this is coincidence or rare association [61].

Tashiro *et al.* have described pancytopenia and subsequent cardiac failure developing in a 46-year-old female TSC patient secondary to hypersplenism [62]. After splenectomy, histology revealed the presence of multiple intrasplenic angiomyolipomas but their relationship to the hypersplenism is unclear.

18.6 Lymphatic Manifestations of TSC

The lymphatic system was first reported to be involved in TSC in 1939 [63] and it is well known that the lymphatic system is commonly involved in females with pulmonary lymphangioleiomyomatosis and less frequently in other individuals with renal AML (Chapter 17). This occurrence reflects the fact that LAM is truly a systemic disease in most patients in whom the respiratory system is most affected due to cyst formation, and patients present for that reason most often [64]. Obstruction of mediastinal lymphatics and/or the thoracic duct likely contributes to chylous effusion in the thorax, seen in 10–20% of LAM patients (Chapter 17) [65–68].

References

1 Tigas, S., Carroll, P.V., Jones, R., Bingham, E., Russell-Jones, D., Powell, M. *et al.* (2005) Simultaneous Cushing's disease and tuberous sclerosis; a potential role for TSC in pituitary ontogeny. *Clin. Endocrinol. (Oxf.)*, **63** (6), 694–695.

2 Nandagopal, R., Vortmeyer, A., Oldfield, E.H., Keil, M.F., and Stratakis, C.A. (2007) Cushing's syndrome due to a pituitary corticotropinoma in a child with tuberous sclerosis: an association or a coincidence? *Clin. Endocrinol. (Oxf.)*, **67** (4), 639–641.

3 Gutman, A. and Leffkowitz, M. (1959) Tuberous sclerosis associated with spontaneous hypoglycaemia. *Br. Med. J.*, **2** (5159), 1065–1068.

4 Davoren, P.M. and Epstein, M.T. (1992) Insulinoma complicating tuberous sclerosis. *J. Neurol. Neurosurg. Psychiatry*, **55** (12), 1209.

5 Kim, H., Kerr, A., and Morehouse, H. (1995) The association between tuberous sclerosis and insulinoma. *AJNR Am. J. Neuroradiol.*, **16** (7), 1543–1544.

6 Boubaddi, N.E., Imbert, Y., Tissot, B., Chapus, J.J., Dupont, E., Gallouin, D. *et al.* (1997) Secreting insulinoma and Bourneville's tuberous sclerosis. *Gastroenterol. Clin. Biol.*, **21** (4), 343.

7 Eledrisi, M.S., Stuart, C.A., and Alshanti, M. (2002) Insulinoma in a patient with tuberous sclerosis: is there an association? *Endocr. Pract.*, **8** (2), 109–112.

8 Root, A.W., Zamanillo, J., Duckett, G., and Sweetland, M. (1986) Gonadotropin-independent isosexual precocity in a boy with tuberous sclerosis: effect of ketoconazole. *J. Pediatr.*, **109** (6), 1012–1015.

9 de Cornulier, M., David, A., and Cohen, J.Y. (1993) Precocious puberty revealing Bourneville tuberous sclerosis. *Arch. Fr. Pediatr.*, **50** (5), 421–423.

10 Ilgren, E.B. and Westmoreland, D. (1984) Tuberous sclerosis: unusual associations in four cases. *J. Clin. Pathol.*, **37** (3), 272–278.

11 Mortensen, L.S. and Rungby, J. (1991) Tuberous sclerosis and parathyroid adenoma. *J. Clin. Pathol.*, **44** (11), 961–962.

12 Yin, W., Zhu, D.M., Wang, D.S., Wang, D.X., Wu, E.H., and Zhu, X.Y. (1984) Parathyroid adenoma with primary hyperparathyroidism and tuberous sclerosis. A case report. *Chin. Med. J. (Engl.)*, **97** (8), 599–602.

13 Bloomgarden, Z.T., McLean, G.W., and Rabin, D. (1981) Autonomous hyperprolactinemia in tuberous sclerosis. *Arch. Intern. Med.*, **141** (11), 1513–1515.

14 Hoffman, W.H., Perrin, J.C., Halac, E., Gala, R.R., and England, B.G. (1978) Acromegalic gigantism and tuberous sclerosis. *J. Pediatr.*, **93** (3), 478–480.

15 Darling, T.N., Skarulis, M.C., Steinberg, S.M., Marx, S.J., Spiegel, A.M., and Turner, M. (1997) Multiple facial angiofibromas and collagenomas in patients with multiple endocrine neoplasia type 1. *Arch. Dermatol.*, **133** (7), 853–857.

16 Grozinsky-Glasberg, S., Franchi, G., Teng, M., Leontiou, C.A., Ribeiro de Oliveira, A., Jr., Dalino, P. *et al.* (2008) Octreotide and the mTOR inhibitor RAD001 (everolimus) block proliferation and interact with the Akt–mTOR–p70S6K pathway in a neuro-endocrine tumour cell line. *Neuroendocrinology*, **87** (3), 168–181.

17 Musat, M., Korbonits, M., Kola, B., Borboli, N., Hanson, M.R., Nanzer, A.M. *et al.* (2005) Enhanced protein kinase B/Akt signalling in pituitary tumours. *Endocr. Relat., Cancer*, **12** (2), 423–433.

18 Zitzmann, K., De Toni, E.N., Brand, S., Goke, B., Meinecke, J., Spottl, G. *et al.* (2007) The novel mTOR inhibitor

RAD001 (everolimus) induces antiproliferative effects in human pancreatic neuroendocrine tumor cells. *Neuroendocrinology*, **85** (1), 54–60.

19 Fukuda, T., Kobayashi, T., Yasui, H., Tsutsumi, M., Konishi, Y., and Hino, O. (1999) Distribution of Tsc2 protein in various normal rat tissues and renal tumours of Tsc2 mutant (Eker) rat detected by immunohistochemistry. *Virchows Arch.*, **434** (4), 341–350.

20 Kenerson, H., Dundon, T.A., and Yeung, R.S. (2005) Effects of rapamycin in the Eker rat model of tuberous sclerosis complex. *Pediatr. Res.*, **57** (1), 67–75.

21 de Tinguy Du Pouet, M., Turpin, G., and Scherrer, H. (1985) Hypothalamic endocrine disorders disclosing Bourneville's tuberous sclerosis. *Presse. Med.*, **14** (10), 599.

22 Perou, M.L. and Gray, P.T. (1960) Mesenchymal hamartomas of the kidney. *J. Urol.*, **83**, 240–261.

23 Verma, B.S. and Tailor, M.H. (1965) Familial tuberous sclerosis: a review with report of three cases. *Indian Pediatr.*, **2** (11), 401–410.

24 Gomez, M. (1979) *Tuberous Sclerosis*, Raven Press, New York.

25 Sareen, C.K., Ruvalcaba, R.H., Scotvold, M.J., Mahoney, C.P., and Kelley, V.C. (1972) Tuberous sclerosis. Clinical, endocrine, and metabolic studies. *Am. J. Dis. Child.*, **123** (1), 34–39.

26 Bereket, A. and Wilson, T.A. (1993) Thyroid dysgenesis and the dysplasia hypothesis in tuberous sclerosis. *Am. J. Med. Genet.*, **47** (3), 417–419.

27 Dicorato, P., Calvanese, A., Maiuolo, A., D'Alessandri, M., Grani, G., Ruggieri, M. *et al.* (2009) Medullary thyroid carcinoma and tuberous sclerosis. *Endocr. Pathol.*, **20** (2), 141–144.

28 Schwarzkopf, G. and Pfisterer, J. (1994) Metastasizing gastrinoma and tuberous sclerosis complex. Association or coincidence? *Zentralbl. Pathol.*, **139** (6), 477–481.

29 Sreenarasimhaiah, J., Armstrong, L.A., Tang, S.J., and Barnett, C. (2009) Pancreatic somatostatinoma and tuberous sclerosis: case report of an exceedingly rare association. *Gastrointest. Endosc.*, **69** (2), 379–381.

30 Verhoef, S., van Diemen-Steenvoorde, R., Akkersdijk, W.L., Bax, N.M., Ariyurek, Y., Hermans, C.J. *et al.* (1999) Malignant pancreatic tumour within the spectrum of tuberous sclerosis complex in childhood. *Eur. J. Pediatr.*, **158** (4), 284–287.

31 Francalanci, P., Diomedi-Camassei, F., Purificato, C., Santorelli, F.M., Giannotti, A., Dominici, C. *et al.* (2003) Malignant pancreatic endocrine tumor in a child with tuberous sclerosis. *Am. J. Surg. Pathol.*, **27** (10), 1386–1389.

32 Noldus, J., Ancker, U., Schafer, H., Conrad, S., and Huland, H. (1994) Multilocular, giant angiomyolipoma of the kidney, adrenal gland and para-aortic lymph nodes. Case report of a 9-year-old boy with tuberous sclerosis. *Urologe. A*, **33** (5), 453–456.

33 Bruneton, J.N., Drouillard, J., Rabin, A., Boedec, R., Broussin, J., and Delorme, G. (1978) Angiomyolipomas (hamartomas) of the kidney. Radiological study apropos of 2 cases. Review of the literature. *J. Can. Assoc. Radiol.*, **29** (4), 252–261.

34 Compton, W.R., Lester, P.D., Kyaw, M.M., and Madsen, J.A. (1976) The abdominal angiographic spectrum of tuberous sclerosis. *AJR Am. J. Roentgenol.*, **126** (4), 807–813.

35 Holtzman, M., Reider-Groswasser, I., and Harel, S. (1983) An unusual association of tuberous sclerosis and adrenogenital syndrome. *Brain Dev.*, **5** (1), 46–48.

36 Stern, J., Friesen, A., Bowering, R., and Babaryka, I. (1982) Multiple bilateral angiomyolipomas of the kidneys in tuberous brain sclerosis in association with pleomorphous pheochromocytoma. *Fortschr. Med.*, **100** (39), 1809–1812.

37 Martin, R.W., 3rd, Rady, P., Arany, I., and Tyring, S.K. (1993) Benign Leydig cell tumor of the testis associated with human papillomavirus type 33 presenting with the sign of Leser–Trelat. *J. Urol.*, **150** (4), 1246–1250.

38 Anderson, A.E., Yang, X., and Young, R.H. (2002) Epithelioid angiomyolipoma of the ovary: a case report and literature review. *Int. J. Gynecol. Pathol.*, **21** (1), 69–73.

39 Guo, H., Keefe, K.A., Kohler, M.F., and Chan, J.K. (2006) Juvenile granulosa cell tumor of the ovary associated with tuberous sclerosis. *Gynecol. Oncol.*, **102** (1), 118–120.

40 Bonetti, F., Martignoni, G., Colato, C., Manfrin, E., Gambacorta, M., Faleri, M. *et al.* (2001) Abdominopelvic sarcoma of perivascular epithelioid cells. Report of four cases in young women, one with tuberous sclerosis. *Mod. Pathol.*, **14** (6), 563–568.

41 Cummings, J.L., Oppenheimer, E.Y., and Hochman, H.I. (1978) Tuberous sclerosis. *Am. J. Dis. Child.*, **132** (12), 1215–1216.

42 Sparling, J.D., Hong, C.H., Brahim, J.S., Moss, J., and Darling, T.N. (2007) Oral findings in 58 adults with tuberous sclerosis complex. *J. Am. Acad. Dermatol.*, **56** (5), 786–790.

43 Hoff, M., van Grunsven, M.F., Jongebloed, W.L., and Gravenmade, E.J. (1975) Enamel defects associated with tuberous sclerosis. A clinical and scanning-electron-microscope study. *Oral Surg. Oral Med. Oral Pathol.*, **40** (2), 261–269.

44 Mlynarczyk, G. (1991) Enamel pitting: a common sign of tuberous sclerosis. *Ann. N. Y. Acad. Sci.*, **615**, 367–369.

45 Russell, B.G., Russell, M.B., Praetorius, F., and Russell, C.A. (1996) Deciduous teeth in tuberous sclerosis. *Clin. Genet.*, **50** (1), 36–40.

46 Hizawa, K., Iida, M., Matsumoto, T., Tominaga, M., Hirota, C., Yao, T. *et al.* (1994) Gastrointestinal involvement in tuberous sclerosis. Two case reports. *J. Clin. Gastroenterol.*, **19** (1), 46–49.

47 Kim, B.K., Kim, Y.I., and Kim, W.H. (2000) Hamartomatous gastric polyposis in a patient with tuberous sclerosis. *J. Korean Med. Sci.*, **15** (4), 467–470.

48 Gould, S.R. (1991) Hamartomatous rectal polyps are common in tuberous sclerosis. *Ann. N. Y. Acad. Sci.*, **615**, 71–80.

49 Goh, S.G., Ho, J.M., Chuah, K.L., Tan, P.H., Poh, W.T., and Riddell, R.H. (2001) Leiomyomatosis-like lymphangioleiomyomatosis of the colon in a female with tuberous sclerosis. *Mod. Pathol.*, **14** (11), 1141–1146.

50 Feriz, H. (1930) Ein Beitrag zur Histopathologie der Tuberosen Sklerose. *Virchows Arch. Pathol. Anat.*, **278** 690–769.

51 Inglis, K. (1950) Neurilemmoblastosis; the influence of intrinsic factors in disease when development of the body is abnormal. *Am. J. Pathol.*, **26** (4), 521–549.

52 Cares, R.M. (1958) The tuberous sclerosis complex. *J. Neuropathol. Exp. Neurol.*, **17** (2), 247–254.

53 Fleury, P., Smits, N., and van Baal, S. (1987) The incidence of hepatic hamartomas in tuberous sclerosis. Evaluation by ultrasonography. *Rofo.*, **146** (6), 694–696.

54 Jozwiak, S., Pedich, M., Rajszys, P., and Michalowicz, R. (1992) Incidence of hepatic hamartomas in tuberous sclerosis. *Arch. Dis. Child.*, **67** (11), 1363–1365.

55 Fricke, B.L., Donnelly, L.F., Casper, K.A., and Bissler, J.J. (2004) Frequency and imaging appearance of hepatic angiomyolipomas in pediatric and adult patients with tuberous sclerosis. *AJR Am. J. Roentgenol.*, **182** (4), 1027–1030.

56 Huber, C., Treutner, K.H., Steinau, G., and Schumpelick, V. (1996) Ruptured hepatic angiolipoma in tuberous sclerosis complex. *Langenbecks Arch. Chir.*, **381** (1), 7–9.

57 Lenci, I., Angelico, M., Tisone, G., Orlacchio, A., Palmieri, G., Pinci, M. *et al.* (2008) Massive hepatic angiomyolipoma in a young woman with tuberous sclerosis complex: significant clinical improvement during tamoxifen treatment. *J. Hepatol.*, **48** (6), 1026–1029.

58 El-Hashemite, N., Walker, V., and Kwiatkowski, D.J. (2005) Estrogen enhances whereas tamoxifen retards development of Tsc mouse liver hemangioma: a tumor related to renal angiomyolipoma and pulmonary lymphangioleiomyomatosis. *Cancer Res.*, **65** (6), 2474–2481.

59 Yang, B., Chen, W.H., Shi, P.Z., Xiang, J.J., Xu, R.J., and Liu, J.H. (2008) Coincidence of hepatocelluar carcinoma and hepatic angiomyolipomas in tuberous sclerosis complex: a case report. *World J. Gastroenterol.*, **14** (5), 812–814.

60 Darden, J.W., Teeslink, R., and Parrish, A. (1975) Hamartoma of the spleen: a manifestation of tuberous sclerosis. *Am. Surg.*, **41** (9), 564–566.

61 Colonna, P., Dimitrov, I., Tordjmann, G., Timsit, G., Messerschmitt, J., and Mussini-Montpellier, J. (1966) Sarcoma of the spleen in the course of Bourneville's tuberous sclerosis. *Presse. Med.*, **74** (9), 447–448.

62 Tashiro, M., Hirose, W., Hanabusa, H., Takahashi, T., Kanki, H., Asaba, Y. *et al.* (2001) Pancytopenia in tuberous sclerosis. *Med. Sci. Monit.*, **7** (3), 444–447.

63 Berg, G.V.G. (1939) Maladie kistique du poumon et sclerose tubereuse du cerveau. *Acta Paediatr. (Uppsala)*, **26** 16–30.

64 Henske, E.P. (2003) Metastasis of benign tumor cells in tuberous sclerosis complex. *Genes Chromosomes Cancer*, **38** (4), 376–381

65 Bustin, F., Gustin, M., Robin, M., Cambier, P., Warling, X., and Chantraine, J.M. (2002) Clinical case of the month. A recurrent chylothorax in Bourneville tuberous sclerosis. *Rev. Med. Liege*, **57** (8), 493–496.

66 Foresti, V., Casati, O., Zubani, R., and Villa, A. (1990) Chylous pleural effusion in tuberous sclerosis. *Respiration*, **57** (6), 398–401.

67 Ioannou, N., Marjani, M.A., and Dunn, E. (1980) Tuberous sclerosis associated with chylothorax. *Conn. Med.*, **44** (5), 279–280.

68 Uozumi, T., Yamashita, Y., Hayakawa, T., Murai, Y., and Kajiki, A. (1983) A case of tuberous sclerosis with pulmonary lymphangiomyoma and chylothorax. *Rinsho. Shinkeigaku.*, **23** (5), 410–414.

69 Folpe, A.L. and Kwiatkowski, D.J. (2010) Perivascular epithelioid cell neoplasms: pathology and pathogenesis. *Human Pathology*, **41**: 1–15.

70 Wagner, A., Malinowska-Kolodziej, I., Morgan, J.A., Qin, W., Fletcher, C.D.M., Vena, N., Ligon, A.H., Antonescu, C.R., Ramaiya, N.H., Demetri, G.D., Kwiatkowsi, D.J., and Maki, R.G. (2010) Clinical activity of mTOR inhibition with sirolimus in malignant perivascular epitheloid cell Tumors: targeting the pathogenic activation of mTORC1 in Tumors. *J. Clin. Oncol.*, **28** (5), 835–840.

Part VI
Family Impact

Tuberous Sclerosis Complex: Genes, Clinical Features, and Therapeutics
Edited by David J. Kwiatkowski, Vicky Holets Whittemore, and Elizabeth A. Thiele

ISBN: 978-3-527-32201-5

19
Impact of TSC on the Family and Genetic Counseling Issues

Vicky H. Whittemore and Janine Lewis

19.1
Introduction

For those families who first hear the words "tuberous sclerosis complex," (TSC) the moment will be forever frozen in their minds. Years after receiving the diagnosis of tuberous sclerosis complex for their daughter, a young couple distinctly remembers the emotions they felt when the physician delivered the diagnosis. Parents relate that they heard the words tuberous sclerosis complex, and then the phrases "rare genetic disease," "no treatment," "tumors in any organ," "possible severe intellectual disability," and "autism spectrum disorder" jumbled together in what followed. They may have been advised not to seek information on the Internet, and cautioned that in some medical textbooks and other sources, it is stated that tuberous sclerosis complex carries a grim prognosis.

Most individuals have never heard of TSC when they receive the diagnosis, so fear of the unknown suddenly looms in front of them.

19.2
Impact on the Family

Caring for an individual with TSC certainly may have an impact on the other family members. Depending on the individual(s) with TSC and the other family members, this may be either a positive or negative impact, and it may also change over time. A study on the impact of a specific diagnosis on outcomes in families with children who have physical disabilities showed that there was a modest effect of diagnosis type on five family outcomes [1]. The outcomes included (1) how parents or guardians rated their child's current health compared to 1 year before the diagnosis; (2) the degree to which the child's physical health caused worry; (3) the degree to which the child's emotional well-being or behavior caused worry; (4) the degree to which the child's health or behavior limited types of family activities; and (5) the degree to which the child's health or behavior interrupted family activities.

Tuberous Sclerosis Complex: Genes, Clinical Features, and Therapeutics
Edited by David J. Kwiatkowski, Vicky Holets Whittemore, and Elizabeth A. Thiele

ISBN: 978-3-527-32201-5

Parents may be concerned about the effect of caring for a child with TSC on their marriage [2]. Parents may become so focused on providing care for their child with TSC that they make their personal lives and relationships, including those with spouses and other children, distant second priorities. Although to a degree this is a normal maternal reaction to an ill child, when carried out to an extreme, it can have clear negative consequences for both spouses and other family members. Couples who have a child with a chronic illness are at special risk for experiencing marital distress, and can benefit from therapies such as emotionally focused therapy [3]. Parents may also worry about the affect on their own health and the other siblings when there is a child with TSC in the family that requires a lot of attention.

In a recent study of a clinic population of children with TSC, about 40% of the children had clinically significant behavioral problems, including symptoms of autism spectrum disorder, inattention, and hyperactivity [4]. Of all the manifestations of TSC, the cognitive and behavioral problems represent the area of greatest concern to parents and caregivers [5]. Nearly 50% of the families reported experiencing significant parenting stress when the child had recurrent seizures, a history of psychiatric diagnosis, low intelligence, and behavioral problems [4]. Clinicians caring for these children should know that behavioral problems may be significant in children with TSC and provide referrals for behavioral intervention. In addition, monitoring of parental stress should be included in the medical management [4]. It has been suggested that clinicians take a "systems" approach to the assessment process by recognizing how the well-being of family members can impact a child with a developmental disability [6].

Effects on the family are even greater when multiple family members have TSC, particularly one of the parents. Families that are aware of the impact of having a child with a chronic disorder can find ways to work together so that the health issues are being attended to and treated as best as possible. Family members may consider family counseling if the child's health issues, emotional well-being, and/or behavior are causing stress and concern. If a family member with TSC has behaviors that make family activities difficult, it may be necessary to find ways in which the family divides their activities so that other siblings can continue to participate in school and outside activities. Interestingly, a study that focused on the families of children with developmental disabilities and the adaptation of their parents, extending from infancy through middle childhood, showed that the child's type of disability predicted trajectories of development in cognition, social skills, and daily living skills [7]. The child's type of disability also predicted changes in maternal (but not paternal) child-related and parent-related stress. Beyond the type of disability, child self-regulatory processes (notably behavior problems and mastery motivation) and one aspect of the family climate (notably mother–child interaction) were key predictors of change in both child outcomes and parent well-being. A different aspect of the family climate–family relations–also predicted change in child social skills. Parent assets, measured as social support and problem-focused coping, predicted change in maternal and paternal parent-related stress, respectively [7].

19.3
Finding Support

Individuals and families each have their own mechanisms for seeking support from others. When there is an individual with TSC in the family, seeking support and information from others with TSC and their families can be invaluable. Learning about how other families cope with family, medical, social, and educational issues can be especially helpful to the parents of a child with TSC. Sharing concerns about growing up and coping with a sibling who has TSC can make a tremendous difference for children and adolescents who have a sister and/or brother with TSC. Knowing that there are other individuals who have the same issues and concerns can make adults with TSC feel as if they are not alone and have friends they can turn to who are understanding and accepting.

Advocacy organizations may provide information and support through some or all of the following:

- Family meetings or gatherings
- Educational activities that also provide an opportunity to network with other families
- Online chat groups or listservs for communicating with other parents, grandparents, adults with TSC, siblings of individuals with TSC, and so on
- Teleconferences where individuals can talk and exchange information about relevant topics important to that group
- Other online resources such as Facebook, Twitter, and so on
- Matching programs through which an individual or family are matched with an individual or family with similar issues and/or experiences

In a survey conducted by Parker [2], some families found that joining a TSC support group helped alleviate their feelings of isolation in caring for their child with TSC. However, many parents had reservations about joining support groups. For parents of a mildly affected child, it could be difficult for them to see children who were more severely affected. Conversely, for parents of severely affected children, some resented seeing families whose children were nearly asymptomatic, happy, and healthy. If an individual or family is not comfortable joining a TSC support group or activity, there are other avenues for support through other family members, friends, religious organizations, therapists and counselors, and numerous other resources.

19.4
Tuberous Sclerosis Complex Organizations and Support Groups

TSC organizations and support groups have been organized in many countries throughout the world. These groups have joined forces to establish Tuberous Sclerosis International, a consortium of the groups so that they may share experiences, information, and research news worldwide. The TSC organizations provide

Table 19.1 Tuberous sclerosis organizations.

Tuberous Sclerosis International Contact: Vicky Whittemore, PhD Tuberous Sclerosis Alliance 801 Roeder Road, Suite 750 Silver Spring, MD 20910 USA Telephone: 301-562-9890 FAX: 301-562-9870 e-mail: info@tsalliance.org Web site: www.tsalliance.org

information about TSC and support to individuals with TSC, their families, health care providers, and educators. In addition, they provide information about the latest TSC research and clinical trials. Some of the organizations organize support group meetings, educational meetings, newsletters, conferences for families and health care providers, physician referrals, matching programs for individuals and families, online chat rooms and/or listservs, physician referrals, advocacy for educational and disability issues, and other information and programs. All of the members of Tuberous Sclerosis International share information and are there to support each other and to assist organizations to form and grow in countries where no support exists. Some of the organizations fund research grant awards to researchers in academic institutions whose research is focused on TSC. The list of all of the members of Tuberous Sclerosis International is too extensive to include here, so a complete list can be obtained by contacting the Tuberous Sclerosis Alliance (see Table 19.1). Information may also be obtained via the Internet.

19.5 Genetic Counseling Issues for Tuberous Sclerosis Complex

There are both genetic counseling and psychological issues to consider and address for individuals who are diagnosed with TSC and for their family members. A referral to a genetic counseling service is recommended for individuals with TSC and/or their family members to allow careful discussion about all the implications of the diagnosis.

19.5.1 Adults with TSC

Health care providers expect that nearly every individual with a TSC mutation will have some signs or symptoms. Most often, symptoms develop before the age of 10. However, since signs and symptoms of TSC can vary widely even within the same family, an individual may have reached adulthood before first learning that they have

TSC. In most cases, an adult is diagnosed with signs of TSC because a child in the family has been diagnosed with TSC, which leads to the evaluation of other family members. For all individuals, the diagnosis of TSC is made on the basis of clinical symptoms and/or by mutation analysis. Approximately 80% of individuals with a clinical diagnosis of TSC will have an identifiable mutation in either the TSC1 or TSC2 gene, found by DNA testing of blood. When an adult is diagnosed with TSC, it has implications for themselves, their family members, and for their children, which significantly adds to the burden of this diagnosis for the individual. Sufficient time needs to be given for the individual to understand and work through all of these issues.

19.5.2 Parents of a Child with TSC

About one third of the time, when a child is diagnosed with TSC, one of the parents also has TSC – a diagnosis that the parent may be unaware of because the symptoms are mild and because neither the parent nor the parent's health care provider has recognized the symptoms. Parents of a child with TSC should have a detailed medical and family history obtained and a clinical evaluation performed for TSC that includes a skin examination with a Woods lamp, retinal exam, brain imaging, and renal evaluation. If a TSC mutation is identified in the child – around 20% of mutations *cannot* be detected yet with a clinical genetic test, then genetic testing can be offered to the parents and other family members to determine their disease status.

Because TSC is an autosomal dominant condition, an affected parent has a 50% chance with each pregnancy of having another child with TSC. If a child inherits the TSC gene – even if the parent's symptoms are mild, the symptoms for the child can range from mild to a more severe presentation; the severity of the symptoms for the child cannot be predicted based on knowing the specific gene mutation.

In two thirds of cases, the parents of an individual with TSC do not show any signs of TSC and do not have a TSC mutation. In 98 or 99% of such cases, the TSC mutation in the child resulted from a spontaneous or *de novo* mutation that occurred only in that individual, and the parents would not be at risk of having any more children affected with TSC. However, there is a slight (1–2%) chance that one of the parents could have some gonadal cells – sperm or eggs cells – that contain a TSC mutation. This condition is called *gonadal mosaicism*, which can only be confirmed if an unaffected parent has a second child with TSC – something that would happen only 1 out of 50–100 times. However, the chance of having gonadal mosaicism cannot be ruled out for any parent who has one child with TSC and no other family history.

19.5.3 Siblings of an Individual with TSC

The diagnosis of TSC has significance for the brothers and sisters of the one diagnosed, as well as for other family members. The risks of siblings having TSC depend on the disease status of the parents. As mentioned above, there is one chance in three that a parent of an affected individual also has TSC. If a parent is affected, the siblings of the first child to come to medical attention and to receive the diagnosis (the proband) have a 50% chance of having inherited the TSC gene. However, owing to wide intrafamilial variability, the symptoms of TSC for brothers and sisters can differ significantly from those of the proband.

If the parents are unaffected, the chance of a sibling having TSC is still 1–2% because of the chance that a parent could have gonadal mosaicism; therefore, the siblings should be offered genetic testing (if the proband's mutation is known) or a clinical evaluation (if the mutation has not been identified).

19.5.4
Family Members of an Individual with TSC

As with the siblings of an individual with TSC, the chance of other family members having TSC depends on the disease status of the parents. If a parent is diagnosed with TSC, then that individual's family members are at risk. The parent should be encouraged to share the information with their other family members so that their family members can make informed decisions about further evaluation for TSC for themselves and for their own families. In some families, health issues are openly discussed, whereas in others it is seldom mentioned. Members of families with open communication styles may be better at discussing issues regarding inherited risks or testing options. Genetic counseling services can often assist families who are uncertain about how to share this information with their families.

19.5.5
Reproductive Options and Decision Making

Prenatal testing options are available to individuals who have TSC and thus are at risk of having an affected child. As mentioned above, with each pregnancy, an individual with TSC has a 50% chance of having a child with TSC. If a child inherits the TSC gene – even if the parent's symptoms are mild, the symptoms for the child can range from mild to severe; the severity of the symptoms cannot be predicted prenatally. Unaffected parents who have a child with TSC have a 1–2% recurrence risk due to the chance of gonadal mosaicism. Providing information about these risks is important for individuals with TSC, asymptomatic at-risk family members, and parents of an affected child *before* they become pregnant to allow time for people to be tested if they are interested and so people can make informed decisions about family planning at their own pace without being pressured by the timeline of a pregnancy. If a pregnancy is already in progress, risk assessment and reproductive options needs to be discussed as early in pregnancy as possible so that parents have the most options available to them. More important, discussion about prenatal testing options always needs to be made in a sensitive, nondirective, and supportive way so that parents can make their own decisions about what is appropriate for their families.

The prenatal testing options for people at risk of having an affected child include molecular genetic testing and fetal ultrasound studies. If a mutation is identified in the parent with TSC, prenatal diagnosis is possible by the analysis of fetal cells obtained from chorionic villi sampling (CVS) between 10 and 12 weeks of gestation or from amniocentesis that is usually performed between 15 and 18 weeks of gestation. Some centers offer early amniocentesis at around 13 weeks of gestation. The results will determine whether the fetus inherited the same TSC mutation as that found in the affected parent; however, the results cannot predict the severity of the symptoms.

Preimplantation genetic diagnosis (PGD) is also available for families when the disease-causing mutation is known. PGD is a technique that can detect the TSC mutation in embryos created through *in vitro* fertilization (IVF) before transferring them to the uterus. Because only unaffected embryos are transferred to the uterus for implantation, PGD provides an alternative to amniocentesis or CVS.

If the disease-causing mutation cannot be identified in the affected parent, families can have multiple high-resolution ultrasounds during pregnancy to look for fetal tumors. Some fetal tumors, such as cardiac tumors, cannot be detected until the third trimester. If a fetal tumor is detected on ultrasound, then the diagnosis of TSC for the fetus is strongly suspected. However, if fetal ultrasounds are normal, the diagnosis of TSC cannot be ruled out. For families who are very concerned about their risk, fetal MRI may be considered to evaluate the fetus for TSC. However, the reliability of this approach and its clinical application in this setting are uncertain at this time.

19.6 Summary

When a member of a family is diagnosed with TSC, it often has a major impact on both the individual and his/her family members. It is important that the individual and the family receive appropriate and accurate information at the time of diagnosis, as well as later on when additional symptoms or manifestations are identified. Local TSC organizations in many countries, including the Tuberous Sclerosis Alliance in the United States, can provide support and up-to-date information for individuals with TSC and their families, caregivers, and health care providers. They can also provide referrals to physicians, genetic counselors, and other individuals who can provide them with guidance and support.

References

1 Eddy, L.L. and Engel, J.M. (2008) The impact of child disability type on the family. *Rehabil. Nurs.*, **33**, 98–103.

2 Parker, M. (1996) Families caring for chronically ill children with tuberous sclerosis complex. *Fam. Community Health*, **19**, 73–84.

3 Cloutier, P.F., Manion, I.G., Walker, J.G., and Johnson, S.M. (2007) Emotionally focused interventions for couples with chronically ill children: a 2-year follow-up. *J. Marital Fam. Ther.*, **28**, 391–398.

4 Kopp, C.M., Muzykewicz, D.A., Staley, B.A., Thiele, E.A., and Pulsifer, M.B. (2008) Behavior problems in children with

tuberous sclerosis complex and parental stress. *Epilepsy Behav.*, **13** (3), 505–510.

5 Prather, P. and de Vries, P.J. (2004) Behavioral and cognitive aspects of tuberous sclerosis complex. *J. Child Neurol.*, **19**, 666–674.

6 Head, L.S. and Abbeduto, L., (2007) Recognizing the role of parents in developmental outcomes: a systems approach to evaluating the child with developmental disabilities. *Ment. Retard. Dev. Disabil. Res. Rev.*, **13**, 293–301.

7 Hauser-Cram, P., Warfield, M.E., Shonkoff, J.P., Krauss, M.W., Sayer, A., and Upshur, C.C. (2001) Children with disabilities: a longitudinal study of child development and parent well-being. *Mongr. Soc. Res. Child Dev.*, **66**, 1–126.

Index

Tuberous Sclerosis Complex: Genes, Clinical Features, and Therapeutics
Edited by David J. Kwiatkowski, Vicky Holets Whittemore, and Elizabeth A. Thiele

ISBN: 978-3-527-32201-5